Landscapes and Hydrology

Christopher W. McVoy
Winifred Park Said
Jayantha Obeysekera
Joel A. VanArman
and Thomas W. Dreschel

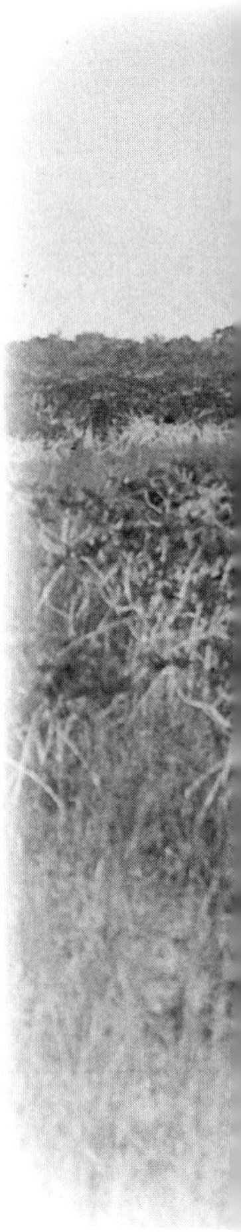

Landscapes and Hydrology of the Predrainage Everglades

UNIVERSITY PRESS OF FLORIDA

Florida A&M University, Tallahassee
Florida Atlantic University, Boca Raton
Florida Gulf Coast University, Ft. Myers
Florida International University, Miami
Florida State University, Tallahassee
New College of Florida, Sarasota
University of Central Florida, Orlando
University of Florida, Gainesville
University of North Florida, Jacksonville
University of South Florida, Tampa
University of West Florida, Pensacola

of the Predrainage Everglades

University Press of Florida
Gainesville
Tallahassee
Tampa
Boca Raton
Pensacola
Orlando
Miami
Jacksonville
Ft. Myers
Sarasota

Printed in the United States of America on acid-free paper

Funding for this research was kindly provided by the South Florida Water Management District, with additional assistance from the Environmental Defense Fund and the U.S. Geological Survey.

16 15 14 13 12 11 6 5 4 3 2 1

Library of Congress Cataloging-in-Publication Data
Landscapes and hydrology of the predrainage Everglades / Christopher W. McVoy . . . [et al.].
p. cm.
Includes bibliographical references and index.
ISBN 978-0-8130-3535-2 (Book/DVD package)
1. Everglades (Fla.)—Environmental conditions—History. 2. Everglades (Fla.)—Surveys.
3. Drainage—Florida—Everglades—History. 4. Wetland hydrology—Florida—Everglades.
I. McVoy, Christopher.
GE155.F6L36 2011
551.4109759'39—dc22 2010042233

Frontis: Diminutive mangroves and hammock with *Paurotis* palm,
Perrine Marl Marsh, 1919. Photo by J. K. Small.

The University Press of Florida is the scholarly publishing agency for the State University System of Florida, comprising Florida A&M University, Florida Atlantic University, Florida Gulf Coast University, Florida International University, Florida State University, New College of Florida, University of Central Florida, University of Florida, University of North Florida, University of South Florida, and University of West Florida.

University Press of Florida
15 Northwest 15th Street
Gainesville, FL 32611-2079
http://www.upf.com

Contents

Supplementary Materials (on DVD)

Historical Observations

Appendices

Illustrations

Plates

Figures

Tables

Supplementary Materials (on DVD)

Appendix Figures

Historical Observation Tables

Appendix Tables

Acknowledgments

Developing a detailed picture of the predevelopment Everglades has been a rewarding but long project. Fortunately the load has been lightened by the project's many faithful supporters, from the original conceivers to those who helped carry it across the finish line. In between, many others have helped, by pointing out obscure historical sources, supporting the research effort, creating illustrations, and critically reviewing the content.

The concept of piecing together a scientifically useful picture of the original Everglades from historical sources is traceable at least back to Aaron Higer, now retired from the U.S. Geological Survey. We thank him not only for showing the way, but also for his great faith in the project and unflagging behind-the-scenes support. We also thank Tim Searchinger, formerly with the Environmental Defense Fund, who aimed the first author at the Everglades, hiring him to "just validate" the so-called Natural Systems Model. Little did Tim know that this decision would eventually result in a book. The late Pete Rosendahl warmly encouraged this project from its inception. Our colleague Martha Nungesser has been steadfast in her belief in and encouragement of this research effort.

Scientists and modelers branching out into historical research need help. We were extremely fortunate in receiving many generous leads into the Everglades historical literature. Through their writings, Junius Dovell, Charles Tebeau and Garald Parker pointed us in the right direction. Others kindly helped in person. Cynthia Plockelman, who for decades built the South Florida Water Management District library, expertly guided us through her special collection of rare historical materials and suggested further contacts. Rebecca Smith and Dawn Hugh of the Historical Museum of Southern Florida lead us to important maps, photographs and original field notes. Gail P. Clement at Florida International University helped with additional sources. Sarah Bellmund, Robert Fennema, Jack Moller, George H. Wedgworth, Dr. Joe Knetsch, and Robert H. Mooney all were generous in sharing their knowledge of relevant historical sources.

As the project grew from a "simple" model validation into a full-blown, independent research study, a number of scientists and researchers reviewed successive versions. Responsibility for the content and any errors of course remain with the authors, but we are grateful for the many detailed, thoughtful comments provided by our reviewers. Their comments sharpened our thinking and greatly improved the final work. A review of the earliest version was organized by the U.S. Geological Survey, with reviewers George Snyder (Everglades Research and Education Center, Univ. of Florida), John Ogden and Bill Robertson (Everglades National Park), Joe Knetsch (Bureau of Survey and Mapping, State of Florida), Wiley Kitchens and Lance Gunderson (University of Florida), Lorraine Heisler (U.S. Fish and Wildlife Service), Joel A. VanArman and Steven M. Davis (South Florida Water Management District), Patrick Gleason (Camp Dresser and McKee), Ronald Hofstetter (Univ. of Miami), and Ben McPherson (U.S. Geological Survey). Review comments on a more extensive interim version were provided by Fred Sklar and Sue Newman (South Florida Water Management District), and by Ben McPherson and Aaron Higer (U.S. Geological Survey), and Kirk W. McVoy (University of Wisconsin). As the project neared completion, a formal review was organized under the guidance of D. Peter Loucks (Cornell University) with reviewers Peter A. Stone (South Carolina Dept. of Health and Environmental Control), Dan Austin (Florida Atlantic University), David L. Cox (David Cox Consulting), Tom Lodge (Thomas E. Lodge Ecological Advisors), and Lance Gunderson (Emory University). Additionally, Michael Duever, Mark Cook and Martha Nungesser, all of the South Florida Water Management District, reviewed portions of the manuscript. The final review, organized by the publisher, was done by Peter A. Stone (South Carolina Dept. of Health and Environmental Control) and William F. Loftus (U.S. Geological Survey, retired).

Illustrations are a critical part of visualizing the Everglades of the 1800s, particularly given how drastically it has been altered. We have been very fortunate to have had extensive and highly talented support. Jane Walters early on recognized the importance of this book and went well beyond the call of duty to produce beautiful illustrations, cheerfully tolerating numerous edits and last minute calls for help. David Williams of Wingin' It Works quickly understood the needs and showed both imagination and great patience in translating concepts into visual reality. Ellen Negley kindly helped us out of a number of graphic binds. Sue Hohner's attention to detail was invaluable for map creation and for smoothing the transition between GIS and graphics. The following individuals created or substantially assisted in creating the illustrations noted.

Jane Walters: Figs. I.2, 3.1, 3.2, 3.3, 3.4, 3.5, 3.6, 3.7, 3.8, 3.9, 3.14, 3.17, 3.18, 4.7, 5.1, 6.2, 11.2, 11.5, and 11.11, and Plates 2, 5, 26, 27, 35, 36, 39, 40, 41, 43, 44, 45, and 46;
Winifred Said: Figs. I.1, 1.11, 4.1, 4.5, 4.6, 4.9, 8.8, 9.1, 9.2, B.1, and D.1, and Plates 2, 3, 4, 5, 10 left, 11, 12, 13, 14, 15, and 16;
Christopher W. McVoy: Figs. 2.1, 3.11, 3.12, 3.13, 4.7, 4.8, 4.9, 5.2, 5.3, 7.3, 8.3, 8.5, 11.4, 11.6, 11.12, and 11.14, and Plates 2, 3, 5, 6, 15, 22, 24, 28, 35, and 36;
Sue Hohner: Figs. 5.1, 11.9, and T.1. and Plates 1, 6, and 8;
David Williams: Figs. 11.3 and 11.7, and Plates 7, 29, and 30;
Nancy Demonstranti: Figs. 11.1 and 11.12, and Plates 25, 31, and 32;
Joel A. VanArman: Figs. 1.5, 1.6, and 1.7, and Plate 9;
Matthew Hinton: Figs. 8.6 and 11.10, and Plates 10 right and 35;
Bernardo Gomez: Figs. 4.7, 11.5, and 11.8;
Oriana Perez: Figs. 2.1, 3.11, and 4.8;
Constance Harbin: Figs. 11.8 and 11.9;
Michael Clay Brown: Plate 3;
Fabiola Santamaria: Fig. 1.4;
Sally Colbert: Plate 23;
Angela Farinelli: Plate 42.

We are grateful to the South Florida Water Management District (SFWMD) for multiple forms of support throughout this project. In addition to providing employment to the authors, the SFWMD made available key geographical information system resources and excellent editors, helped create technical illustrations, provided financial support at critical times, and generously underwrote the color plates. The dedication to intellectual integrity shown by the SFWMD during this project has been extremely gratifying. Christopher W. McVoy and Winifred Park Said are also particularly grateful for the SFWMD's support of their extensive fieldwork in the present day Everglades. The field knowledge gained greatly assisted in our understanding of the predrainage system. Fred Sklar, Director of the Everglades Division, not only harbored the first author but provided invaluable diplomacy, editing and encouragement. Terry Petrucha successfully guided production of many of the illustrations. Julie (Pennington) Clark kindly edited an early version of the first three chapters. Dean Powell, Seán Sculley, and ShiLi Miao were important shepherds, keeping the research and writing efforts on track. Fabiola Santamaria patiently built the databases of references and township surveys. Michael Manna reviewed and revised Appendix Q. Malak Ali kept the illustrations organized and along with Fabiola and Sue Hohner helped prepare the manuscript for submission. We thank Shannon Estenoz of the SFWMD Governing Board for her support and encouragement. The first author also thanks the Environmental Defense Fund and John Vecchioli and Irv Kantrowitz of the U.S. Geological Survey for employment during the early phases of the project.

The first author's writing has been at times less than concise. We thank our editors for having been able to see past this, and for their help in freeing the underlying essence. John Byram was instrumental in bringing the manuscript to print (and DVD). Hands down winner in reducing the printed word count! His support and good cheer considerably lightened preparation of the manuscript for submission. Editor Carol Wanta was an inspiration, helping us see a broader audience and role for this book, suggesting most of the present structure, and pushing the writing to a much higher standard. Eagle-eyed copy editor Sally Antrobus read every page, claimed ignorance of all issues scientific, yet was extremely helpful (and gracious) in catching numerous oversights on our part. The book is measurably stronger for her work. Finally, Marthe Walters of the University Press of Florida kept the book on track throughout the production process.

The first author is grateful for many types of support that made this effort possible. My parents, Kirk and Hilda, inspired an empirical approach as well as a love and curiosity for the natural world. Botanist Hugh H. Iltis was an early and inspiring role model. Charlie Hall instilled a systems ecology approach to thinking across disciplines and to applying science to societal issues. Suresh Rao and Jeff Wagenet taught me transport processes in porous media and further sharpened my thinking. During the initial phases of the research, Rosa Durando and Stella and "Ric" Rossi kindly took me into their homes in between commutes to New York City. Soraya Esposito and Oriana Perez tolerated long hours and divided attention. My friend and mentor Bharata opened my eyes and heart to new directions, and never once doubted that we would see the book in print. And a loving thank you to my dear wife Kimberly—a constant support, always sure that this was the right thing to do. She has been a rock during the long journey.

All of the authors thank our families, colleagues and friends for their support of this effort. We are especially thankful to all who have worked and continue to work, in many different ways, for the restoration of the Everglades. May this book help illustrate what it once was. Finally, we thank those firsthand observers of the Everglades—the Native Americans, explorers, surveyors, botanists, agriculturalists, and early settlers—who, little knowing that their words might still be read more than a century later, took the trouble to record what they saw around them even when it had little or nothing to do with their immediate tasks.

Foreword

Though thick enough to be one, this work is not a history book. It includes a great deal of history, but it goes well beyond that. In my opinion, it is a *diagnostic ecohistory* of the Everglades. You are already familiar with another type of diagnostic science; we call it medicine. The *diagnostic ecohistory* you now hold in your hand, and especially the way we will use it for Everglades restoration, closely resembles the practice of medicine.

When we are sick, a familiar sequence begins. We go to the doctor and write down our medical history and our current status. Then we sit down with the diagnostic scientist, the doctor, who reviews our history and current status and then begins asking specific questions about the changes occurring in our body and the possible reasons for them.

This is when the real diagnostic science begins. The doctor combines, in his head, all the knowledge from his years of medical training, all his experiences with other patients, and everything he knows about the patient, and mentally starts putting it together, looking for cause and effect. As he does this, he starts coming up with hypotheses that might explain the patient's problem.

And even though the doctor cannot experiment on the patient to test his hypotheses (his sample size is one!), we have faith in his ability to help. We let the doctor take the first step, knowing that we may have to change course slightly as we learn more.

This is what we need to do with the Everglades, a patient that also needs serious healing. Here, too, we have only a sample size of one. Here, too, many types of experiments may not be practical. And for the Everglades, a place that has not "been herself" for more than a century, it is even difficult to obtain the "medical history," that is, a description of her original, healthy condition. (The ecologist sees the beauty of the system, so I always refer to the Everglades as "she"; also, she has the ability to tolerate pain, which is necessary.)

To heal our patient we need to know what is wrong with her. To know what is wrong, we need to know what "right" looked like. We need to know which things, if changed, would give us back an Everglades system.

This book, this *diagnostic ecohistory*, does all that. It provides the "medical history" for our patient, helping us understand her original appearance and the inner workings that produced that appearance. By also tracing the course of her afflictions, the canals and levees that have worn her down, it also helps us make a correct diagnosis, suggesting hypotheses that can explain the changes have occurred. Most importantly, this book tells us which components are needed to make a marsh into an Everglades. If we do nothing we will still have a marsh. But we won't have an Everglades, a very important point.

This brings us to the next phase, restoration. We have to define what restoration is. Some things cannot be changed, we recognize that. At the same time, if we change other things, the critical ones, they could give us back an Everglades system.

What is Everglades restoration? You find out in this book what the components are: connectivity, water depths, sheet flow, water quality. But it is more. Technically, we can put the water wherever we want it. But that isn't what we want, we want a system that is the Everglades, that would recognize herself as her former healthy self, a system that is self-repairing. Make the changes at the upstream end. You repair the quality of water going in at the top and I feel that quality will come back very quickly. The same concept exists with water quantity. Fix upstream and let the system improve itself, instead of fixing individual areas. This patient is very complex, so you must make the conditions right, then let her heal herself as much as possible.

We say that she should be restored not just relative to the water, but to all the components, to define the new Everglades like the old Everglades. We want her to be fully restored to a functional Everglades, less than the original extent, but a functional Everglades and then we measure the changes. We use the tools provided by this book, this *diagnostic ecohistory* of the Everglades. This is what we do in our lives, our careers, and our bank accounts: aim in the right direction, watch over time, and keep checking: are we going forward or back? Here we do the same, only with a restored Everglades. The characteristics discussed in this book are the diagnostics that let us determine whether restoration is moving in the right direction.

So who should use this book? Anyone involved in ecosystem restoration: engineers, scientists, planners, and decision makers. For ecosystems in other parts of the world, it can be a template for identifying the original predevelopment conditions. For the Everglades, it is the reference of record, the encyclopedia of what makes the Everglades the Everglades. I have no doubt that it will become *the* diagnostic tool for understanding what makes this patient sick and for measuring her progress toward recovery.

Aaron Lee Higer
U.S. Geologic Survey, retired, 1960–2010
Presently affiliated with Geosciences Dept.,
Florida Atlantic University

Preface

A Journey through the Everglades

In the nineteenth century the Everglades was surrounded by forest on all sides but the north. For most explorers this meant it was a place that one "came upon," with a vast vista suddenly spreading out before the eye: "an opening appeared, letting in a flood of daylight, and we suddenly burst into the pathless Everglades" (Willoughby 1898a, p. 104).

This description and others are from firsthand observations made a century or more ago. The earlier of these observers saw what was then an undisturbed Everglades, prior to man-made drainage of either the Everglades or the sources of water that fed it, Lake Okeechobee and the Kissimmee River Valley. At that time, during the middle of the nineteenth century, most of the Everglades was a vast, regionally structured peatland formed in a subtropical climate. It included a nearly imperceptible slope that was nevertheless sufficient to keep the water always in motion: "limpid" and "not stagnant," as the early accounts emphasized. It truly was an enormous, exceedingly flat, and seemingly limitless place. Only sawgrass, sloughs, and scattered tree islands were visible to the south, while farther north, just unbroken tall dense sawgrass, "as far as the eye could see."

The peat soils of the Everglades filled the bedrock basin, 40 miles (60 km) wide, out to the tree lines on the western and eastern borders (plate 2). This same peat formed the southern shoreline of Lake Okeechobee, 70 miles (110 km) long (plate 6). Water flowed from Lake Okeechobee over this peat shoreline directly into the Everglades. The water covered the full 40-mile width of the Everglades, flowing slowly southward as a broad, shallow sheet, reaching outlets in Biscayne Bay and the Gulf of Mexico more than 100 miles (160 km) downstream. The slope, a drop of a bit over 20 feet (6 meters) along 100+ miles (160+ km) or 3 inches to a mile, while almost invisible, was still adequate to create the apparently critical distinction between lake and "River of Grass" (Douglas 1947).

Depending upon the traveler's direction of approach, reaching the edge was often marked either by relief that the immediate struggle had ended or by dismay at the new struggles ahead if the traveler chose to proceed onward into the Everglades. Coming westward from the Atlantic Coast or northeastward from the Gulf of Mexico, explorers usually paddled, often strenuously, up one of the coastal rivers. These were typically enclosed by forest, so explorers noted the sudden widening of the view when they finally reached the Everglades, as well as the silent, suddenly calm water:

> Just at sundown we arrived at the end [head] of the [New] river. Here the water was pouring from the Everglades through a narrow channel and running so swiftly it taxed our strength to the utmost to paddle the old canoe through. When we reached the still water of the Glades we paused to eat. (Pierce 1970, p. 158)

> The approaches to the Everglades are singularly beautiful. . . . The ideal approach is by water, and the Everglades rivers are not matched for beauty anywhere else in the world. . . . As one ascends the river, and the fresh water of the Glades overcomes the brackish tidal water, the mangrove gives way to its next of kin, the cocoa-plum, a tree of similar habit of growth. Farther up, the cypress fills the shore-line, and the lilies come into view. The whole voyage is a scene of shifting greens—the dark olive of the mangrove, the lemon-like foliage of the cocoa-plum, and the lighter green of the growing cypress, touched into life by vivid sunshine, making a scene of unique beauty. This beauty is enhanced by the "dark water," as it is called, which flows out of the Glades. . . .
>
> . . . Seven miles up the New River, and three miles up the Miami, the rivers open suddenly into the vast expanse in which they have their source; and the wide reaches of the "grass-water" seem to drink up the sunshine only to give it form again in waving grass and gleaming water, in swaying vine and star-bright lily and orchid. (Dix and MacGonigle 1905, p. 526)

Though relatively still when compared to the swiftly flowing rivers, the water of the Everglades was not at all stagnant, a situation that was frequently noted, with some surprise, by explorers navigating farther west into the Everglades: "everywhere, and at all seasons of the year, the water in the Glades is clear, pure, and, though sometimes warm, palatable, without the least suggestion of staleness or stagnancy. One of the most singular conditions of the region is that the water seems to move in one mass, with a general trend from the northeast in a southerly direction toward the outlets by which it finds its way into Biscayne Bay and the Gulf of Mexico" (Dix and MacGonigle 1905, p. 521).

Approaching the eastern side of the Everglades, explorers encountered either the Marl Marshes, discussed later, or more commonly, the Ridge and Slough landscape (plate 2). This landscape formed a patterned mix of three main elements: long, linear, slight ridges covered with dense sawgrass; "channels" or "leads" of water, often covered with water lilies; and slightly elevated tree islands. The channels and leads, now called "sloughs," allowed for some degree of navigation:

> Seen in perspective when the water is low, the Everglades resemble far-reaching prairies, with here and there distant clumps of trees outlining islands. Even when the water is high it is not prominently seen, as the interminable growth of tall [saw] grass hides it all, save in the numerous little channels which wind aimlessly about, sometimes leading into a closed pocket, sometimes broadening into open spaces abloom with pond-lilies. This level, prairie-like effect has a peculiar charm for the eye—something like that of the level yet living monotony of the sea. It suggests silence, but not stagnation; repose, but not death. Explorers all bear testimony to the impulse to adventure into this spreading plain of grass and water—a little farther, and still a little farther; always seeking that which lies just beyond the horizon. (Dix and MacGonigle 1905, p. 525)

But passage was not always easy, especially if one's route happened to head crosswise, rather than parallel to the linear ridges and sloughs: "there is no obvious method of forcing a passage. The region is not exactly land, and it is not exactly water. That is, you cannot travel by land, because there is water there, now shallow, but again rather deep. And you cannot travel by water,—at least not freely and at will,—because there is land there, or, rather, high, close-grown, saw-edged grass, which as effectually bars the progress of a boat. An amphibious motorcar with scythe-bearing attachment would appear to be the only kind of machine that might reduce this natural fortress, and that machine has not yet been constructed" (Dix and MacGonigle 1905, p. 513).

Travelers entering the northern Everglades from the east encountered an arm of the Ridge and Slough landscape (plate 2), parallel to what is now the L-8 canal (fig. I.1). In contrast, travelers entering the northern Everglades from the west encountered a very different landscape, the Sawgrass Plains. Instead of the mix of water lily sloughs, sawgrass ridges, and tree islands encountered in the Ridge and Slough landscape farther south and east, travelers came eastward out of the woodlands to find a continuous mass of sawgrass, the Sawgrass Plains landscape. The descriptions from surveyors measuring out the townships and ranges across the Florida peninsula echo the vastness seen on the east side, but are more workaday, perhaps a reflection of their profession but perhaps also of the less inviting expanse of continuous, impenetrable sawgrass: "On the west edge of dense Sawgrass marsh. Beyond this point the Sawgrass extends S.E.-E-NE & N.W. in an unbroken and dense marsh with two feet or more of water at all times" (Tannehill 1871-T42 R31; see list of Township Survey References, page 321); "Beyond 80 chains—dense sawgrass at all times overflowed & worthless" (Tannehill 1871-T42 R32).

Unlike the explorers who found arrival at the edge of the Everglades a welcome relief from the strong currents of the eastern coastal rivers, the surveyors arriving at the western edge of the northern Everglades abandoned further progress as "impracticable":

> 40.00 [chains] 1/2 Mile. Made mound & pits in edge of impracticable Sawgrass bordering . . . the Everglades. The sawgrass stretches as far as the eye can see, forming a perfectly level horizon eastward. Tuesday March 19th [1872]. Relinquish line. (Apthorp 1872-T46 R34, p. 39)
>
> I find the Eastern portion [of this Township] to be one vast sea of overflowed land, as far as the Eye can reach to the Eastward and is known as the Everglades it being impracticable to survey on account of the water and sawgrass. (Stearns 1874-T47 R34, p. 264½)

Prior to the onset of man-made drainage, as Everglades water levels rose and fell each year under the influence of seasonal rainfall and inflows from Lake Okeechobee, the large northern peat landscape called the Sawgrass Plains alternated between being inundated and being boggy—damp peat but without surface water. Farther south, the predrainage microtopography of the larger Ridge and Slough landscape, which gave rise to descriptions of the Everglades as half land, half water, also was subject to the ebb and flow of seasonal moisture. During the wet season, water levels rose to cover the peat of the ridges and, for short periods, the peat surface of many tree islands as well. By the end of the dry season, water levels typically dropped below much of the ridge peat, but water generally still remained within the sloughs, perpetuating the half-land, half-water landscape. It is likely that this year-round persistence of water in sloughs was a critical component of predrainage Everglades ecology, especially for aquatic species. Exchanges—of water, of organisms, of chemical constituents—between ridges and sloughs were likely also important.

The preceding passages, echoed by many, are glimpses back through time that paint a picture of the Everglades as it existed before the first man-made major alterations. The chapters that follow provide detailed descriptions of the predrainage Everglades, gleaned from a plethora of historic sources, including notes from explorers' expeditions and surveyors who traveled and worked in the Everglades during the early to mid-nineteenth century.

Introduction

This study was initiated to support hydrologic and ecological restoration of the Everglades, a heavily drained ecosystem. Given the absence not only of predrainage hydrologic time series measurements but also of predrainage ecological studies, it seemed important to assemble the best possible sources to yield a qualitative and quantitative picture of original, "natural" Everglades conditions. A synthetic study in historical ecology was needed, combining historical predrainage information with modern tools such as remote sensing and geographical information systems. Several factors suggested that quantifying the predrainage hydrology of the Everglades might be especially important: (1) the climate is one of strongly seasonal, yet variable, rainfall with the associated annual potentials for both flood and drought; (2) the absence of readily available locations for water storage, other than the remaining present-day Everglades and Lake Okeechobee; and (3) the direct adjacency of three large, potentially competing users of water, two of which also need flood control: a rapidly growing urban population (6,000,000 and growing), a large agricultural area, and the remaining Everglades (fig. I.1; plate 2b). Finally, the competing demands for water in South Florida are sufficiently strong that it has frequently been tempting to cite selectively only historical sources that support a particular interest. We strove in this study to overcome such temptations by examining the full historical record exhaustively, with the hope of producing an objective view of this time period.

It also seemed important to not be daunted by the absence of traditional time series measurements of predrainage stages, flows, water quality, etc. We were conscious that the lack of such measurements could mean an a priori dismissal of an ecological history study as merely anecdotal. However, to us it seemed wiser, especially in the face of one of the larger, if not the largest, ecological restoration projects ever undertaken, to subscribe to the opposite view. We chose to assume that ample and sufficiently precise information would be available from multiple firsthand observers. Further, we assumed that enough consistencies or inconsistencies would emerge to provide a measure of the reliability and accuracy of the resultant predrainage picture. It seemed a responsibility to comb the widest variety of available sources of direct and indirect information, gleaning the maximum possible amount of scientifically useful predrainage knowledge. We brought to this task the perspectives of a systems ecologist and soil physicist, a botanist, and a hydrologist.

The specific objectives of this study were to:

1. Map and characterize the vegetation, soils, regional topography, microtopography, and hydrology of the predrainage landscapes of the Everglades;
2. Quantitatively estimate predrainage water depths and hydroperiods for each landscape;
3. Map and characterize the landscapes bordering the Everglades sufficiently to understand their individual relationship to overall Everglades hydrology;
4. Map predrainage patterns of Everglades water flow;
5. Document changes that occurred within the Everglades between the time of first drainage (1880s) and time of first system-wide, scientific mapping (1940s);
6. Make available to the reader as much as possible of the "data" we used to achieve the previous objectives, e.g., photographs, maps, and essential portions of historical accounts.

Scope

Most maps of southern Florida from the nineteenth and early twentieth centuries include lines labeled as "Border of the Everglades" (plate 2a). The Everglades was originally visibly distinct, and the edge was easily recognizable. This Everglades extent was the focus of this study. Additional attention was directed to the bordering landscapes, particularly to the relation between water levels there and within the Everglades, and to understanding the flows of water in and out of the

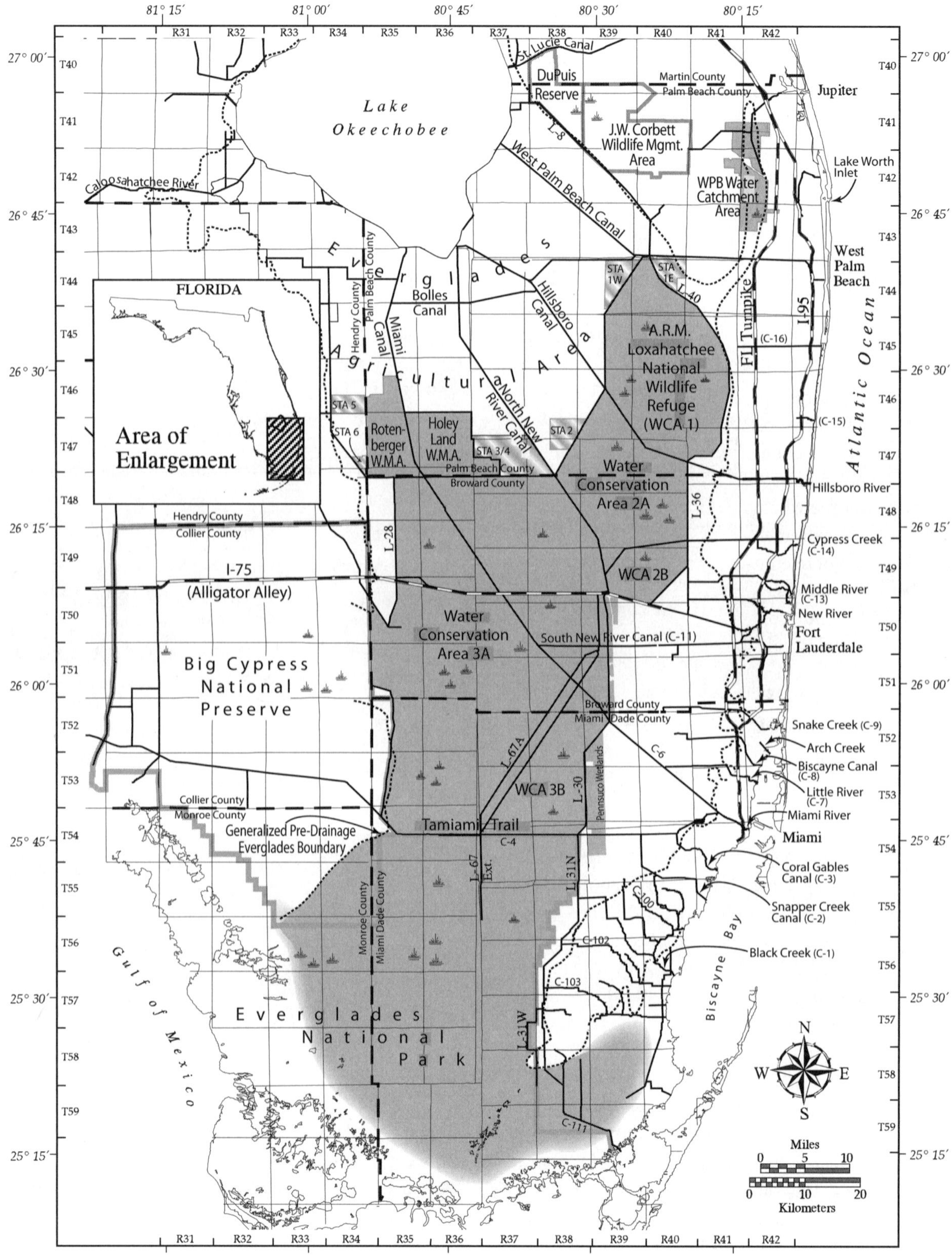

Figure I.1. Location of major canals, levees, and land units present in South Florida in 2010. Gray shading indicates remaining portions of the Everglades. Diagonal shading in Townships T44, T46, and T47 indicates planned or constructed Stormwater Treatment Areas (STAs), designed to remove phosphorus from water to be delivered to the Everglades.

Everglades. However, the coastal mangroves were not investigated.

The time periods of focus were primarily the 1800s and, secondarily, the early postdrainage period extending into the 1940s. Later drainage continuing into the 1960s, in and around the Everglades Keys (plate 2a), was generally not addressed. The ecological picture of the Everglades developed in this study may be thought of as applying to the mid-1800s—that is, 1850 plus or minus a few decades. The focus on the mid-1800s is natural. The numbers of recorded observations of the Everglades started increasing in the 1830s, and by the 1880s the predrainage period was over, as large-scale drainage works began altering Everglades hydrology. We make additional use of postdrainage observations from as late as the 1940s. These early postdrainage observations are, in some cases, similar enough to the predrainage ones to be useful supplements. Additionally, mapping studies conducted over the whole Everglades in the early 1940s proved to be an extremely useful tool for locating geographical points of reference and identifying the context for predrainage sources.

Ecologically, the scope of this study included the physical and biological, particularly vegetation, and generally neglected chemistry. This emphasis was dictated largely by the kinds of information gleaned from historical sources. Knowledge of predrainage water and soil chemistry would be quite helpful, but unfortunately neither early explorers nor military commanders nor land surveyors recorded such data. Information on predrainage microbiology would also have been useful, but again the historical record, either directly or by inference, seems to be quite silent on this aspect. The historical record of predrainage fauna is also not extensive, and its applicability is limited somewhat by animal mobility.

Physically, we focused on the wetland soils associated with different portions of the Everglades; on regional topography, which changed substantially as drainage brought about oxidation and subsidence of the organic peat soils; on microtopography associated with the patterned peatland portions of the Everglades; and on hydrology.

Regarding hydrology, our attention was drawn to the original water depths, for which the historical record proved quite rich, providing detailed descriptions of the locations of inflows and outflows, the climate-driven annual rise and fall in water depths, and the types and directions of surface water flow.

Our overall focus was primarily at the scale of what we refer to as "landscapes"—the recognizable physiographic regions of distinct vegetation, soils, microtopography, and hydrology in the Everglades that extended over tens of miles. An interest in landscapes meant attention to the scale of important variation within each landscape. This was quite extensive within the apparently very uniform Sawgrass Plains but was reduced to tens or hundreds of feet within the Ridge and Slough landscape to take into account the pattern of ridges, sloughs, and tree islands. For the most part we used English units of measurement, not SI (International System of Units), because nearly all original sources used these units.

Rationale

The research reported here is an exercise in historical ecology. It is an attempt to paint a picture, using scientific interpretation as quantitatively as possible, of the main landscapes of the predrainage Everglades as they would have been seen during the middle of the nineteenth century. The resulting picture is our interpretation of the historical information; in essence, a hypothesis. Unlike hypotheses from other fields of science, it is impossible to test this hypothesis, which would require direct observations of the original system. Yet it is perhaps insightful to quote Marguerite Yourcenar, the first woman to be elected to the Académie Française, as she reflected upon researching her remarkable work *Memoirs of Hadrian*: "which is not to suggest, as is too often done, that historical truth is never to be attained, in any of its aspects. With this kind of truth, as with all others, the problem is the same: one errs, more or less" (Yourcenar 1986, p. 276).

Though the task Marguerite Yourcenar set herself—to enter the mind of a genius, to speak in his voice as he approaches death—was infinitely greater than ours, her approach to the preparatory scholarship, as she describes it, bears uncanny similarities to ours:

> The rules of the game: learn everything, read everything, inquire into everything, while at the same time adapting to one's ends the Spiritual Exercises of Ignatius of Loyola, or the method of Hindu ascetics, who for years, and to the point of exhaustion, try to visualize ever more exactly the images which they create beneath their closed eyelids. Through hundreds of card notes pursue each incident to the very moment that it occurred; endeavor to restore the mobility and suppleness of life to those visages known to us only in stone. When two texts, or two assertions, or perhaps two ideas, are in contradiction, be ready to reconcile

them rather than cancel one by the other, regard them as two different facets, or two successive stages, of the same reality, a reality convincingly human just because it is complex. Strive to read a text of the Second Century with the eyes, soul, and feelings of the Second Century; let it steep in that mother solution which the facts of its own time provide; set aside, if possible, all beliefs and sentiments which have accumulated in successive strata between those persons and us. And nevertheless take advantage (though prudently, and solely by way of preparatory study) of all possibilities for comparison and cross-checking, and of new perspectives slowly developed by many centuries and events separating us from a given text, a fact, a man; make use of such aids more or less as guide-marks along the road of return toward one particular point in time. Keep one's own shadow out of the picture; leave the mirror clean of the mist of one's own breath; take only what is most essential and durable in us, in the emotions aroused by the senses or in the operations of the mind, as our point of contact with those men who, like us, nibbled olives and drank wine, or gummed their fingers with honey, who fought bitter winds and blinding rain, or in summer sought the plane tree's shade; who took their pleasures, thought their own thoughts, grew old, and died. (Yourcenar 1986, p. 276–277)

While our focus was on the reconstruction of landscapes rather than on an unusual human being, we too sought to "learn everything, read everything, inquire into everything," to make notes on each firsthand source, to commit to memory all place names, to overlay each map, and to "try to visualize ever more exactly" the details of landscapes already a century and a half old. We too compared multiple sources, cross-checking for consistencies (very common), seeking to understand contradictions (very rare). Later, secondary studies were consulted with some circumspection and compared with our own developing images. We too strove to "leave the mirror clean" of our own mists; how well we succeeded is for the reader to judge. To assist in such judgment, we have included as many as possible of the relevant portions of the source materials we consulted (see DVD).

Chapter 1 thoroughly examines the sources we used, but a word on approach is appropriate here. While it is true that we did "inquire into everything" in the sense of not excluding any types of sources, it is also true that we did not "read everything." We read voraciously, omnivorously, but not systematically. Thus we have looked at sources as varied as newspaper articles, business letters, maps, military histories, the minutes of state agency meetings, land surveyors' field notes, hunting guides, transcribed oral histories, and more. Over 900 sources have been cataloged and annotated. With the exception of land surveys, our approach to finding sources was generally more relational than systematic, following leads backward from one to the next, pursuing the firsthand ones most vigorously. We did not, however, systematically read, for instance, all the Miami newspapers for any time period, relying instead on leads from other sources to point toward the most relevant articles. At some point, the sources making up this web began converging, with new suggestions turning up as already examined or as repetition of the content of an already examined source. This convergence was reassuring, leading us to believe that we probably examined perhaps 90 to 95 percent of the existing historical sources likely to contain scientifically useful information regarding the predrainage Everglades.

A Brief History

Dating of basal peats suggests that the Everglades are about five thousand years old. Organic soil (histosols, or simply "peat") began forming as wetland plant remains accumulated within a bedrock trough (fig. I.2). By the 1800s, and perhaps before that, peat had accumulated to the point that the water surface on top of the peat was about even with the elevation of the bordering mineral ground surface (Wright 1912, p. 17). Accumulation to this elevation may reflect a point of equilibrium; further accumulation and increased height above the surroundings likely would have increased rates of peat oxidation. At its thickest, near the southern shore of Lake Okeechobee, the peat accumulation was 10–12 feet (3.0–3.7 m) deep by the late 1800s. The peat in this area may have accumulated to sufficient height to have formed a dam, allowing Lake Okeechobee waters to rise to higher levels (Gleason et al. 1984, p. 301). Interestingly, although there are probably too few cores to draw definitive conclusions, the available studies of peat soil cores suggest that the vegetation seen in the large peat landscapes of the Everglades in the 1800s may have been present during much of Everglades history. Cores from the Sawgrass Plains landscape studied by Dachnowski-Stokes (1930) suggested that sawgrass (*Cladium jamaicense*) was present throughout most of the length of the undated cores (fig. 6.2), except for layers of amorphous peat that

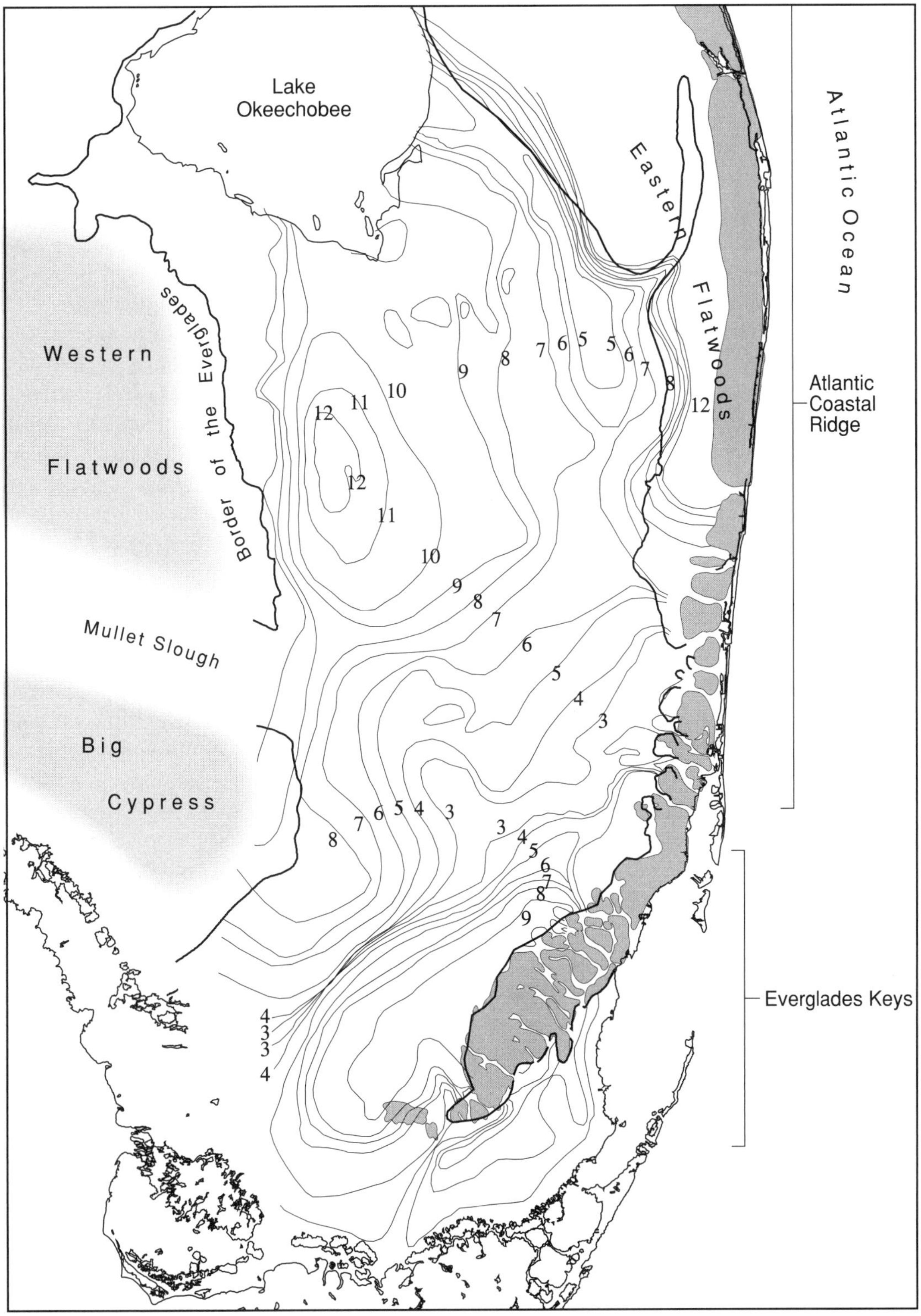

Figure I.2. The Everglades basin and bordering uplands. Contours show bedrock elevation in feet above Punta Rassa datum (Jones et al. 1948e, supplemented by Parker et al. 1955). Eastern and western shading indicate slightly higher bordering landscapes; eastern areas delineated by Leach et al. (1972).

occurred at similar elevations, and presumably during similar time periods.

In contrast, soil cores taken within the Ridge and Slough landscape and studied by Cohen and Spackman (1984), Gleason et al. (1994), and Willard and Holmes (1997) show alternating layers of sawgrass peat and slough (water lily) peat. This suggests that while the exact locations, and widths, of individual sloughs and ridges may have shifted over time, the presence of a landscape of alternating sloughs and ridges has existed for as much as a millennium (Bernhardt et al. 2004, p. 12; Bernhardt and Willard 2009, p. 10).

Starting in 1884 engineers began drawing water from Lake Okeechobee as a first step toward draining the Everglades. That year, therefore, marked the distinction between the "predrainage" and "postdrainage" Everglades, terms that occur frequently throughout this book. Although the historical record is somewhat unclear, it appears that the efforts of Hamilton Disston's company, the first to use mechanized means to dredge canals, were in fact successful in lowering Lake Okeechobee water levels, perhaps by as much as three to five feet below normal, for a number of years, possibly until as late as the mid-1890s. Such a reduction would certainly have affected Everglades hydrology.

About 1900–1905 lake levels appear to have risen again, and the State of Florida initiated a second wave of drainage activity focused on dredging canals cutting northwest to southeast through the Everglades. Those four major canals were intended to drain both the Everglades and Lake Okeechobee and appear to have been very effective. The first of these, the North New River Canal, was opened in 1912; the last, the West Palm Beach Canal, was opened in the early 1920s. Chapter 3 recounts the severe hydrologic and ecological effects of these canals.

By the 1940s concerns about overdrainage of the Everglades, inadequate drainage of the agricultural lands close to Lake Okeechobee, and the need for flood control after the enormous flood of 1947 prompted initiation of the extensive federal water control project, the Central and Southern Florida Project for Flood Control and Other Purposes. With most of the construction occurring during the 1950s and '60s, this project expanded the canal system, added levees, and created the system of impounded Water Conservation Areas in place today.

The levees, which were built as part of the water control system implemented by the Central and Southern Florida Project, exerted mixed ecological influence on the remaining Everglades. On the one hand, they helped raise water depths closer to predrainage levels; on the other, they created wedges of water depth with some areas too shallow and others too deep. Additionally, the levees essentially eliminated sheet flow, the movement of water across the whole surface of the landscape, "as if in one mass."

By the 1990s momentum to restore the Everglades had mounted sufficiently that the U.S. Congress authorized a review of the federal Central and Southern Florida Project for Flood Control and Other Purposes with a focus on restoration of Everglades hydrology. In 2000 Congress approved the Comprehensive Everglades Restoration Plan, directing it to be adaptive to new knowledge and to be based on scientific information.

Present-day conditions differ greatly from predrainage conditions. More than half of the Everglades has been drained for more than a century, with a large northern portion farmed (Everglades Agricultural Area, EAA) and a large eastern portion urbanized (plate 2b). Where once the whole southern shoreline of Lake Okeechobee flowed freely into the Everglades throughout much of the year, those flows are now blocked by the 30-foot-high Hoover Dike, which protects towns and the farmed northern portion from the historical seasonal inundation. The remaining portions of the original Everglades are owned by the State of Florida (north of Tamiami Trail, fig. I.1) or by the federal government (Everglades National Park, south of Tamiami Trail). While the federal portion has a strong mandate for environmental protection and preservation, the state portions upstream from it have multiple-use mandates and are referred to as "Water Conservation Areas" (fig. I.1).

The five Water Conservation Areas are impoundments separated by levees. The flow of water between these Water Conservation Areas and thence into Everglades National Park no longer occurs as surface water "moving in one mass" across the whole landscape but, instead, is through severely concentrated point sources controlled by water management gates. Under the multiple-use mandates, water depths within the Water Conservation Areas can be lowered to respond to urban or agriculture water supply needs or raised in response to flood control needs. In addition to obstructing the flow of water across the landscape, the levees that impound the Water Conservation Areas also create wedge-shaped profiles of water depth: generally shallower than predrainage depths toward the upstream end of each Conservation Area and generally deeper toward the downstream ends. Canals further alter water movement across the whole landscape by creating low resistance paths that bypass or "short-circuit" overland flow.

Drainage, elevated influxes of nutrients, compartmentalization, and altered water depths have all affected the remaining Everglades. Chapter 3 examines some of these changes in detail. Severe early drainage caused large-scale oxidation, subsidence, and outright burning of peat soils, with losses of as much as 10 feet in elevation in the northern portions of the Everglades. Influxes of phosphorus to the canal water have created an expanding area of cattail (*Typha* spp.) invasion, eliminating natural vegetation and altering the peat microtopography; that is, the elevation differences between ridges and sloughs. Peat microtopography has also diminished, or even vanished, in other locations not strongly affected by phosphorous influxes. The originally strong peatland patterning formed by ridges and sloughs has been blurred, if not eliminated, by the combined influence of altered water depths and greatly reduced water flows caused by the mandated Water Conservation Area impoundments. Similarly, the elevation of some tree islands above the surrounding sloughs appears to have decreased, which exacerbates the water depth "wedge effect" and affects tree survival. Within Water Conservation Area 2A, the outlines of former tree islands are still visible, but almost all tree species have disappeared.

Despite these problematic changes, a substantial portion of the remaining Everglades can be restored to close to original conditions, provided that the original driving forces of hydrology and water chemistry are restored at the upstream border, and provided that surface water is once again allowed to flow slowly along the full length and breadth of the landscape.

1

Sources and Methods

Ecological characterization of a region ideally begins with field surveys. As firsthand observation of the predrainage Everglades is no longer possible, we relied by necessity on observations previously recorded by others. To accomplish this task, we examined a variety of historical sources including the descriptions, maps, measurements, routes, and photographs of early observers such as explorers, soldiers, surveyors, and scientists. As many sources as possible were located, cataloged, and annotated.

By comparing these sources with historical and modern mapping, it was possible to reconstruct the nature and extent of the various landscapes present within the Everglades during the 1800s and prior to mechanized drainage (before 1882). Information and concepts from soil science, plant ecology, hydrology, and environmental history contributed to the landscape reconstructions. Thorough cross-checking of the historical sources revealed consistency among the diverse accounts, maps, and measurements that, in turn, provided confidence in the accuracy of the reconstructions. The value of thorough cross-checking among multiple sources and source types cannot be overemphasized.

Numerous maps, historical and more recent, were enlarged or reduced to a common scale and overlain for comparison of spatial patterns. In almost all cases, patterns were easily recognized, even between maps produced with very different technologies and as much as a hundred years apart. For example, the Marl Transverse Glades, natural cuts between the higher-lying Everglades Keys, were drawn on township plat maps by John Jackson in 1847 (fig. 1.1). These same cuts are clearly recognizable on a drainage map (e.g., fig. 1.8) drawn seventy years later by John King (1917a), and on a soil map (plate 11) published one hundred years later (Jones et al. 1948a).

At the scale of the entire Everglades basin, overlays of maps of bedrock topography, ground surface topography, soils, and vegetation (e.g., fig. I.2, plates 11, 13, 14 and 16) revealed strong similarities in spatial pattern. Combined with the soil surveyors' descriptions of individual soil types, the overlays provided an understanding of the different soils of the Everglades and insights into the impacts of drainage upon them. With these insights, the soil map of the Everglades Drainage District from the 1940s (Jones et al. 1948a) became a key tool for interpreting historical observations of the predrainage Everglades.

Historical observations from narrative sources such as descriptions of expeditions, government township survey notes, etc., were overlaid on the 1940s soil map,

Figure 1.1. Sample township plat map, surveyed in 1847. The northwest portion of the township was part of the Everglades (Rockland Marl Marsh). Shapes in southern and eastern portions of township depict Everglades Keys; that is, higher-lying "islands" or "keys" of pine woods (triangular pine tree symbols) and the source of "Dade County pine." Narrow strips of "Sawgrass Prairie" between the Keys were lower-lying Marl Transverse Glades, now sites of drainage canals (cf. fig. I.1). See figure 1.8 and plate 4 for surrounding context. Source: Jackson (1847-T55 R39); "Sawgrass Prairie" and "Everglades" labels enlarged.

then evaluated within the context suggested by the surrounding soils. Similarly, the observations were compared with township survey plat maps from the 1840s and 1870s, which often indicated vegetation types (such as "pine [forest]," "cypress," "sawgrass," "Everglades") and showed locations of topographic highs and lows, Seminole settlements, and natural drainages. The use of maps to locate and provide context for the individual observations helped to delineate more accurately the predrainage landscapes and to estimate hydrologic characteristics for each.

The landscape names adopted in this study, with few exceptions, follow the original names used by authors prior to the 1950s. Similarly, the soil nomenclature of the Jones et al. (1948a, e) soil map is followed (appendix O), recognizing that it differs from contemporary soil taxonomy, but also noting that this soil classification was developed at the scale of the entire multicounty Everglades basin and that the names tend to reflect closer ties to predrainage Everglades conditions.

Overview of Primary Sources

A diverse range of source materials was consulted. These materials included vegetation, soil, and topographic mapping studies from the 1940s; government township surveys; accounts of expeditions into the Everglades; early maps; narrative accounts by persons acquainted with the predrainage Everglades; early photographs; construction surveys and profiles; hydrologic, climate, and soil records; and scientific studies (predrainage as well as postdrainage). Each category of source material was analyzed differently, as discussed in the following sections.

With regard to the selection of historical materials, some comment is appropriate. The approach might best be described as "omnivorous": any source encountered that might contain firsthand observations of the predrainage Everglades was carefully read. Leads to these firsthand observations were obtained opportunistically, initially from secondary sources such as Dovell (1947a), S. M. Davis et al. (1994), and Gleason et al. (1984); later from inspection of historical collections (especially the Historical Museum of Southern Florida and the P. K. Yonge Library at the University of Florida); and as the study progressed, from conversations with numerous individuals and historians knowledgeable about the Everglades.

The sources examined might be divided loosely into two types: "professional" sources, such as surveyors' notes, records of exploring or military expeditions, maps, and early land suitability assessments; and "cultural" sources, such as hunters' or anglers' accounts, newspaper articles, and accounts from early settlers. We researched these two types of sources differently due to differences in their accessibility. Maps, township surveys, early scientific studies, and accounts of cross-Everglades expeditions were searched systematically; that is, we strove to identify and obtain essentially all available examples in each of those categories. In contrast, we researched cultural sources less systematically. All specific leads encountered were followed, but we did not exhaustively examine periodicals, oral histories, or early settlers' histories for any particular time period. While the absence of systematic searching might present a possible source of bias, in practice it is likely that few relevant sources were omitted. The duration of the project, the authors' contacts with historians, and extensive examination of primary sources mentioned in secondary sources meant that we examined hundreds of cultural sources. While there is no way to know the number of sources not encountered, it seems likely to be small.

All sources examined contributed to the overall predrainage picture developed, but those that could be more precisely identified in both space (e.g., by landscape) and time (e.g., by pre- or postdrainage period and by time point within the dry season) were weighted more heavily. Surprisingly, it proved possible to locate many sources quite specifically, both in time and space, particularly once knowledge of early place names was developed. Part way through the study, a "saturation phenomenon" was increasingly observed: as we examined additional sources, they tended to repeat previously obtained information rather than providing new or significantly different information. While those additional sources did not often result in new insights regarding predrainage conditions, they did often provide useful corroboration of other sources as well as a useful indication that a large proportion of the predrainage literature probably had been examined.

Two types of information may have been underutilized in this study: oral histories and archaeological studies. Oral histories were examined only indirectly, through those already published or used in secondary sources. Oral histories, especially from the Seminole and Miccosukee tribes, may yield additional information, especially if they include clues that could reliably source the observations to predrainage times and to specific locations.

Archaeological studies of the Everglades, such as those reviewed by Griffin et al. (2002), were examined only minimally in this study. Archaeological information older than about 1800 would precede the time period of interest, 1840–1880, but later archaeological information might shed additional light on predrainage water depths, flora, and fauna. To do that, it would be important to establish whether the archaeological materials had been transported from distant locations. We did, however, examine paleoecological studies of vegetation history based upon the analysis of soil cores.

With the possible exception of the influence of underutilization of oral histories and archaeological sources, it

Table 1.1. Chronology of Everglades drainage activities (left column) and key sources used in this study (right column)

Drainage Activity/ Relevant History	Year	Information Source
	1775	B. Romans, *A Concise Natural History of East and West Florida*
	1823	C. B. Vignoles, *Observations upon the Floridas*
	1823	J. Knetsch, "William Cooley Explores the Everglades"
Second Seminole War	1835–42	
	1837	J. L. Williams, *The Territory of Florida: Or Sketches of the Topography . . .*
	1838–41	Col. Harney expeditions (Anonymous 1840, 1841)
U.S. Government resolution to reclaim Everglades	1845	G. MacKay township surveys
	1847	J. Jackson township surveys
	1848	B. Smith, *Report on Reconnaissance of the Everglades*
Swamp and Overflowed Lands Act passed	1850	
Internal Improvement Fund established	1855	
	1856	J. C. Ives, *Military Map of the Peninsula of Florida South of Tampa Bay*
	1870s	M. Williams township surveys
H. Disston drains large area in Kissimmee Valley	1880s	
Canals dredged from Lake Okeechobee into Caloosahatchee River	1882–83	
	1883	*Times-Democrat* expedition (Hopkins; Peoples and Davis; A. P. Williams)
	1885–86	W. Mickler township surveys
	1892	Ingraham expedition (Marchman; Moses)
	1892	J. Kreamer, *Map of Hic-po-chee and Okeechobee Sugar Lands*
	1897	Willoughby expedition, H. L. Willoughby, *Across the Everglades*
Everglades Drainage District created	1905	
North New River Canal begun	1906	
Miami and Hillsboro canals begun	1910	
	1907	J. T. Stewart, *Report on Everglades Drainage Project in Lee and Dade Counties, Florida*
	1911	U.S. Congress, *Everglades of F1orida* (Senate Document 89)
Miami Canal locks opened[a]	1911–13	
	1911–18	Numerous township surveys completed for the Trustees of the Internal Improvement Fund
North New River Canal completed	1912	
Last uninterrupted natural overflow from Lake Okeechobee	1912	
	1913	J. W. Harshberger, *Phytogeographic Map of South Florida*
Hillsboro Canal completed (approx.)[a]	1915	Baldwin and Hawker, "Soil Survey of the Fort Lauderdale Area, Florida"
	1917	J. W. King surveys and reports
West Palm Beach Canal completed[a]	1920	
Tamiami Trail and Canal completed	1928	
	1940	First aerial photos of the entirety of the Everglades (U.S. Department of Agriculture–Soil Conservation Service)
	1943	J. H. Davis, *Vegetation Map of Southern Florida*
Impoundment of Water Conservation Areas by levees	1954–63	
	1948	Jones et al. soils map (with *Soils, Geology and Water Control in the Everglades Region)*
Levee 30 constructed; southward flow pushed west	1953	
Levee 67 constructed; southward flow pushed west	1963	

Note: a. Opening dates for canals are sometimes not well defined as openings may initially have been partial, followed by later improvements, and/or full opening.

seems unlikely there was a systematic bias in the choice of sources consulted. The quantity and diversity of sources examined, along with the frequent similarities across sources, suggest that possible systematic biases probably have not greatly influenced the conclusions of this study.

The timeline shown in table 1.1 gives an overview of the timing and sequence of the main sources used and their relation to the timing of drainage milestones. The major phases of the Everglades chronology include early descriptions of the Everglades recorded prior to 1840; expeditions, surveys, and maps that were conducted and/or prepared from 1840 to 1910; construction of canals to the north and southwest of Lake Okeechobee in the 1880s; construction of major Everglades canals between 1906 and the 1920s; and a second major period of levee and canal construction primarily during the 1960s and 1970s. Systematic scientific investigation of the Everglades began in the middle of the twentieth century.

Mapping Studies from the 1940s

The decade 1938–1948, here referred to as the 1940s, is a key period in the scientific investigation of the Everglades. Three major efforts, each distinguished by extensive fieldwork and a synoptic analysis of the Everglades as a whole, occurred during this period. John H. Davis Jr. of the Florida Geological Survey described, classified, and mapped the vegetation of southern Florida (1943a, b). A multiagency team of researchers from the University of Florida Agricultural Experiment Station, the U. S. Department of Agriculture–Soil Conservation Service (USDA-SCS), the U.S. Geological Survey, and the Everglades Drainage District surveyed both surface and bedrock elevations (Jones et al. 1948d), classified and mapped the soils (Jones et al. 1948a, e), and mapped (in simplified form) the natural vegetation or human land use (Jones et al. 1948a). The U.S. Geological Survey, under the direction of Garald Parker and numerous colleagues, conducted extensive hydrological studies, particularly in the southern Everglades (Parker et al. 1955).

These three studies made use of the earliest set of aerial photographs of the entirety of the Everglades (USDA-SCS 1940), a key source of information that covered the whole of the Everglades and the Atlantic Coastal Ridge. Earlier aerial photographs of the coastal portions of southern Florida exist (U.S. Coast and Geodetic Survey 1928a, b, c) but do not include the interior (main body) of the Everglades. Many predrainage landscape patterns are still visible in the 1940 U.S. Department of Agriculture–Soil Conservation Service (USDA-SCS) photographs despite hydrologic changes and some degree of urban or agricultural development. The index for the USDA-SCS aerial photos was compiled in early June 1940, implying that the area was flown during the dry season, sometime between January and May 1940. Rainfall in 1940 and in 1939 was average; 1938 was one of the sixth driest of the 100-year record. The photos, therefore, reflect normal dry season conditions directly preceded by an average year.

The soil and land cover maps (plates 11, 14) produced by Jones et al. (1948a), and the vegetation map (plate 13) produced by J. H. Davis (1943b), are particularly important to the research presented here and are referenced repeatedly. They are important because of (1) the timing of the field research in relation to the chronology of drainage and drainage effects and (2) the researchers' approach to classification of vegetation communities and soil types.

The Jones et al. and the J. H. Davis mapping efforts were conducted sixty years after managed drainage began and approximately twenty-five years after completion of the four major canals: North New River, Miami, Hillsboro, West Palm Beach canals. Therefore the maps represent an already altered ecosystem. They are nevertheless very useful, for three reasons. First, at the time of mapping in the 1940s, maps of water districts (Jones et al. 1948b) and aerial photographs (USDA-SCS 1940) indicate that agricultural and urban land use of southern Florida was still restricted mainly to uplands, leaving much of the Everglades undeveloped and covered with some form of native vegetation, albeit disturbed. Second, the extensive system of levees that would later impound the Everglades into a series of basins had not yet been constructed. Third and most important, Everglades vegetation and soils were still responding to the long-term hydrological stress imposed by drainage. The lag between the onset of drainage and the change in soils and vegetation is not sharply defined and, depending on the characteristic temporal scale of the variable for change, may require decades or longer (Gunderson 1994). For those areas with longer lag times, this means that a map made thirty years after drainage began may still contain useful clues regarding predrainage conditions. The process of adjusting (hindcasting) those portions of the Jones et al. soil map (1948a) affected by drainage is discussed in more detail in chapter 4.

The second reason for the importance of the J. H. Davis (1943b) and Jones et al. (1948a) maps is their approach to classification. Both classification systems were clearly based on consideration of the Everglades as an entire ecosystem (J. H. Davis 1943a; Jones et al. 1948e).

Field observations from all parts of the Everglades and surrounding uplands were evaluated to develop the descriptions of vegetation communities and soil types, from the western border with the Big Cypress Swamp to the eastern border along the Atlantic Coastal Ridge, from north of Lake Okeechobee to the mangrove fringe along the southern coast. Unfortunately, subsequent mapping efforts have frequently lacked this ecosystem-wide perspective.

Soil Map (1948)

Because vegetation responds more quickly to hydrological change than do soils, soil information was weighted more heavily throughout this study as a basis to estimate predrainage conditions. As a result, we relied extensively upon the Jones et al. soil map (1948a). The detailed soil map (1948a, shown much reduced in plate 11), a more general version showing agricultural capabilities (1948c), a topographic map of the same extent (1948d), and an associated report, *Soils, Geology, and Water Control in the Everglades Region* (1948e) arose out of a major research effort conducted in the Everglades from 1938 to 1948. The multiagency and multidisciplinary effort involved the U.S. Dept. of Agriculture–Soil Conservation Service (A. R. Stephens 1943; Evans and Allison 1942; Gallatin and Henderson 1943; J. C. Stephens 1943), the Florida Geological Survey (J. H. Davis 1943c), the U.S. Geological Survey (Parker and Hoy 1943; Ferguson 1943), the Everglades Drainage District (Bestor 1943), the University of Florida Agricultural Experiment Station (Allison 1943; Henderson 1942; Neller 1943), and the Everglades Fire District (Bender 1943).

The Jones et al. soil map (1948a) encompassed the former Everglades Drainage District—specifically, all of Broward County; most of Palm Beach and Dade counties; the eastern portions of Collier and Hendry counties; and portions of Monroe, Glades, Highlands, Okeechobee, Martin, and St. Lucie counties. The map consists of thirty-eight sheets at a scale of 1:63,360 (1 inch = 1 mile), which are labeled as "complying with the national map accuracy requirements." A combination of aerial photography at a scale of 1:40,000 (USDA-SCS 1940) and extensive field surveying provided the basis of the Jones et al. soil map:

> Land surface elevations were determined to 0.1 foot at 330-foot intervals, and depths of peat soil were measured by probings at intervals of 660 feet. Approximately 2,000 miles of levels were run by Soil Conservation Service engineers. (Jones et al. 1948e, p. 51)
>
> The intensity of the [field survey] coverage varied greatly, according to nature of the land, existing use, and probable capability for use. In the areas of peat and muck soils, the land now cultivated was investigated in every field where necessary, to locate significant boundaries. On the open sawgrass plains, east-west lines were run every 6 miles and supplementary north-south lines were run wherever additional information appeared to be necessary. . . . Borings into the peat and marl were made with a 2-inch auger every half mile. The samples obtained were examined and were preserved for further observation and analysis wherever the material differed appreciably from that obtained in nearby borings. (p. 59)

Digitization of the map in 1995 for the present study and digital comparison with more recent data confirmed the high degree of horizontal surveying accuracy.

Map information collected by the Jones et al. surveyors concentrated on "soil conditions that influence the use and management of the land" and consisted of soil type (sixty-five categories), soil depth (four categories), ground slope (three categories), and vegetation cover (seven categories) or land use (five categories). The collected data formed the basis of the soil map (plate 11) as well as a map of vegetation ("land cover") for those areas still undeveloped in the 1940s (plate 14). Soil types were characterized as to texture and structure of the surface soil and subsoil, organic matter content, pH, permeability, and type and depth of underlying material. Soil depth was also measured and separately mapped.

Soils were classified according to the system prevalent at the time (USDA 1938), with factors specific to the Everglades noted by Henderson (1942). McCollum et al. (1976) discussed the relation between the soil types distinguished in Jones et al. (1948a, e) and in more modern soil classifications (such as USDA-SCS 1970; McCollum et al. 1978) for the Palm Beach County portion of the area. Although classification concepts have changed somewhat since the 1940s, the Jones et al. soil map (1948a) still stands as a valuable comprehensive description of the Everglades region, *as it was in the early 1940s*, and it remains the only soil map covering the entire Everglades basin.

While this soil map is a very accurate depiction of 1940s soil conditions, the severity of postdrainage changes experienced by the organic soils of the Everglades prior to the 1940s (chapter 3) indicates that it is not appropriate to consider this map an accurate depiction of predrainage soil conditions. As one of the tools for reconstructing predrainage conditions, the Jones et al. soil map (1948a) was therefore both extremely useful and at times challenging. Comparison of the map with information from other time periods suggested that for some locations, postdrainage hydrology had changed the soil *type*, but left the *shape* of the soil polygons essentially unaltered; in other locations, both type and shape were greatly altered. Successful use of this map for this study therefore required careful examination, taking into account position in the landscape, the type of soil, and site-specific drainage history.

For the purposes of this study, the sixty-five soil types identified by Jones et al. (1948a) included more detail than was relevant. The sixty-five soil types were therefore aggregated into fourteen soil groups (table 1.2). The aggregation was based on similarity of the soil profiles, topography, and location of the soils within the Everglades region. Visual comparison of the original (plate 11), and the grouped map (plate 15), suggests that most of the spatial detail was preserved. Three general physiographic regions were readily discernible: wetlands, uplands, and areas of intermediate hydrology. Identification of wetland soils (Okeechobee mucks, Okeelanta peaty muck, Everglades peats, Loxahatchee peats, Gandy peats, Ochopee marls, and Rockland area and Perrine marls) and of upland soils (the upland sands and Rockdale complexes) was relatively straightforward.

The hydrologically intermediate soils (Hialeah mucky marl, Davie mucky fine sands, Davie fine sands, and wet sands) include soils that were originally of intermediate hydrology and also soils altered by drainage

Table 1.2. Soil groups defined in this study and their geographic relation to postdrainage vegetation types and physiographic areas

Soil Group[a]	Constituent Soil Types[b]	Correponding Vegetation Types[c]	Corresponding Physiographic Areas[d]
Okeechobee mucks	20, 21	Custard-apple Zone	—
Okeelanta peaty muck	30	Willow and Elderberry[e]	—
Everglades peats	10, 11, 12, 14, 50, 51, 53	Saw-grass Marshes (medium dense to sparse) and Saw-grass Marshes (dense)[f]	Peat marshes
Loxahatchee peats	40, 41, 42, 43	Sloughs, Ponds, Lakes (with aquatic plants)[g]	Peat marshes (ridge-slough area)
Gandy peats	60, 61	Tree-islands, Bay Tree Forests	Peat marshes (islands)
Hialeah mucky marl	71	Saw-grass Marshes (with wax myrtle thickets)	Marl glades
Davie mucky fine sands	83, 83S	Wet Prairies, Saw-grass Marshes (with wax myrtle thickets)[h]	Sandy prairies
Davie fine sands	82, 82S	Wet Prairies[i]	Sandy prairies
Wet sands	80, 81, 84, 86	Pine Flatwoods Forests	Sandy prairies
Upland sands	85, 87, 87S, 94, 95P, 96, 98	Pine Flatwoods Forests	Sandy pinelands
Ochopee marls	72, 72S, 73, 73S	Cypress Forests, Marsh Prairies, Southern Everglades	Western marl glades
Perrine marls	74, 74S, 74V, 74Y, 74P	Southern Coast Marsh Prairies	Marl glades, mangrove swamp
Rockland	106	Marsh Prairies, Southern Everglades	Rocky glades
Rockdale complexes	99, 100	Miami Rockland Pine Forests	Rocky pinelands

Sources: Groupings from this study; soil numbers from Jones et al. (1948a, e), vegetation types from J. H. Davis (1943a, b), and physiographic regions from Gallatin et al. (1958, pp. 5–9).

Notes: a. Soil groups defined by aggregation of the soil types shown in second column. Plate 15 illustrates the groupings.

b. Numbered soil types are from Jones et al. (1948a); for detailed descriptions see appendix O, this study. Plate 11, digitized from Jones et al. (1948a), shows all 65 soil types and can be compared with the groupings shown in plate 15. The following minor soil types from Jones et al. were excluded from this study: 62, 63, 70, 74T, 75–78, 88–92, 95, 97, 101–105, 107.

c. Column shows the main vegetation types identified on J. H. Davis's (1943b) vegetation map that matched the soil groupings; determinations based on overlay. Additional minor areas of other vegetation types that fell within certain soil groups are indicated in notes below.

d. Column shows physiographic areas, identified by Gallatin et al. (1958) and Leighty et al. (1954) on the basis of associations of vegetation, soil, and hydrology, that matched the soil groupings. Matches determined by overlay and were generally fairly close.

e. Okeelanta peaty muck also includes some areas within the Custard-apple Zone.

f. Everglades peats also includes some areas within several other types: Saw-grass Marshes (with wax myrtle thickets); Saw-grass Marshes (with abundant ferns and cattails); Willow and Elderberry; and Slough, Ponds, and Lakes (with aquatic plants).

g. Loxahatchee peats also includes areas of Saw-grass Marshes (medium dense to sparse) and of Marsh Prairies, Southern Everglades.

h. Davie mucky fine sands also includes areas of Slough, Ponds, and Lakes (with aquatic plants) and of Saw-grass Marshes (medium dense to sparse).

i. Davie fine sands also includes areas of Saw-grass Marshes (medium dense to sparse); Saw-grass Marshes (with wax myrtle thickets); and Miami Open Pine Forests.

j. Wet sands also includes areas of Cypress Forests; Wet Prairies; Saw-grass Marsh (with wax myrtle thickets); and Freshwater Marsh (outside the Everglades).

k. Upland sands also includes areas of Saw Palmetto or Dry Prairie.

l. Ochopee marls also includes areas of Saw-grass Marshes (medium dense to sparse).

m. Perrine marls also includes areas of Tree Islands, Bay Tree Forests, and Cypress Forests.

n. Jones et al. (1948a, e) did not consider the Rockland area to be a soil type, but earlier information suggests that greater water depths and more soil were present prior to drainage. Rockland also includes areas of Saw-grass Marshes (with wax myrtle thickets).

and oxidation so that they artificially resemble intermediate soils. These "artificially intermediate" soils occurred especially on the western and eastern edges of the Everglades basin, where the open Everglades met the adjoining forested uplands. The soils in those areas were most severely affected by drainage. Changes in the intermediate soils are discussed in chapter 3; the conceptual reversal of these changes to produce a hindcast map of the soils as they likely appeared in the 1850s is discussed in chapter 4.

The rightmost two columns of table 1.2 illustrate the similarity in spatial distribution of the fourteen soil groups to the postdrainage distribution of main vegetation categories identified by Davis (1943a, b) and to postdrainage physiographic regions identified by Gallatin et al. (1958). The similarity in pattern can also be seen by comparing plate 15 (soil groups) with plate 13 (vegetation) and with figure 4.9 (physiographic regions—similar to those drawn by Gallatin et al. 1958). The patterns seen in all three cases reflect a combination of the original predrainage Everglades landscapes and the postdrainage spatial changes imposed by altered hydrology. As such the patterns represent a starting point, but not an ending point, for understanding the components of the predrainage Everglades.

Vegetation Map (1943)

The bulletin *The Natural Features of Southern Florida* (J. H. Davis 1943a) provides a thorough analysis of the landscapes and vegetation present in the 1940s. The well-known vegetation map (J. H. Davis 1943b) included in this bulletin is a map of the types and approximate distribution of the vegetation present at the time of mapping. This map is a depiction of 1940s postdrainage conditions rather than an attempt to draw the original predrainage vegetation. The only exceptions were for the areas where the native vegetation had already been cleared for agricultural or urban land use, particularly the Custard Apple Swamp (where the dominant tree was *Annona glabra*, also known as pond apple) originally present along the southern shore of Lake Okeechobee, and parts of the Eastern Flatwoods and the Atlantic Coastal Ridge. For these specific areas, Davis estimated the predrainage vegetation. However, the estimates appear to have been informal and not precise.

J. H. Davis either did not have the resources to conduct a spatially precise mapping effort or did not feel it was wise to do so. His map therefore differs noticeably from the contemporary Jones et al. (1948a) soil map. The locations and shapes of Davis's vegetation polygons are only approximate, limiting the usefulness of overlays to the regional scale. For example, where in the Jones et al. soil map (1948a) specific tree islands can be easily and unambiguously matched to present day islands through overlay in a geographical information system (GIS) (see, e.g., plates 31, 32), this cannot be done at all with the Davis (1943b) vegetation map. While this reflects a different mapping style than that used by Jones et al. (1948a), it does not in any way diminish the value of the Davis map as an insightful analysis of the different types of vegetation present in and around the Everglades in the 1940s and of their approximate spatial relationship.

As already noted, it is important to reiterate that Davis mapped the vegetation that was present in the early 1940s. Due to the severe impact of pre-1940 anthropogenic drainage (chapters 2 and 3), the Davis map therefore cannot be considered as a representation of predrainage vegetation.

Government Township Surveys

Federal and state township surveys and their accompanying field notes were an invaluable resource for this study, in fact meriting a separate list of references, see page 321. These data, obviously spatially referenced and dating primarily to the 1840s, 1870s, and 1910s, greatly helped in extending backward to predrainage times the view of the Everglades presented by the 1940s soil and vegetation maps.

Townships are (within measurement error) six-mile-by-six-mile (9.7 by 9.7 km) parcels of land with the four corners surveyed into place and aligned on a nominally north-south grid. The townships are referred to as "Township X South Range Y East" (usually abbreviated as T54 S R42 E, or simply T54 R42); X and Y refer to the number of townships away from a zero point in Tallahassee, Florida. The exterior boundaries of the townships usually were surveyed first and separately. The survey crew walked and marked each half mile (0.8 km) and mile (1.6 km) along each of the four six-mile-long boundaries. Sometimes conditions necessitated something other than walking, as indicated by the following excerpt from a 1911 survey:

> This is about western edge of the cypress . . . abandoned line here [middle of Section 34] to be resumed in Glades Boats. Resumed this line some months later. (Ensey 1911-T44 R41; South boundary of township)

As interest in settling an area increased, a separate surveying contract was typically granted to "subdivide" the

6	5	4	3	2	1
7	8	9	10	11	12
18	17	16	15	14	13
19	20	21	22	23	24
30	29	28	27	26	25
31	32	33	34	35	36

Figure 1.2. Standard numbering for the thirty-six 1-mile-square sections of a township.

township. These subdivision surveys crossed the township at one-mile (1.6 km) intervals, marking the corners of all thirty-six of the one-square-mile "sections." Numbering of the sections was standardized (fig. 1.2).

Field notes for the surveys included descriptions of the general character of the vegetation traversed during each mile as well as names of the specific "witness trees" used to locate the markers placed at each half-mile point on the surveyed one-mile sections. In the open Everglades the absence of witness trees was also indicated; this helped locate the transition between the bordering forests and the Everglades. "Pits and mounds" were sometimes used to mark the half-mile points if posts could not be set. Indications in the notes of conditions being, for example, "too wet for pits and mounds" provided additional information on water conditions at the time of the survey.

A change of survey instructions issued by the Trustees of the Internal Improvement Fund in approximately 1910 ensured that many of the surveys commissioned after that year included additional information on soil type, soil depth, and water depth. The surveys done during the 1910–1920 period provide exceptionally detailed information concerning Everglades landscapes. In many townships the location of each transition between slough and ridge, ridge and slough, ridge to channel, channel to island, etc. was explicitly measured (fig. 1.3). We used this information to help determine the predrainage extent of the Ridge and Slough landscape (chapter 8) and to quantify average widths of the ridges, sloughs, and channels that were characteristic of that landscape (appendix C). The presence of Ridge and Slough landscape within specific townships was determined through a semi-quantitative analysis of vegetation types encountered along the six-mile (9.7 km) transects traversed by the surveyors (appendix B; chapter 8). Similar tables were prepared for transects extending from the Ridge and Slough landscape into the bordering eastern uplands to understand this transition better (appendix D).

The observations recorded in the township survey notes provide a spatially objective assessment of the landscape. The locations of the sample points occur on a predetermined, fixed grid, without a bias that would emphasize any particular landscape or water depth. This objectivity, and the fact that it is combined with accurate knowledge of the observed locations and often with specific dates, make the township surveys an invaluable resource. It is worth noting that while spatial errors exist within surveys, for the purposes of our study, they are irrelevantly small. Cheves (2007), for instance, describes township T41 R37 along the eastern shore of Lake Okeechobee as one of the "top ten GLO [General Land Office] nightmares in the country." While this township clearly presents important legal and surveying problems, the underlying blunder of 10 chains (660 feet; 200 m) is for our purposes a reassuring indication that even the "worst case" spatial errors are far too small to affect our analysis.

Comparison of the township plat maps with the Jones et al. soils and vegetation maps (1948a) was useful for investigating changes in landscapes that occurred between the time of the township surveys and the 1940s. A special citation style was adopted in this research to identify specific surveys uniquely. This was necessary not only because the same surveyor often surveyed multiple townships within one year but also because the same township was often surveyed several times by different surveyors. Each survey is cited in this research with the surveyor's name, year, and township and range numbers; thus Jackson 1847-T56 R38 refers to John Jackson's survey of Township 56 S Range 38 E in 1847 (township survey references are listed separately, following the book's main reference list). Our analyses of the township surveys included the following aspects:

1. Creation of a comprehensive collection of Everglades region township plat maps reduced to a standard size (8.5 × 11 inches) for reference and comparison. This was helpful for identifying the various surveyors, years, and times of year the sur-

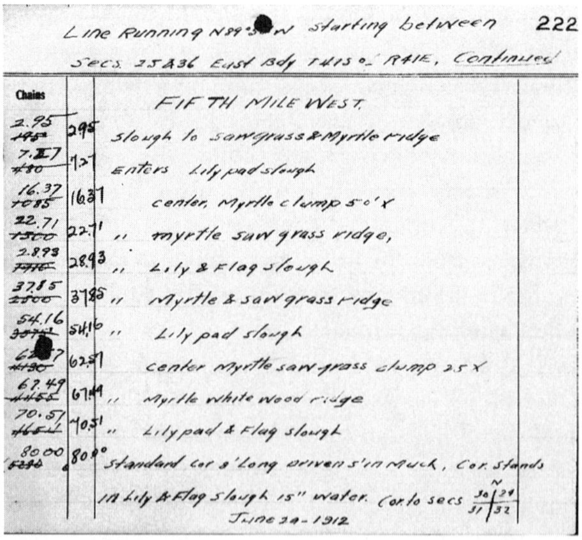

Line Running N89°...W starting between 222
Secs. 25 & 36 East Bdy T47S R41E, Continued

Chains		
		FIFTH MILE WEST.
2.95	295	Slough to sawgrass & Myrtle ridge
7.17	717	Enters Lily pad slough
16.37	1637	center, Myrtle clump 50' x
22.71	2271	" myrtle saw grass ridge,
28.93	2893	" Lily & Flag slough
37.85	3785	" Myrtle & saw grass ridge
54.16	5416	" Lily pad slough
62.?7	62?7	center Myrtle saw-grass clump 25 x
67.49	6749	" Myrtle white wood ridge
70.51	7051	" Lilypad & Flag slough
80.00	8000	Standard cor a Long driven s'in muck, Cor. stands in Lily & Flag slough 15" water. Cor. to secs 30 29 / 31 32
		June 24-1912.

Figure 1.3. Sample page from the field notes for the survey of Township 47 South, Range 41 East. Entries show distances in chains (1 chain = 66 feet = 20.1 m; 80 chains = 1 mile = 1.6 km) to the different vegetation types encountered while surveying a 1-mile westbound segment between Sections 25 and 36. Alternating sloughs and sawgrass ridges reported are representative of the Ridge and Slough landscape. Source: Ensey (1911-T47 R41).

Table 1.3. Township surveys in and around the Everglades: Surveyors, locations and years worked (see also fig. 1.4 and the list of Township Survery References, page 321)

Surveyor	Year	Locations
G. MacKay[a]	1845	Atlantic Coastal Ridge: T41–43[b] R42[b]; T44 R41–43; T45 R42–43; T46–47 R41–43; T48 R41–42; T49 R42–43; T50–52 R42; T53 R41–42; T54–55 R40–41; T56 R38–39
A. H. Jones	1845	Northeast edge of Everglades and East: T41 R37–43; T42 R39–43; T43–44 R40–43
J. Jackson[c]	1847	Southern Everglades, East: T55 R39–41; T56 R38–40; T57 R38–39; T58 R38
J. Jackson[c]	1860	Northern Everglades, West: T40–41 R31–32; T42 R30; T43 R31
W. J. Reyes	1855–56	Northeast edge of Everglades and East: T40 R37–40, 42–43; T41 R37–39, 43; T42 R39
W. J. Reyes	1858	Everglades, East: T42 R40, 42–43; T43 R40, R42–43; T44 R41–43
W. L. Apthorp	1872	Northern Everglades, West: T43 R31–32; T44 R32–33; T45–47 R33–34
J. D. Tannehill	1871	Northern Everglades, West: T40 R31–33; T42 R31–32; T42 R30–32
M. A. Williams[d]	1870–72	Atlantic Coastal Ridge: T45 R41–43; T46 R41, 43; T47 R43; T48–49 R42–43; T50–51 R42; T52 R41–42
M. A. Williams[d]	1874–75	Western Everglades: T48–49 R33–34; Atlantic Coastal Ridge: T53 R42; T57–58 R39–40, T59 R38–39
S. G. Hamblen	1873	Northern Everglades, West: T43–44 R31–32; T44 R33; T45–46 R33
W. Mickler[e]	1885–86	Southern Everglades, West: T52–58 R35–36; T53–57 R33–34; T55 R32, 37
W. C. Sollie	1884	Southern Everglades, West: T50 R33; T51 R34–35; T52 R33–34
J. O. Fries	1898	Eastern Everglades: T50 R40–41; T51 R41–42
J. T. Hancock	1907	Northern Everglades, West: T40 R32-R33
J. W. Newman[f]	1908	Eastern Everglades: T50–51 R41; T54 R40
F. C. Elliot[f]	1911–12	Northern Everglades, West: T43 R32–34; T44–46 R33–34; T47 R34
R. F. Ensey[f]	1911–12	Eastern Everglades: T44–52 R41; T45 R39; T46 R40; T50 R40; T52 R42; T54 R40
W. J. Kackley[f]	1912	Eastern Everglades: T51 R40–41
L. D. Franklin[f]	1912	Eastern Everglades: T45–49 R41
L. D. Franklin[f]	1914	Northern Everglades, West: T43 R35; Eastern Everglades: T50 R39–40; T51 R38–40
T. E. Frederick[f]	1912	Eastern Everglades: T52–53 R40
T. E. Frederick[f]	1914	Northern Everglades, West: T44–45 R35–36
C. R. Horne[f]	1914	Northern Everglades: T44–45 R37, T44 R39–40
O. A. Hardin[f]	1914–16	Northern Everglades, West: T42–43 R36–39; T43 R34; T44 R33; T44–46 R34–36; T47 R35–36
J. King[g]	1917	Eastern Everglades: T54 R37–41; T55 R39–41
L. B. Merriam[f]	1918	Northern Everglades, West: T41 R32; T42 R32–34; T43 R34
So. Eng. Co.	1918	Eastern Everglades: T54 R37–39
N. B. Broward[f]	1918	Eastern Everglades: T50–51 R38; T47–50, 52–53 R39; T49–50 R40

Notes: a. Additional information provided in MacKay (1846), Jackson (1847), and Knetsch (1994).

b. T = Township, R = Range.

c. Additional information provided in Knetsch (1992) and Krome (1904).

d. Additional information provided in Knetsch (1993).

e. Mickler (1859; 1861; 1885).

f. Survey for the Trustees of the Internal Improvement Fund.

g. Not a formal township survey.

veys were conducted and, additionally, for identifying regional patterns such as the location of the predrainage "Edge of the Everglades," which was marked on many plat maps.
2. Creation of a comprehensive collection of field notes for the surveys.
3. Comparison of information obtained from the field notes for numerous townships concerning vegetation, soil, and water depth with that given on the Jones et al. soils and vegetation maps (1948a) and on the J. H. Davis (1943b) vegetation map. This helped develop an understanding of the patterns of soil and vegetation change at particular locations over time.
4. Tabulation of water depths as classified by landscape, location, and date. These were recorded in appendix A, organized by landscape. The township surveys used in this study are shown in figure 1.4. Table 1.3 indicates when and where in southern Florida the various surveyors worked.
5. Tabulation of historical observations of direction of water flow (appendix S).

Major Expeditions

Accounts of a number of expeditions into the Everglades prior to, or shortly after, drainage are available in published literature (table 1.4). Typically these expeditions were able to paddle, pole, push, or in some cases even sail between three and twelve miles (5–19 km) per day. At each night's camp an account of the day's travel was recorded. These records often included direct or indirect descriptions of water depths, the plants seen, changes in topography or landscape, and occasionally animals seen or hunted. The accounts generally also included estimates of the distances and courses (directions) traveled each day. In this study we reconstructed the expedition routes using these daily distances and courses, along with the known starting points. The known endpoints provided a degree of check on the accuracy of the route reconstruction. The accuracy of the locations of the night camps appeared to be within a few miles (3–5 km) of the likely actual locations, sufficiently accurate to provide useful insights at the scale of the Everglades as a whole.

Tables were prepared of the daily quotes regarding water depths, substrate (soil and rock), vegetation, and topography observed along each segment of the expedition. These tables, together with the reconstructed routes, essentially constitute informal predrainage "ecological transects" across the Everglades. Through overlay of the routes on maps of vegetation (J. H. Davis 1943b) and soils (Jones et al. 1948a), and on the hindcast landscape map produced by this study (plate 5), the transect observations were compared with estimated predrainage landscapes and with postdrainage vegetation and soils. We used any similarities or differences to correct landscape boundaries on the hindcast map, to improve the definitions of the different landscapes, and as a source of quantitative water depths. Appendices E through N present analyses of each expedition and its relevance to the determination of the nature of the predrainage landscapes. Figure 1.5 shows an example of information derived from one of the expeditions, in this case a cross section (profile) of the soil across the Everglades.

These expeditions were never designed or intended to be "ecological transects." This, and the richness of the information they contain, suggest that it is prudent to consider possible sources of bias associated with their use. Two obvious sources might be the time of year and the choice of route. It would not be surprising if both were chosen to allow travel in atypically deep water depths. Perusal of table 1.4 shows that while a number of expeditions occurred in November–December—that is, in the early dry season when water levels would still have been high—a surprising number also occurred quite late into the dry season, in March and April. For almost all expeditions, the dates are known and can be used to interpret the expedition's observations. Plate 10 left does suggest deliberate choice of routes; most were through the Ridge and Slough landscape, for the most part avoiding the Sawgrass Plains and the Marl Marshes. Within the Ridge and Slough landscape, the expedition accounts themselves suggest a preference for sloughs over ridges—natural, considering the dense, uninviting sawgrass on ridges—but also suggest that expeditions traveling primarily in the transverse, east to west or west to east direction nevertheless crossed a good number of ridges. In short, the issue of bias should not be neglected, but careful reading of a number of different expedition accounts can be expected to give a fairly complete picture of predrainage Ridge and Slough landscape conditions.

Early Maps

We reproduced the southern Florida portions of the early maps listed in table 1.5, adjusting them to similar scales. We then compared the maps for useful spatial information, such as descriptions of landscape and terrain; landmark points; location and nature of the edge of the Everglades; routes across the Everglades; locations of Seminole villages, fields, burial grounds, and travel routes; locations of important tree islands; and

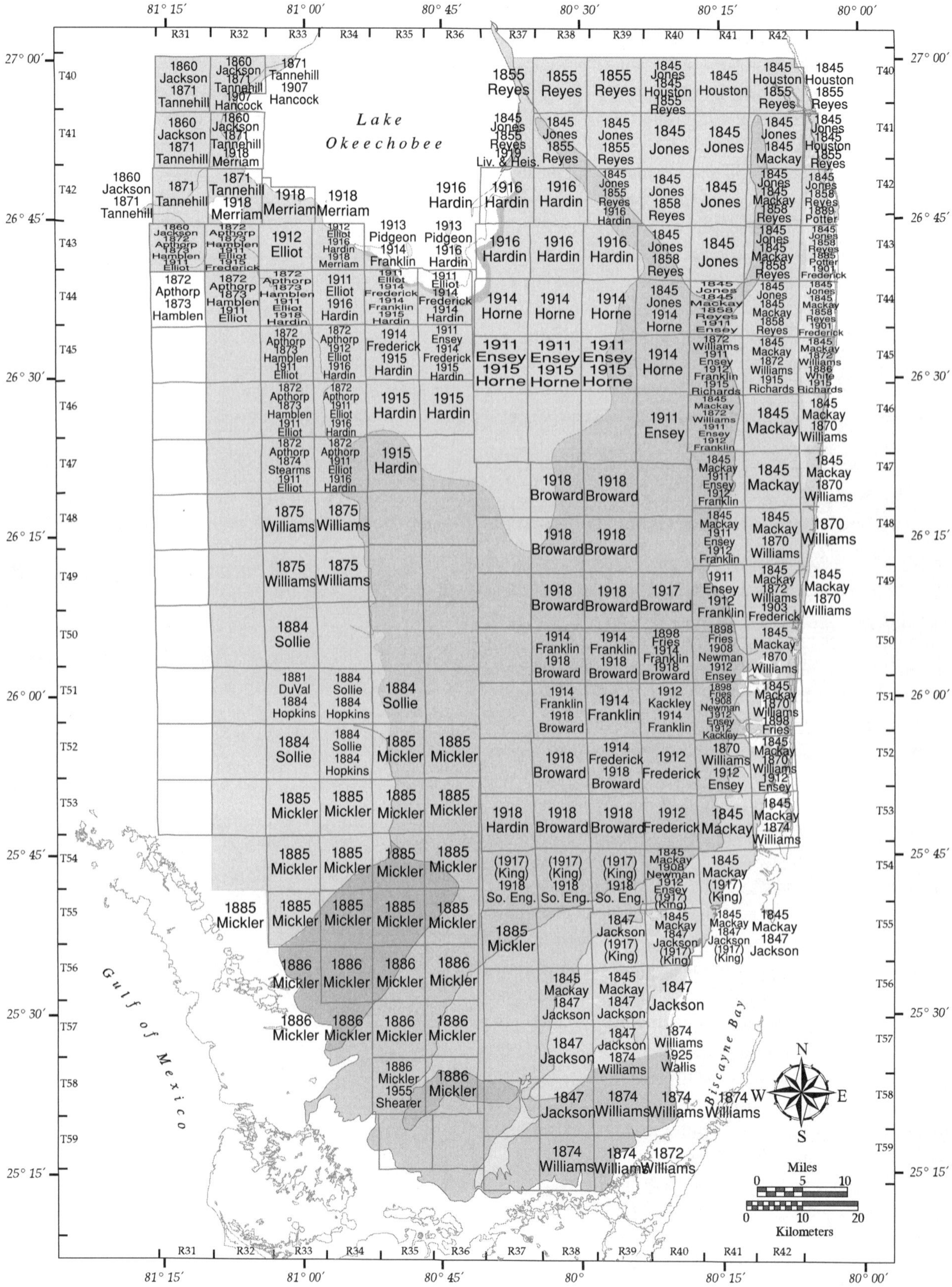

Figure 1.4. Township surveys studied during this research, showing year surveyed and name of surveyor. See also table 1.3 and the list of Township Survey References, page 321.

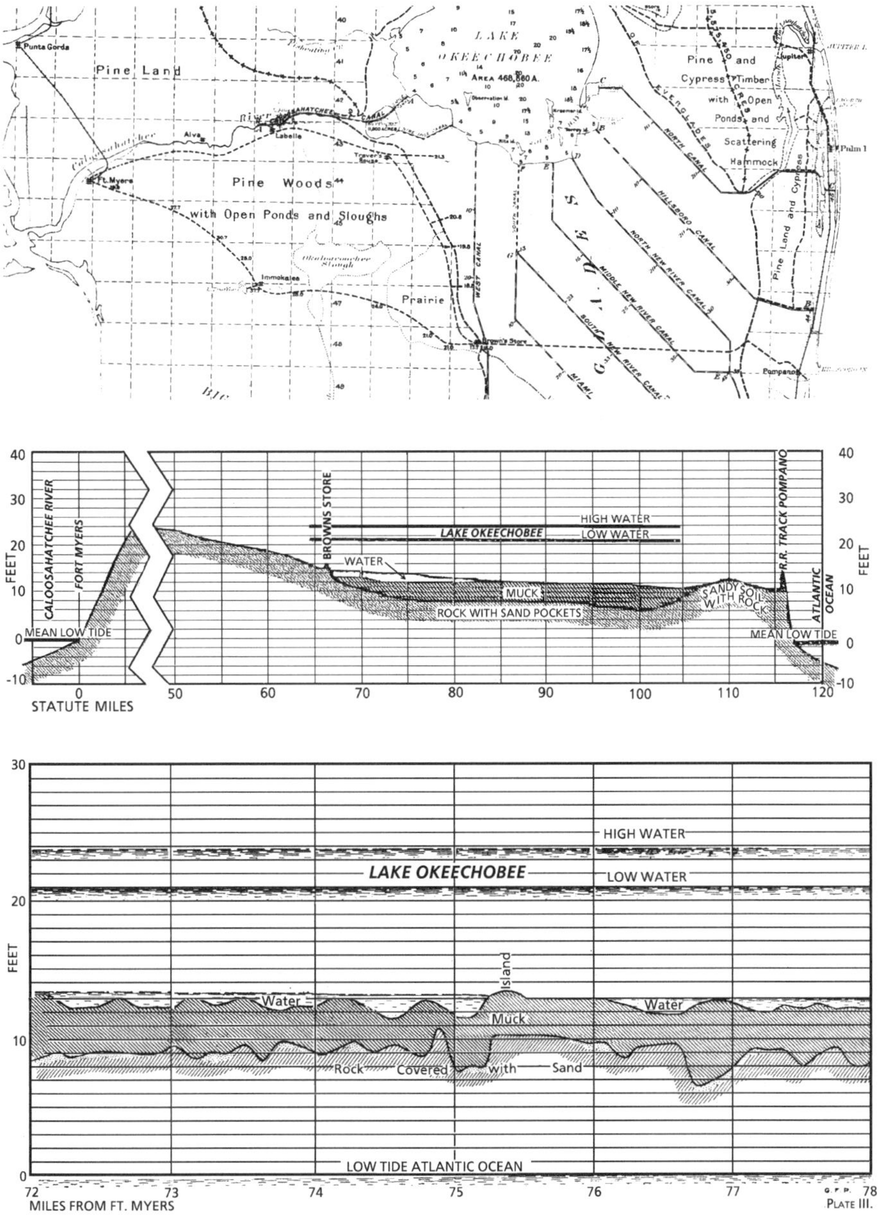

Figure 1.5. General (middle) and detailed (bottom) profile views of the soils of the Everglades as measured during the 1907 expedition from Brown's Store to Pompano (top, Pompano at lower lower right). See also appendices K and L. The expedition crossed in approximately Township 48, passing mostly through the Ridge and Slough landscape, with a short central portion through the Sawgrass Plains. The undulations in the peat ("muck") surface shown for miles 72–78 most likely correspond to alternating sawgrass ridges and sloughs. Sources: Soil profiles, Wright (1909c, plate III); map, Wright (1909b).

depths of organic ("muck") soil. The resultant familiarity with a collection of diverse maps from different time periods was useful for locating specific features mentioned in other sources. Such features can be hard to locate on modern maps because they no longer exist, the names have changed, or the features have become unrecognizable. Examples of Everglades landmarks and features no longer found on modern maps include the 13-Mile Canal, Brown's Store, Sam Jones' Seven Islands, Big City Island, and Rio Ratones, to name a few.

Several maps merit particular attention. *The Military Map of the Peninsula of Florida South of Tampa Bay* (Ives 1856b) stands out as a useful and remarkably detailed map of the predrainage Everglades region (fig. 1.6). Ives's map includes considerable phytogeographical information, showing symbols for pine, palmetto, oak, cypress, and "koontee" (coontie, *Zamia pumila*) plants as well as hatching to indicate the following landscape types that can still be recognized today: sawgrass, swamp, marsh, scrub, wet prairie, dry prairie, wet hammock, and dry hammock. Spatial accuracy of the map is good along the Atlantic Coastal Ridge, the Everglades Keys (Miami Rock Ridge), and Biscayne Bay, while shapes and sizes of some features are distorted (e.g., Long [Pine] Key, Cape Sable, and Lake Okeechobee). Despite these distortions, Ives's map has earned

Table 1.4. Early expeditions into the Everglades

Appendix[a]	Leader	Month/Season	Year	Route/Location	Method of Transport	Citation(s)
N	W. Cooley	Fall	1827	North fork of New River to northwest and back	horseback	Knetsch (1989, p. 40)
N	W. Cooley	Fall and winter	1827	South fork of New River to Rio Ratones (Snake Creek) and Biscayne Bay	"whale boat"	Knetsch (1989, p. 41)
N	W. Cooley		1827–1836	New River to 15 miles west	boats	Knetsch (1989, p. 41–42)
N	W. Cooley and Capt. Powell	Summer and fall	1836	South fork of New River to 30 miles west	boats	Knetsch (1989, p. 42)
	Col. W. Harney	April	1838	One of the Marl Transverse Glades (T55 R40)	on foot	Sunderman (1950, pp. 26–27)
E	Col. W. Harney	Dec. 4–15	1840	Miami River to Big Cypress to Harney River	16 cypress canoes	Anonymous (1840); *St. Augustine News* (1841)
F	Col. W. Harney	Jan. 1–10	1841	Fort Dallas (Little River) to Fort Lauderdale; Everglades	4–5 large canoes	A. M. Brooks (1880, pp. 245–255)
N	Capt. M. Burke	Oct. 9–14	1841	Fort Dallas to Chikikos Island to Prophet's Landing	canoes	Sprague (1848, pp. 333–334)
N	Lt. J. T. McLaughlin	Nov. 16–24	1841	Lake Okeechobee to Loxahatchee Slough to Biscayne Bay	canoes	Sprague (1848, p. 334)
N	Maj. T. Childs	December (8 days)	1841	Fort Dallas to Prophet's and Waxy Hadjo's landings and back to Fort Lauderdale	canoes	Ives (1856a, pp. 22–23); Sprague (1848, pp. 348–355)
N	Lt. J. T. McLaughlin	Dec. 10–15	1841	Prophet's Landing to cypress [forest] at head of Loxahatchee Slough	canoes	Sprague (1848, p. 358)
N	Capt. Wade	December (2 days)	1841	Loxahatchee Slough to Fort Lauderdale	canoes	Ives (1856a, pp. 16–18)
G	Lt. J. Rodgers	Feb. 12–22	1842	Fort Dallas to Rio Ratones to Lake Okeechobee	16 cypress canoes	Preble (1842)
G	Lt. J. Rodgers	April 2–10	1842	Lake Okeechobee to Fort Lauderdale	16 cypress canoes	Preble (1842)
N	B. Smith	Sept. 6–9	1847	Into the Everglades from Little River	"batteau," 21 by 4 feet beam, 5 inch draught	Smith (1848b, pp. 119–122)
N	B. Smith	Sept. 10–18	1847	Miami River to Prophet's Island and back	"bateau," same as above	Smith (1848b, pp. 122–127)
N	Capt. S. Dawson	March 1–7	1855	Miami River 27.5 miles west and back; "43 miles . . . measured along the route"	canoes, poled	Ives (1856a, pp. 21–22) Gaby (1993, p. 129)
N	Capt. S. Dawson	June	1855	Miami River 53 miles to Prophet's Landing and back	canoes	Ives (1856a, p. 22); Dix and MacGonigle (1905, p. 516)
N	M. A. Williams	March 21–24	1874	Township 57, Range 39	2 canoes	Williams 1874-T57 R39
N	Pierce and Bradley	February (estimated)	1882 (est.)	New River to Snake Creek to Biscayne Bay	"Indian canoe"	Pierce (1970, pp. 157–160)
	J. Henshall	Feb. 22–25	1882	New River to Pine Island and back	canoe + sail	Henshall (1883, pp. 52–53)
N	Pierce and Spencer	Spring	1882 (est.)	New River to Snake Creek to Biscayne Bay	"factory canoe"	Pierce (1970, pp. 165–167)
H	Maj. A. P. Williams (*Times-Democrat*)	Nov. 10–Dec. 6	1883	Lake Okeechobee to Shark River	5 canoes + 2 larger provision "batteaux"	Wintringham (1964); Hopkins (1884b, pp. 104–108)
N	Quimby	November–December	1888	Biscayne Bay to Snake Creek to New River	"big Indian canoe"	Pierce (1970, pp. 228–229)

Appendix[a]	Leader	Month/Season	Year	Route/Location	Method of Transport	Citation(s)
I	J. Ingraham	March 17–April 5	1892	Fort Shackleford to Miami River	cypress skiff + canoes	Moses (1892); Church (1892)
N	Griswold	February–March	1895 or 1896	Everglades Keys west; New River west	canoe	Griswold (1896)
J	Lt. H. Willoughby	Jan. 8–23	1897	Shark River to Miami River	2 canvas canoes	Willoughby (1898a, b)
	Seminoles	?	1890s	Big City Island to Pine Island	canoes	Fries 1898-T50 R40–41
N	A. W. Dimock	?	1907	Harney River to Miami River	powerboat	Dimock (1907)
K	Brett, Chadwick and others	Feb. 8–April 4	1907	Brown's Store to Pompano	canoes	J. T. Stewart (1907, pp. 62–63); Wright (1909c, pp. 156–159)
L	Stewart	March 19–24	1907	Fort Lauderdale to Pine Island (T50 R41)	canoe, poled	J. T. Stewart (1907, pp. 46–51)
M	J. Harshberger	June	1912	Lake Okeechobee to Fort Lauderdale (North New River Canal)	steamboat	Harshberger (1914, pp. 166–172)
N	J. King	February–March	1917	Miami River to Shark River	"Glades skiff"	Larned (1917; 1918a–h); King (1917c, f)

Note: a. Additional details for major and some minor expeditions are included in the Appendix shown in the first column.

enduring respect from later users in the field (Knetsch 1995).

John Harshberger, a plant ecologist, prepared the earliest map (1913) specifically delineating Everglades vegetation (fig. 1.7). While largely undifferentiated within the interior of the Everglades, the map does address the borders and adjoining uplands in detail. An early agricultural land capability map by James M. Kreamer (1892) was an important source for determining the predrainage extent of sawgrass and depths of sawgrass-peat soils in a large area southwest of Lake Okeechobee (plate 9).

In 1915 the soil scientists Mark Baldwin and H. W. Hawker surveyed upland soils near Fort Lauderdale and in a six-mile-wide (9.7 km wide) band along the full length of the North New River Canal, creating the first detailed soil map of southern Florida (Baldwin and Hawker 1915). The combination of high quality fieldwork and particular timing (shortly after the onset of drainage) made this study an extremely important complement to the Jones et al. (1948a, e) soil study. Many similarities between the two studies were found in the coastal ridge area east of the Everglades. The soil classification of the upland sands found there does not change greatly when exposed to anthropogenic drainage. However, it was the differences between the two studies with regard to the organic soils of the Everglades that helped elucidate an understanding of the soil changes caused by drainage (chapters 2 and 3).

A detailed but unpublished map, originally prepared for drainage planning (King 1917a), provided useful information regarding the Ridge and Slough landscape near the present Tamiami Trail and the adjoining portions of the Everglades Keys landscape. King mapped the Everglades Keys and Marl Transverse Glades (fig. 1.8) independently of the township maps; however, his map indicates very similar spatial patterns. King's map also provides one of the earliest graphic indications of tree islands oriented parallel to the northeast-to-southwest direction of water flow.

In working with all these maps, we found it was helpful to overlay them at appropriate scales on more modern maps or on georectified satellite imagery. Even though changes in land use and in the organic soils have greatly altered the appearance of southern Florida, overlays frequently helped reveal traces of the locations of predrainage boundaries and thus helped clarify the nature and extent of postdrainage changes to the Everglades.

Narrative Accounts

Narrative accounts did not necessarily yield specific information in terms of water depths or landscape boundaries, but they often helped supplement information from other sources. Descriptions of vegetation, when included, provided a way to identify the landscapes described by the authors. Similarities in landscape and/

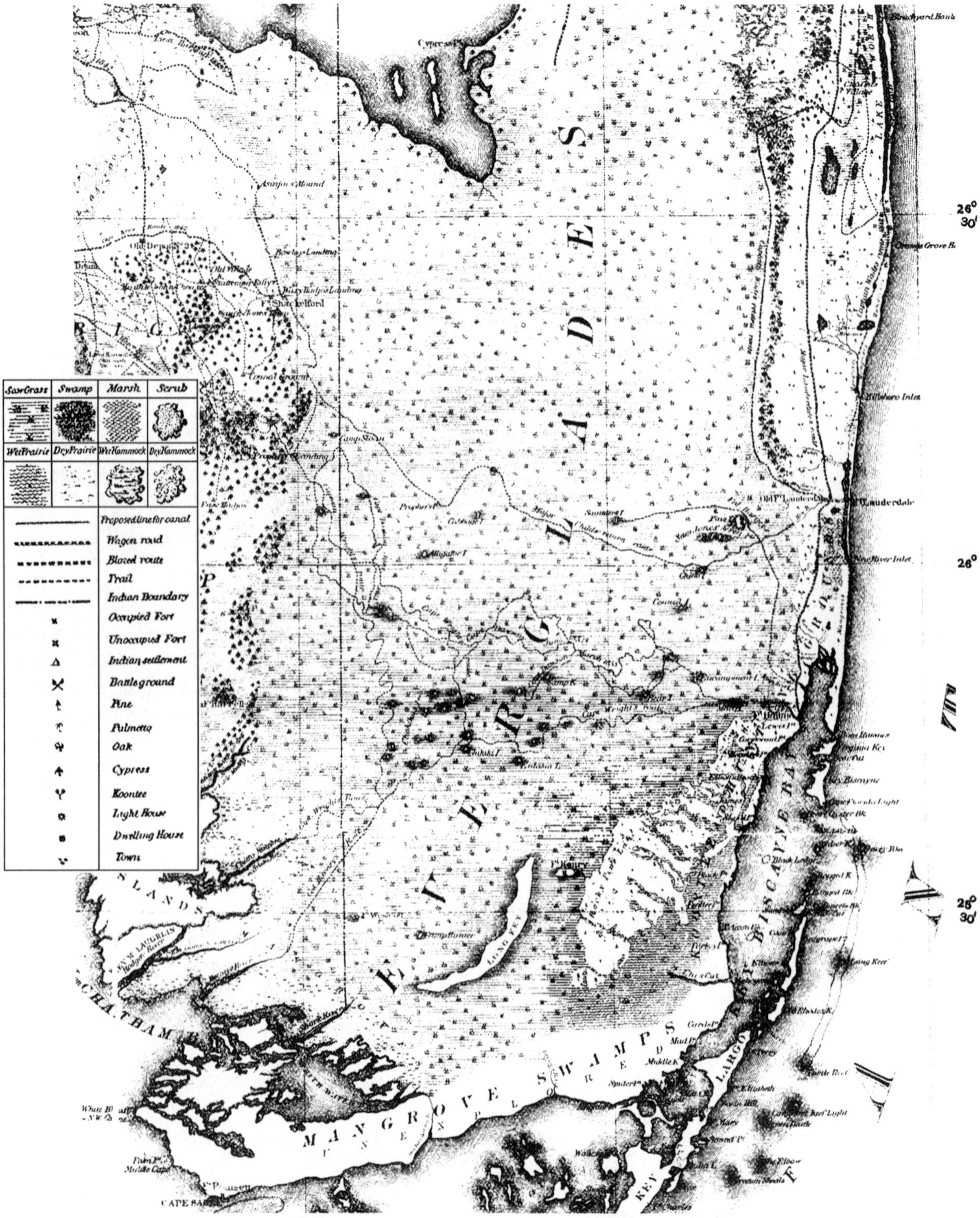

Figure 1.6. Everglades portion of the *Military Map of the Peninsula of Florida South of Tampa Bay*, drawn in 1856. Note the white area of Everglades Keys ("Rocky Pine Land") separated by thin, darker Marl Transverse Glades south of Fort Dallas (Miami). The southeast side of Everglades Keys (Perrine Marl Marsh in this study) was mapped as wet prairie. Farther north, between the Miami and New rivers, the Everglades directly bordered the narrowest portion of the (discontinuous) Atlantic Coastal Ridge. North of Fort Lauderdale the eastern Everglades bordered a strip of cypress. Reproduction of Ives (1856b).

Table 1.5. Early maps of southern Florida examined in detail for this study

Title	Year	Comments	Citation	Plate or Fig. number
Map of the Seat of War in Florida	1839	"Passage for small boats across the Everglades as reported by the Indians"	MacKay and Blake (1839)	
The State of Florida	1846	Many tree islands, Seminole landings on edge marked U.S. Bureau of Topographical Engineers (1846)		
A Plat Exhibiting the State of the Surveys in the State of Florida	1853	Landings marked; accuracy limited	Florida Surveyor General's Office (1853)	
Map of the State of Florida	1856	Routes indicated: Child's, Dawson's, Wright's	U.S. Bureau of Topographical Engineers (1856)	
Military Map of the Peninsula of Florida South of Tampa Bay	1856	See text, this chapter	Ives (1856b)	1.6
Map of Hic-Po-Chee and Okeechobee Sugar Lands	1892	Clear delineation of sawgrass, slough area, and pineland on northwest edge of Everglades	Kreamer (1892)	pl. 9
Florida East Coast RW [Railway]	1903	Details of Everglades keys south of Miami; Perrine Marl Marsh vegetation	Krome (1903m)	
Official Map of Canals	1908		Trustees of the Internal Improvement Fund (1908b)	
Map Showing Proposed Plan for Draining the Florida Everglades	1908	Stewart's west to east survey route	Wright (1909b)	1.5
Map of the southern portion of the peninsula of Florida showing the Everglades Drainage Canals	1911	Canal status on May 1, 1911	U.S. Congress (1911, plate VI, opposite p. 200)	
Map of Everglades Drainage District	1911	Muck depths, dredge locations	Ensey and Elliot (1911)	
State of Florida South of latitude 28° 30′	1913	Recommended canals	Florida Everglades Engineering Commission (1914, map between pp. 12–13.)	
Phytogeographic Map of South Florida	1913	See text, this chapter	Harshberger (1913)	1.7
State of Florida	1915	State of canal completion	Hassan (1915)	
Soil Survey of the Fort Lauderdale Area, Florida	1915	First soil map based on detailed field survey	Baldwin and Hawker (1915)	
Map showing results of examination of the Tamiami Trail Lands	1917	First known map drawing sloughs and sawgrass ridges	King (1917b)	8.2
Contemplated Location and Dimension of the Main Drainage Canals in the Southern Drainage District	1917	Tree island orientation and edge between Miami Rock Ridge and Everglades	King (1917a)	1.8
Location Map of a Portion of Dade Co. Florida	1920	Detailed depiction of Peat Transverse Glades and intervening high ground	Love (1920)	

or hydrologic descriptions given by different authors helped extend the time periods for which a recognizable landscape was known to occur. Key accounts include Vignoles (1823b), J. L. Williams (1837), Smith (1848a, b), William Cooley in 1851 (in Knetsch 1989), Ives (1856a), U.S. Congress (1911), and Larned (1917). Charles B. Vignoles (1823b) gave quite detailed descriptions of plant communities along the uplands east of the Everglades. Vignoles's plant classification scheme is similar to that found in Matson and Sanford (1913b) and in the Harshberger phytogeographic map and description of vegetation (1913, 1914).

John Lee Williams (1837) had frequent contact with the Seminoles and through them developed a good understanding of the large directional tree islands in the vicinity of the present Tamiami Canal. Buckingham Smith's well-known lyrical description of the Everglades (within the Ridge and Slough landscape) was partially

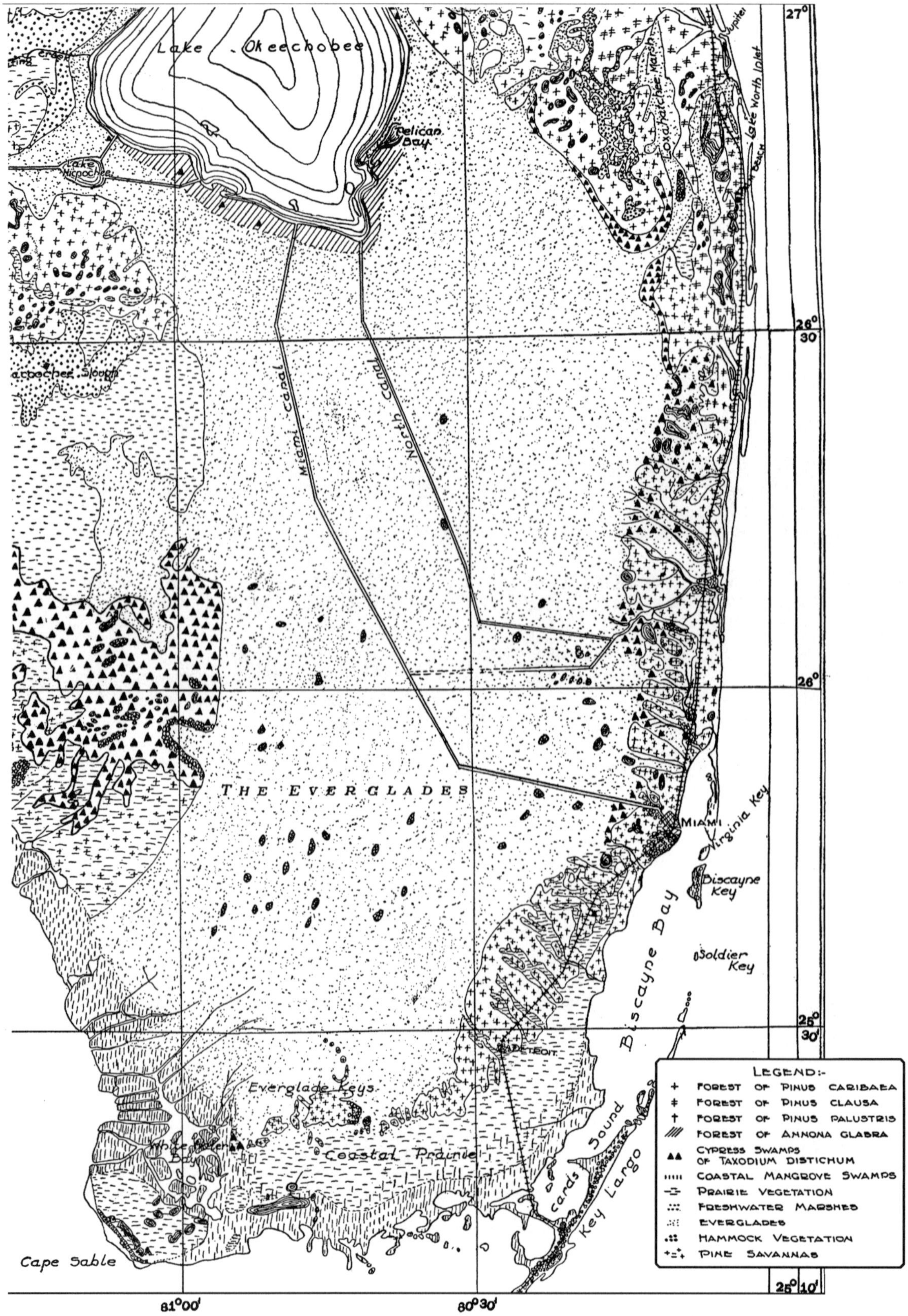

Figure 1.7. Everglades portion of the *Phytogeographic Map of South Florida* drawn by the plant ecologist John Harshberger. Note the coastal rivers extending into the Everglades and the distinct edge between the Everglades and bordering uplands to the east. Reproduction of Harshberger (1913).

reprinted in U.S. Congress (1911). In spite of his account being strongly in favor of drainage of the Everglades, Smith drafted an inspired description of the then undisturbed Everglades. Observations detailed in letters sent to Smith from individuals with experience in the Everglades were also used for background information. William Cooley was a settler who lived on the New River from 1823 to 1836. His letter, reprinted in Knetsch (1989), gives remarkable detail concerning the eastern Everglades. The memoir prepared by Lt. Joseph C. Ives (Ives 1856a) to accompany his well-known map contains many quotes from firsthand observers of the Everglades. W. Livingston Larned (1917) provided a serialized account of the engineer and naturalist John King's trip from Miami to Florida Bay.

Photographs

Although photographs can be invaluable for characterizing and recognizing predrainage Everglades landscapes, several practical limitations were encountered. For photographs to serve as fully qualified sources, the location and date of each had to have been identified by the photographer. This was rarely done. Nevertheless, most photographs were easily rated for presence or absence of water and, in some cases, even approximate locations and dates were still helpful. Photographs have proven to be particularly useful for identifying landscape types and the boundaries between them. We examined photographs from various sources, primarily the following: Willoughby (1898a), U.S. Congress (1911), Harshberger (1914), Baldwin and Hawker (1915), King (1917e, f), Parker (1940), Brown (1943), collections of the Historical Museum of Southern Florida, and the John K. Small collection in the State Archives of Florida.

Other Surveys, Profiles, and Studies

Linear surveys associated with the building of canals, railroads, and roads (Tamiami Trail) were used in addition to government township and expedition-related surveys as sources of information concerning soil and/or water depths (table 1.6). Unfortunately, field notes were not located for all of the surveys listed.

In many of these surveys the soil surface and the underlying bedrock surface were measured and subsequently drawn as profiles (fig. 1.9). In some cases either the survey location or the survey legend permits identification of the soil material: sand, marl, or peat (most often referred to as "muck"). These profiles were used to estimate soil depths at the time of the survey and then

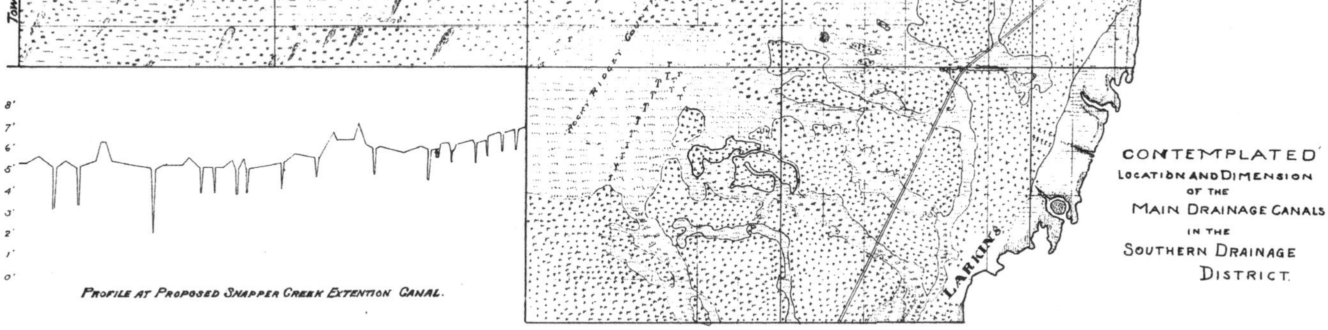

Figure 1.8. "Contemplated Location and Dimension of the Main Drainage Canals in the Southern District," 1917 blueprint map. Note the teardrop shape and northeast-southwest orientation of tree islands (Townships 53–54, Ranges 37–39) within the Ridge and Slough landscape, and the narrow Marl Transverse Glades extending southeastward between the Everglades Keys in T55 R39–40 and the southeast corner of T54 R40. Contemplated canals were the Tamiami Canal (1 mile, 1.6 km south of north boundary of T54; built there) and Snapper Creek Extension Canal (1 mile, 1.6 km north of south boundary of T54; not built). Reproduced from King (1917a).

Table 1.6. Construction surveys and soil profiles used in this study

Location	Survey Type/Surveyor	Citation(s)
Townships 42–44, Ranges 32–36	Agricultural survey including muck depths	Kreamer (1892)
South of Miami	Florida East Coast Railroad survey	Krome (1903m; see also 1903l; 1904)
West into Everglades from U.S. township surveys of Atlantic Coastal Ridge	Agricultural and drainage reconnaissance surveys	Rose (1898, pp. 455–456)
13 (18?) miles west from head (falls) of Miami River	Levels run by John Newman for drainage planning	Stewart (1907, p. 46)
West Palm Beach to Lake Okeechobee	Leveling survey in 1913 by George B. Hills	Florida Everglades Engineering Commission (1914, p. 9)
Major canals	Canal construction surveys	Florida Everglades Engineering Commission (1914)
North New River Canal	Agricultural soil survey	Baldwin and Hawker (1915)
Township 54 Ranges 37–38	Soil depth survey	King (1917d)
Tamiami Trail and Snapper Creek	Proposed canal survey and soil profile	King (1917g)
Eastern Tamiami Trail	Canal and road survey profile	Southern Engineering & Construction Co. (1918)
Western Tamiami Trail	Canal and road survey profile	Crabtree (1921)
Everglades Drainage District	Muck depths along transects	Ensey and Elliot (1911)
Townships 51–54, Ranges 39–40	Map of peat thickness	U.S. Army Corps of Engineers (1954, plate 5)

to compare these graphically with current soil depths (figs. 3.12, 3.13). (An uncertainty may arise with this approach if probing, rather than augering, was used to estimate soil depths, because an estimated depth reported as peat or muck may in fact have included marl or sand.)

Although there are apparently no published scientific studies representing work conducted in the Everglades before construction of the first drainage canals in 1880, several key observational studies were completed sufficiently early in the history of drainage to provide useful information. John Stewart worked during 1907 for the U.S. Department of Agriculture, Drainage Division. During the first half of that year he traveled extensively throughout southern Florida, personally inspecting different parts of the Everglades and interviewing local residents with experience in the area (Stewart 1907). His observations and commentary provide a landscape ecology perspective on the Everglades in the few years just before completion of the four main canals—the North New River, Miami, Hillsboro, and West Palm Beach canals.

Samuel Sanford was a geologist and field naturalist who surveyed possible routes for a railroad from West Palm Beach south to Miami and on to the Keys before he transferred to the Florida Geological Survey. His descriptions of the Everglades in Matson and Sanford (1913b), particularly of the adjacent Eastern Flatwoods and the Atlantic Coastal Ridge, are thorough. In addition to his phytogeographic map (Harshberger 1913) previously mentioned, plant ecologist Harshberger produced a detailed report (Harshberger 1914) with specific descriptions of Everglades landscapes. Similarly, the soil map of Baldwin and Hawker (1915) was accompanied by a report with landscape descriptions of use to this study, particularly of the Sawgrass Plains and the Ridge and Slough landscapes.

The botanist/plant ecologist Roland Harper also worked for the Florida Geological Survey. He was acutely aware of the changes being caused by drainage and by agricultural or urban development. Harper's detailed descriptions of the vegetation communities of the Everglades reflected his concern that some might disappear permanently in the face of development (Harper 1927, pp. 75, 124). Others shared this concern. In *From Eden to Sahara: Florida's Tragedy* botanist John K. Small (1929b) reflects an awareness of the Everglades as an ecosystem under assault by overdrainage. Small collected botanical specimens in southern Florida over a number of years. Although his technical publications primarily addressed flora at the species level, his numerous articles provided material for landscape characterization, particularly in the Everglades south of Long Pine Key (see chapter 9).

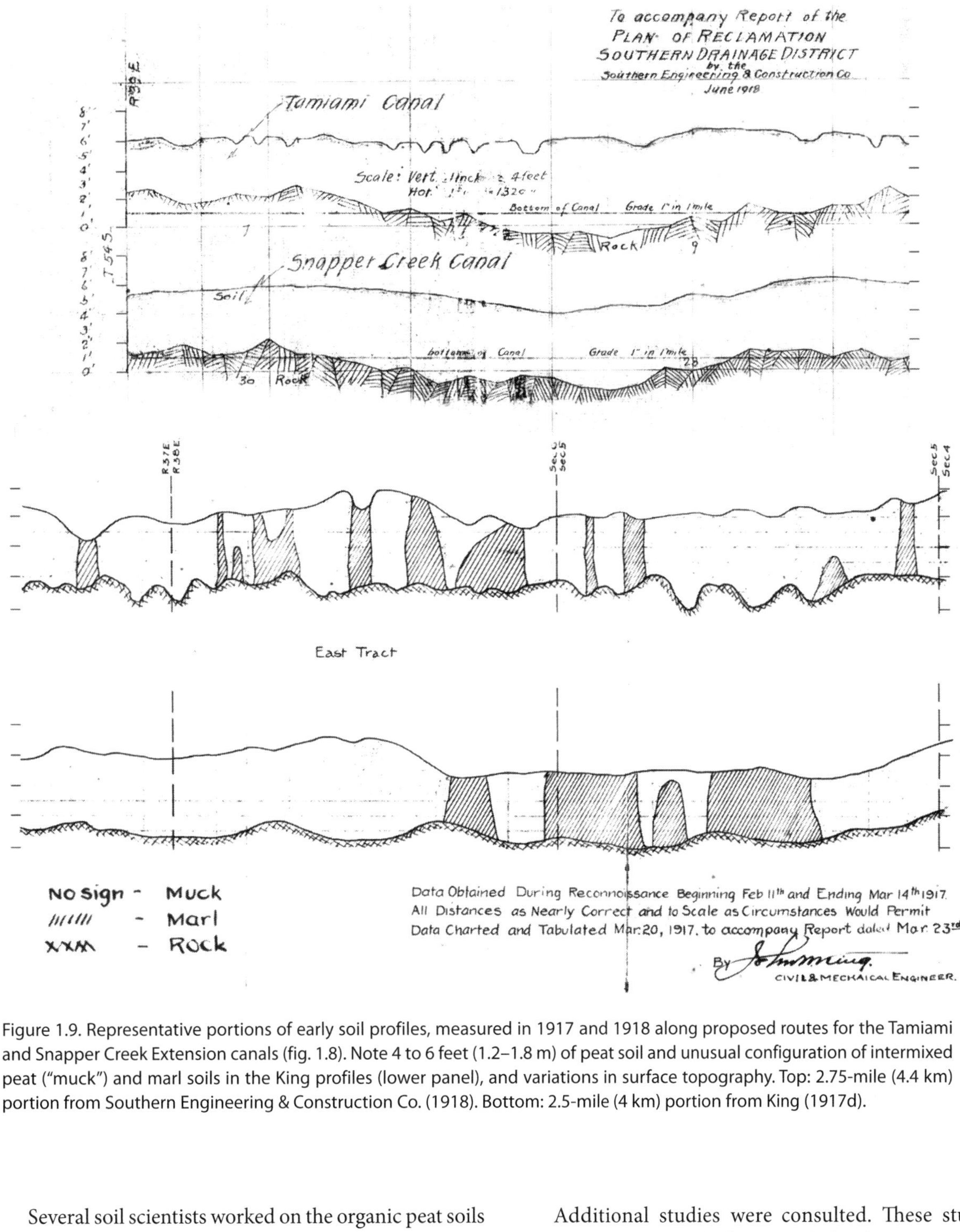

Figure 1.9. Representative portions of early soil profiles, measured in 1917 and 1918 along proposed routes for the Tamiami and Snapper Creek Extension canals (fig. 1.8). Note 4 to 6 feet (1.2–1.8 m) of peat soil and unusual configuration of intermixed peat ("muck") and marl soils in the King profiles (lower panel), and variations in surface topography. Top: 2.75-mile (4.4 km) portion from Southern Engineering & Construction Co. (1918). Bottom: 2.5-mile (4 km) portion from King (1917d).

Several soil scientists worked on the organic peat soils of the northern Everglades. Dachnowski-Stokes took many soil cores within what later became the Everglades Agricultural Area (Dachnowski-Stokes 1930). B. S. Clayton conducted long-term experiments to measure rates of peat subsidence as a function of water table height (e.g., Clayton et al. 1942; Clayton 1943; Stephens and Johnson 1951a, b).

Additional studies were consulted. These studies were conducted well after Everglades drainage was established, but the authors generally brought a long-term outlook to their work as well as an interest in predrainage conditions. These studies include Dovell (1947a, b), Robertson (1953), Egler (1952), Loveless (1959b), Craighead (1971), Tebeau (1973), Alexander and Crook (1973, 1975), Tropical Bioindustries (1990), and various studies by Dan Austin and his graduate students (Austin 1976; Austin et al. 1977; Steinberg 1980).

Rainfall Records

A quantitative analysis of the influence of weather variations, long-term climate changes, and multiyear and multidecadal climate cycles on the historical hydrologic observations gathered during the course of this study was beyond the scope of our work. Additionally, such an analysis would be limited by the scarcity of rainfall data available prior to approximately 1890.

We did, however, analyze the available rainfall data to distinguish wet and dry years. The resulting graph (fig. 1.10) was used to place particular hydrological observations from the historical record in context. Data were extracted from the South Florida Water Management District and National Oceanic and Atmospheric Administration databases. The first database includes information from Mitchell and Ensign (1928) and from the Division of Water Survey and Research, Florida State Board of Conservation (1954).

For each rainfall station we calculated a long-term, mean annual rainfall amount based on all data available for that station. Using that long-term mean, the deviation of each year's annual rainfall from the mean was calculated. Finally, for each year, and across all stations, the average of the deviations was calculated for all of the available rainfall stations. For earlier years, fewer stations contributed to the calculated deviation because of differing periods of records.

Summary

During the course of this study, a diverse range of source materials was consulted: vegetation, soil, and topographic mapping studies from the 1940s; government township surveys from the mid- to late 1800s; accounts of expeditions into the Everglades; early maps; narrative accounts by persons acquainted with the predrainage Everglades; early photographs; construction surveys and profiles; scientific studies; and rainfall records. Figure 1.11 charts the flow of this information through its assimilation, iterative refinement, and synthesis. The process of mentally reconstructing the predrainage Everglades through logical synthesis of all of these sources might be referred to as "forensic ecology" and is analogous to archaeological reconstruction of past cultures on the basis of the remaining physical evidence.

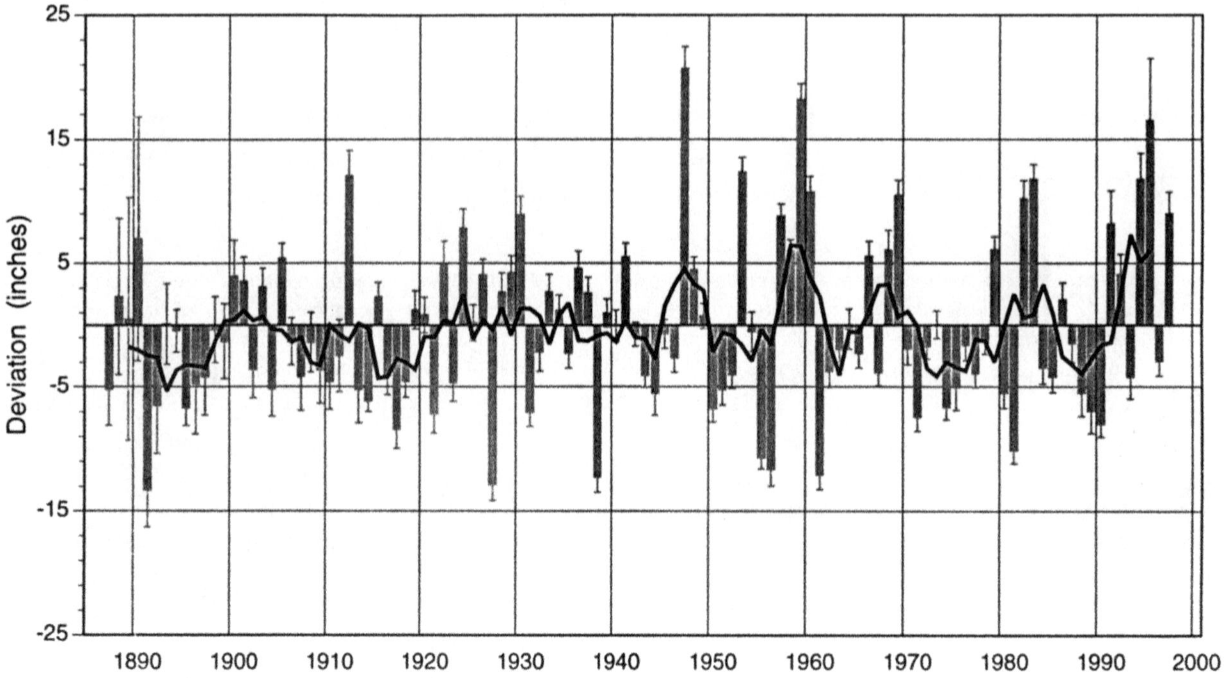

Figure 1.10. Variability of annual rainfall in South Florida, 1887 to 1997. Deviations shown for each year are the mean of deviations of all rainfall stations available for that year (see text). The solid line is the five-year running average.

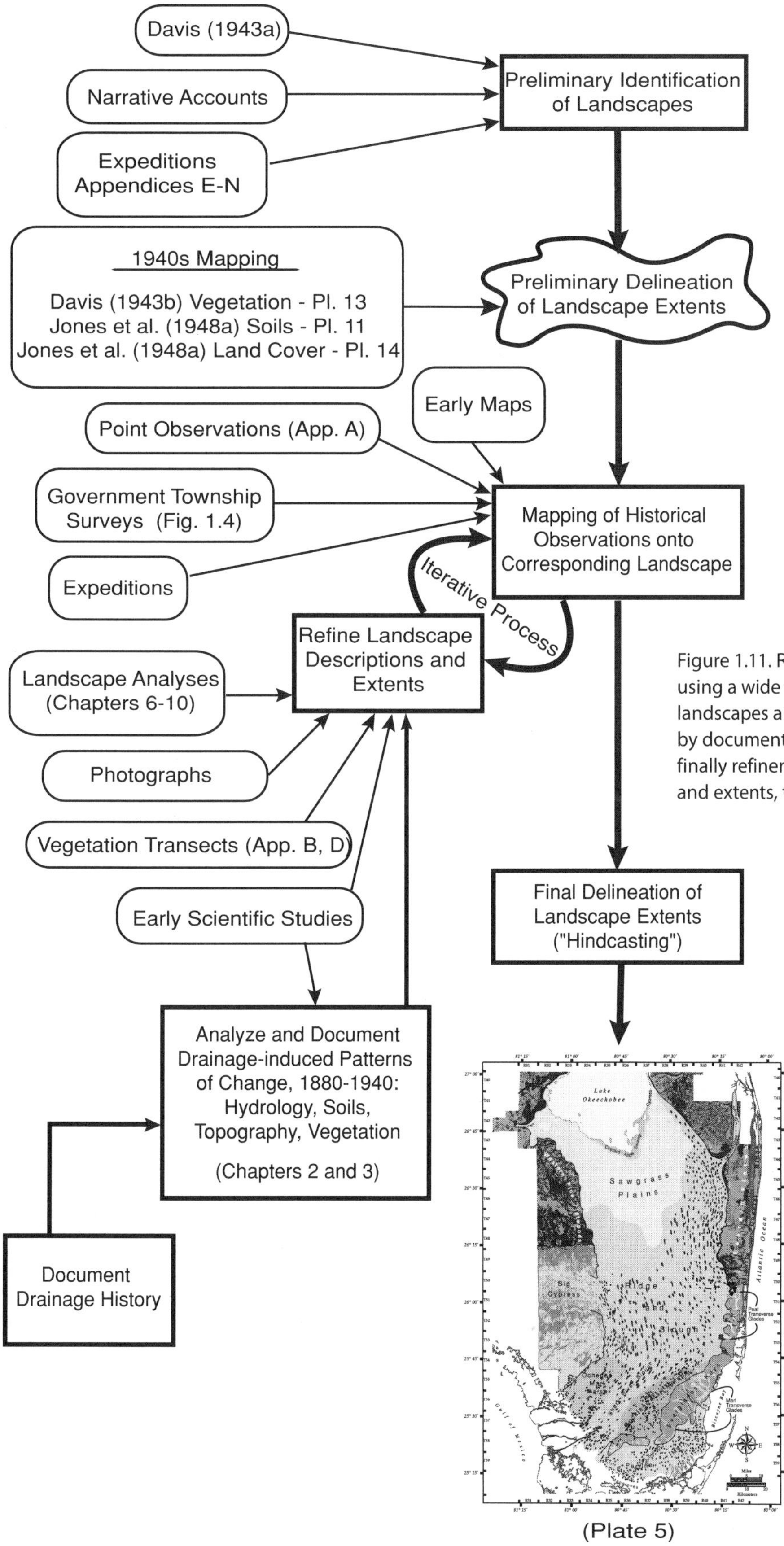

Figure 1.11. Research flow chart. Steps included using a wide range of sources to identify landscapes and delineate their extents, followed by documentation of the drainage history, and finally refinement and "hindcasting" descriptions and extents, to arrive at the map shown in plate 5.

2

Phases of Everglades Drainage, 1880s to Early 1950s

The 1940s were a turning point in the environmental history of the Everglades. First, key scientific studies were conducted that resulted in the earliest detailed, field-verified maps of the entire region: the soil and land use maps of Jones et al. (1948a, b, c, d, e) and the vegetation map of J. H. Davis (1943a, b). Those maps together provide an excellent picture of the region—at a particular point in time, and at a particular point in the course of the drainage-induced alterations to the Everglades. This chapter clarifies the history of drainage and the related changes that preceded the spatially comprehensive maps of the 1940s, allowing the picture they present of the Everglades to be placed in proper context.

Second, as indicated in this chapter, it had become apparent by the 1940s that the first water management tool, drainage, needed to be supplemented with a second one: water conservation (Mayo 1940, p. 82; Johnson 1974, pp. 174, 183). By the 1940s the many years of drainage afforded by the canals dredged in the 1880s and 1910s had drastically affected the Everglades. Large portions of the Everglades had dried out, so much so that even the peat soils had burned. The efficacy of the canal system at removing water had been demonstrated for some time; in fact, even before the 1940s, there had been growing concern that in some areas, the canal system was too effective. It was suggested that drainage should be focused only on the parts of the Everglades intended for agricultural use. There also was growing awareness that the relationship between water levels in the Everglades and other water resources of the region had been underestimated. As the level of the water table in the Everglades fell, water tables elsewhere around the region also fell, allowing salt water to intrude into coastal wells. These considerations, combined with the need for flood control after the major flood of 1947, led to the creation of the large, joint federal-state sponsorship of the Central and Southern Florida Project for Flood Control and Other Purposes (C&SF Project) in 1948.

We examined historical and early scientific sources to determine the time sequence of Everglades drainage between the 1880s and the 1940s and to identify the physical and ecological changes that took place as a result. Changes in the landscape were dramatic, rapid, and extensive. Enormous areas that had originally been covered by surface water much or all of the year became areas of exposed peat soil with water tables three feet (1 m) below the ground surface—as much as 9 feet (3 m) below in some areas (Everglades Drainage District, Operations Department 1928; Clayton et al. 1942, pp. 70–71). Sheet flow disappeared from large areas, recurring only under particularly wet conditions. Subsidence and decomposition of soils was widespread (Clayton 1943, p. 118). Stephens and Johnson (1951a, p. 191) noted that "the soils within the [640,000 acres of the Everglades Agricultural Area] lost approximately forty percent of their original volume . . . since drainage operations began." In addition to microbial oxidation, muck (peat) fires took a great toll. Florida Commissioner of Agriculture Nathan Mayo stated that large areas, "in some instances over two hundred square miles in a single body, have been denuded of all peat and muck by fires" (Mayo 1940, p. 81; see also Bender 1943, p. 150). The topographic changes were so great near Lake Okeechobee that "about 250,000 acres . . . that formerly sloped away from [Lake Okeechobee] now slope toward it" (J. H. Davis 1943a, p. 119, fig. 26; Lloyd 1929, p. 76). Not surprisingly, vegetation communities were substantially altered (Baldwin and Hawker 1915, pp. 752–753; J. H. Davis 1943a, p. 261). In short, many of the most severe changes to the Everglades had already occurred prior to the mapping activities of the 1940s and prior to initiation of the extensive C&SF Project in 1948.

Understanding those drainage effects provides a means to understand the landscapes better as they were mapped in the 1940s, as well as to understand their original predrainage condition better. It must be noted that, valuable as Davis's vegetation map and the soil map of Jones and colleagues have been, the soil, land use, and vegetation they mapped was the product of nearly sixty years of reduced water levels and hydroperiods. As a consequence, despite being the earliest comprehensive maps of the Everglades, those maps cannot be used as de facto estimates of predrainage conditions in the Everglades. However, when combined with other scientific and historical evidence, they provide a key springboard for estimating the extent and characteristics of the original predrainage Everglades.

Anthropogenic changes in the landscapes of the Everglades may be understood most clearly by distinguishing three phases of drainage history (Light and Dineen 1994): the Lake Okeechobee phase (1882–1896), the muck canal phase (ca. 1906–1930), and the impoundment phase (ca. 1950–1975). Construction of the impoundment phase was completed by about 1975; however, hydrologically the phase continues to the present (2010). The first two phases lowered water depths throughout the Everglades. The focus during those times was primarily on drainage; few levees or elevated barriers to flow were constructed. During the Okeechobee phase, Hamilton Disston's company dredged canals to connect Lake Okeechobee directly to the Caloosahatchee River. Flows to the river increased, and it appears that lake levels were successfully lowered for a period of time. Lowering of the lake would have reduced inflows into the Everglades (chapters 5 and 11).

The muck canal or state canal phase was built mostly by the State of Florida. A central feature of this phase was the dredging of four major canals running south and southeast through the Everglades to the Atlantic Coast. As indicated later in this chapter, these canals, initially operated with minimal controls, had dramatic affects on Everglades hydrology.

The third or impoundment phase began around 1948 and extended into the 1970s (table 2.1). The work was done as part of a joint state and federal effort, the Central and Southern Florida Flood Control Project. While this project created some new canals and improved existing ones, the main focus was on construction of many miles of levees to create impoundments known as Water Conservation Areas. This phase grew out of an increasing recognition of the "[water management] benefits of functions other than drainage and reclamation," and also out of a related sense that there was "an urgent need for the development of a comprehensive plan [for water control]" (Allison et al. 1944, p. 1). As intended, the impounded Water Conservation Areas generally raised water levels relative to those present during the excessive drainage of the muck canal phase, bringing them closer to predrainage water depths. Ecologically, this improved the average water depths within the impounded areas north of Tamiami Trail, but at four costs: (1) increased fluctuations in water depth due to use of the areas for storage, (2) virtual elimination of free-flowing sheet flow, (3) creation of water depth "wedges," and (4) unnatural variation of flows to areas south of the Tamiami

Table 2.1. Approximate dates of initial construction and opening for service of major hydraulic works affecting the Everglades

Canal	Initiated	Opened[a]	Citation(s)
Three Mile Canal, Lake Hicpochee to Lake Flirt	1882?	1884	Meigs 1879; Harney (1884)
Small, local truck farming canals related to railroad	>1896	[1897?]	Jones et al. (1948e)
North New River Canal	1906	1912	Interbureau Committee (1930); Clayton (1936)
South New River Canal	1906–1909	1913	Marston et al. (1927); Fla. Everglades Eng. Comm. (1914)
Miami Canal	1910	1913[b]	Marston et al. (1927); Fla. Everglades Eng. Comm. (1914)
Hillsboro Canal	1910	[1915?]	Marston et al. (1927); Jones et al. (1948e)
West Palm Beach Canal	1913–1917	[1920?]	Marston et al. (1927)
Caloosahatchee Canal	1915	1925[c]	Marston et al. (1927); Herr (1943)
Saint Lucie Canal	1916	1926	Marston et al. (1927); Elliot (1927)
Tamiami Trail and Canal	1916	1928	Tamiami Trail Commissioners (1928)
Lake Okeechobee South Shore Levee[d]	1921	1926	Marston et al. (1927); Parker et al. (1955)
Lake Okeechobee Levee[e]	1932	1938	Parker et al. (1955)

Notes: a. Opening dates of canals are less well defined since improvements, redredging, widening, etc. often continued for a number of years.

b. Dupuis (1954b), a doctor who settled in the Lemon City area north of Miami in 1898, gives 1911 as the date of opening for the Miami Canal. Gaby (1993) states March 1912.

c. According to Herr, chief engineer of the Okeechobee Flood Control District, 1929–1943+: "The Caloosahatchee Canal had been in existence prior to that time [1925] but its capacity was small and it had little effect on Lake elevations" (Herr 1943). A second project increasing its capacities was completed in 1938.

d. This was a low muck levee. It was seriously breached during the hurricanes of 1926 and 1928.

e. "This is a much more solid levee constructed by the U.S. Army Corps of Engineers with a top elevation of 32–34 feet above mean sea level" (Herr 1943).

Trail. The wedges arose from the difference between the water slope, which tended toward zero within each impoundment, and the ground slope, which in these areas retained much of the original predrainage fall. The elevation drop within each impoundment is sufficient to create water depth differences of 1 to 3 feet (0.3–0.9 m), resulting in strong upstream-downstream gradients in water depth and plant community (see also chapters 8 and 11).

We distinguish the different drainage phases to clarify both their timing and their effects. Water depths as well as water flows were altered. The spatial patterns of vegetation, soils, and topography mapped in the 1940s were the combined result of the original predrainage patterns modified by the drastic changes associated with the first two drainage phases. The landscape conditions seen at present are even more complex, due to the third drainage phase (impoundment), which created and rewetted the Water Conservation Areas but did so under essentially nonflowing conditions. The original predrainage patterns, the conditions mapped in the 1940s, and the conditions at present (2010) all differ considerably from one another. Attempts to relate them to one another must take into account all phases of Everglades water management history.

Lake Okeechobee Phase

The chronology and results of Hamilton Disston's efforts to lower Lake Okeechobee water levels are documented in detail in chapter 5 and are only summarized here. Chapter 5 suggests that under predrainage conditions Lake Okeechobee was directly and strongly connected to the Everglades, with water passing from the lake into the Everglades during and after the wet season of most years, along 70 miles (110 km) of the southern shoreline. Early designers of drainage efforts were well aware of this connection, so they also knew that to drain the Everglades for agricultural use, the lake levels would first have to be lowered (fig. 2.1). Even as late as 1929, when Lake Okeechobee had already been lowered several times and substantially, the critical role of the lake was still being emphasized, as in this report from an agronomist to the U.S. Army Corps of Engineers: "The control of Lake Okeechobee is of first importance because of the fact that the overflow of Lake Okeechobee adds an immense amount of water to the Everglades in addition to the water naturally collected from its own heavy rainfall during the rainy season" (Lloyd 1929, p. 75). Although no systematic stage records are available and the narrative record is not unanimous, as far as can be determined from contemporary firsthand observations, Disston's efforts in the 1880s do appear to have been successful in lowering lake levels for a period. Even a 1- or 2-foot (0.3 or 0.6 m) lowering would have reduced outflows from the lake greatly, with correspondingly great alterations to Everglades hydrology (chapters 5 and 11).

By about 1903 to 1905, when interest in draining the whole of the Everglades was renewed, Lake Okeechobee water levels had apparently risen again, closer to predrainage levels, and outflows into the Everglades had resumed. Although circumstances are not clear, the rise was likely related to the decline of Disston's companies after his death in 1896 (Knetsch 2006, pp. 204–205). Nevertheless, the fifteen or so years during the late 1800s when outflows from Lake Okeechobee into the Everglades were apparently reduced likely had a lasting influence.

Muck Canal Phase

The next period of drainage activity, referred to in Light and Dineen (1994) as the "Cut 'n Try Era," began about 1905 and continued into the early 1930s (table 2.1). This period has also been referred to as the muck canal phase because the four major canals built during this phase—the North New River, Hillsboro, Miami, and West Palm Beach canals—were created for the most part simply by dredging "muck" (that is, peat soil) out of the Everglades. Throughout much of Everglades literature, "muck" and "peat" are used interchangeably to describe the peat soils that were originally 85 percent to 90 percent organic material and were present over much of the Everglades. (Separately, there were true mucks, containing about 50 percent organic matter, along the southeastern portion of the Lake Okeechobee shoreline, corresponding approximately to the extent of the Custard Apple Swamp described in chapter 7.) The origin of these canals through muck dredging had hydrological and ecological implications. Only short portions where the canals crossed the Eastern Flatwoods or the Atlantic Coastal Ridge were dredged through sand or rock. Within the Everglades, the spoil banks were simply peat soil, and that soil disappeared quickly through subsidence, oxidation, fire, and physical sloughing. For example, the caption to figure 4 in Evans and Allison (1942, p. 39) reads: "A ten to twelve foot dike built of peat burned to a depth of a foot or more beneath the natural ground surface." With the disappearance of the peat spoil banks, over time these canals became simple rimless cuts, level with the surrounding landscape (fig. 2.2).

Hydrologically, the absence of associated levees

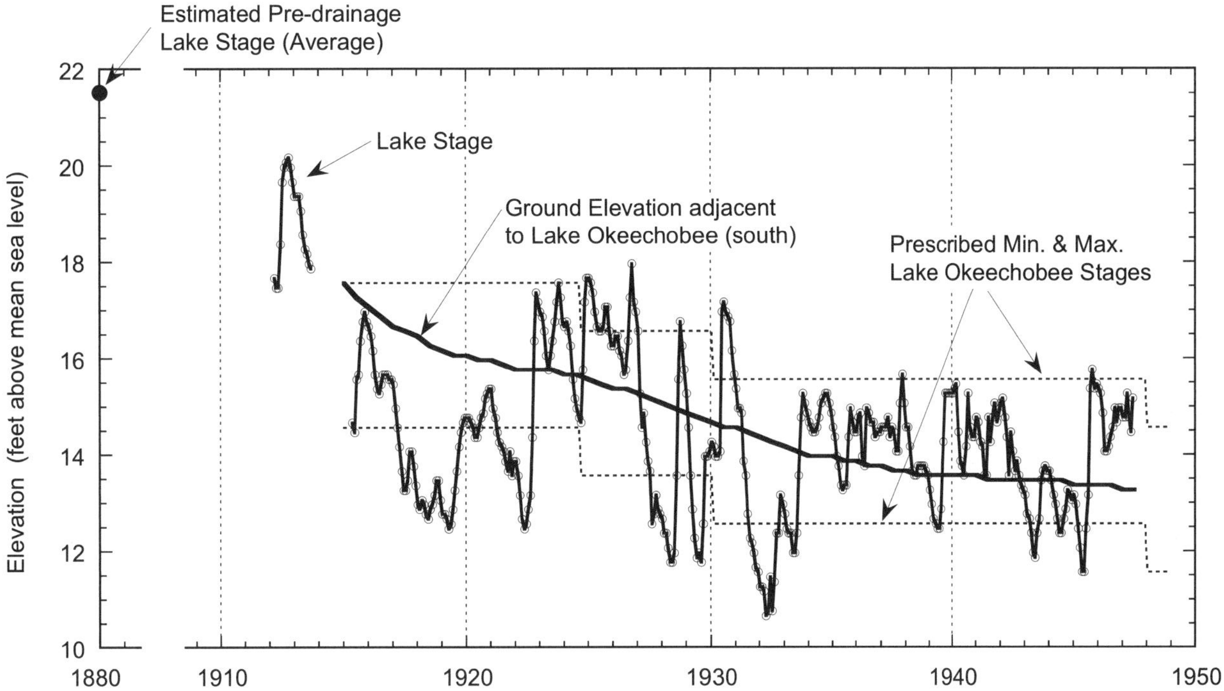

Figure 2.1. Partial history of water levels in Lake Okeechobee and ground elevations of the peat soil near the southern border of the lake. Note especially the 4- to 5-foot (1.2 to 1.5 m) drop in ground elevation and in average lake level between 1880 and 1915. After 1915 the decline in elevations of the prescribed (managed) lake minimum and maximum levels approximately paralleled the declining ground surface elevation. Sources: 1912 to 1913, Florida Everglades Engineering Commission 1914; 1915 to 1947, L. Johnson (1947).

Figure 2.2. View of the North New River Canal, January–March 1915. Note the absence of levees and closeness of water level to land surface. Photo from Baldwin and Hawker (1915; fig. 1 in plate 23).

meant that the muck canals could drain water directly from the landscape along the full length of their passage through the Everglades; that is, not only the agricultural areas near Lake Okeechobee but also the "wild lands" not yet in use farther downstream: "The broad sloughs that once carried the main flow of surface water [sheet flow] are now intercepted by the canals and their subsidence valleys, quickly draining surface waters to the sea" (Allison et al. 1944, pp. 6–7). When conditions were such that all of the intercepted sheet flow could be carried off by the canals, sheet flow "shadows" (areas of reduced flow) would have been created on the downstream side of the canals. Additionally, downstream sheet flow shadows would, of course, also have been present whenever the canals eliminated surface flow by drawing water tables below the soil surface. At the other extreme, under higher flow conditions the absence of levees meant that water could flow *across* the canal surface (e.g., Parker et al. 1955, pp. 332–333, 370). Boat captain Lawrence Will indicated that the cross flow in the 1910s could be strong enough to cause substantial problems for early boat captains on the North New River canal: "Near the Fifteen Mile Post [measured from south end of canal, now very close to southern tip of Water Conservation Area 2B; Section 5 of Township 50 Range 40], no banks rose on either side [of the canal] and here, during flood times, the southward flowing 'Gulf Stream of the Everglades' swept across the canal's current, throwing the biggest boats off course" Will (1965, p. 86). As discussed in the next section, the lack of levees and the unrestricted drainage of both the agriculturally used and the unused wild lands (i.e., the then remaining Everglades) later became central issues in the planning of the impoundment phase.

In addition to the muck canals, the major drainage-related structures built during this period were the St. Lucie Canal, the levees around the southern shore of Lake Okeechobee, and the borrow canal and levee associated with the Tamiami Trail highway. The falls at Fort Thompson on the Caloosahatchee River, first blasted in 1881 by Captain Fred Menge for Hamilton Disston, were again cleaned out in 1909 by the dredge *Caloosahatchee*, improving navigation to Lake Okeechobee and likely increasing outflows (Will 1965, p. 48). Aside from navigation, the focus of these efforts was twofold: the lowering of water levels in Lake Okeechobee to eliminate outflows into the Everglades (J. A. Williams et al. 1911, p. 197; Wright 1912, p. 29), and the drainage of the Everglades themselves to make land available for agriculture. The muck canals, like the Caloosahatchee Canal, increased flows in approximately the original, natural flow direction. In contrast, the St. Lucie Canal connecting Lake Okeechobee to the Atlantic Ocean created a completely new, man-made watercourse; it did not follow a natural channel or flow direction or even a natural slope.

The four major muck canals, the North New River, Hillsboro, Miami, and West Palm Beach canals, were part of a larger set of proposed canals, apparently envisioned to follow the original slope(s) of the Everglades toward the original outlets. The previously named four canals followed the southeasterly slope of the northern and eastern Everglades, connecting with the coastal rivers that drained Everglades water into Biscayne Bay and the Atlantic. A second, more southerly set of canals that were never built would have followed the southwesterly slope, draining into Florida Bay (Florida Everglades Engineering Commission 1914, p. 14; Elliot 1927, p. 64; fig. 11.14). The construction of Tamiami Trail, which appears to have diminished southwesterly flows, and plans for an Everglades National Park may both have been contributing reasons for not building the second set of southwesterly canals.

Dredging of three of the four major muck canals, the North New River, Hillsboro, and Miami, began in 1906–1910 and was completed by 1915. The West Palm Beach Canal was likely finished within the following five years. Contemporary accounts give a sense of the dramatic and rapid effectiveness of the muck canals. The following observations from the newly dredged eastern ends of North and South New River canals were made in the fall of 1907:

> That the canals are a success and are reclaiming the land as the dredges progress, is thoroughly established. The canals reduce the water level from the surface to a point six feet below the surface of the ground as shown by the water in the canal, and the land for a mile on either side of the canal is entirely reclaimed, and is practically ready for preparation for cultivation, and the general influence of the drainage reaches to a much greater distance than one mile. (Jennings 1907, p. 122)
>
> . . . The positive truth is that the superintendent finds it necessary to keep a sufficient quantity of water in the canal to float the dredge, while in front of the dredge the water is pouring over the front of the canal and falling six feet over a perpendicular dam to the water level of the canal and thus going on to the ocean.
>
> The result is that the reclamation of the land is fully demonstrated. We walked for a distance of half a mile or more along an Indian trail or canoe route through the saw grass, where twenty days ago the Indians traveled with their boats and canoes, the water having all

> been drawn off from this territory by the cutting of the canal, thus lowering the water level. (Jennings 1907, p. 125)

William S. Jennings's 1907 description of the drastic and rapid shift from Seminole canoe route to dry land walking path is corroborated by other accounts. In 1882 James Henshall navigated the same area during the dry season (February–March), sailing a Seminole cypress canoe from the headwaters of the New River to Pine Island (plates 41, 43; Henshall 1884, pp. 104–106). In 1898 surveyor John O. Fries noted multiple canoe landings on Sam Jones' Seven Islands (Long Key) and canoe routes between them and Pine Island and Big City Island (Fries 1898-T50 R40, 1898-T50 R41, 1898-T51 R41, 1898-T51 R42; see list of Township Survey References, page 321). In contrast, ten years later and a year after Jennings's description, surveyor John Newman drew a plat map of the same area, showing more than 800 parcels (Newman 1908-T50 R41). Two hundred and seventy of the 800 parcels were already marked as sold.

Similarly dramatic changes were reported farther south a few years later when the lower end of the Miami Canal was opened: "When the Miami Canal locks were opened in the early part of 1911, garden vegetables in the edge of the Everglades died, and some of the driven wells, 25 feet deep, went dry as far east as the Florida East Coast Railway, a distance of four miles" (Dupuis 1954b, p. 43). Dupuis' description certainly should not be interpreted as an indication that the regional water table dropped by 25 feet (8 m)—a physical impossibility. However, Dupuis was an early settler in Miami (1898 in Lemon City), a doctor, and later a promoter of drainage in the Little River area and is likely a credible source. Mechanisms and specific effects aside, it seems appropriate to conclude that opening of the Miami Canal did in fact rapidly cause some form of visible changes to the Everglades.

A simple, approximate comparison helps one to comprehend the dramatic influence of the muck canals on Everglades water depths, not only at their initial opening but also over the subsequent decades. Early canal discharge measurements do not appear to exist, but data collected by the U.S. Geological Survey (table 2.2) indicate that during the 1941–1946 period, the four main muck canals together discharged an average of 1.7×10^6 acre-feet (2.1×10^9 m^3) of water per year. Assuming those canals drained this water from the Custard Apple, Sawgrass Plains, and Ridge and Slough landscapes—that is, from 2.2×10^6 acres (8.9×10^9 m^2; see table 4.2)—then this corresponds to an average annual discharge of 0.77 feet or 9 inches (23 cm). During the 1940s, when inflows from Lake Okeechobee into the Everglades were negligible (Parker et al. 1955, p. 561), the other principal components of the Everglades water budget were net precipitation (rainfall minus evapotranspiration) and sheet flow (surface runoff) to the southwest through Shark River Slough. The sheet flow component, an outflow from the area in question, is not well known, but net precipitation can be reasonably estimated. Parker et al. (1955, p. 561) estimated an average of 51 inches (130 cm) of rain per year during 1940–1946. Assuming that 43–46 inches (109–117 cm) of this returned to the atmosphere as evapotranspiration (German 2000, p. 47), then annual net precipitation for the period would have been 6–9 inches (15–23 cm). The canal discharge outflow of 9 inches (23 cm) would therefore correspond to roughly 100–150 percent of the net precipitation inflow; a very substantial influence on the water budget and, of course, an outflow that did not exist under natural, predrainage conditions. The dramatic reductions in water level and the associated ecological changes in the Everglades observed during the muck canal phase appear to be a logical consequence of a dramatic alteration in the water budget; in short, "the drainage worked."

As the northern and central portions of the Everglades were being drained by the four major muck canals, the canal and levee associated with the Tamiami Trail highway (U.S. 41) were lowering water depths in the southern Everglades. Tamiami Canal differed from the muck canals in that its position and alignment had

Table 2.2. Annual canal discharge, 1940–1946 (units = 1,000 acre-feet)

Canal	1940	1941	1942	1943	1944	1945	1946	Mean
West Palm Beach	854	1109	842	502	459	575	741	726
Hillsboro	342	583	414	83	97	149	173	263
North New River	401	638	472	148	94	178	231	309
Miami	479	599	545	299	227	277	456	412
Total	2076	2929	2273	1032	877	1179	1601	1710

Source: Parker et al. (1955), tables 35, 37, 39, 47.

been determined almost exclusively by transportation needs, not by the natural slope of the landscape. Unlike the muck canals, which mostly ran approximately parallel to the predrainage flow direction (and slope), the Tamiami Canal ran almost exactly perpendicular to the flow direction. Lowering of water levels in the southern Everglades was probably more the result of partial blockage by the Tamiami Trail roadbed than the result of drainage by the Tamiami Canal.

Captain James Jaudon and other promoters of the Tamiami Trail had extensive land interests located in several townships of Everglades land south of the proposed Tamiami Trail and hoped to drain the lands for farming (King 1917c, d, e, g). Blockage and diversion to the sea of the southward flow of Everglades water was an explicit goal of the Tamiami Trail promoters: "The idea actuating the Dade County Commissioners was that the drainage of the Everglades would be promoted by the construction of the proposed road, because it was the plan to dig a canal and use the rock excavated from the canal for the road bed. The canal would constitute a waterway of value in draining the adjacent lands, and the drainage thus effected would enhance their value to the State" (Tamiami Trail Commissioners 1928, p. 13).

J. H. Tatum, a Miami resident, publicly opposed the road plan because of a concern for flooding upstream from the proposed road. Tatum was apparently unconvinced that the borrow canal to be dug on the north side of the roadbed levee would successfully divert to the sea all the water coming from the north. At a public meeting on September 8, 1915, "Tatum claimed that the building of the road across the Everglades would flood Dade county [north of the proposed Tamiami Trail], and he was unalterably opposed to spending money to turn a deluge of water upon the valuable lands of this county" (Tamiami Trail Commissioners 1928, p. 14).

Tatum's concerns apparently were well founded. Only a few years after initiation of dredging and levee construction in 1916, upstream flooding was indeed noted in Townships 52–53 Range 39 due to blockage of southward flow across Tamiami Trail. In a written response to a 1923 complaint by the Pennsylvania Sugar Company ("Pennsuco"), F. C. Elliot, chief engineer of the Everglades Drainage District, agreed that the Tamiami Trail acted "as a continuous dam across the Everglades preventing the natural flow of water and jeopardizing [by flooding] the lands East and Northwest along the Tamiami trail and Miami Canal" (Elliot 1923, p. 45).

Despite this official recognition of the problem by Elliot, blockage of flow by the Tamiami Trail was still reported seven years later: "Federal Highway No. 94, locally known as Tamiami Trail, has been built across the 'lower glades.' . . . It serves to check, to a substantial degree, the further southward movement of those waters to their ultimate outlet at the southern end of the State" (Okeechobee Flood Control District of Florida 1930, p. 28).

Reduction of flow to the southern Everglades continued until sometime between 1930 and 1939, when conveyance under the Tamiami Trail roadbed was most likely increased. In 1939 the U.S. Geological Survey began measuring discharge through lateral canals extending under the Tamiami Trail (Parker et al. 1955, p. 10). Leach et al. (1972, p. 150) include a table giving locations of bridges along Tamiami Trail. The irregular spacing of the bridges and canals suggests that they were specifically sited to align with the original sloughs of the Ridge and Slough landscape: "The greater discharge [compared to eastward or westward flow] from the [Tamiami] canal is through openings under the highway along its south bank into the natural sloughs leading to the tip of the Florida peninsula" (Allison et al. 1944, p. 17).

In considering the influence of Tamiami Trail and the Tamiami Canal, it is helpful to recall that both were constructed after the muck canals were completed; that is, both have always functioned in the downstream hydrologic shadow of the muck canals as well as in the shadow of reduced inflows to the Everglades from Lake Okeechobee. A significant exception to this occurred during the great 1947 flood, when water levels rose so high and over such a large area (fig. 11.8; see also Johnson 1974, pp. 135–149), that the canals were overwhelmed. To a degree this flood reproduced, and in places likely exceeded, some aspects of the predrainage Everglades hydrology. At this time water actually flowed *over* Tamiami Trail, as told by Fred Dayhoff to the botanist Taylor Alexander:

> [After] the 1947 hurricane, . . . flood waters as deep as a foot were still running over the Tamiami Trail. Not yet 8 years old, [Fred] and his brother sat on opposite fenders of their 1937 Packard, . . . as their father drove slowly. Fred recalled that giant whirlpools marked the submerged north ends of the culverts that passed under the road, and that the floodwater over the road was swift. With their rifle, they took turns shooting bass that were running over the road against the current. Upon each hit, the other brother would jump off and attempt to grab the fish. They had to be quick because the current was fast enough that the fish would be quickly carried into the deep water over the sawgrass south of the road (Lodge 2004, pp. xxxi–xxxii).

The overflow and the swiftness of the current, apparently present across the full width of the Everglades ("ended . . . eight and a half miles westward from the Tamiami Trail's 'Fortymile Bend'"; Lodge 2004, p. xxxii) shed light on possible predrainage sheet flow conditions. A statement by Lamar Johnson, then assistant chief engineer of the Central and Southern Florida Flood Control District, is pertinent: "While there is no data available on the flood flow through the Everglades, early descriptions support the belief that water levels in relation to ground surface were little different than those that prevailed in 1947" (Johnson 1950, p. 5). At the same time, Dayhoff's observations of sufficient water depths for shooting fish cannot be assumed to be identical to predrainage conditions, due to an altered flow cross section. Vertically, the roadbed would have reduced the flow cross section. Horizontally, the flow cross section in 1947 would partially resemble predrainage conditions, as the L-67A and L-67C levees (fig. I.1), which block flow across Tamiami Trail in Ranges 37 and 38, had not yet been constructed. However, some of the southeastward flows that would originally have discharged through the coastal rivers (see chapters 10 and 11) probably were at this time diverted south and southwest by the north-south levees already present in Dade and Broward counties, thus increasing the discharge across Tamiami Trail.

Completion of other drainage projects during the muck canal phase likely shifted some of the burden of lowering water levels in Lake Okeechobee away from the muck canals. These projects include completion of the St. Lucie Canal, enlargement of the Caloosahatchee Canal in the mid-1920s, and construction of a muck levee along the south shore of Lake Okeechobee. Ecologically, completion of the muck levee may not have had a strong effect as lowered lake levels had previously eliminated most outflows into the Everglades: "The 1912 overflow was the last uninterrupted natural discharge from the Lake [Okeechobee]" (Johnson 1950, p. 4).

By 1940 the works of the Lake Okeechobee and muck canal phases had resulted in hydrologic isolation of the Everglades from Lake Okeechobee. These works, which included completion of the muck canals, completion of the St. Lucie and Caloosahatchee canals, and the construction of Tamiami Trail, all had strongly affected the Everglades and resulted in lowered water tables throughout the Everglades basin.

An obvious but sometimes overlooked effect of lowering the water table below the surface of the Everglades was the elimination of sheet flow. The mechanistic relationship between the presumably low velocities of predrainage sheet flow and the highly directional landscape patterning of the Everglades is still unknown, but it appears likely that some aspect of water flow affected the patterning. Elimination of sheet flow over much of the Everglades during most of the muck canal phase almost certainly played a role in the postdrainage alteration of the original patterns.

Impoundment Phase

The impoundment phase of Everglades water management history refers to the extensive joint federal and state project initiated in 1948, the C&SF Project. Strictly speaking, the timing of this phase (constructed primarily during the 1950s and 1960s; Light and Dineen 1994) places it beyond the scope of this chapter, but the phase is discussed here because the motivations leading up to it shed considerable light on the effects of the muck canal phase on the Everglades. In addition, chapter 8 makes use of the transition from the muck canal phase to the impoundment phase to help estimate predrainage vegetation within the Ridge and Slough landscape. From an Everglades perspective, one of the dramatic results of the C&SF Project was the transformation of 1,340 square miles (860,000 acres; 350,000 ha) of "unused land" of the Everglades into three giant "conservation pools" (Central and Southern Florida Flood Control District 1957, p. 25); that is, Water Conservation Areas 1, 2 and 3 (fig. I.1). While the motivation for constructing these pools was not primarily for preservation of natural areas, the de facto result was partial reversal of the extreme drainage of the muck canal phase. The increases in water depths that accompanied creation of the "pools" (see appendix R), in turn protected at least some aspects of the patterned portions of the original Everglades. At the same time, as pointed out by botanist Taylor Alexander and discussed further in chapter 11, "[In work done for the Department of Interior in the early 1970s] the already known conditions in the Water Conservation Areas were verified, namely the effects of too little water in the northern portions and too much flooding in the southern ends—a simple outcome of impoundment in a sloping landscape" (Lodge 2004, p. xxxiv). In addition, creation of the impoundments marked the end of unrestricted sheet flow.

The origin of the impoundment phase (i.e., the C&SF Project) is often ascribed to the great flood of 1947, popularly symbolized in the so-called "Weeping Cow" report of that year (Everglades Drainage District 1947). While that flood and the photographs in the report certainly catalyzed approval of the project, recognition of the need for a more comprehensive water management plan, one that focused on aspects other than drainage alone, had been building for a number of years. The original drainage plans of the nineteenth and early twentieth centuries had simply called for draining the entire two million acres (810,000 ha) of organic Everglades soils. The efforts to lower and dike Lake Okeechobee as well as the dredging of the muck canals through the Everglades reflected this intent. However, soon after these steps came into operation, concerns began emerging that a different approach was needed. A report by the Flood Control District stated: "[Around the time of the 1947 flood] it finally became apparent that the whole problem of proper water supply and control had been aggravated by uncoordinated efforts at improved drainage. . . . Nature had been disrupted. In turn, man had not sufficiently planned to restore the balance" (Central and Southern Florida Flood Control District 1957, pp. 18–19). A number of years earlier, prior to 1940, the state commissioner of agriculture, Nathan Mayo, had been similarly critical: "Heretofore the plan of reclaiming the Everglades has been essentially an effort to get rid of water rather than a plan to use the water intelligently. Excess water from the Kissimmee watershed and Lake Okeechobee has been hurried through canals to the ocean and the Gulf, while large areas of unused land [in the Everglades] have been allowed to dry out and burn up for lack of that water" (Mayo 1940, p. 82).

As early as 1927 an "Engineering Board of Review" had recognized that soil subsidence was indeed a significant issue, despite previous engineering estimates that it would not be. The lowering of the recommended high and low control levels for Lake Okeechobee from 19 and 16 feet respectively (as recommended by the Randolph Commission in 1913) to 17 and 14 feet was seen as "necessary because the subsidence of muck soils adjacent to Lake Okeechobee already has reached 3 feet to 4 feet in various places and is still increasing, whereas the Randolph Commission [in 1913] estimated that the subsidence would be less than 8 inches" (Marston et al. 1927, pp. 48–49). The Engineering Board of Review felt this subsidence would continue: "We believe that a further general subsidence of three feet below the present ground surface for present depths of 6 feet or more of muck may be expected in any tract of the Everglades lands within a period of five to fifteen years following the construction of the arterial canals through the tract, and we recommend that the arterial canals be designed in anticipation of such subsidence" (Marston et al. 1927, p. 75). Two years later Major General Lytle Brown of the Office of the Chief of Engineers shared the concern regarding soil loss, foreshadowing a soil thickness issue that soil scientists would address in the 1940s: "Subsidence, if not checked, may bring the land surface so close to the water table or to the underlying rock as to make cultivation impracticable" (Brown 1929, p. 51).

To reduce the tendency of the peat soil to oxidize, the Engineering Board of Review called for maintaining water tables as high as possible in the peat soils, primarily through the construction of levees along the canals' banks. These levees along both the arterial and the main canals (Marston et al. 1927, p. 14) would prevent drainage of unused portions of the Everglades, instead focusing the drainage on the farmed lands: "Control of the ground water table . . . will, in our opinion, best be accomplished . . . by retaining upon all undeveloped lands as fully as practicable their wet weather accumulation of water. The latter provision would be accomplished by constructing and maintaining the spoil banks of the arterial canals as continuous levees . . . preventing the flow of water into the canals from all adjacent lands excepting those tracts subjected to actual development" (Marston et al. 1927, p. 80). An agronomist from the Mississippi Agricultural Experiment Station agreed with the need for "continuous levees along both sides of each canal to retain the water in the Everglades and prevent subsidence from excessive drainage" (Lloyd 1929, p. 77). The Engineering Board of Review expected that such levees, and the associated retention of water, "will be of definite importance in preventing fires, in checking both subsidence of undeveloped lands and their growing up with objectionable vegetation [e.g., willow and myrtle], in the general maintenance of a satisfactory dry weather ground water table, and in provision of a substantial increase in canal capacity for emergency use" (Marston et al. 1927, p. 80). A decade later the state commissioner of agriculture took Marston et al.'s (1927) suggestion to retain water on the unused lands a step further, urging that in addition, excess water from other areas be brought in and stored on the unused areas: "Large areas in the Glades will not be placed in cultivation for several decades, if not longer. Unless protected from fire, the soil on them will in the meantime be destroyed. It is, therefore, recommended that excess water from Lake Okeechobee be conveyed to unoccupied portions of the Glades for the purpose of raising the water table in the soil, or even flooding the surface, in order that it cannot burn" (Mayo 1940, p. 82).

It is noteworthy that all of the urgings during the 1920s through 1940s to refine and spatially differentiate drainage of the Everglades were based only on differences in land usage. During this period the focus was on agriculture, and all of the Everglades was still considered potentially drainable and arable. Spatial distinctions were based only on whether the land was already "in use" at that time. There were few discussions of the Everglades as wetlands or natural areas, of value in and of themselves. This began to change under increasing influence of soil studies. These did not specifically emphasize the Everglades as wetlands, but they did indirectly open the door to consideration of non-agricultural land uses. With the onset of systematic mapping of Everglades soils in 1939 (Jones et al. 1948a, e), the rationale for geographical differentiation of drainage intensity gained a more mechanistic basis. By the early 1940s soil scientists had linked three relevant observations: (1) farming in the Everglades would require field-scale ditching; (2) the organic soils of the Everglades were subject to compaction by dewatering and by oxidation; and (3) it was not economically feasible to cut field ditches into the bedrock underlying the organic soils. Together, these observations dictated that for field drainage to work, the postsubsidence soil thickness must be great enough to accommodate the depths of the ditches. Soil scientists therefore concluded that only those soil types that could be expected to have sufficient thickness after subsidence "warranted intensive development and cultivation" (Evans and Allison 1942, p. 44). Using this criterion, the Everglades were differentiated into "(1) potentially cultivable lands; [and] (2) areas that have no immediate agricultural value" (C. K. Davis (1943, p. 116). Significantly, all of the Loxahatchee peats—that is, all of the soils associated with the "ridge and slough lands [landscape]" (C. K. Davis 1943, p. 116; see also chapter 8)—were excluded as being subject to "extreme shrinkage" (Jones et al. 1948c; Evans and Allison 1942, p. 45). Within the prized Everglades peats of the Sawgrass Plains (chapter 6), only the deep and very deep phases were considered of agricultural value (Evans and Allison 1942, p. 44). Altogether, the net result of these soil studies was a great shift in the goals for draining and cultivating the Everglades. The target area decreased by a factor of five, from the majority of the Everglades (i.e., the two million acres or 810,000 ha of organic soils) down to only about 400,000 acres (160,000 ha). In terms of the original predrainage landscapes, the drainage and agricultural focus shifted away from the Ridge and Slough landscape and concentrated instead on the Sawgrass Plains and Custard Apple landscapes.

With clearer understanding of the physical, soil-related reasons why the Ridge and Slough landscape would not, and should not, be destined for agricultural use, water management goals for this landscape also shifted. It became more clearly recognized that even though the Ridge and Slough areas had not been developed or used, they had in fact already been overly drained by the canals and as a result had suffered serious soil loss from peat fires (Mayo 1940, p. 84; Allison et al. 1944, p. 23–24; Bellamy 1947, in Huser 1989) as well as from microbial decomposition induced by drainage (Allison et al. 1927, p. 37; Clayton 1943, p. 119). As a result, one of the goals of the impoundment phase was to reverse drainage of the Ridge and Slough areas. The chief drainage engineer for the U.S. Sugar Corporation wrote: "It is commonly thought desirable that the large areas of idle lands [i.e., the Ridge and Slough landscape] should be restored, as nearly as possible, to their original condition of inundation in order to control fires and encourage wild growth and normal soil building processes again" (Bestor 1942, p. 96). The motivation for rehydrating the overdrained Ridge and Slough areas was not only to benefit the areas themselves; it also reflected increasing agricultural and urban needs. Increases in the fraction of the Sawgrass Plains landscape under cultivation created a greater need to move excess water out of that landscape and into the agriculturally unusable Ridge and Slough landscape: "The expansion of the Everglades Agricultural Area also tended to create a new problem on the flood side. As more land was put into cultivation, less land [within the Sawgrass Plains landscape] was available as a 'dumping ground' for excess water. The result was more serious and frequent flooding" within that landscape (Central and Southern Florida Flood Control District 1957, p. 20). Increasing urban populations in the coastal area created additional demand for water storage and also for raised Everglades water stages to counteract saltwater intrusion into coastal wells: "During the dry years with the resulting dehydration of the glades and the intrusion of salt water into the coastal areas, it became apparent that water conservation was just as much or more a necessary function of any plan as drainage" (Central and Southern Florida Flood Control District 1957, p. 19).

All of the preceding factors suggested a need to change the way the major muck canals provided drainage for the Everglades: "It became apparent also that the very structures [canals] designed to drain certain areas and to protect them in time of flood were also depriving them of necessary moisture during other periods" (Central and Southern Florida Flood Control District 1957, pp. 19–20). As the chief drainage engineer of the Everglades Drainage District later wrote in his memoirs, an advisory committee's report (Allison et al., 1944) proposed "that the present canal system in the Everglades be reserved for the primary service of lands in agricultural use at the time [i.e., the farmed portions of the Sawgrass Plains]. All other wild lands [i.e., the Ridge and Slough landscape] would be denied service from the existing drainage system, and rainfall would be retained on them for wildlife and recreational use, at the same time retarding soil subsidence" (Johnson 1974, p. 175). This proposal met with approval; however, these particular canals had been designed and constructed at an earlier time when the goal of Everglades water management was focused on draining the entire extent of the Everglades. As a result, when goals shifted to increasing the drainage of the Sawgrass Plains (traversed by the upper reaches of the canals), and simultaneously decreasing drainage of the Ridge and Slough landscape (lower canal reaches), there was no physical way to do this. The original muck spoil banks had oxidized and subsided away (e.g., fig. 2.2), so any surface water present in the adjoining Ridge and Slough landscape was automatically drawn into the canals, decreasing their efficiency at draining the upstream farmed areas.

An important aspect of the C&SF Project, therefore, was to provide the lower reaches of the muck canals with levees—as proposed thirty years earlier by Marston et al. (1927). With the levees in place, the canals would be able to transport water drained from the upstream agricultural areas efficiently downstream through the Ridge and Slough landscape without draining this agriculturally unused landscape. It was also recognized that with additional placement of several stand-alone levees not associated with existing canals, a system of impoundments ("conservation areas") could be created: "The dual purpose of retarding runoff from unused land into overloaded canals will largely be accomplished by the same levees and dykes that enclose the conservation areas, thereby increasing the efficiency of the present system in service of lands in agricultural use" (Allison et al. 1944, p. 39). These levees compartmentalized more than 60 percent of the then remaining Ridge and Slough landscape into a series of three large impoundments, designated as Water Conservation Areas. Water would be held in these impoundments for flood control and water supply: "south and east of the agriculturally usable lands in the Everglades, three giant conservation pools totaling 1,340 square miles are being provided by encircling levees. These pools are designed to impound waters [drained] from the [upstream] agricultural areas of the glades and from parts of the east coast region. Spillways will connect the pools, and outlets will be provided into Everglades National Park at the south end of the basins" (Central and Southern Florida Flood Control District 1957, p. 25). Lamar Johnson, the chief drainage engineer of the Flood Control District at that time, reminds us of the original intent. His postconstruction description of the intent closely matches Mayo's preconstruction intent (Mayo 1940, p. 82; see previous pages): "The original purpose [of the Water Conservation Areas] was a place to retard runoff of direct rainfall and to pump other excess flood waters for storage. Benefits such as recreation and wildlife preservation were secondary objectives" (Johnson 1974, p. 183).

The thinking prevalent during the impoundment phase of Everglades drainage, as well as the rationale behind creating the Water Conservation Areas—that is, creation of control structures to reduce the "wasteful" loss of fresh water into the ocean—are both well illustrated in a related comment by Lamar Johnson. In a report he prepared for Everglades National Park in 1950, while chief drainage engineer of the Flood Control District, Johnson "analyzed the park's water problem" (Johnson 1974, p. 209). He proposed what he thought would be a "practical solution" to help the park meet its acknowledged responsibility to re-create a balance between salt and fresh water: "[several] knee-high overflow dikes about six miles apart across the Everglades flow way [i.e., across Shark River Slough]" (Johnson 1974, p. 209; see also Johnson 1950, pp. 11–12). These presumably would have created a series of stepped-down, pooled terraces that would have held back water within Shark River Slough. The chief engineer's apparent surprise and not a little indignation at the "horrified opposition" of Everglades National Park officials to his proposal illustrates his focus on the "usefulness" of water and hence on water storage:

> The recommendation to use the water that they [Everglades National Park] do get more efficiently has been completely ignored, . . . With the increasing demand for fresh water in Florida, water should be retained on the land wherever possible and practicable. It is inexcusable to allow fresh water to flow into the sea with no attempt to retard the runoff. Florida saw too much of that practice in the all-out drainage period [i.e., the muck canal phase] of its history. The park people know this, yet the park sits there like a fledging egret on its nest, mouth open and squawking, waiting to be fed. (Johnson 1974, p. 211)

This focus on storing and conserving water is very likely the same focus that motivated conversion of 40 percent of the Everglades still present in 1940 into the three impounded Water Conservation Areas. Such a focus, especially on the part of a drainage engineer, is easily understood in the context of the preceding wasteful muck canal phase, during which the canals unnecessarily drew water off lands that were not even in agricultural use.

At the same time, Johnson's emphatic comment and its implicit disregard for the natural hydrology of the Everglades are surprising given Johnson's background and knowledge. Lamar Johnson possessed considerably more than an ordinary engineering understanding of the Everglades. He first came to the Everglades with his father in 1921; he took pride in being able to pole a "gladeboat . . . with the dexterity of a Seminole" (Johnson 1974, pp. 98–99); he surveyed extensively in the Everglades (by gladeboat; Johnson 1974, pp. 94–112); he helped raise an early and influential alarm regarding subsidence and peat loss (Stephens and Johnson 1951a, b); and he was concerned about the vegetative changes he saw in the Everglades during his lifetime (Johnson 1958, pp. ii, 29).

As an engineer Johnson certainly understood that water within impoundments would pool; that is, the water surface would tend to become level. Johnson was aware of Everglades topography (e.g., Stephens and Johnson 1951a, b), particularly the regional slope in the downstream direction, so he would have been able to foresee that the combination of sloped ground with pooled water would create "wedges," or gradients in water depth, shallower upstream and deeper downstream. He would also have been able to foresee that over the long length of the Water Conservation Areas, the magnitude of the depth gradient would be significant. Even more curiously, the following intriguing suggestion, made two decades after the Water Conservation Areas were created, clearly indicates that Johnson understood that these wedges and other impoundment effects would not have been present in the original Everglades:

> In my opinion, extensive studies should be made to determine the benefits, if any, of making the pumped inflow [to the Water Conservation Areas, especially WCA 1] move in a sheet flow over the whole area, rather than letting most of it flow down the perimeter channels as it does at the present time. This would probably increase the total storage by distributing the flow over the total area, holding the pool surface on a gradient. People, even those who should know, are inclined to forget *that the historic flow through the Everglades was for long periods after the wet season at approximately the same depth throughout the length of the floodway* [i.e., water surface on a gradient parallel to the ground surface gradient]. The slope of the land was very gentle and the vegetation offered enough resistance to cause the water to flow very slowly. (Johnson 1974, p. 186; emphasis added)

In view of Johnson's in-depth understanding of both pre- and postdrainage Everglades hydrology (see also chapter 11), it is hard to understand his criticism of the park's opposition and his somewhat dismissive comment that "the Park Service does not interfere with nature by doing anything artificial" (Johnson 1974, p. 209). One can only speculate that the concept of storing water for "useful purposes" thoroughly dominated the thinking of the time, and hence outweighed, in his mind, any importance of maintaining the original flow way hydrology—that is, the sheet flow, the original water surface gradient, and undiminished freshwater discharges to Florida Bay.

Clearly the Water Conservation Areas were intended to store water and to do so at water depths greater than those of the preceding muck canal phase. It is less clear, however, how the newly impounded water depths would compare to those originally present in the predrainage Everglades. As it turns out, for practical reasons, vegetation played a strong role in establishing regulation schedules for the Water Conservation Areas, providing an indication of the approximate depths along with illustrating that impoundments necessarily include a range of depths. Chief Engineer Lamar Johnson noted: "The conservation areas were originally planned to store *more* water than they presently do. They generally were planned to be operated at higher stages. The questions of hurricane windtides [seiches] and seepage losses resulted in an interim levee height and lower temporary operating stages" (Johnson 1974, p. 185). The U.S. Army Corps of Engineers assumed that large areas of open water with long wind fetches would increase the danger of seiches, while stands of emergent vegetation would dampen them. The Corps therefore asked several scientists (Givens 1957; J. H. Davis 1957; Andrews 1957) to comment on the likely types of aquatic or emergent vegetation that could be expected in Water Conservation Area 1 under a regulation schedule of 12.5–15 feet (3.8–4.6 m) above mean sea level or one of 14–17 feet (4.3–5.2 m): "The most pertinent problem presented me by the Corps of Engineers was the possibility of drowning enough vegetation to make an open pool of water that might have waves during storms that would top or injure the levees at their present height" (J. H. Davis 1957, p. C-2).

Chapter 8 (Ridge and Slough landscape) and appendix R discuss in detail the vegetation changes predicted by J. H. Davis, Givens, and Andrews as well as later actual observations after enclosure of the Water Conservation Areas. In general, the vegetation suggests that the midpoint of the impounded water depths was similar to but somewhat shallower than predrainage water depths, with areas of former sloughs becoming wet prairies of emergent vegetation and former sawgrass ridges becoming invaded by shrubs, especially willow and myrtle.

The impoundment phase affected the impounded Water Conservation Areas most directly, primarily by increasing average water depths relative to the muck canal phase and by decreasing sheet flow. In addition, as discussed further in chapter 11 (see also plate 30), the impoundment phase caused significant upstream-downstream water depth gradients to develop within each Water Conservation Area. Within the area that became Everglades National Park, downstream from Water Conservation Areas 3A and 3B, the effects of the impoundment phase were more complex and more variable over time, being especially dependant on the particular policies of water release from Water Conservation Area 3A. Very generally, however, the effects might loosely be described as the opposite of those experienced by the Water Conservation Areas. As Everglades National Park was an unimpounded area, sheet flow within the park was able to continue unimpeded, *provided* that a continuous discharge of upstream water was available at Tamiami Trail and *provided* that one was sufficiently far from the distorting effects of the point source inputs (i.e., the four S-12 gates along Tamiami Trail in Ranges 35 and 36; see fig. I.1). The absence of impounding structures within Everglades National Park appears also to have averted the development of water depth gradients. How the impoundment phase changed water depths within Everglades National Park relative to the muck canal phase is less clear. Theoretically, the presence of upstream impoundments increased the availability of water relative to the muck canal phase; however, how actual releases compare is less clear and likely has varied over time.

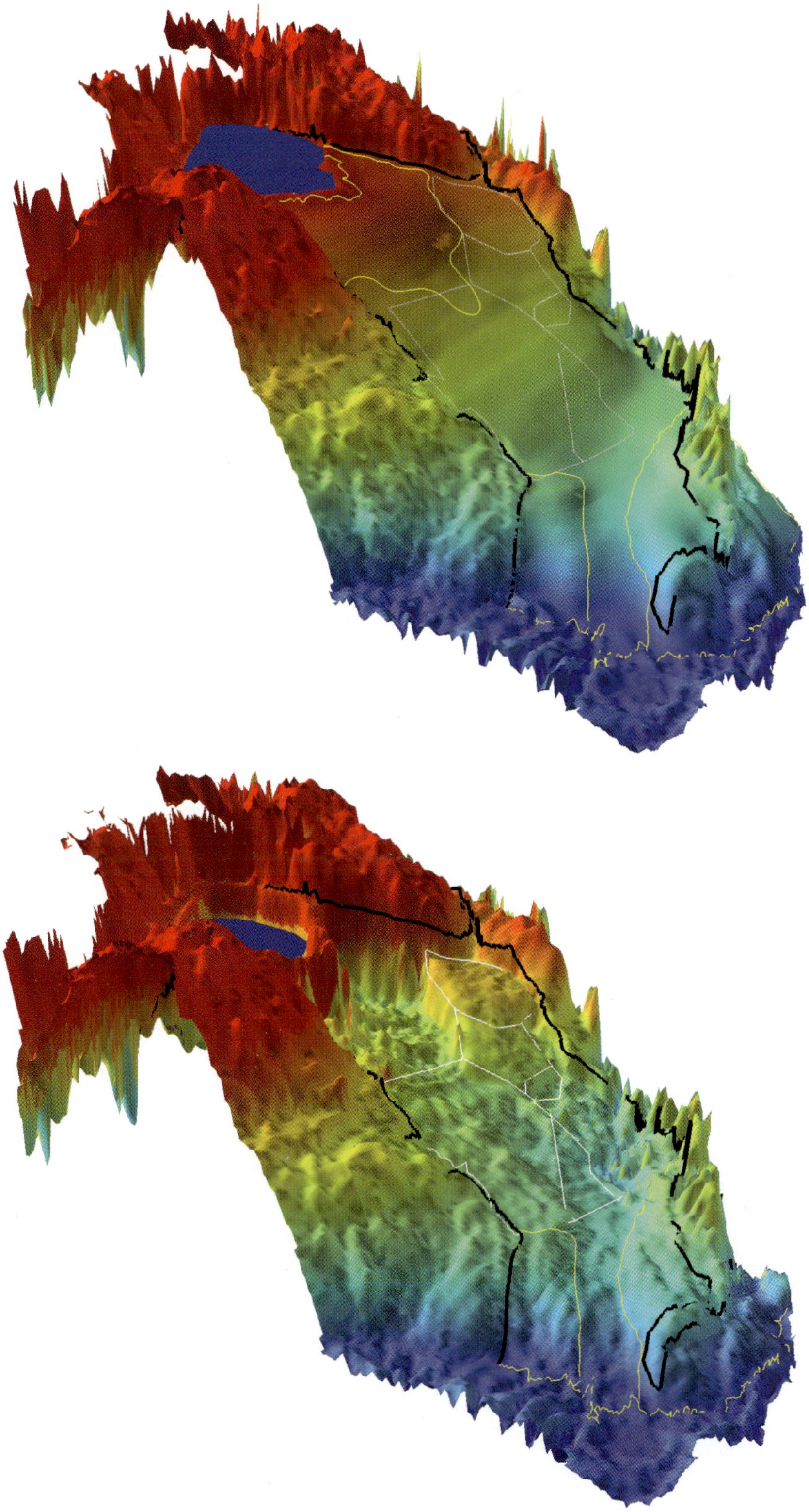

Plate 1. Three-dimensional views of the predrainage Everglades (top) and current extent (bottom). Vertical exaggeration: 9000×. Black lines = Everglades boundary; yellow lines = landscape boundaries; white lines = current Water Conservation Areas (shown for reference). Predrainage Lake Okeechobee is drawn at 21 feet (6.4 m) above NGVD 29. For visibility, current Lake Okeechobee is drawn at 17 feet (5 m) with dike at 21 feet (6 m); the top of the dike in actuality is at 35 feet (11 m), and lake level varies between about 10 and 17 feet (3–5 m) above NGVD 29. Sources: Predrainage contours from plate 6; current levels from South Florida Water Management Model (SFWMD, Hydrologic Systems Modeling Division 1998).

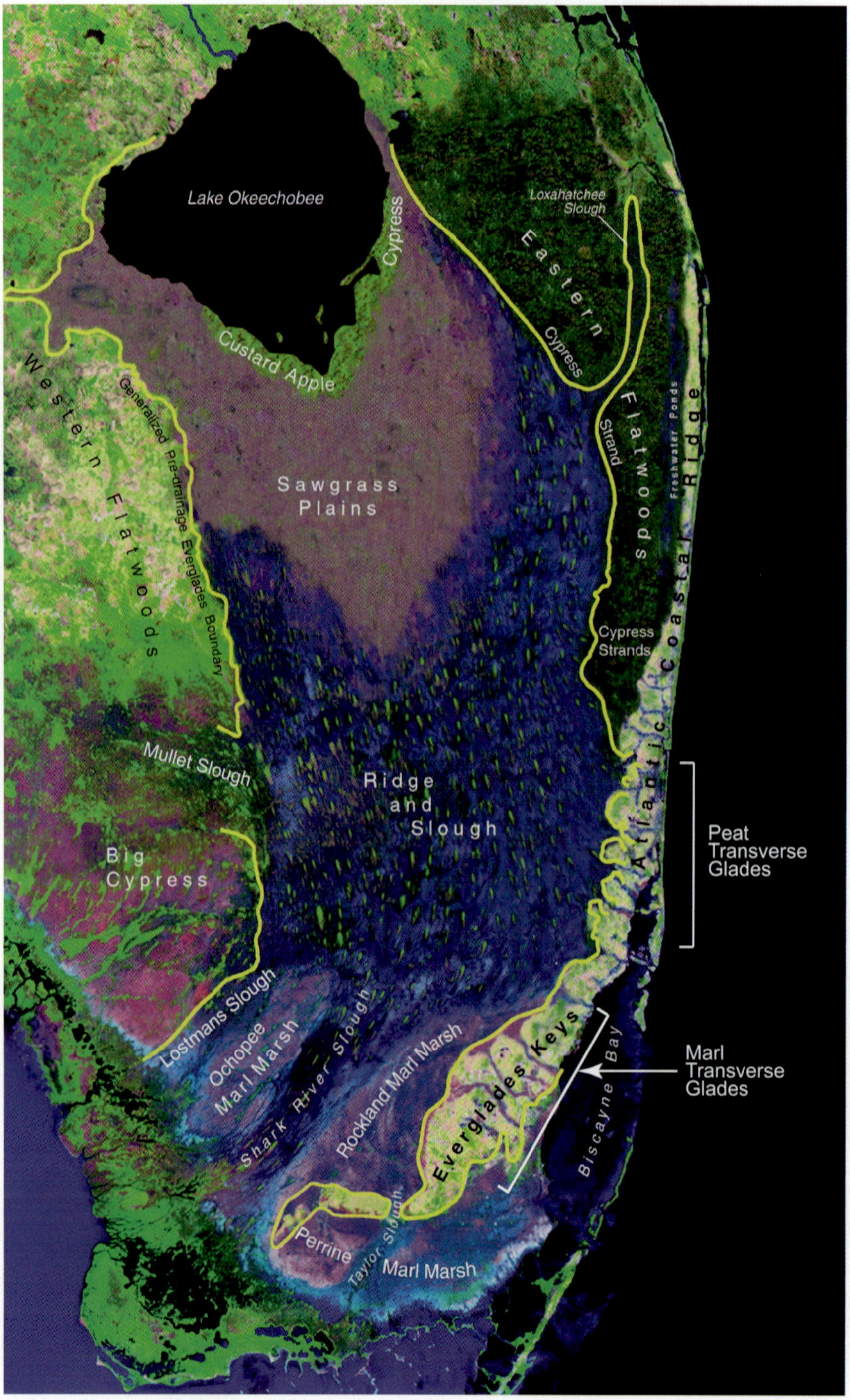

Plate 2a. Landscapes of the Everglades: Reconstructed "satellite image," ca. 1850s. Yellow lines show original border of the Everglades. The image was created by mapping appropriate present-day (1994) remnant visual signatures onto appropriate predrainage landscape extents.

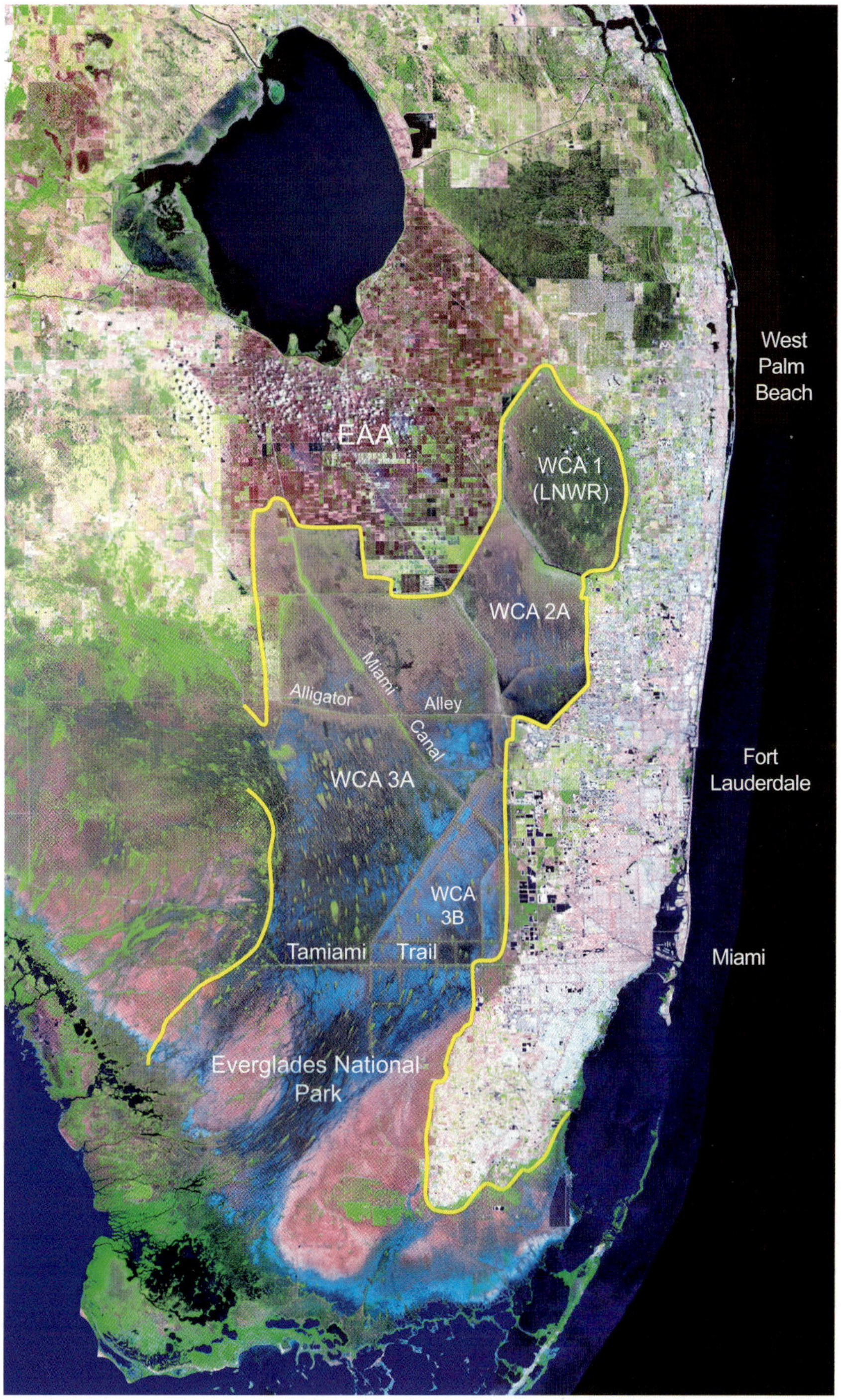

Plate 2b. Landscapes of the Everglades: Actual satellite image, 1994. Yellow lines indicate remaining portions of the Everglades (2010). The image is from Landsat TM data, digitally processed by the South Florida Water Management District.

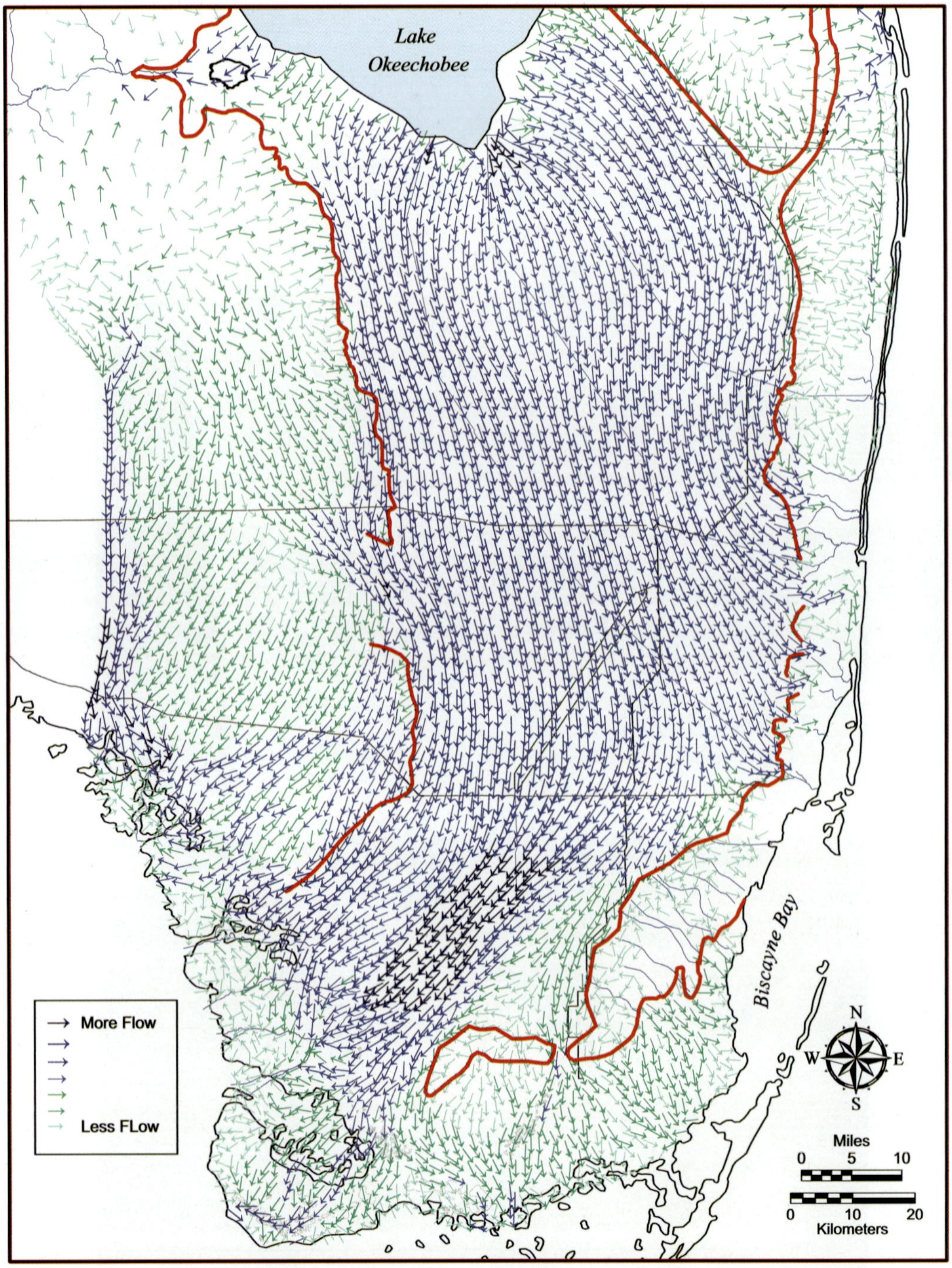

Plate 3a. Hydrology of the Everglades: Predrainage surface water flows, ca. 1850s. Arrow size and color reflects flow magnitude. Selected postdrainage levees and roads shown for orientation. Source: Natural System Regional Simulation Model v3.3 (Said and Brown 2010). Model developed by the Hydrologic and Ecosystems Simulation Modeling Department, South Florida Water Management District.

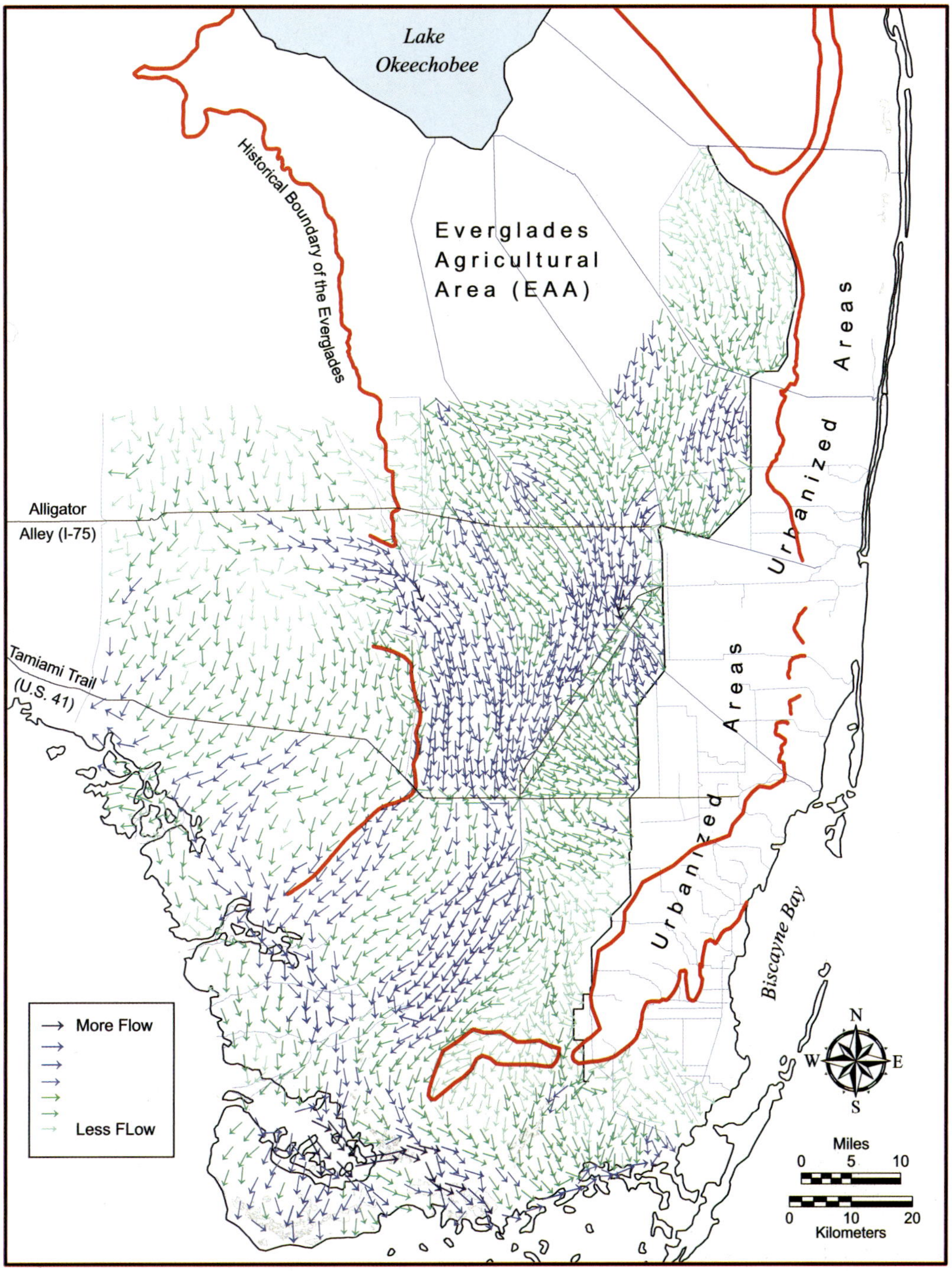

Plate 3b. Hydrology of the Everglades: Current surface water flows, 2010. Arrow size and color reflects flow magnitude (same scale as plate 3a). Source: Glades-LECSA Model (Senarath et al. 2008, 2010; see also Lal et al. 2005; South Florida Water Management District 2006). Model developed by the Hydrologic and Ecosystems Simulation Modeling Department, South Florida Water Management District.

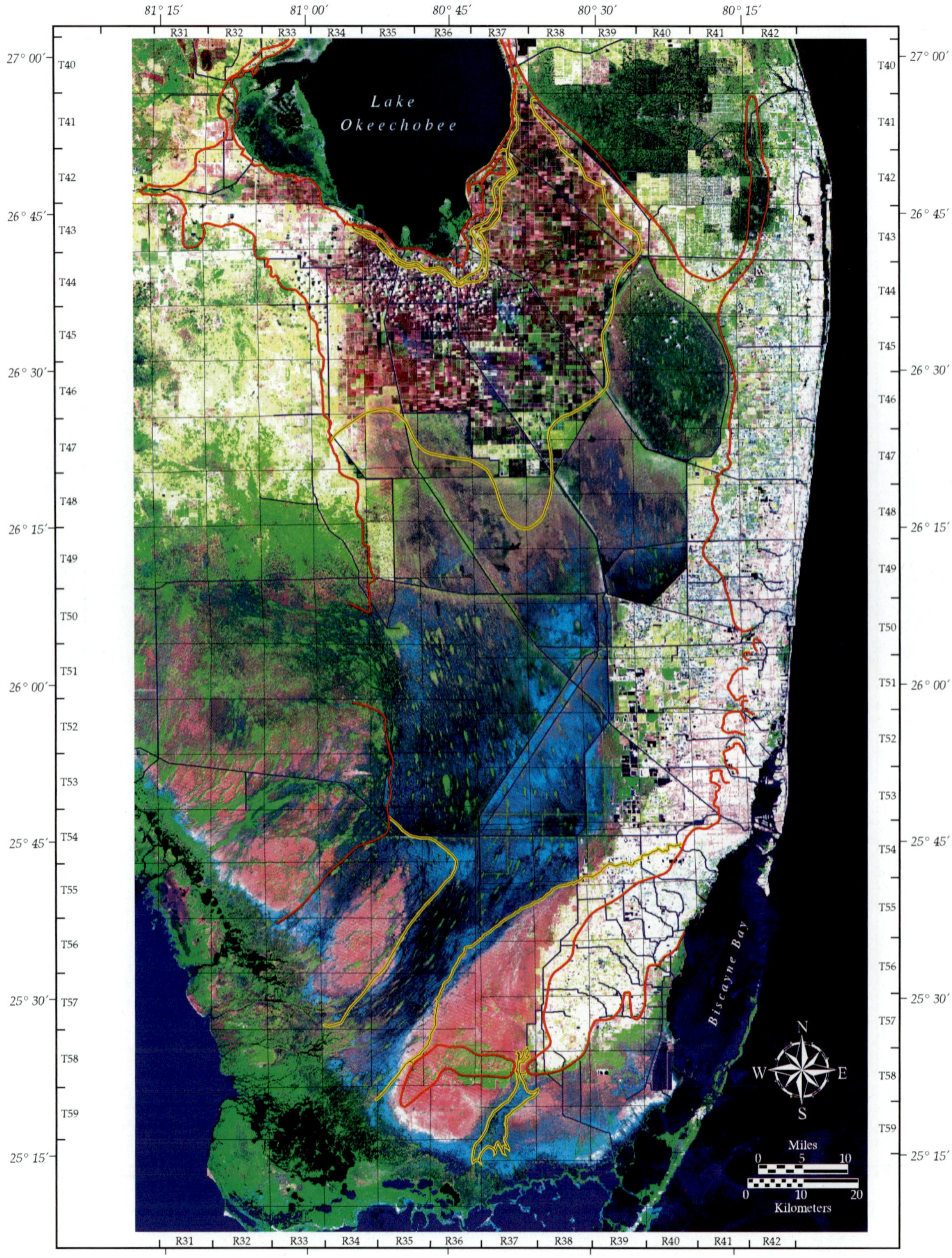

Plate 4. Satellite image of southeastern Florida, April 1, 1994. Red lines mark the predrainage border of the Everglades; yellow lines mark predrainage landscape boundaries. Shown for orientation are the township range grid (black) and 1990s canals (blue; see also fig. I.1). Landsat TM imagery, digitally processed by the South Florida Water Management District. Area N of Township 44 (approximate) from March 29, 1993.

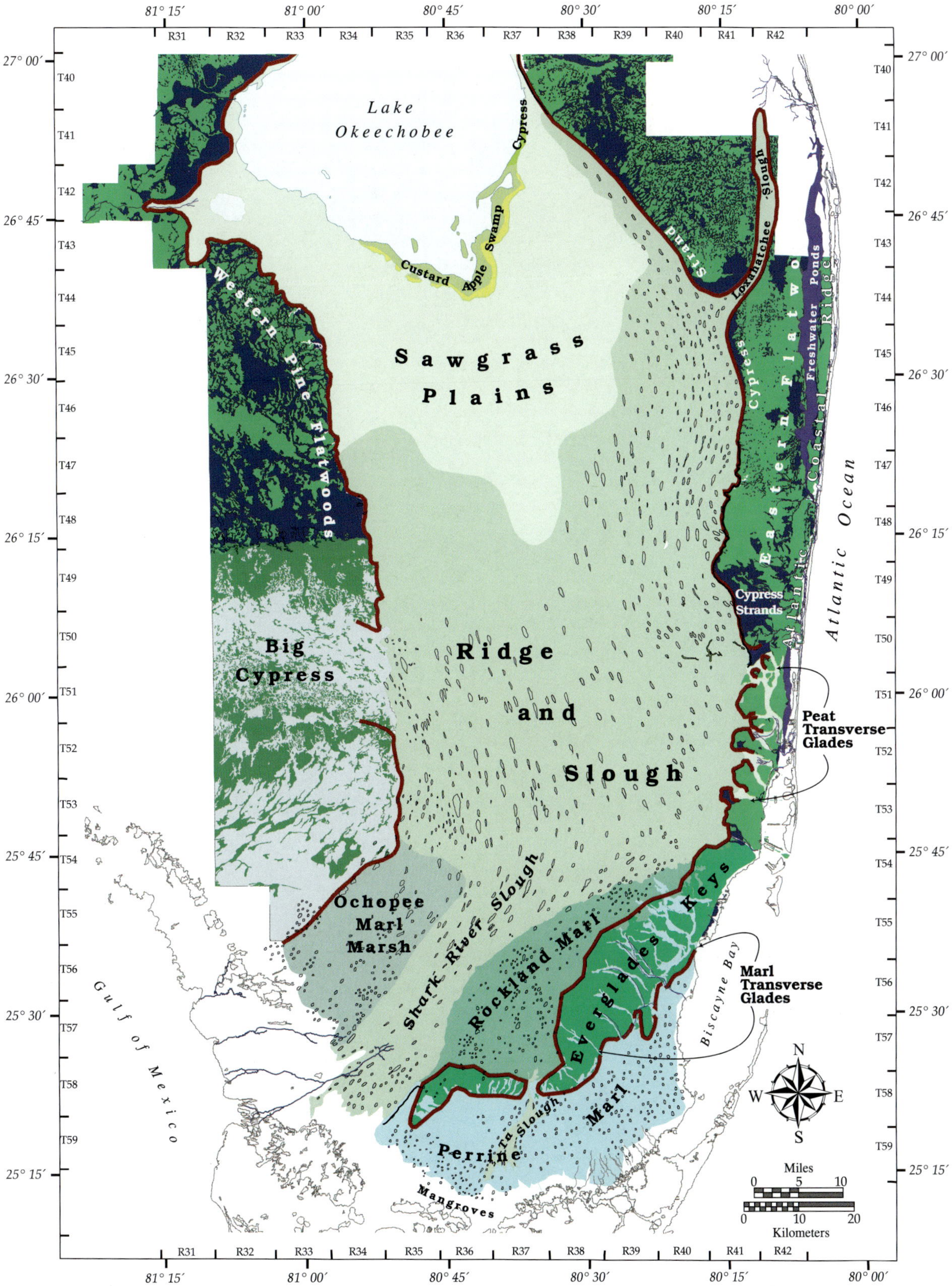

Plate 5. Landscapes of the predrainage Everglades and bordering areas, ca. 1850. Source: This study.

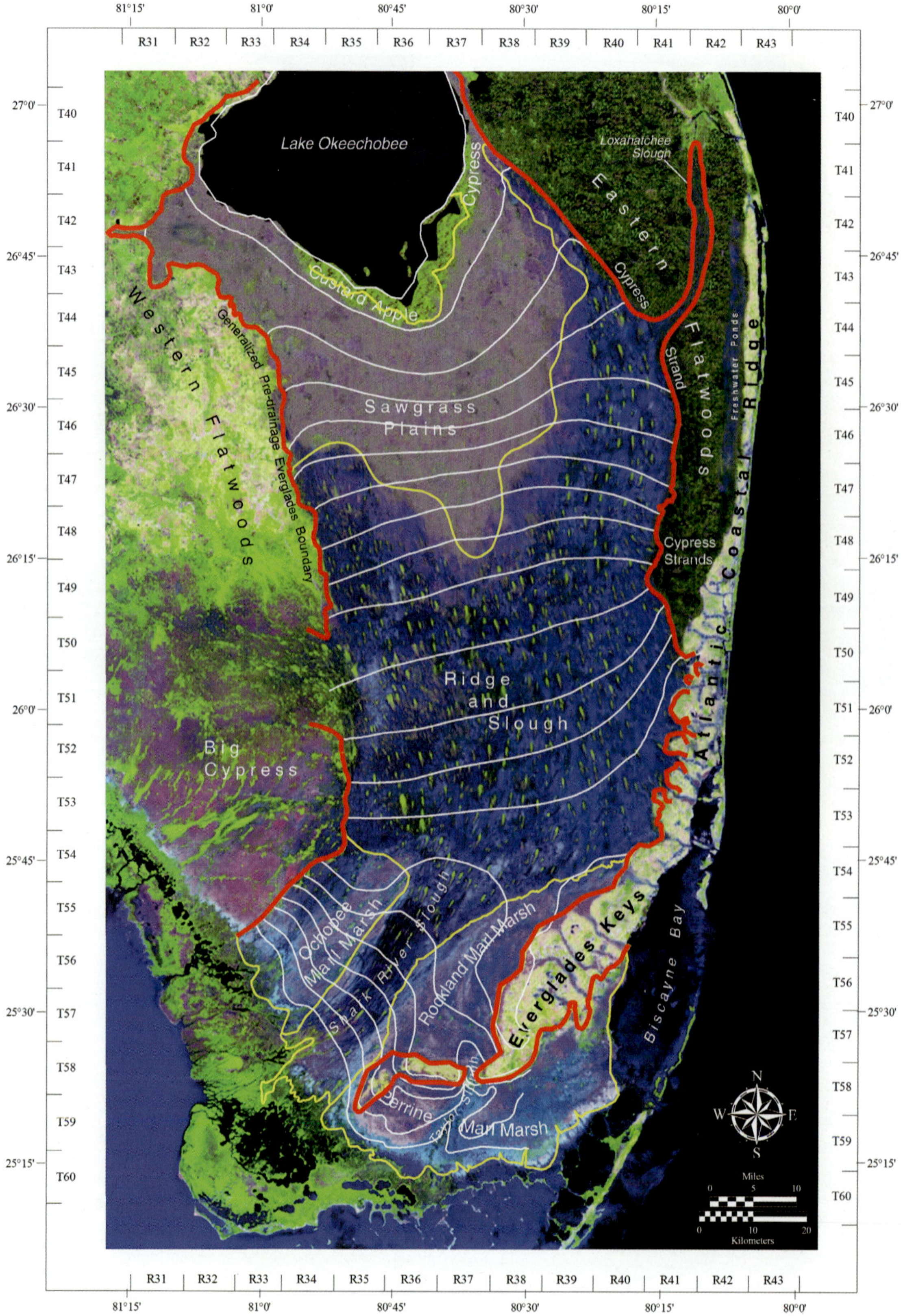

Plate 6. Predrainage Everglades estimated topographic contours (white lines; 1 foot to 20 foot; 0.3 m to 6.1 m above NGVD 29), overlain on a reconstructed "satellite image." Fine-scale topographic variation (microtopography) is not shown. The shoreline of Lake Okeechobee was about 21 feet (6.4 m) above NGVD 29. Red lines mark the predrainage border of the Everglades; yellow lines show predrainage landscape boundaries. Source: Chapter 11.

Plate 7. Artist's rendering of the predrainage Everglades, in oblique aerial view, "exploded" to show substrate. Cuts illustrate the relation of peat surface to ridges, sloughs, tree islands, and the different predrainage landscapes.

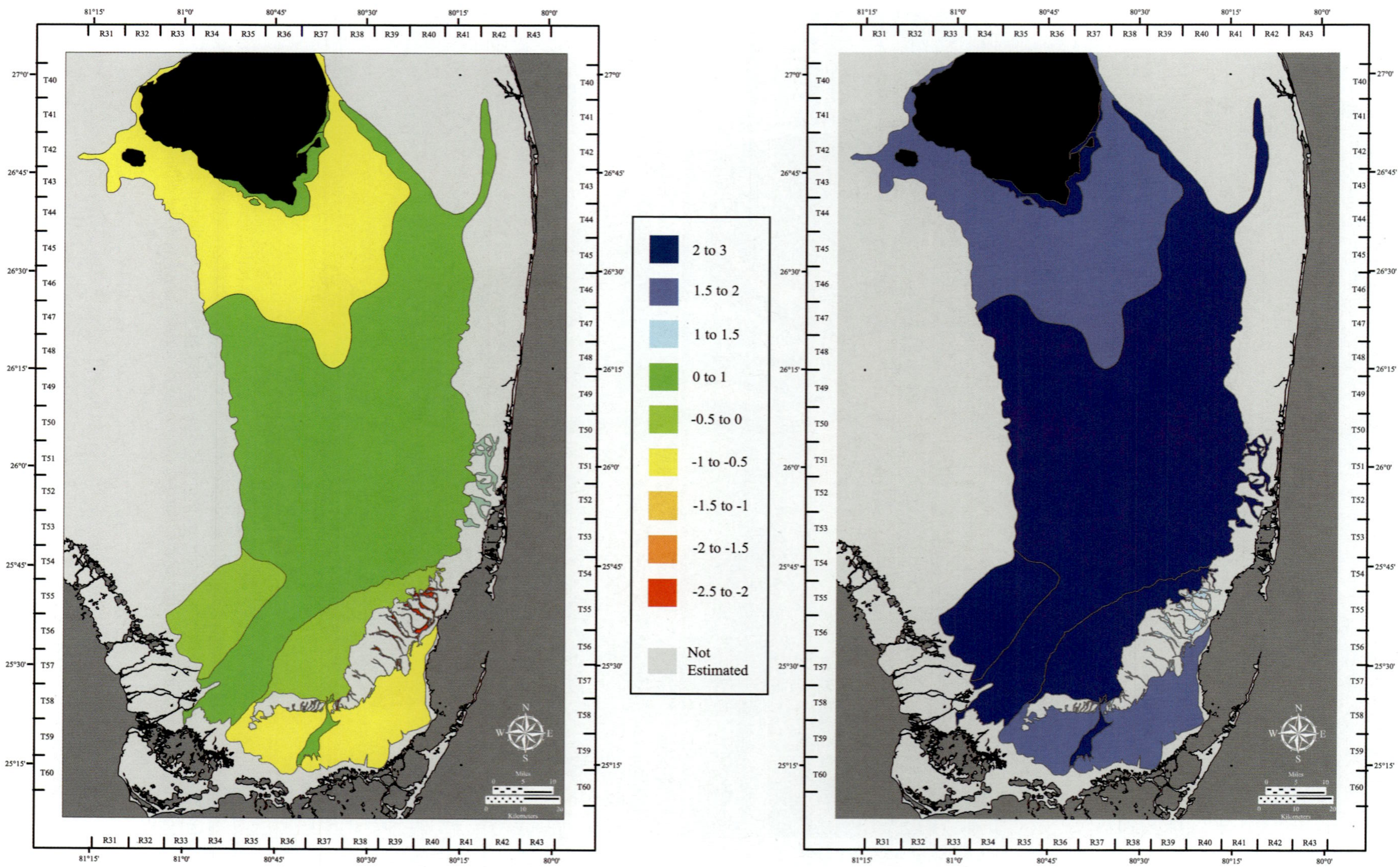

Plate 8. Water depths in the predrainage Everglades: Long-term average annual low, at the end of the dry season (left), and long-term average annual high, at end of the wet season (right). Annual lows and highs occur, on average, in mid-May and mid-October. Source: Table 11.4.

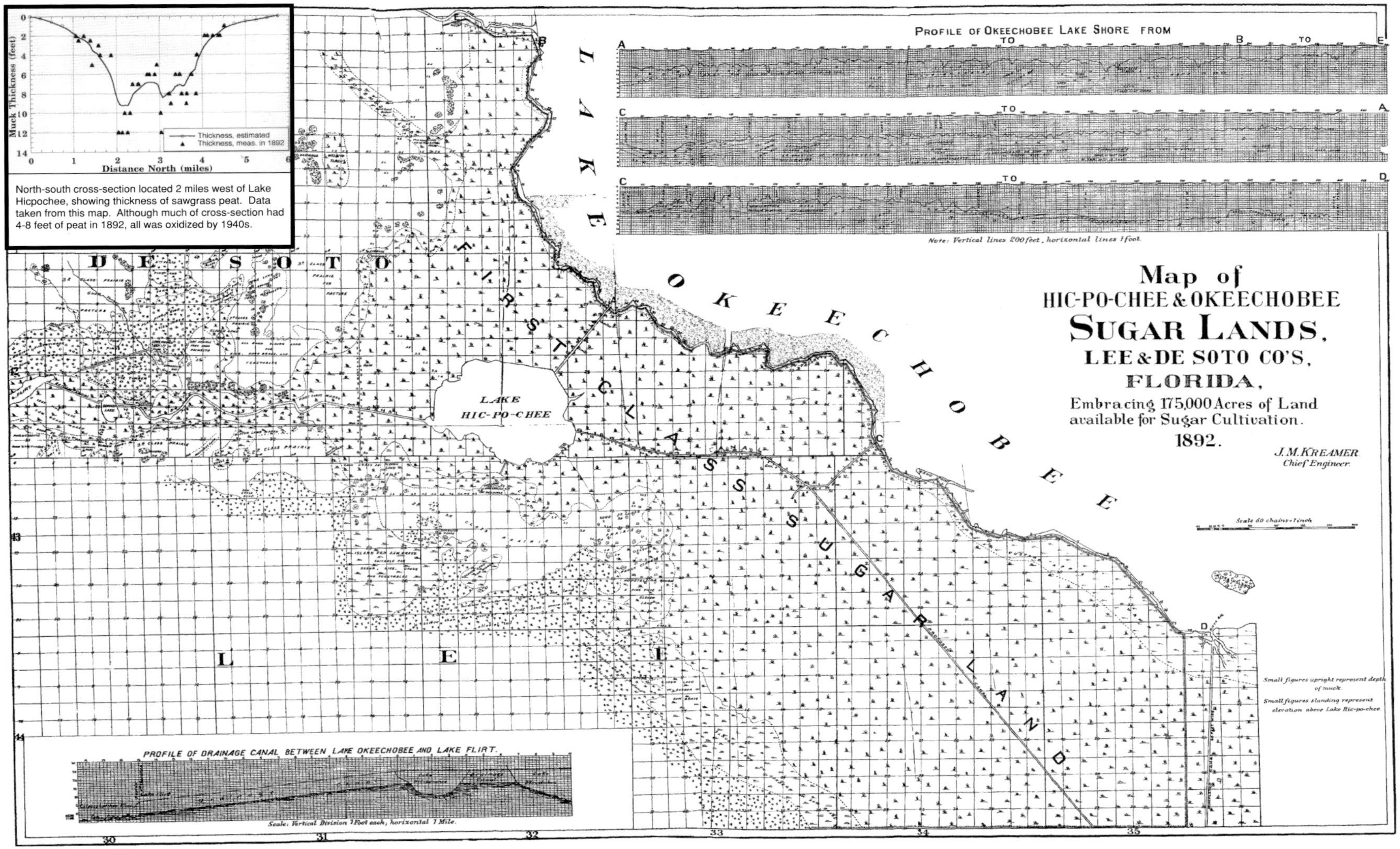

Plate 9. *Map of Hic-Po-Chee and Okeechobee Sugar Lands* prepared in 1892 by J. M. Kreamer for the Atlantic and Gulf Coast Canal and Okeechobee Land Company. Muck (peat) depths are shown along the Lake Okeechobee shoreline, the Three- and Nine-Mile canals, and two marsh transects. The inset (top left) graphs peat thicknesses noted on the map for a north-south transect two miles west of Lake Hicpochee. Reproduced from Kreamer (1892).

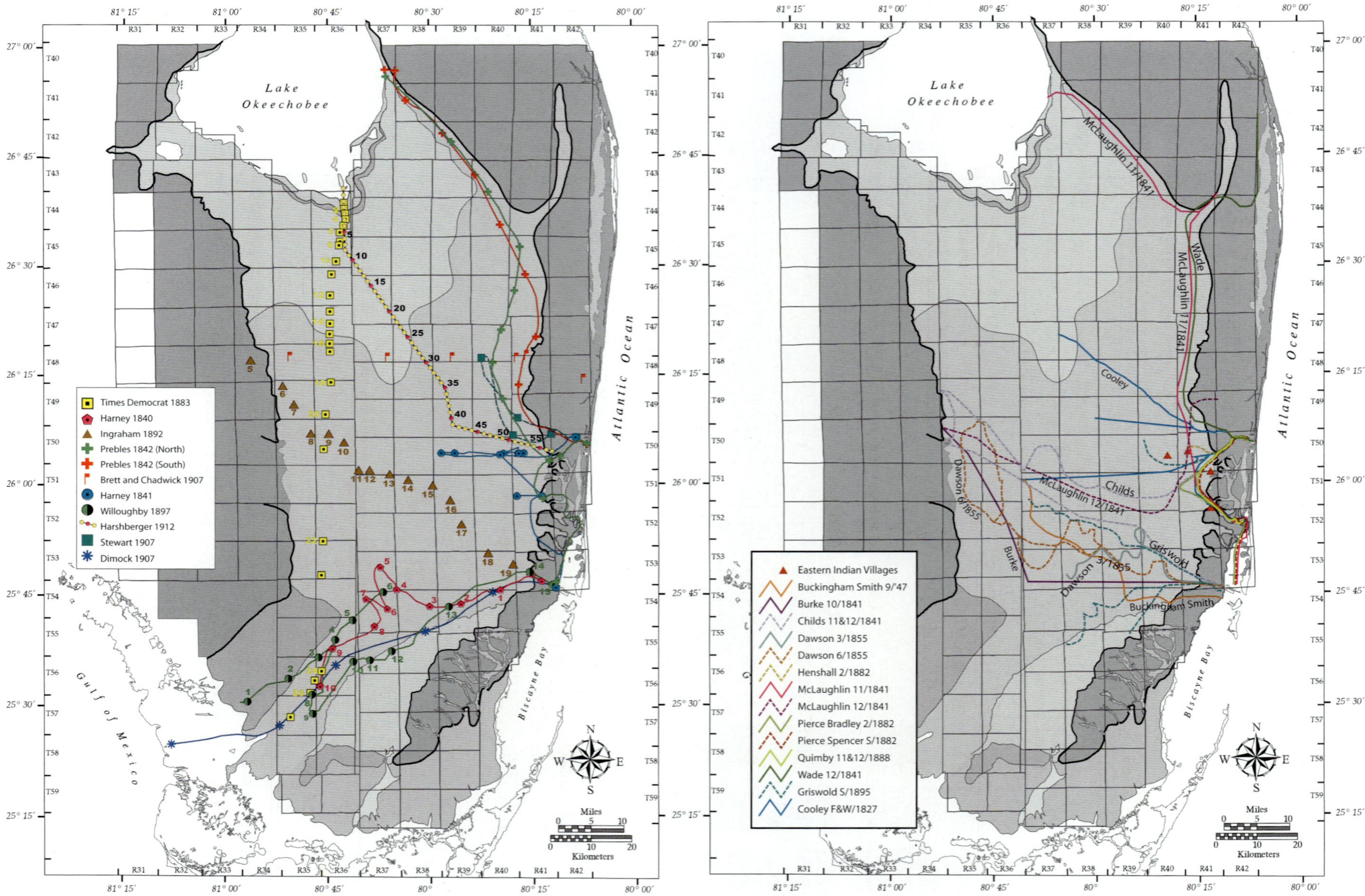

Plate 10. Routes of expeditions in the Everglades, 1827–1912. Left map indicates the routes of major expeditions, right map the routes of less-documented expeditions. Firsthand accounts exist for these routes (see appendices E through N). Overnight camps are numbered where possible. Note the predominance of routes within the Ridge and Slough landscape. Legend shows chief of party and year (plus month or season where possible).

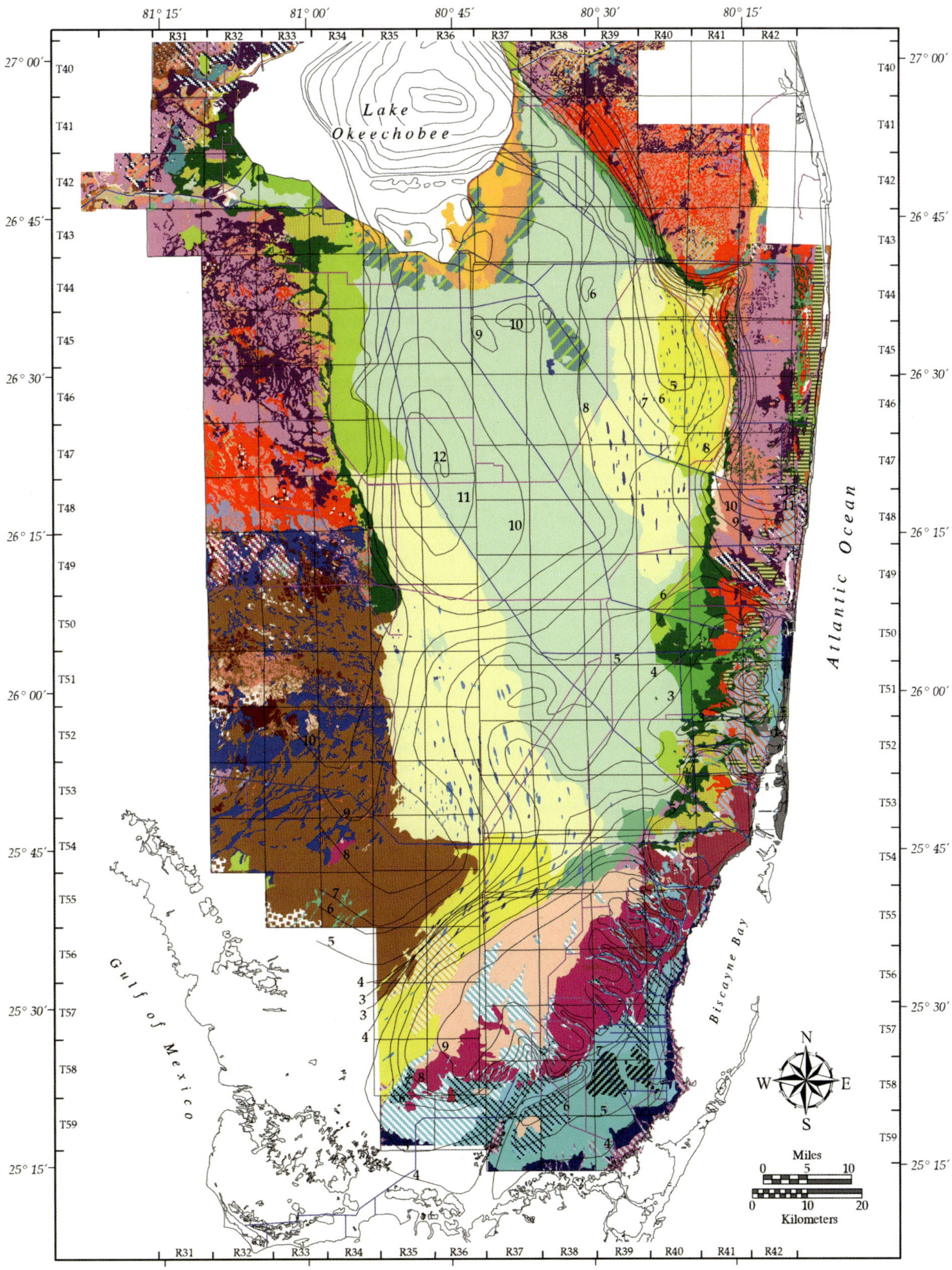

Plate 11. Soils map of the Everglades Drainage District, ca. 1940. See plate 12 for legend. Overlays: contours of rock floor of the Everglades, township range grid, and 1990 canals. Soils digitized from Jones et al. (1948b); rock floor digitized from Parker et al. (1955).

Soils

- Everglades peat
- Everglades peat over shallow marl
- Everglades peat over shallow sand
- Everglades peat over deep sand
- Okeechobee muck
- Okeelanta peaty muck
- Loxahatchee peat
- Loxahatchee peat over shallow sand
- Loxahatchee peat over shallow marl
- Loxahatchee peat over deep sand
- Everglades peaty muck
- Everglades peaty muck over shallow sand
- Everglades peaty muck over deep sand
- Gandy peat
- Gandy peat over sand or marl
- Brighton peat
- Istokpoga peat
- Hialeah mucky marl
- Ochopee marl
- Ochopee marl, shallow phase
- Ochopee fine sandy marl
- Ochopee fine sandy marl, shallow phase
- Perrine marl
- Perrine marl, shallow phase
- Perrine marl, very shallow phase
- Perrine marl (peat substratum)
- Perrine marl, shallow phase (peat substratum)
- Perrine marl, tidal phase
- Copeland fine sandy loam, shallow phase
- Keri fine sand
- Manatee fine sandy loam
- Parkwood fine sandy loam
- Arzell fine sand
- Charlotte fine sand
- Davie fine sand
- Davie mucky fine sand
- Delray fine sand
- Immokalee fine sand
- Pompano fine sand
- Broward fine sand
- Broward fine sandy loam
- Broward-Ochopee complex
- Felda loamy fine sand
- LaBelle loamy fine sand
- Palmdale fine sand
- Palmdale loamy fine sand
- Sunniland loamy fine sand
- Leon fine sand
- Dade fine sand
- Palm Beach fine sand
- St. Lucie fine sand
- Rockdale fine sand-Limestone complex
- Rockdale fine sandy loam-Limestone complex
- Alluvial soils undifferentiated (poorly drained)
- Coastal beach
- Made land
- Mangrove swamps (undifferentiated soil materials)
- Mine pits and dumps
- Rockland
- Swamp (undifferentiated soil materials)

General Vegetation

- Scrub Forests
- Inland Swamps
- Cypress Forests
- Cypress-heads or domes
- Pine Flatwoods Forests
- High Pine Forests
- Miami Rockland Pine Forests
- Miami Open Pine Forests
- Mangrove Swamps
- Mangrove Swamps and Salt-water Marshes
- Salt-water Marshes
- Salt Prairie (Cape Sable area)
- Fresh-water Marshes (outside the Everglades)
- Wet Prairies
- Saw-palmetto or Dry Prairies
- Southern Coast Marsh Prairies
- Coastal Beach and Dune Vegetation

Main Everglades Vegetation

- Saw-grass Marshes (medium dense to sparse)
- Saw-grass Marshes (dense)
- Saw-grass (with wax-myrtle thickets)
- Willow and Elderberry Zone (mostly cultivated)
- Custard Apple Zone (mostly cultivated)
- Saw-grass marshes (with abundant ferns and cat-tails)
- Sloughs, Ponds, and Lakes (with aquatic plants)
- Tree Islands, Bay Tree Forests
- Tree Islands, Hammock Forests
- Marsh Prairies, Southern Everglades

Cover or Land Use

- Cropland
- Orchards and groves
- Urban areas
- Grassland
- Idle land
- Sawgrass
- Myrtle and bay
- Cypress
- Hardwoods and cabbage palmetto
- Pines and saw palmetto
- Mangrove
- Mixed grasses and water plants

Legends for plates 11, 13, and 14

Clockwise from left:
Plate 11, Soils (Jones et al. 1948b)
Plate 13, Vegetation (Davis 1943b)
Plate 14, Land Cover/Land Use (Jones et al. 1948e)

Plate 12. Legends for plates 11, 13, and 14.

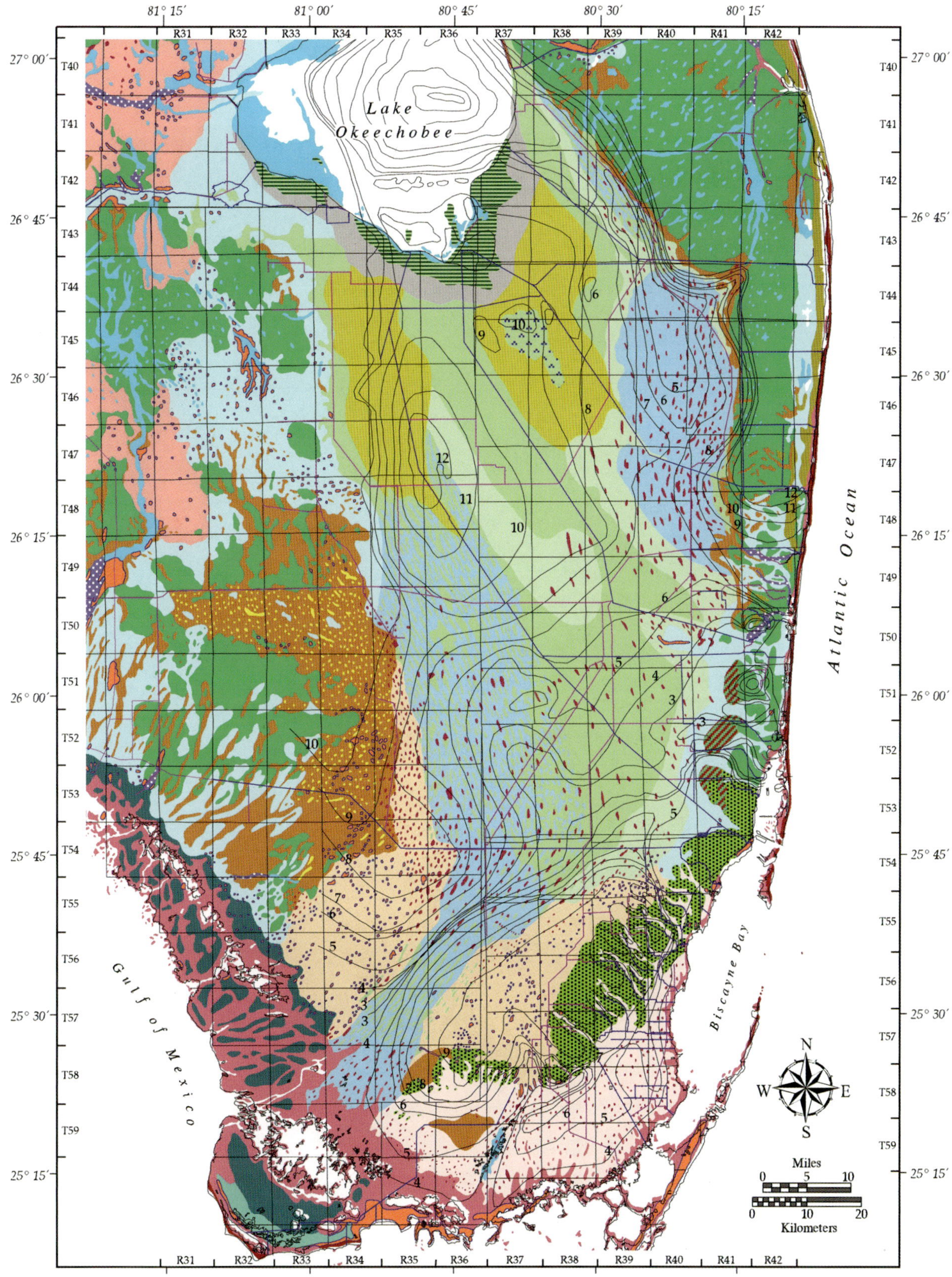

Plate 13. Postdrainage vegetation map of southern Florida, ca. 1940. See plate 12 for legend. Overlays: contours of rock floor of the Everglades, township range grid, and 1990 canals. Vegetation digitized from Davis (1943b); rock floor digitized from Parker et al. (1955).

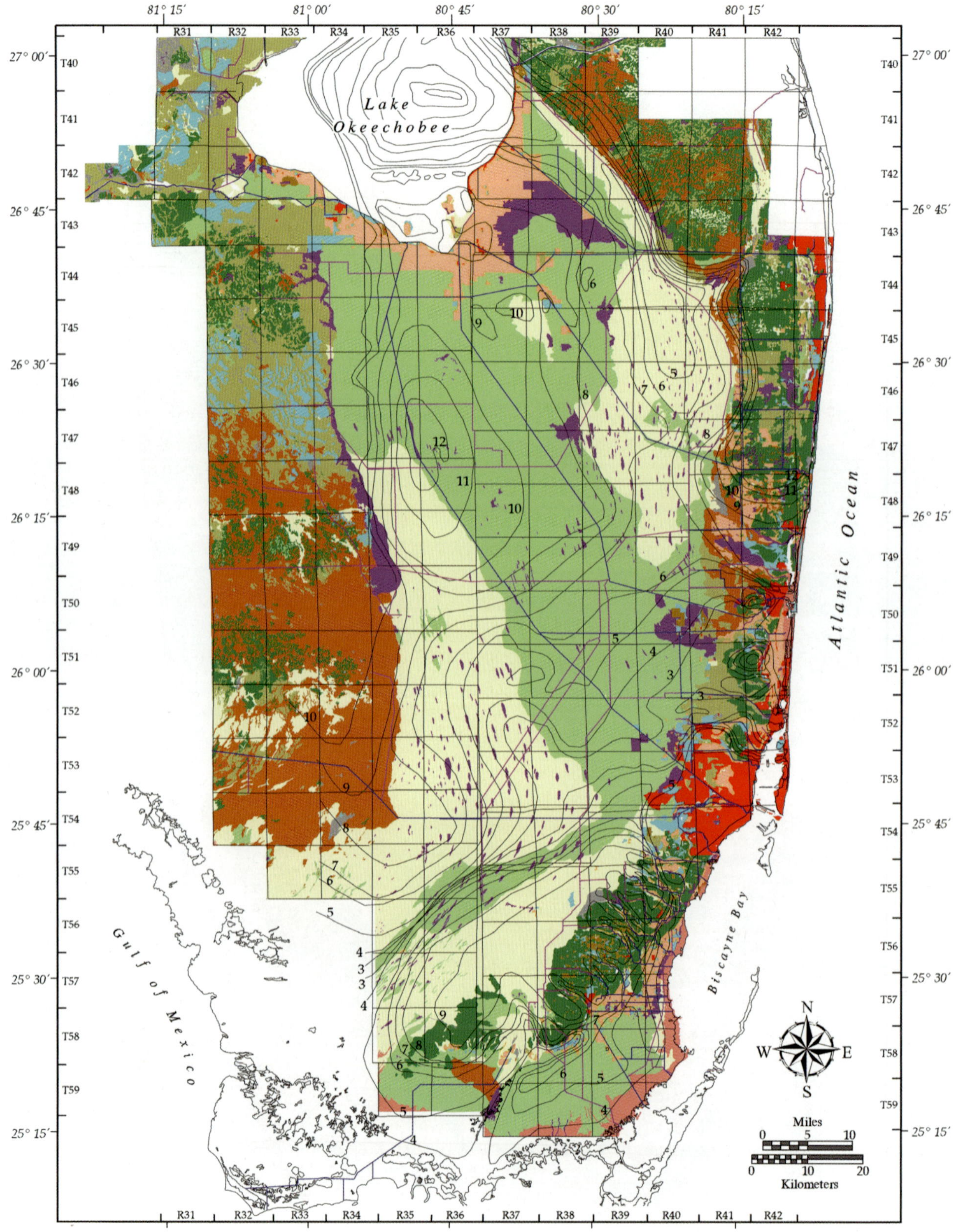

Plate 14. Map of "land cover" or "land use" of the Everglades Drainage District, ca. 1940. See plate 12 for legend. Overlays: contours of rock floor of the Everglades, township range grid, and 1990 canals. Land cover/land use digitized from Jones et al. (1948b); rock floor digitized from Parker et al. (1955).

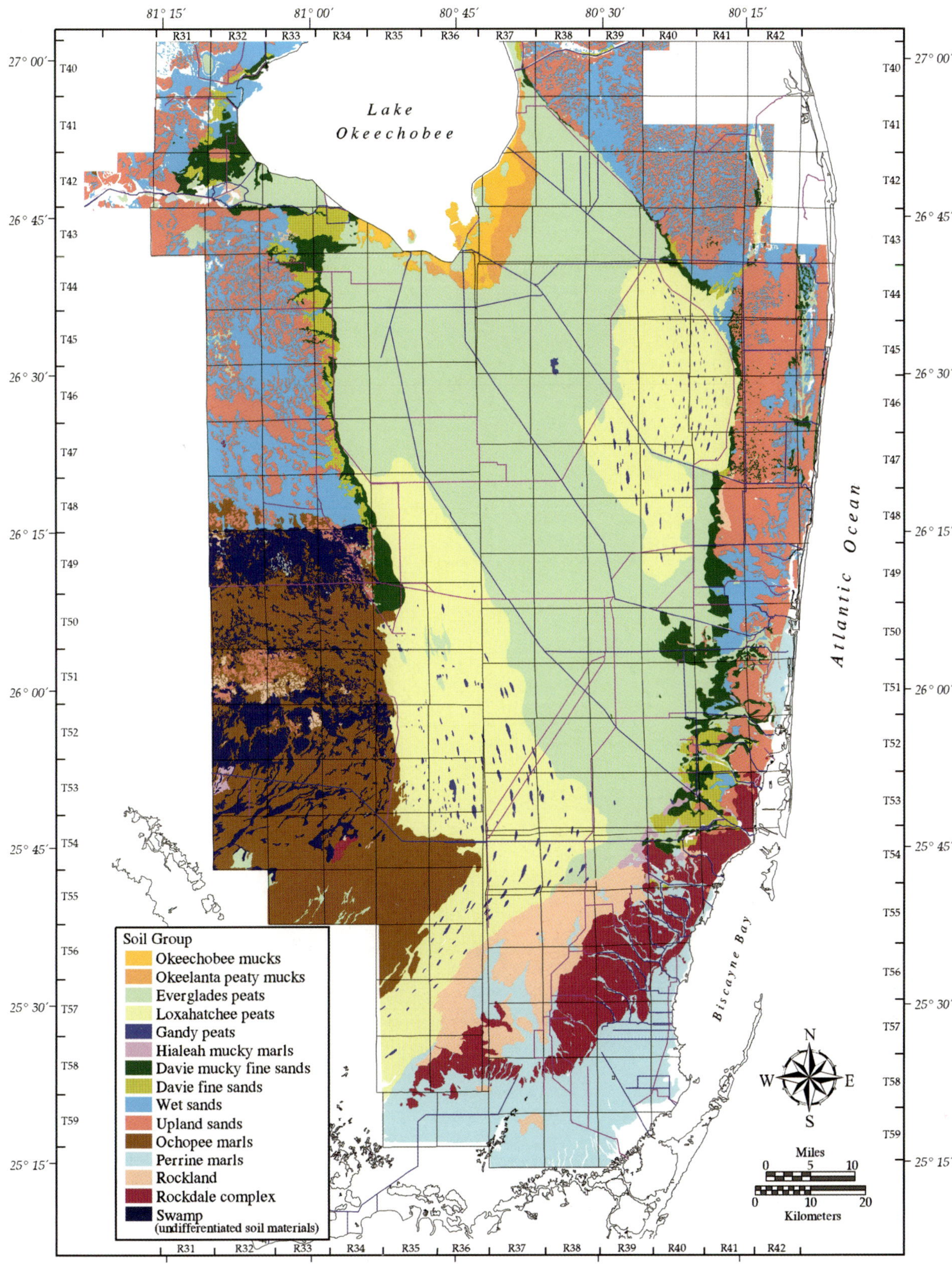

Plate 15. Generalized soils map of the Everglades Drainage District, ca. 1940. Overlays: canals present in 1940 (blue), canals built 1940–1990s (purple), and township range grid. Soils digitized from Jones et al. (1948b).

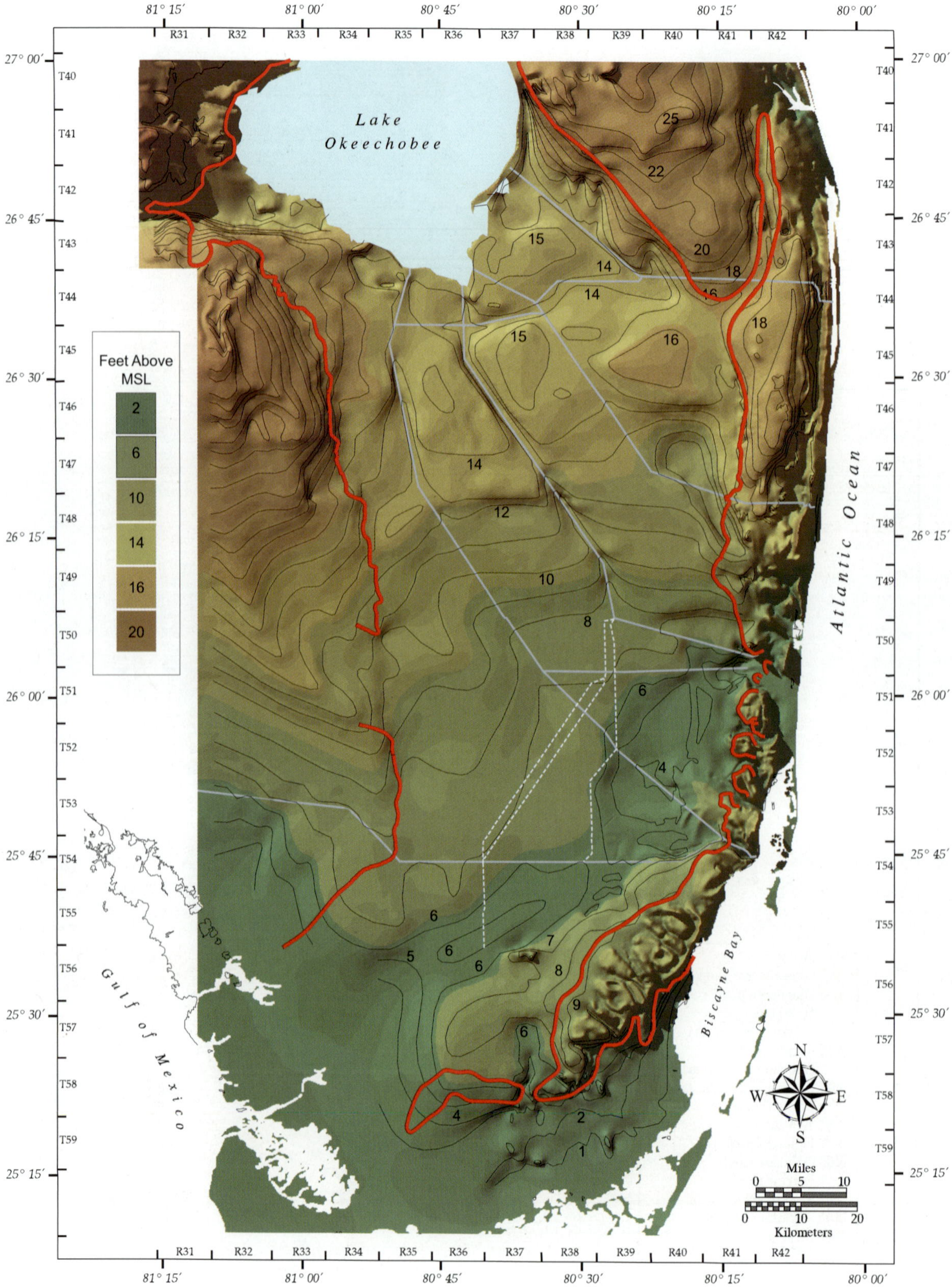

Plate 16. Shaded relief map of land surface of southeastern Florida in 1940. Note "subsidence valleys" along canals (see also figs. 3.6, 3.7) and "subsidence crater" surrounding the southern reach of the Miami Canal (see also figs. 3.9, 3.12). Dashed lines are canals/levees not present in 1940; shown for location only. Data for the Everglades and Everglades Keys from Jones et al. (1948e), with additional data for the Atlantic Coastal Ridge from U.S. Army Corps of Engineers (1960a [map]) and U.S. Geological Survey quadrangles (contours on coastal ridge and Everglades Keys omitted for graphic clarity).

Plate 17. Ridge and Slough landscape in February 1917. Note the sharp, distinct edges on most sawgrass ridges, and the open water of sloughs, with few emergent plants. Top: NE Shark Slough area, "511 . . . same point [Section 24 on its Western boundary, Range 37 E. Township 54 South, . . . showing evident 'lakey' conditions], taken in westerly direction." Bottom: SW corner WCA 3B, "512 . . . at or near Southeastern corner of section 5, range 37 E. Township 54 South . . . lakelike conditions with flowing water." King (1917f).

Plate 18. Ridge and Slough landscape in February 1917. Sawgrass ridges with tree islands in background. Top: NE Shark Slough, "507 . . . on top of Tableland near Range line 37/38, Township 54, looking West." Bottom: NE Shark Slough, "508 . . . view showing method of effecting passage [through sawgrass ridge] from one slough to another, corner of Section 29, Range 38 East, Township 54 South." King (1917f).

Plate 19. Ridge and Slough landscape in February 1917. Note water lilies in sloughs in the midground of both photos. Top: NE Shark Slough, "503 . . . members of party locating numbers on stakes of Tamiami Trail." Bottom: NE Shark Slough, "505 . . . view of Central Slough basin, taken in the Southwestern portion of Section 20, Range 38 E. Township 54 S., looking in an Easterly direction." King (1917f).

Plate 20. Ridge and Slough landscape in February 1917. Distinct sloughs surrounded by tall sawgrass ridges. Note the unidentified light-colored edge exposed at base of sawgrass ridges. Both photos taken about 1 mile north of Tamiami Trail, but approximately 15 miles apart. Top: Middle south end of WCA 3B, "504 . . . Central Slough basin, taken in Section 6 Range 38 E., Township 54 South, looking South" King (1917f). Bottom: SW corner of WCA 3A, "No. 1 . . . approach to coastal hammock [far background] at Northern edge of Hopkins Tract [Section 10 Township 54 Range 35, estimated]" King (1917e).

Plate 21. Painting of the Everglades (Ridge and Slough landscape), 1932. This is most likely a view from Tamiami Trail. Note the alternating sawgrass ridges and open water sloughs. Such birds feeding in Ridge and Slough landscape in the 1930s likely reflects reduced water depths due to several decades of uncontrolled drainage (chapters 2, 3). The original caption reads: "Herons in the Everglades. Vast numbers of herons of several species feed in the Everglades, and at certain seasons they may be seen to advantage from the motor highways. The birds shown are: Upper left, an adult and two immature (white) Little Blue Herons. Upper center, an American Egret; upper right, a single Snowy Heron. Lower right, three Louisiana Herons, two adult and one (the center bird) immature. The standing bird is a Ward's Heron." Painting reproduced from Howell (1932).

Plate 22. Oblique aerial and ground views of portions of the Ridge and Slough landscape considered to resemble its predrainage condition most closely. Dark areas in the oblique view are sawgrass ridges; light areas are sloughs. Sawgrass ridges form the background, sloughs the foreground in the lower two photographs. Photographs by Christopher W. McVoy, taken in Water Conservation Area 3A, 2005.

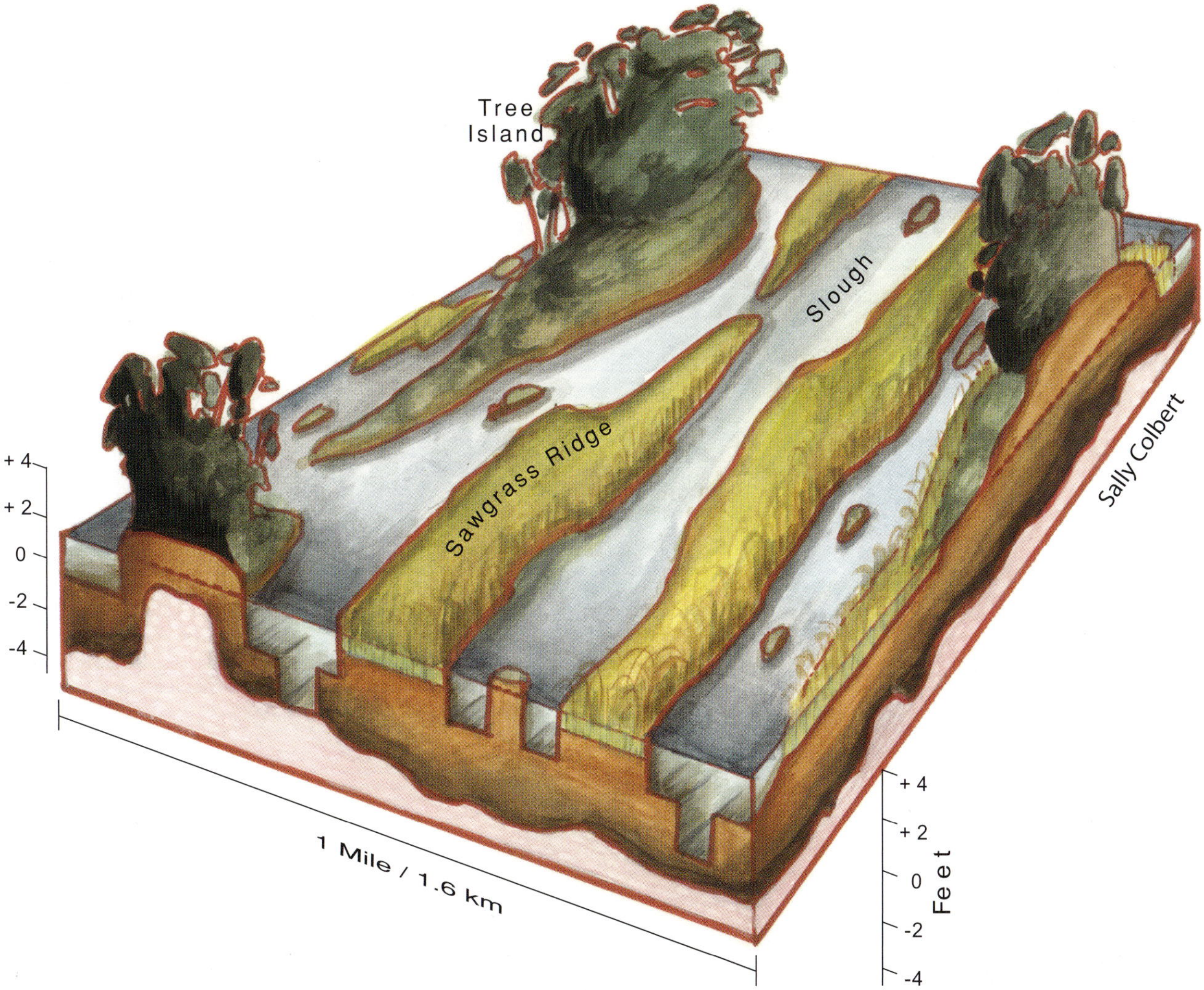

Plate 23. Scale model of typical block (1 × 1.5 miles) of the predrainage Ridge and Slough landscape. Vertical exaggeration: 200×. Vegetation is not to scale. Sloughs would have been either open water or partially covered with water lilies (not shown at this scale). Note relation of peat elevations, shown in cross section, to the spatial pattern of vegetation types.

Miami Canal

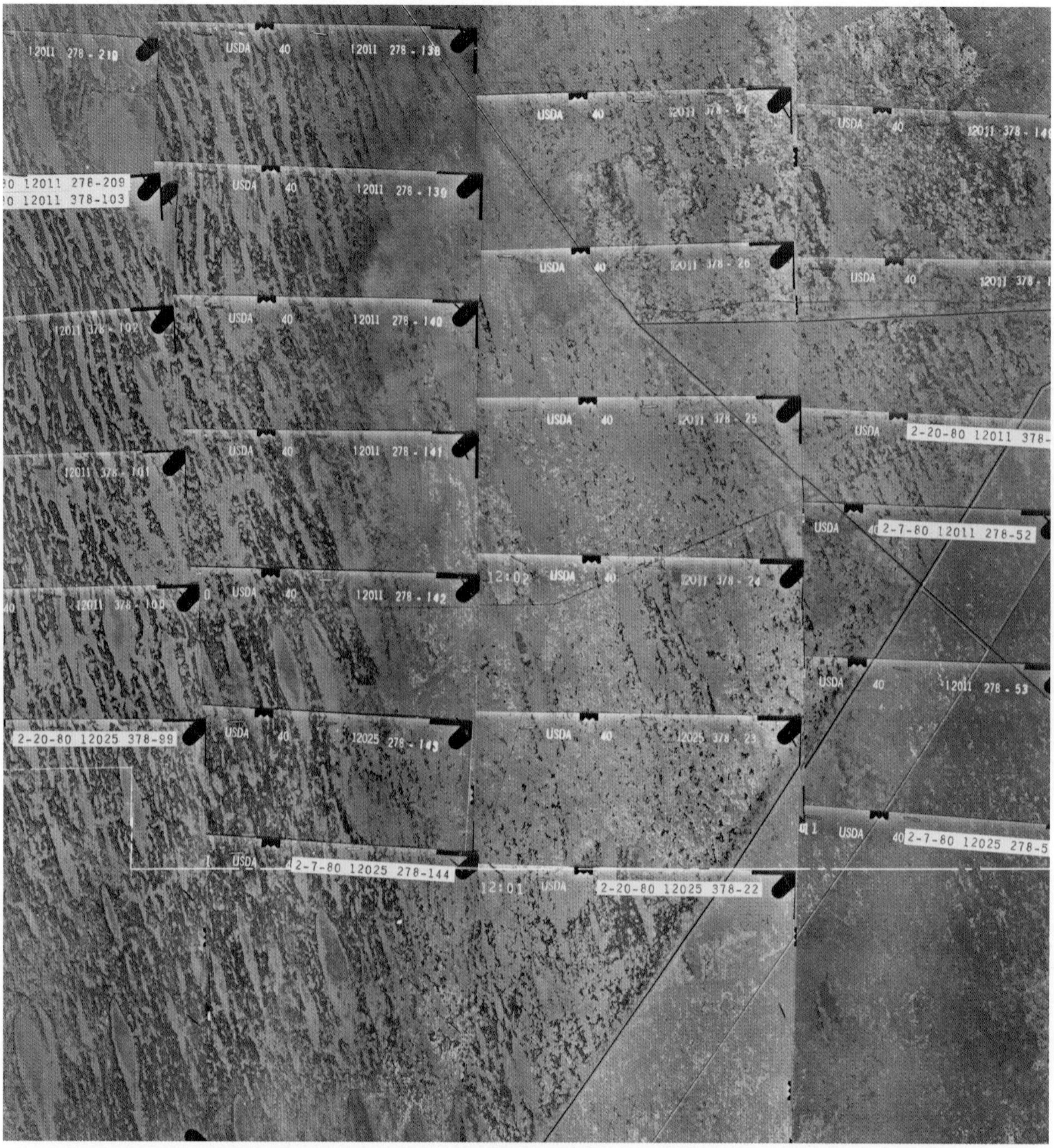

L-67A Canal
and Levee

Plate 24. Aerial photomosaic of Ridge and Slough landscape, January–February 1980. The mosaic depicts the east side of Water Conservation Area 3A, approximately Township 51 Range 38 (see plate 4 for location). Light gray indicates sawgrass ridges or tree islands; black indicates deeper water of sloughs. Note at left the clear linear pattern of distinct, elongated sawgrass ridges and sloughs, similar to oblique aerial images from Matlack (1939). Sawgrass ridges cover less than 50 percent of landscape. In the right half of the image, note similar directionality, but a much less distinct pattern, and apparent expansion of sawgrass coverage. The fine-scale pattern suggests that the sawgrass has expanded laterally from original sawgrass ridges, displacing sloughs. Sloughs have likely been invaded by wet prairie species. Note also that although the directionality appears approximately constant across whole image, the spatial pattern differs between the compartments formed by canals and/or levees. Source: U.S. Department of Agriculture (1980).

Plate 25. Satellite image of Water Conservation Areas 3A and 3B and northern Everglades National Park, April 1994. Aligned reddish ovals are strand tree islands; green generally indicates sawgrass; and dark blue or purple areas are generally sloughs. Note the alignment of sawgrass ridges parallel to tree island alignment in the area south of I-75 and west of the Miami Canal (see fig I.1 for road and canal names). Landsat TM imagery, digitally processed by the South Florida Water Management District.

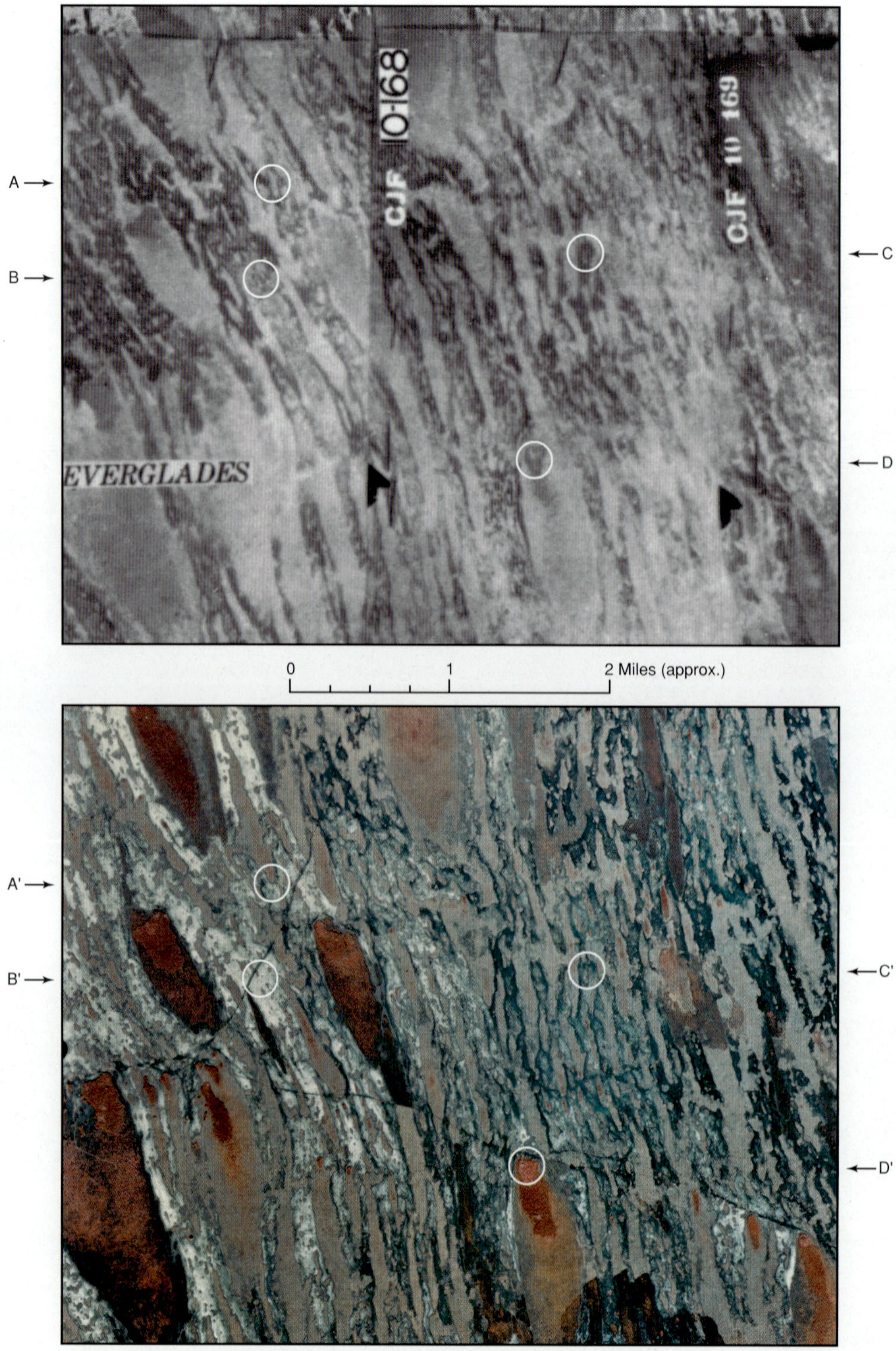

Plate 26. Aerial photographs of Ridge and Slough landscape, 1940 (top) and 1990 (bottom). The area is in Water Conservation Area 3A, west side, 7 miles south of I-75. Gray indicates sawgrass ridges (A' marks the southern tip of a ridge); white (B') or darker blue (C') indicates sloughs; and red indicates woody vegetation (D') of tree islands. Note the remarkable similarity in landscape pattern, including fine detail. Sources: U.S. Department of Agriculture (1940) and South Florida Water Management District (1991).

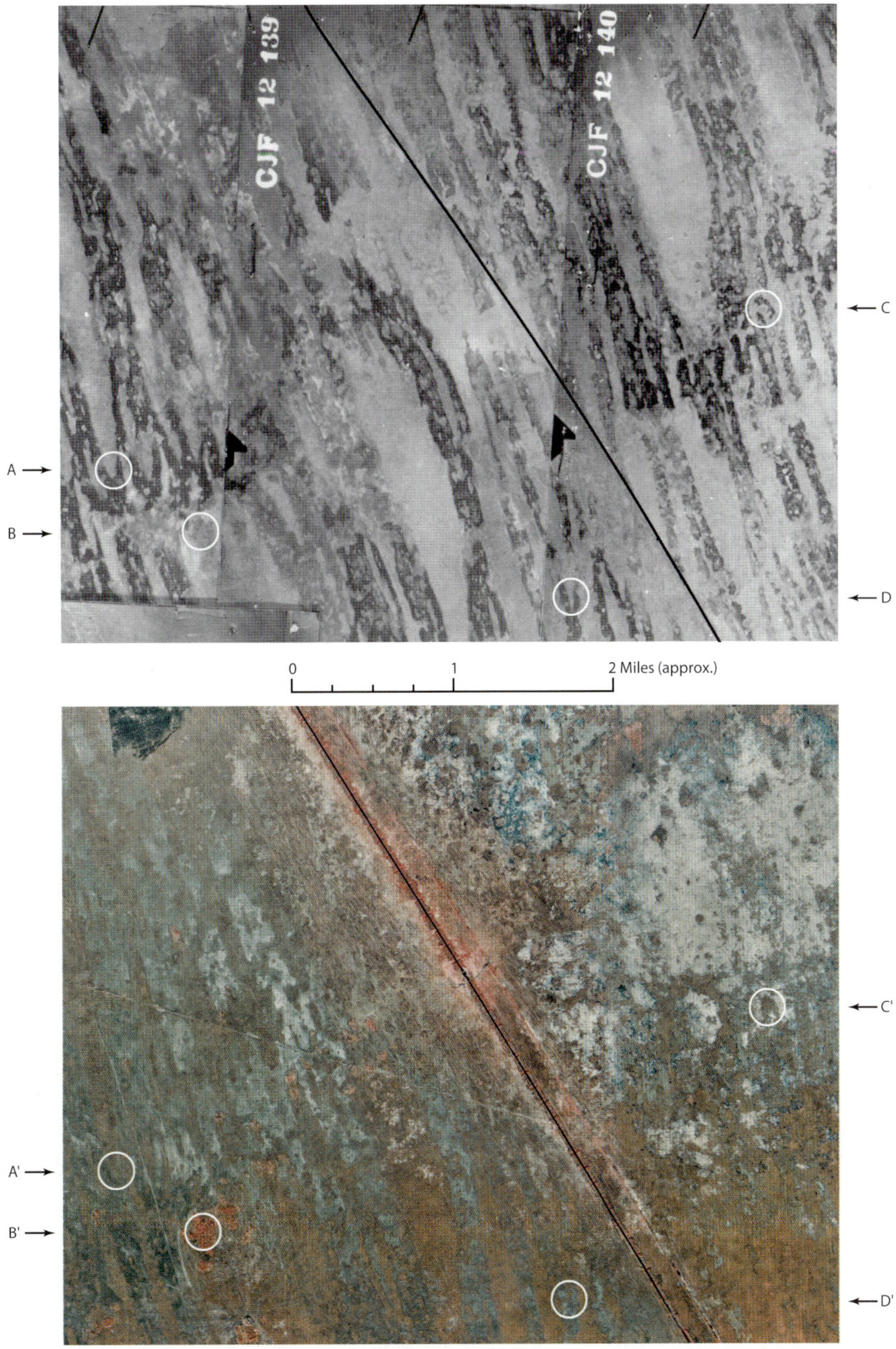

Plate 27. Aerial photographs of Ridge and Slough landscape, 1940 (top) and 1990 (bottom). The area is in Water Conservation Area 3A, west side, 4 miles north of I-75. Colors are as in plate 26. A, C, and D indicate tips of sawgrass ridges (1940); B' is woody vegetation (1990). "Al Henschel's camp" is at bottom right of 1990 photo. Note partial disintegration of the landscape pattern west of the Miami Canal and almost complete disappearance east of the canal. Sources: U.S. Department of Agriculture (1940) and South Florida Water Management District (1991).

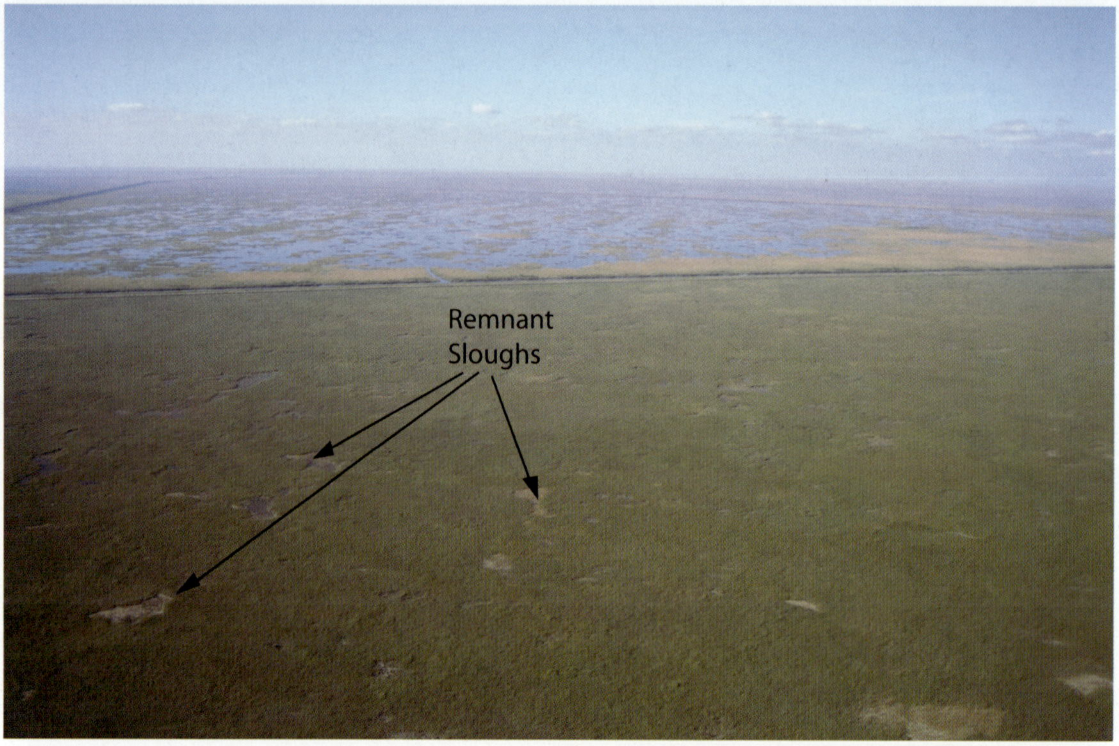

Plate 28. Preserved (top) and degraded (bottom) portions of the remnant Ridge and Slough Landscape. Top: Water Conservation Area 3A, showing sawgrass ridges and aquatic sloughs with open water or water lilies. Whitish edge lines are wrack bands of dead vegetation, mostly sawgrass, deposited by Hurricane Wilma. Bottom (foreground): "The Pocket," between the L-67A and L-67C levees. Here, as in Water Conservation Area 3B, sawgrass ridges have greatly expanded, leaving only remnant sloughs. Background behind the L-67A levee is Water Conservation Area 3A. Compare these with 1917 images (plates 17–20) and with expedition descriptions (appendices E–N). Photographs by Christopher W. McVoy, 2005.

Plate 29. Drainage effects on the Ridge and Slough landscape. This landscape originally covered half of the Everglades. Cross sections show average annual high and low water levels under predrainage (top) and current (2010; bottom) conditions. Current conditions are drawn from northern WCA 3A (Shark River Slough in Everglades National Park is similar); predrainage conditions are based on appendix R. Predrainage sloughs retained water throughout most years, favoring aquatic species. Current sloughs frequently dry out, creating dense wet prairie plant stands, reducing ridge (and tree island) elevations through peat oxidation, and periodically eliminating aquatic fauna or forcing them into alligator holes or canals.

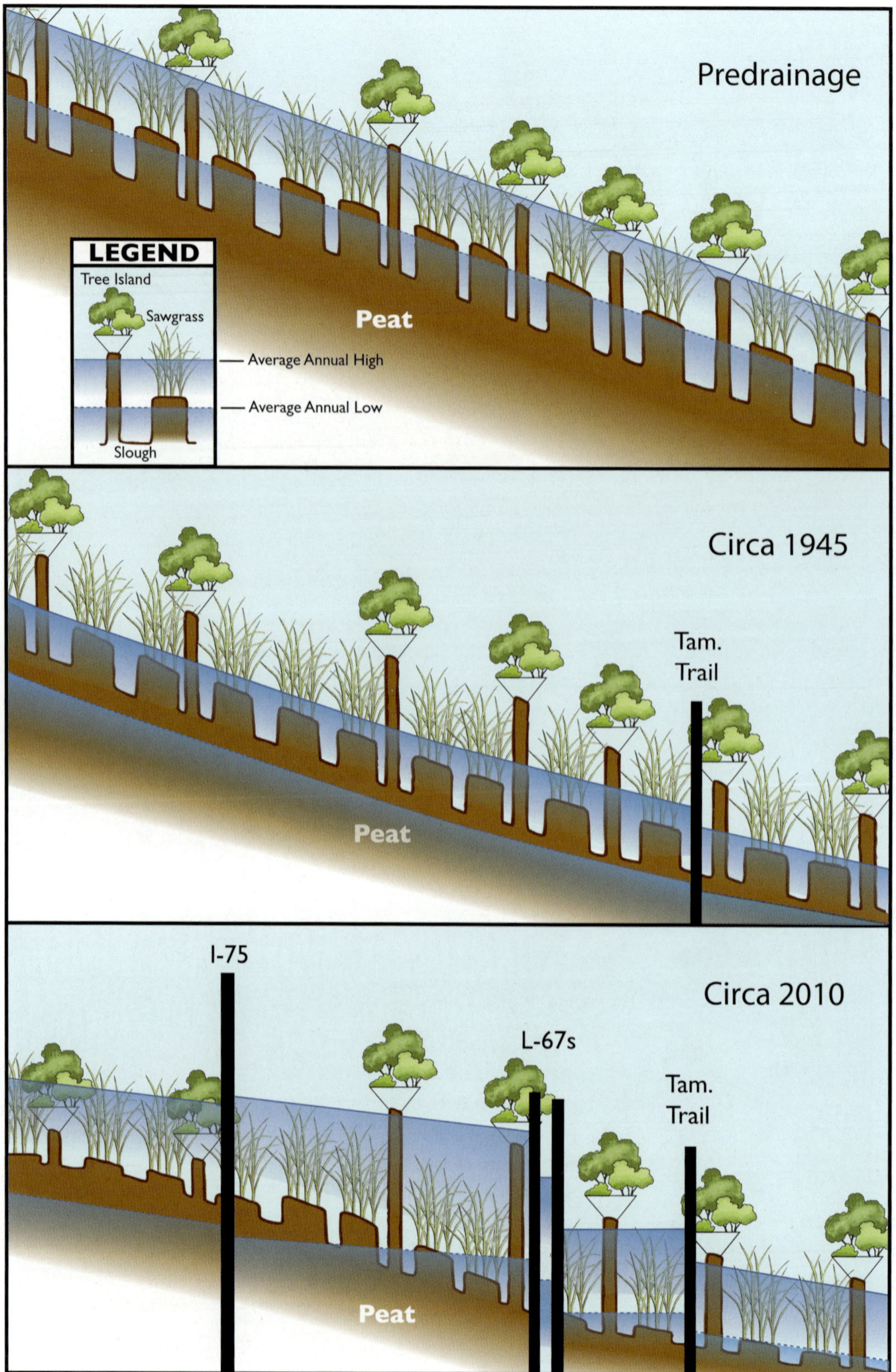

David Williams/Wingin' It Works

Plate 30. Water and ground surface along a transect through the Ridge and Slough landscape, three time periods. The transect follows a (curved) flowline from the northern edge of Water Conservation Area 3A, through WCA 3B and into Everglades National Park (Shark River Slough). Peat elevations and water depths are to scale; vegetation is not. In the predrainage system, the water surface was parallel to the ground surface; under current impounded conditions, water "wedges" develop, deepest close to the L-67s and Tamiami Trail, becoming shallower with distance upstream. Source: Chapter 11.

Plate 31. Strand tree island position and orientation in central Ridge and Slough landscape, 1940 and 1994. Base map is Landsat TM imagery, digitally processed by the South Florida Water Management District (April 1994). Overlain white polygons indicate areas of "Gandy peat" soil or "myrtle and bay" land cover, both digitized from Jones et al. (1948a).

Plate 32. Strand tree island position and orientation in northeastern Ridge and Slough landscape, 1940 and 1994. Base map is Landsat TM imagery, digitally processed by the South Florida Water Management District (April 1994). Overlain white polygons indicate areas of "Gandy peat" soil or "myrtle and bay" land cover, both digitized from Jones et al. (1948a).

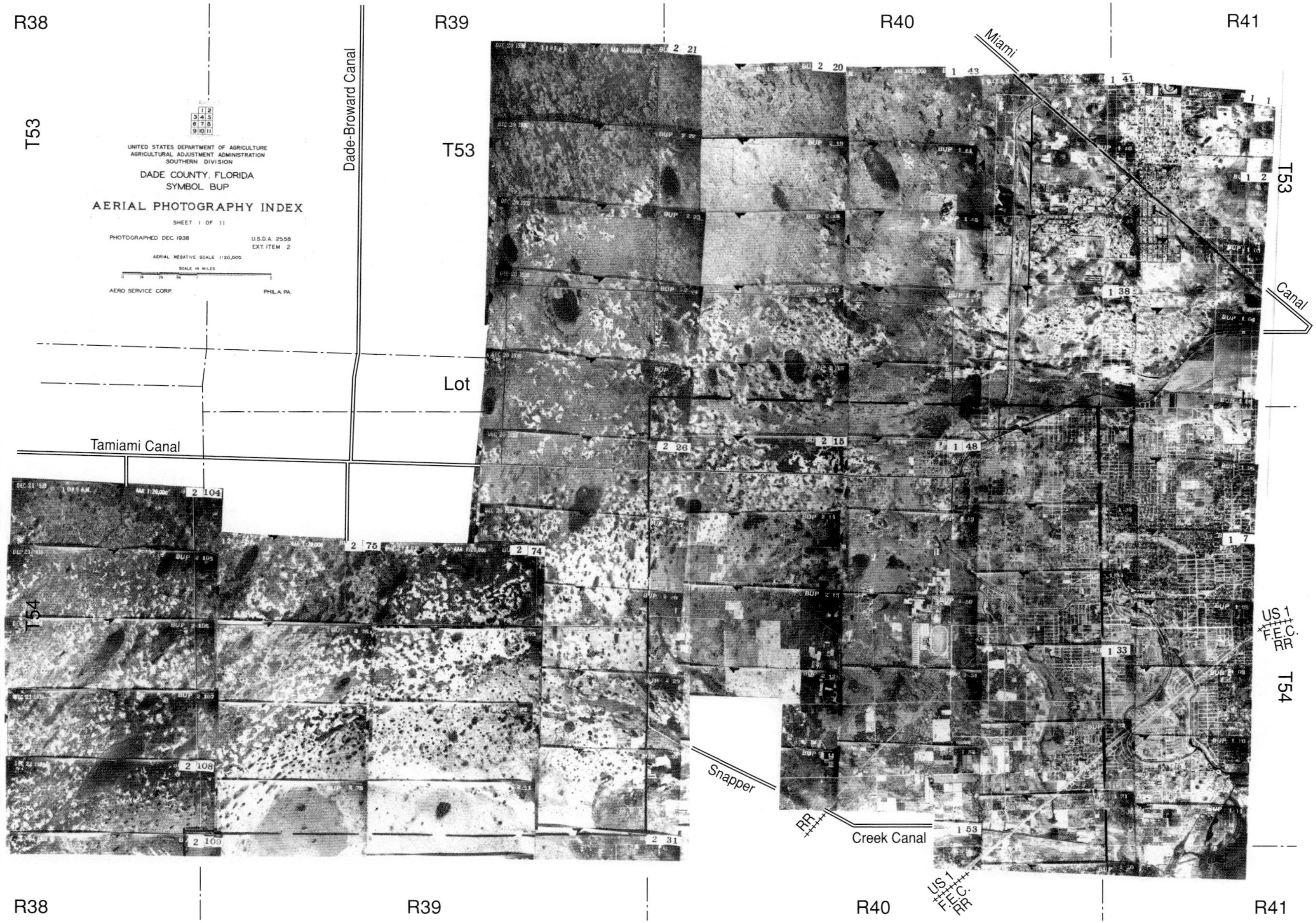

Plate 33. Aerial photomosaic, of Ridge and Slough landscape near head of Miami River, 1938. Tree islands oriented NE-SW in Township 54 Ranges 38–39 are part of Northeast Shark Slough, draining southwest. Tree islands oriented NW-SE in T53 R40 are part of drainage southeastward to Miami River. Shift in direction can be seen in R39, from SSE in T53 to SSW in (western) T54. Source: U.S. Department of Agriculture (1938).

Plate 34. Aerial photomosaic, 1964, of Taylor Slough, a separate portion of the Ridge and Slough landscape. The wettest and widest portion is the lower left quadrant. Compare pattern there of linearly aligned, ovate patches of higher ground with other aerial images of the Ridge and Slough landscape and also with the Little River Peat Transverse Glade in 1925 (plate 46). U.S. Geological Survey (1964).

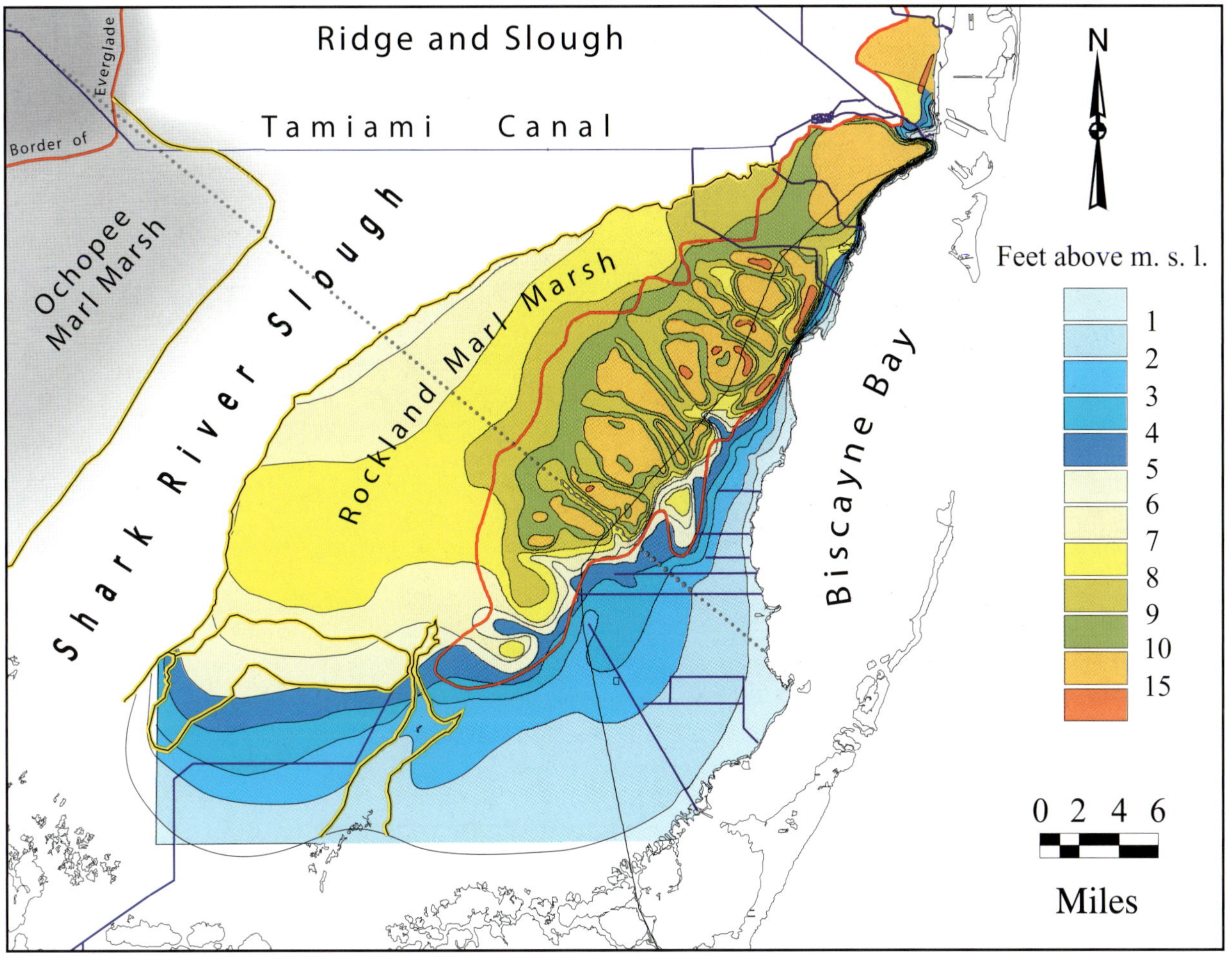

Plate 35. Contour map of the Everglades Keys (> 10 feet elev.; orange) and the Marl Transverse Glades (8–9 feet elev.; olive green). Note the ledge at 9-foot elevation (dark green) along the western edge of the Keys, 1 foot higher than most of the Marl Transverse Glades. See also photographs of Everglades Keys (fig. 10.5) and Marl Transverse Glades (figs. 10.4, 10.6), and predrainage maps from 1847 and 1856 (figs. 1.1, 1.6). Dotted line indicates location of cross section shown in figs. 4.7 and 11.5. Sources: Landscape extents, this study; contours, U.S. Army Corps of Engineers (1960a).

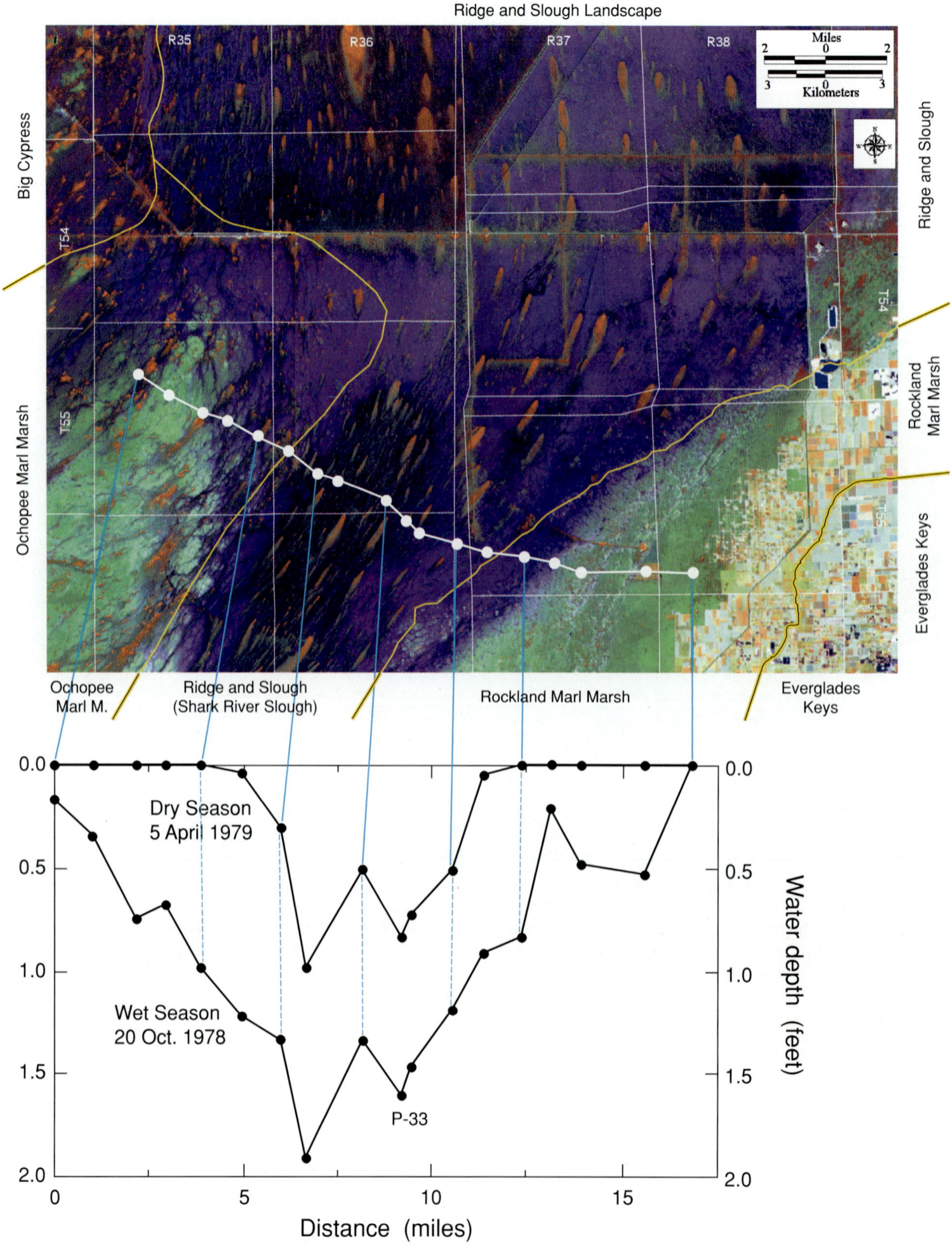

Plate 36. Postdrainage water depths along transect across Shark River Slough and the bordering marl marshes. The concave pattern reflects current postdrainage ground and bedrock surface topography. See also figure 11.5. Sources: Measured depths and locations from Rosendahl and Rose (1981); Landsat TM imagery, digitally processed by the South Florida Water Management District (April 1994).

Plate 37. Ochopee Marl Marsh landscape, February 1917. Top: "No. 2 . . . disintegration of sloughs and out crop of Lossman's River limestone—location Section #1, Range 35 [possibly 34] East, Township 55 South." Bottom: "No. 4 . . . breaking up of sloughs in Central Eastern border of Hopkins tract [east side of T55 R34] showing drying conditions of Everglades and the coastal hammock in the distance." Source: King (1917e).

Plate 38. Aerial photomosaic of three landscapes, 1940: Sawgrass Plains (far left), Ridge and Slough (center), and Eastern Flatwoods (upper right). The border of the Everglades extends diagonally southwest from upper left, with Loxahatchee Slough extending north-northeast to upper right as an "arm" of the Everglades. The West Palm Beach Canal (C-51) crosses slightly above the middle of the image. Within the Eastern Flatwoods, note the mosaic of wet prairie ponds (dark, roughly circular) and surrounding pine and cypress woods (lighter gray). Additional detail is visible in plate 39. Source: U.S. Department of Agriculture (1940).

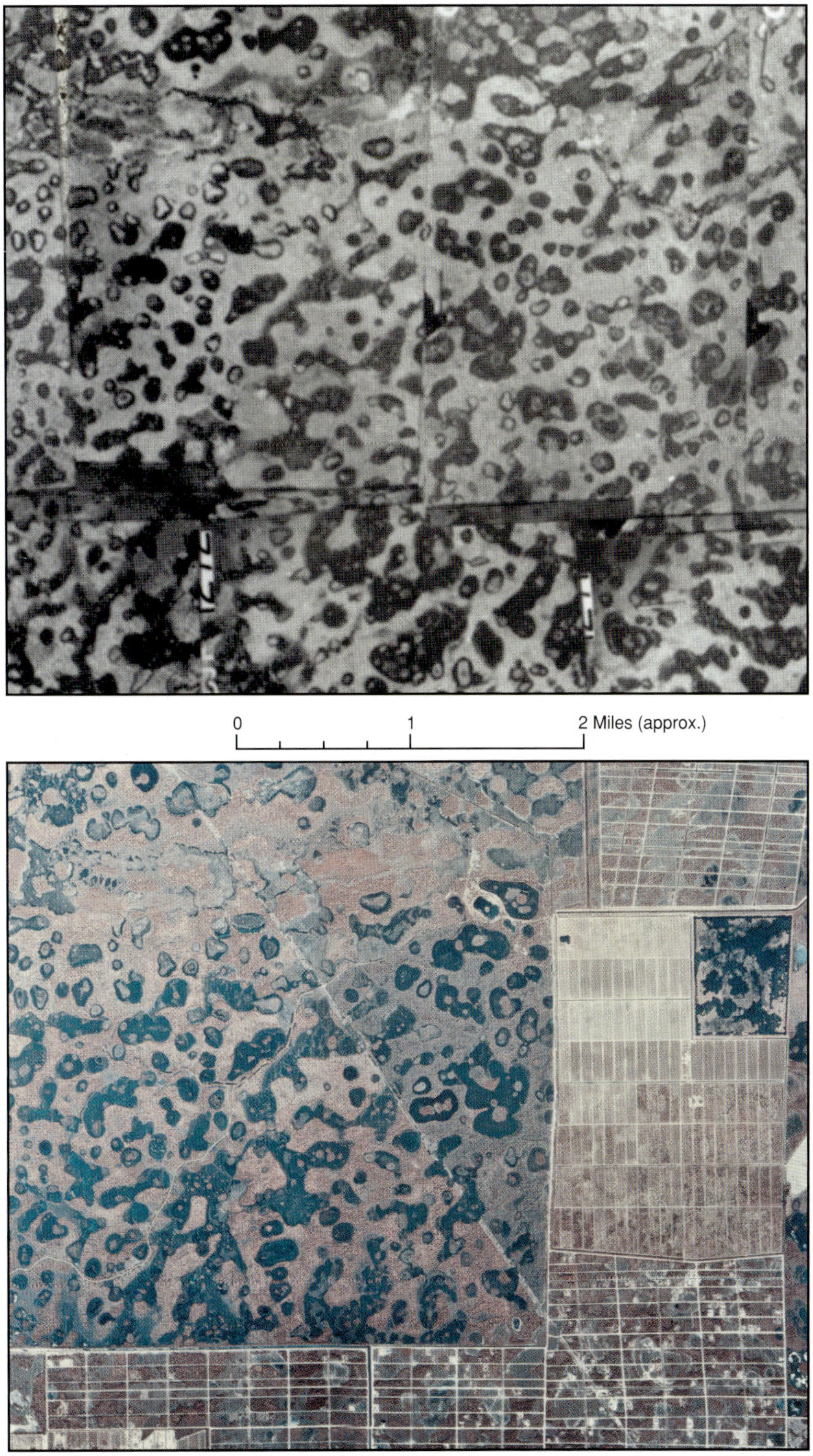

Plate 39. Aerial photographs of the Eastern Flatwoods in 1940 (top) and 1991 (bottom), showing a mosaic of ponds (dark, roughly circular) within a matrix of pine and cypress woods. Landmarks are Hungryland Slough across the top and the Corbett Wildlife Management Area youth camp entrance at upper right of undeveloped area. In the 1991 image, note evidence of impoundment (wedge-shaped area bordered by powerline) and disintegration of pattern under deep impoundment (far right rectangle). Contrast with plates 24 and 26. Sources: U.S. Department of Agriculture (1940) and South Florida Water Management District (1991).

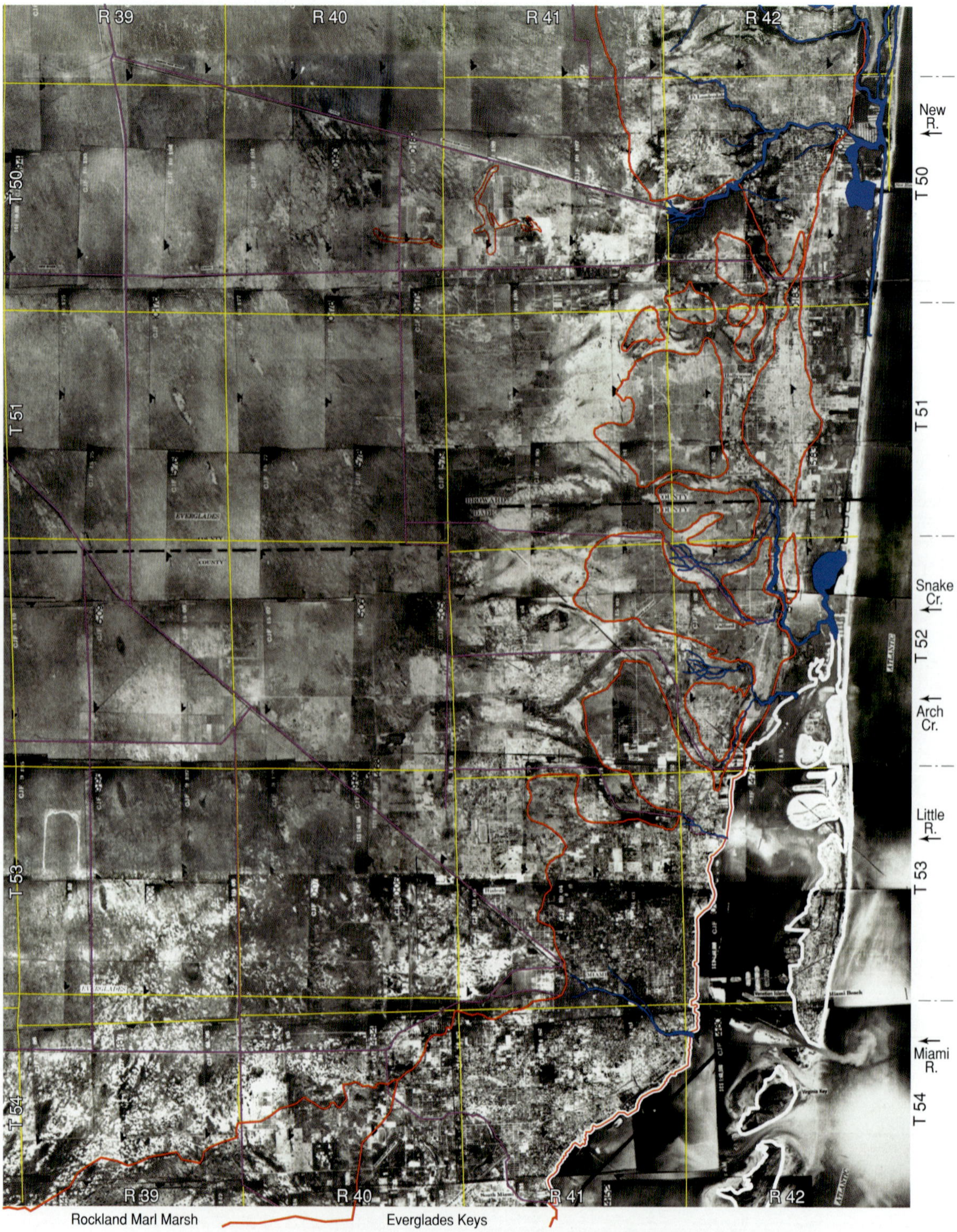

Plate 40. Aerial photomosaic, 1940, of eastern Everglades and Atlantic Coastal Ridge, New River (Fort Lauderdale) to Miami River region. Red polygons, taken from the soil map (Jones et al. 1948a), are islands of high ground forming the discontinuous Atlantic Coastal Ridge. Peat Transverse Glades and coastal rivers are visible between the islands. The 1- to 2-mile-wide whitish area immediately west of islands (R41) is most likely a sand subsoil, exposed when overlying peats oxidized. Compare with the predrainage maps of the same area (plates 41, 43). Source: U.S. Department of Agriculture (1940).

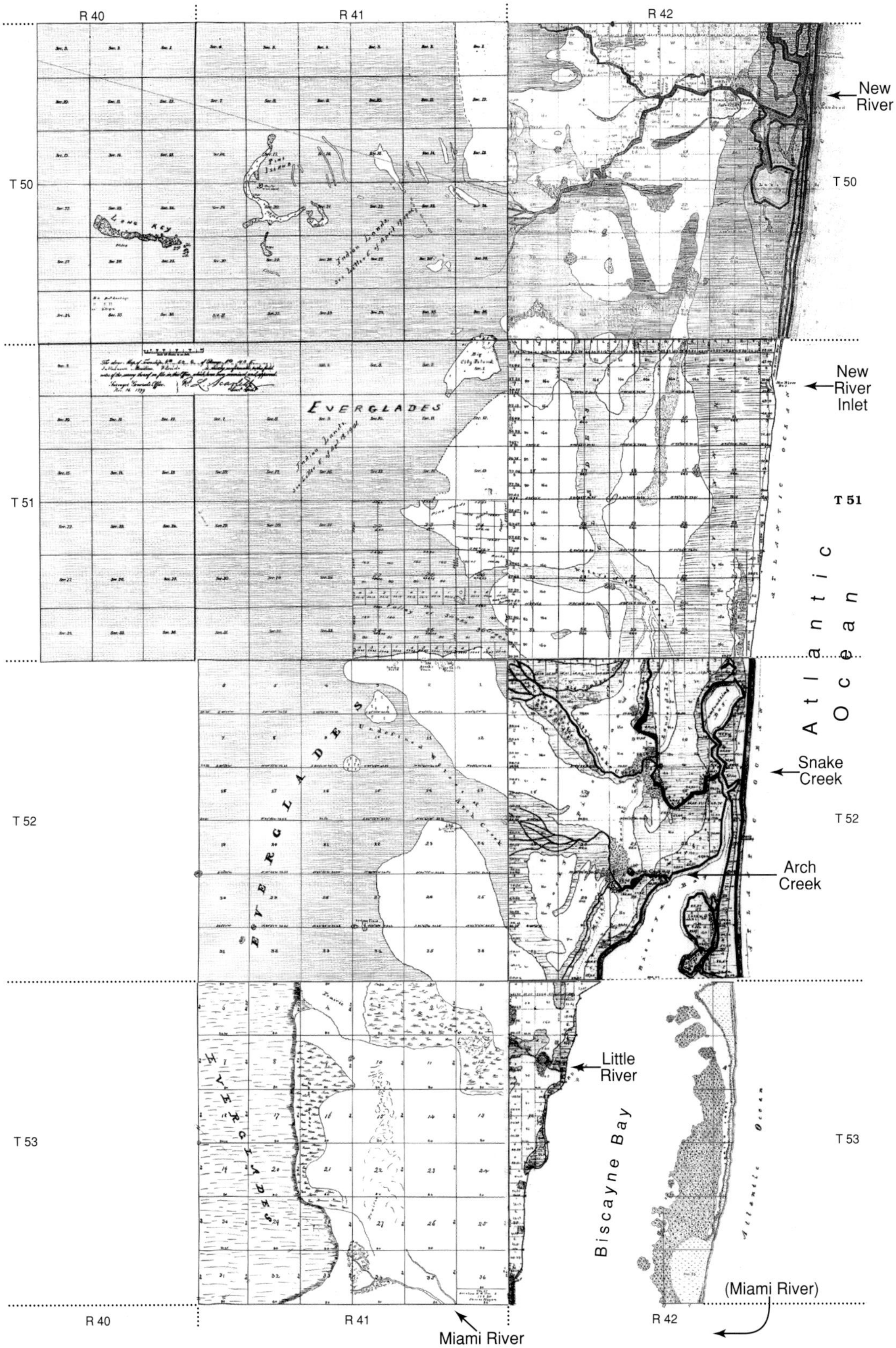

Plate 41. Predrainage landscape map of eastern Everglades and Atlantic Coastal Ridge, New River (Fort Lauderdale) to Miami River region. White areas are islands of high ground ("pine and palmetto"), forming the discontinuous coastal ridge. Shaded areas between islands are Peat Transverse Glades. The Everglades (shading in R40, R41) directly border the ridge of islands and merge continuously with the transverse glades. Sources: Ten township plat maps, surveyed 1845–1898, see figure 1.4.

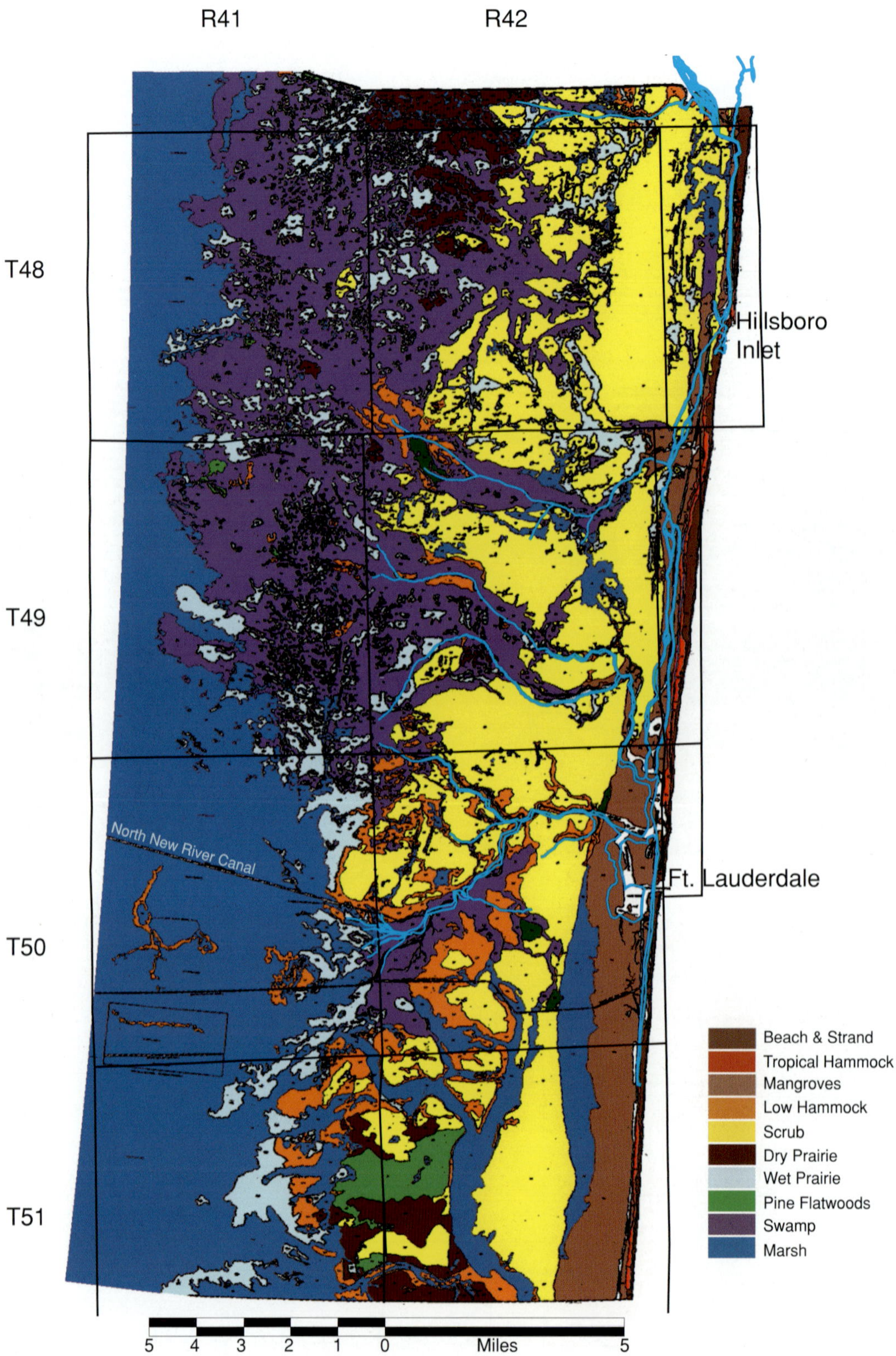

Plate 42. Estimated natural vegetation of Broward County, 1940. Overlaid on 1940 imagery are township boundaries (black) and predrainage rivers (blue); geo-referencing is approximate. Note extensive areas of cypress swamp draining from the Everglades eastward to the coastal rivers: Hillsboro Creek, Cypress Creek, North and South forks of Middle River, and New River. See also plates 40 and 41. Redrawn from Steinberg (1980).

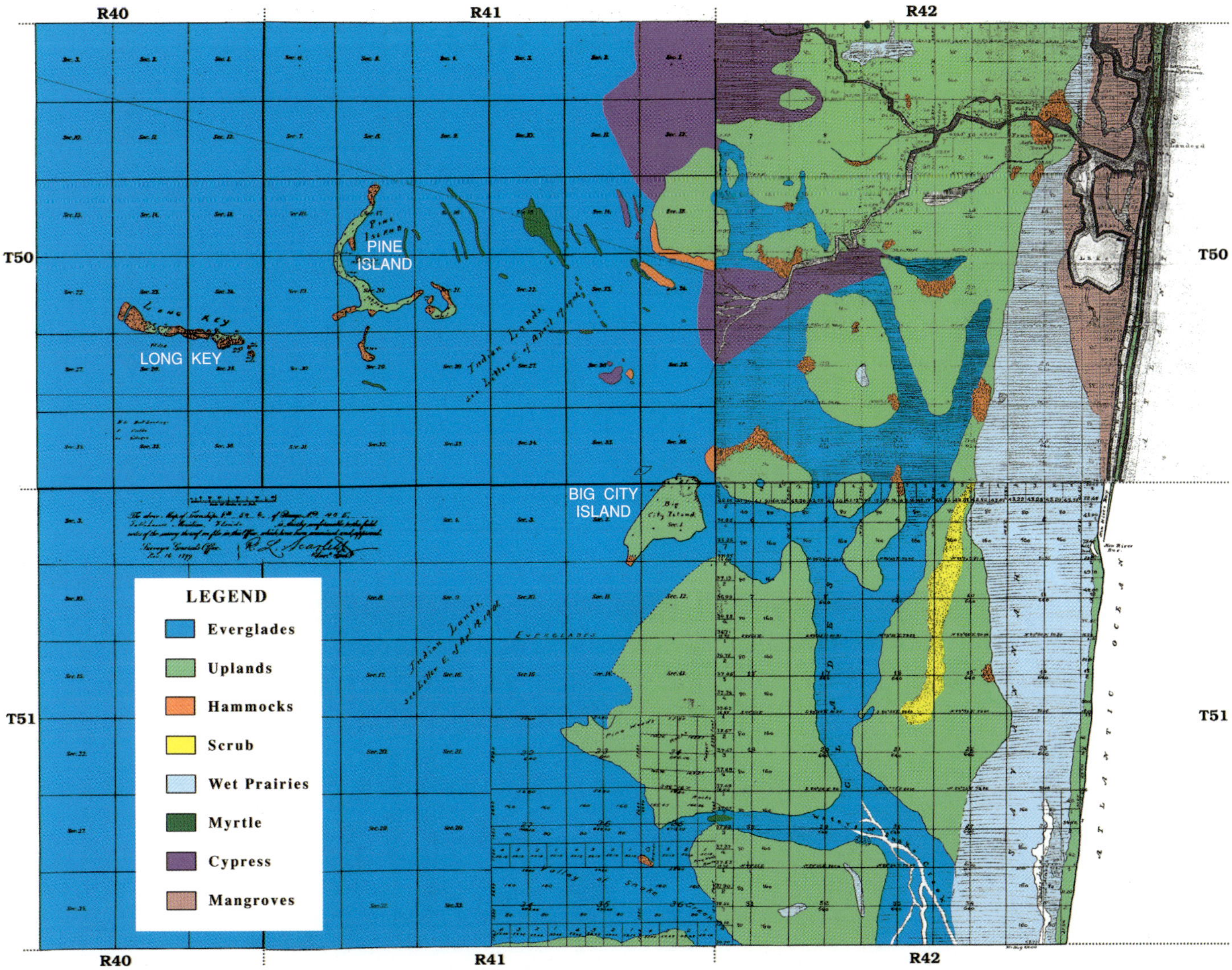

Plate 43. Pine Island, Big City Island, and Sam Jones' Seven Islands (Long Key) in the predrainage Everglades near Fort Lauderdale, ca. 1898. Black lines are the original township survey plats, 1870–1908; colors are vegetation categories indicated in survey field notes. Eastward, the Everglades extended into the Atlantic Coastal Ridge as the Peat Transverse Glades. Redrawn from township survey plats: Fries (1898-T50 R40, 1898-T50 R41 1898-T51 R42); Newman (1908-T50 R41, 1908-T51 R41); Williams (1870-T50 R42, 1870-T51 R42).

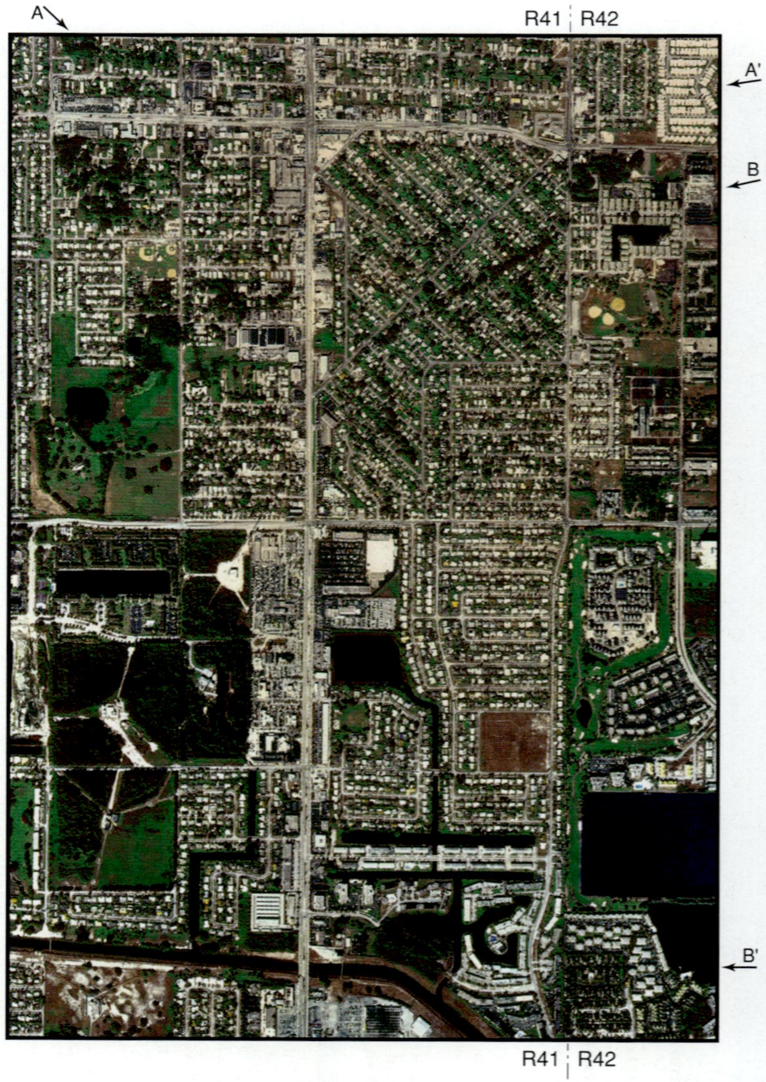

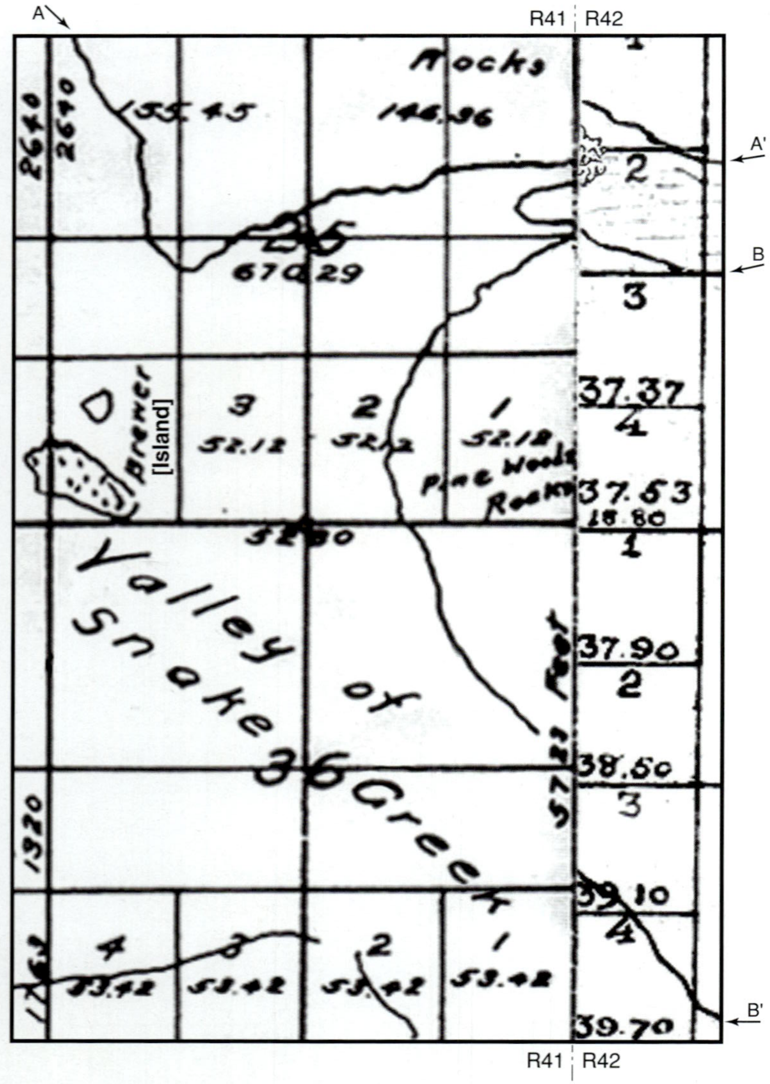

Plate 44. Edge of the Everglades and two Peat Transverse Glades near the mouth of Snake Creek, 1908 (right) and 1994 (left). (southeast quarter of T51 R41, portion of R42). In the 1994 photo the northern portion of a line of oak trees is still visible, as are oak trees of the eastern half of a former tree island and scattered oaks of the western half. Compare with 1947 and 1955 photos, plate 45; for regional context, see plates 40, 41. Chitto Tustenuggee (Snake Warrior's) Island within the Everglades is labeled "Brewer Island" near the western border of the township plat. Sources: Aerial photograph from 1994, South Florida Water Management District; survey map, Newman 1908-T51 R41 and Williams 1870-T51 R42.

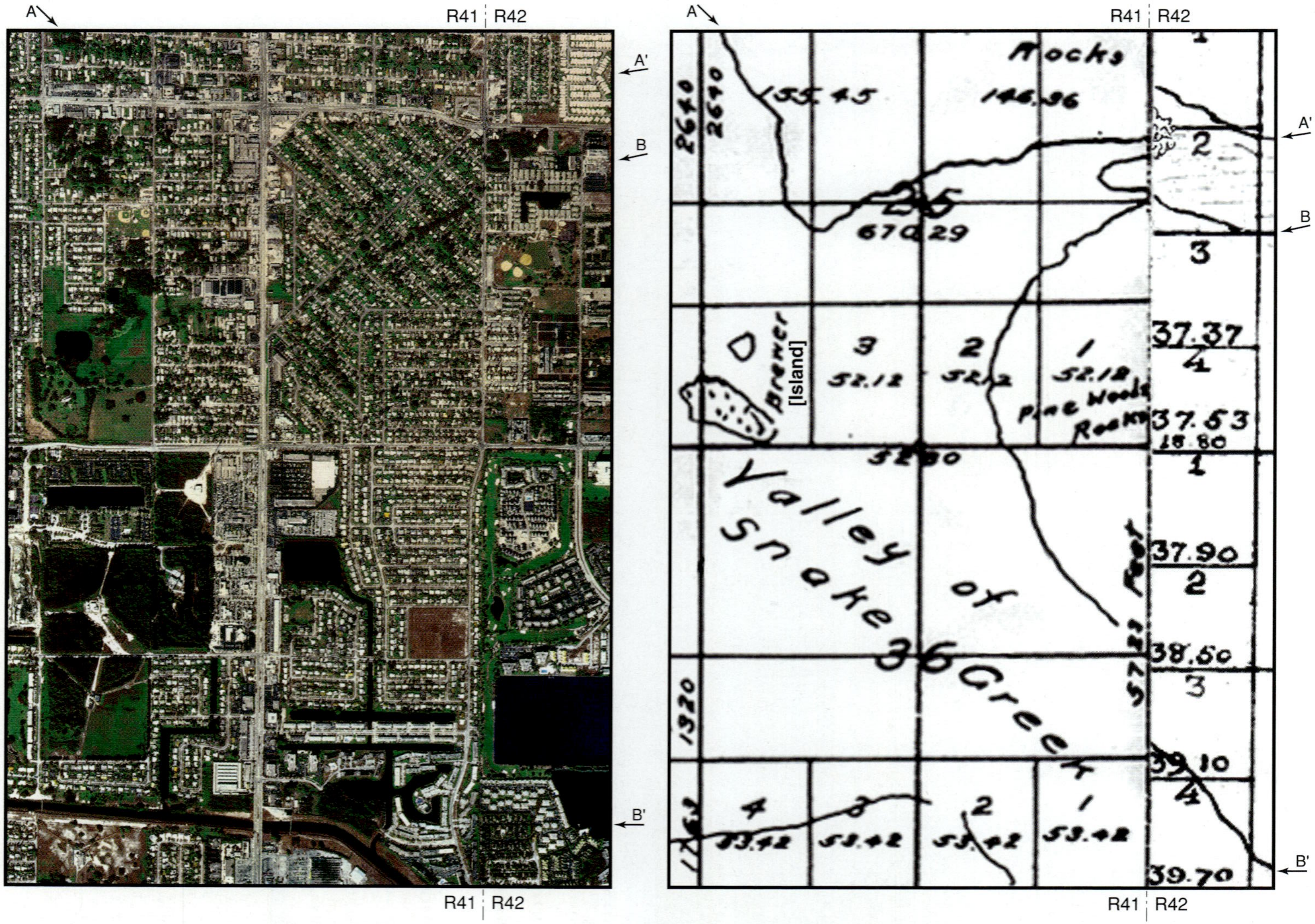

Plate 45. Aerial photos, 1947 (left) and 1955 (right), of edge of the Everglades and two Peat Transverse Glades, Snake Creek area (same extent as plate 44). The upper transverse glade (A' and B) includes a tree island in the 1947 photo. By 1955 the west half of the tree island has been developed, but scattered oaks are left standing. The lower transverse glade (B' and Snake Creek Canal) drains southeast. Two distinctive lines of (postdrainage) oak trees mark the sharp border between the Everglades and pineland, curving across the top (A–A') and at right (B–B'). Compare the oak line with the pinelands edge surveyed in 1908 (plate 44).

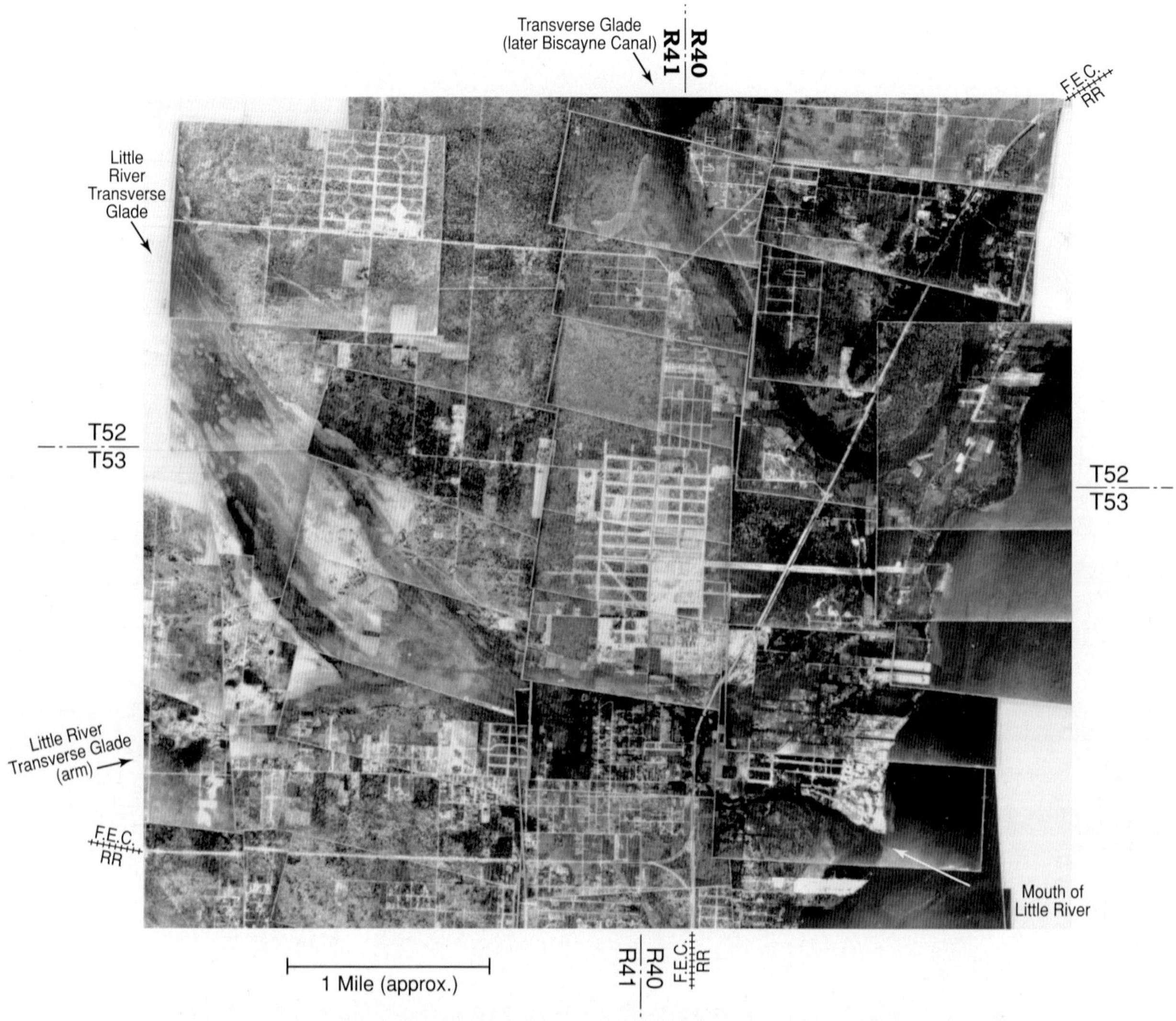

Plate 46. Aerial photomosaic of the Little River Peat Transverse Glade, 1925. Note the directional, aquatic texture of the transverse glade. Street grid development is confined to the higher coastal ridge. In 1845 there was sufficient outflow to operate a coontie mill on the Little River draining the transverse glade.

3

Effects of Everglades Drainage, 1880s to Early 1950s

Hydrologic Changes

The available quantitative and narrative data were examined to assess the degree of hydrologic change caused by the Lake Okeechobee and muck canal phases of mechanized drainage. Those sources were also used to estimate how quickly the hydrologic changes occurred. While the number of distinct quantitative sources is small, each includes a substantial number of data points, giving a good indication of spatial and temporal trends. Qualitative sources provide a means to extrapolate the quantitative data to larger areas. Two main sources of qualitative data were found: observations from agricultural researchers working in the northern Everglades and observations from surveyors and biologists traversing the southern Everglades. Some additional information was available from the agriculturally used areas encroaching westward into the Everglades from Miami and from the pinelands southwest of Miami. Indirect information was also available from groundwater changes observed along the coastal ridge, particularly the progress of saltwater intrusion. Generally, the sources all suggest significant lowering of water levels over large areas as early as the 1920s and earlier in specific areas.

Figures 3.1, 3.2, and 3.3 help to illustrate the magnitude of changes between predrainage and postcanal (1920s–1930s) conditions. Figure 3.1 graphs 350 field measurements of water depths (distance above or below the ground surface) made in the Everglades during the dry season of 1915, only three years after the opening of the North New River Canal (Baldwin and Hawker 1915, p. 753). The earliness of this remarkable dataset suggests that the measurements may still show some resemblance to predrainage conditions, while also reflecting the influence of lowered Lake Okeechobee stages (and the associated reductions in lake discharge into the Everglades) as well as direct drainage of the Everglades by the canal itself. Both the map and graph in figure 3.1 clearly indicate that drainage had already substantially depressed the water table along the first 10 miles (16 km) of each end of the canal. Water depths in the middle reaches of the canal were likely also somewhat reduced. The measured water depths from the middle reaches—about 0.5 feet (15 cm) in the Sawgrass Plains landscape and about 1 foot (30 cm) in the Ridge and Slough landscape—therefore constitute a strong source of information regarding

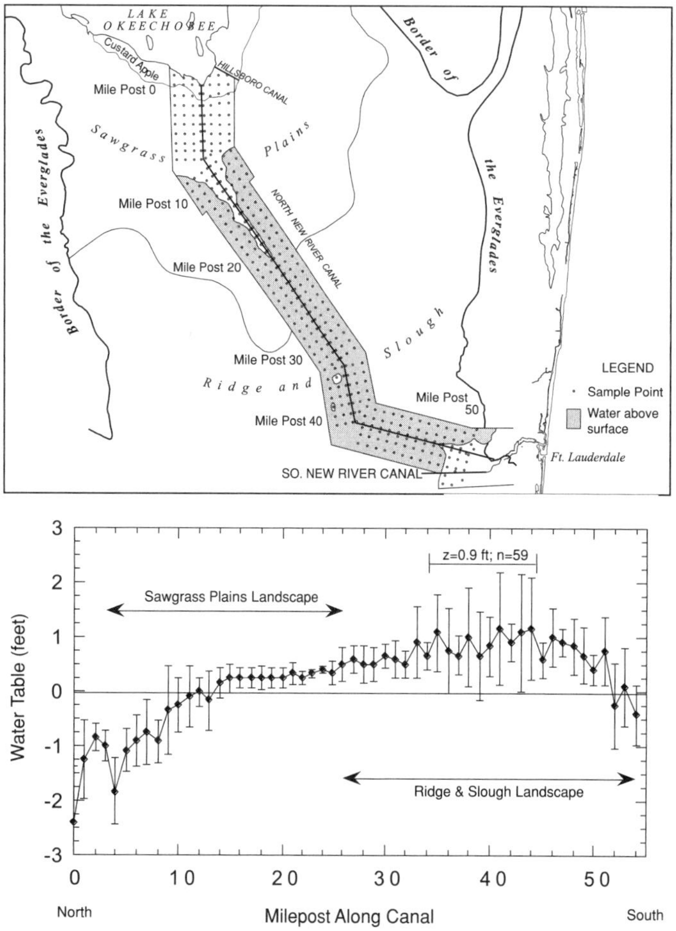

Figure 3.1. Surface water ponding depths along the North New River Canal in 1915, three years after canal opening, measured during the dry season, January 22 to March 16, 1915. Top: Location of sample points and area of ponding. Bottom: Measured depths and standard deviations. Note differences between landscape types. Map and data are from Baldwin and Hawker (1915).

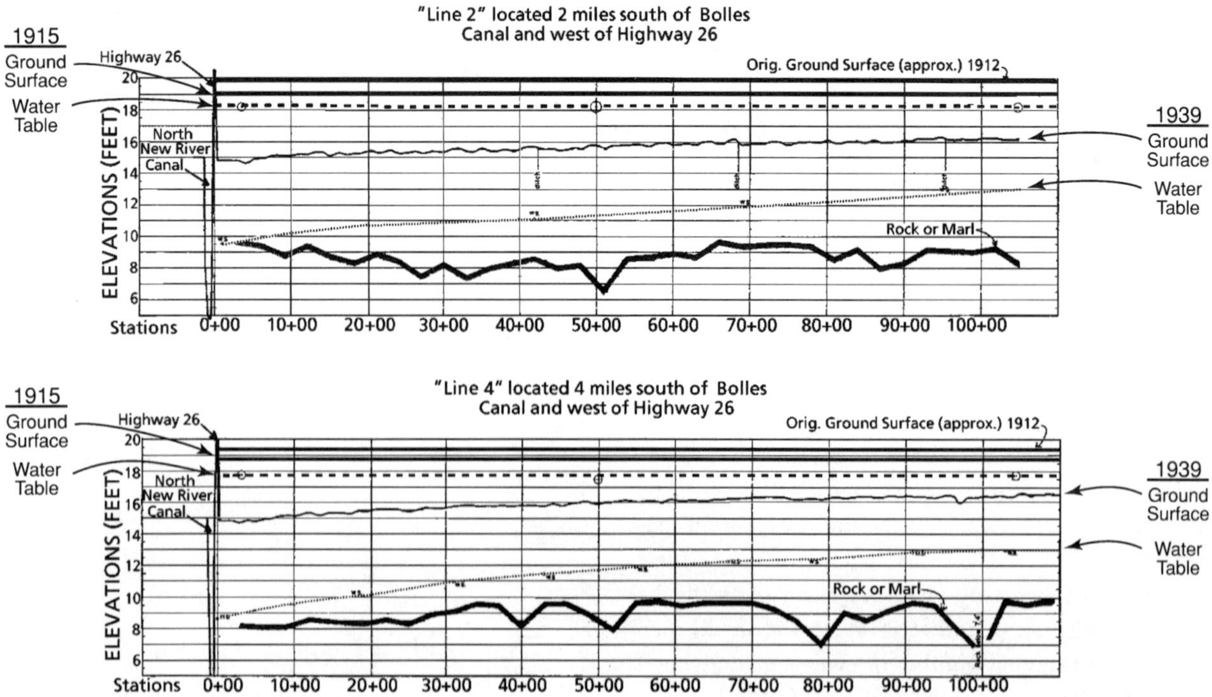

Figure 3.2. Water table and ground surface elevations along two transects in the Sawgrass Plains area during February–March of 1915 and 1939. Both were dry years. Line 2 (upper panel) is shown as transect D_1 in figure 3.9; line 4 as D_2. Sources: Bedrock, 1912 ground surface, and 1939 elevation data, Clayton et al. (1942); 1915 ground surface levels estimated using subsidence data from Clayton et al. (1942) and Stephens and Johnson (1951a); 1915 water table data, Baldwin and Hawker (1915; see also fig. 3.3).

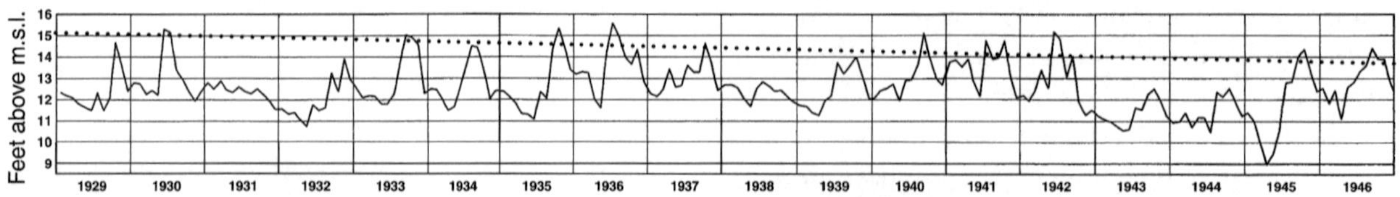

Figure 3.3. Monthly mean stage (1929–1946) of the Hillsboro Canal at Shawano (T45 R38, Section 13). Parker et al. (1955) considered these stages to be representative of the surrounding Everglades (see text). Note that the water table was below ground surface (dotted line) during almost all of this period. Source: Redrawn from Parker et al. (1955).

predrainage dry season (February–March) water depths in these landscapes (see plate 5 for landscape locations and chapters 6 and 8 for landscape descriptions). The combination of the influence of drainage and below average antecedent rainfall (Parker et al. 1955, pp. 36–37; Simpson 1920; p. 121; see also fig. 1.10) suggests that these measurements constitute an informal lower boundary for estimating typical predrainage water depths for this time of year. The greater water depths present in the Ridge and Slough landscape compared to the Sawgrass Plains are consistent with many predrainage descriptions of these two landscapes. The greater variability of the water depths measured in the Ridge and Slough area likely reflects the greater topographic variation (i.e., the microtopography associated with ridges and sloughs; see chapters 8 and 11) originally present in this landscape.

Figure 3.2 compares the 1915 conditions shown in figure 3.1 with measurements made in 1939, a quarter century after the opening of the major canals. The measurements were made in the spring (January to March) in both cases, and precipitation for the two years was approximately comparable. The level of the ground surface, as well as the water table, near mileposts 6 through 8 of the North New River Canal, had declined dramatically: the surface of the peat soil by 3 feet (1 m) and the water table by about 7 feet (2 m). Instead of being above ground as under predrainage conditions, by 1915 the water table was depressed along the first 10 or so miles (16 km) at either end of the canal; however, by 1939, the water table had been drawn down fully halfway between the subsided peat surface and the underlying bedrock.

Figure 3.3 shows seventeen years of water level measurements from the Hillsboro Canal at Shawano. This point was located at mile marker 17 of the Hillsboro Canal in Township 45 Range 38, between Sections 13 and 14, a point within the Sawgrass Plains landscape about 5 miles (8 km) northwest of the border with the Ridge and Slough landscape. Garald Parker considered these measurements to be a useful indication of hydrologic conditions in the surrounding area, in this case the formerly inundated marsh: "Records of water levels in the Everglades are extremely useful in showing the nature and the trends in water conditions over the immediate surrounding area. When such a record is collected on a canal, the stages usually are representative of a large area. For this reason, the record of daily stage of Hillsboro Canal at Shawano, collected since January 1, 1929, is worthy of special attention because it is the longest continuous period of record known in the Everglades" (Parker et al. 1955, p. 371). Two important conclusions regarding the degree and timing of drainage can be drawn from figure 3.3. First, water in the canal apparently was below the adjacent ground surface most of the time, often dropping to 3 feet (1 m) below the ground surface during the dry season. The water table in the soil away from the canal, of course, would have been higher, following a concave shape ending at the next canal. Interestingly, figures 3.4 through 3.7 suggest that a similar concave shape of the peat ground surface developed during the course of drainage, transforming the initially extremely level ground surface into a "pillow" configuration, with low points at each canal. This type of transformation, with ground surface shape tending to conform to the water table shape, is an expected result of the subsidence rate being proportional

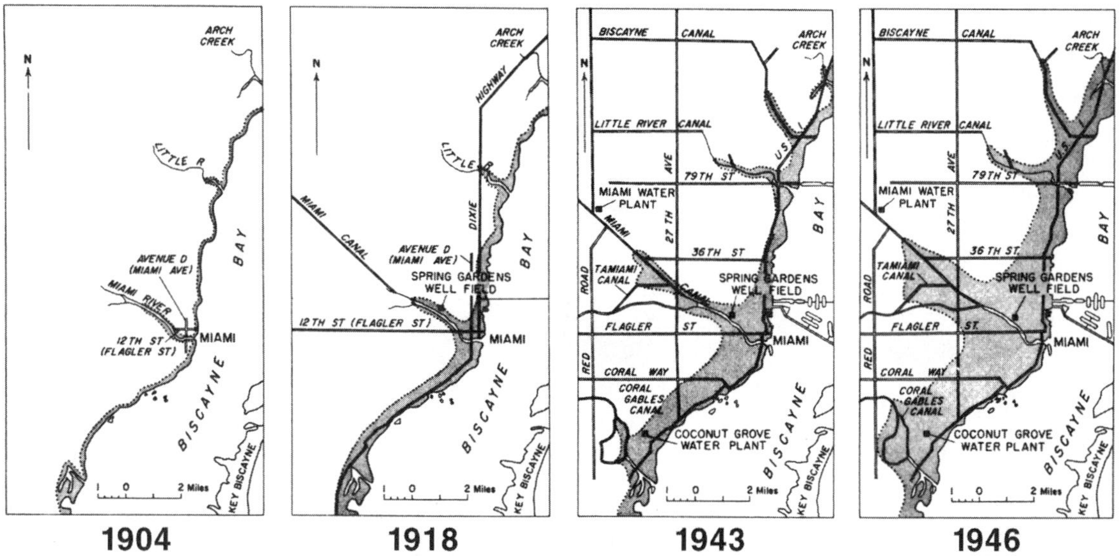

Figure 3.4. Intrusion of salt water into the Biscayne Aquifer, Miami area, as water levels in the Everglades were lowered by canal drainage from 1904 to 1946. Reproduced from Parker et al. (1955).

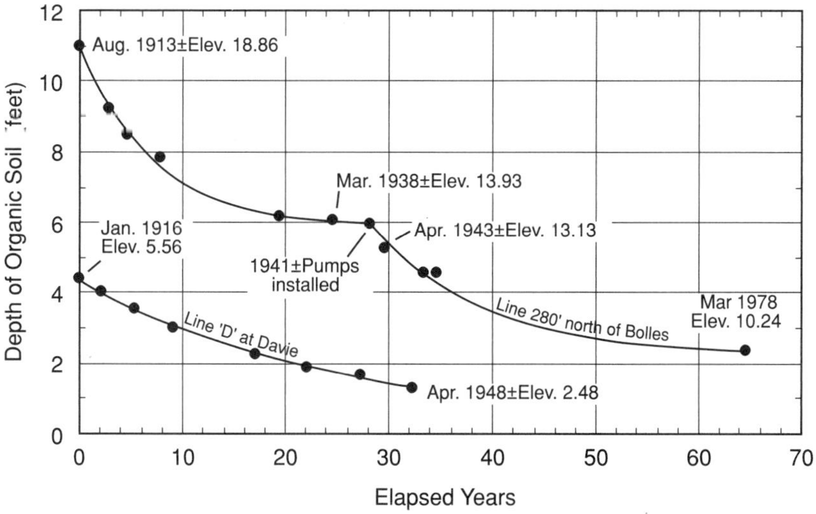

Figure 3.5. Subsidence of Everglades organic soils, 1913–1978. Survey lines were established in 1913 and 1916 by the U.S. Department of Agriculture. "Survey line 280' north of Bolles Canal at Okeelanta" is located in T44 R36, Section 36 along transect C and north of transect D in figure 3.9. "Line 'D' at Davie" is transect E in figure 3.9. Sources: Stephens and Johnson (1951a); Shih et al. (1979a, b, c).

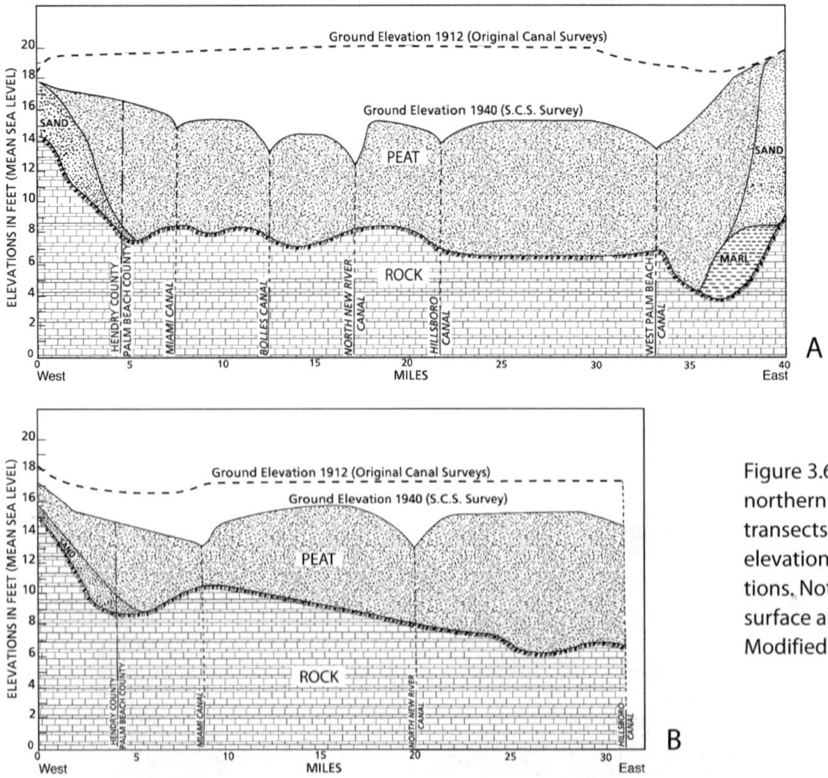

Figure 3.6. Subsidence of organic soils in the northern Everglades, 1912–1940. The location of transects A and B is shown in figure 3.9. The 1912 elevations are likely similar to predrainage elevations. Note the change from convex to concave surface and subsidence valleys along each canal. Modified from Stephens and Johnson (1951a).

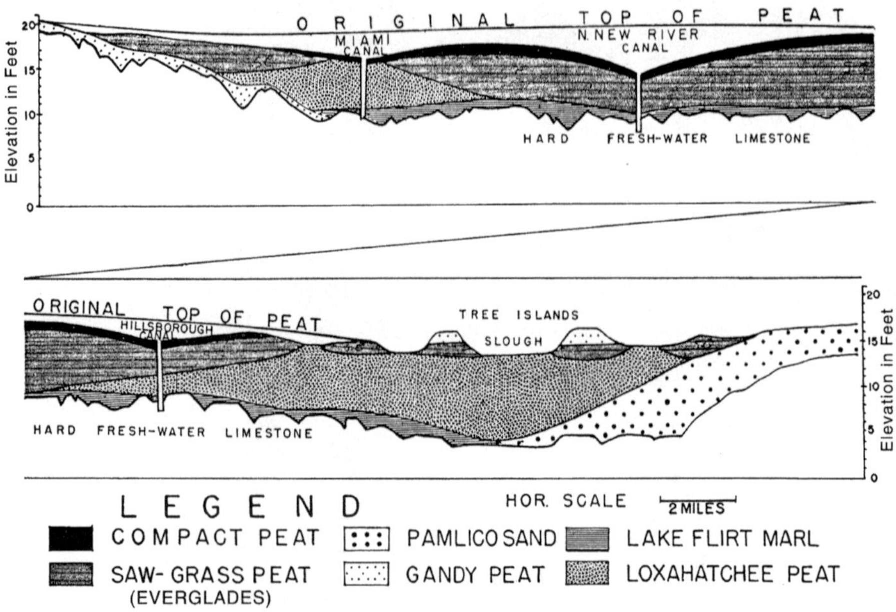

Figure 3.7. Subsidence of peat soil in the northern Everglades, 1882 to approximately 1940. Note the black surficial layer of postdrainage compact peat above the sawgrass (Everglades) peat. This black surficial layer was a defining characteristic used to recognize the Everglades peat mapped by Jones et al. (1948e). The location is shown in figure 3.9 as transect C. Modified from J. H. Davis (1946).

to the depth of the water table (Clayton 1943, p. 120; Stephens and Johnson 1951a, p. 218). The development of this configuration, with subsidence having occurred even at the locations furthest from canals, establishes that water tables throughout the entire area must have been drawn below ground surface for a substantial fraction of the year. Apparently, drawdown patterns resembling those shown in figure 3.3 for a canal were present over wide areas. The Sawgrass Plains and likely parts of the Ridge and Slough landscape would have been affected.

Figures 3.6 and 3.7 and other measures of soil subsidence help clarify that the canals drew water tables below ground surface over wide areas. Figure 3.3, even though the data are for water levels in a canal, helps establish the timing of the canals' effects. (As an aside, the ability of the canals to draw down marsh water levels was, of course, enhanced by reductions or elimination of outflows from Lake Okeechobee; see chapter 5.) Figure 3.3 indicates that the water levels do not show a declining trend over the eighteen-year period 1929–1946. The dip in 1943 to 1945 is most likely a reflection of the 1942–1943 and 1944–1945 droughts (Parker et al. 1955, pp. 36–39), rather than part of a trend. The absence of a declining trend suggests that the dramatic drawdown of the water table had already been effectively completed prior to 1929; during the 1929–1946 period, the canals were simply maintaining the previously drawn down levels.

Soil scientists and drainage engineers working in the northern Everglades (Clayton 1936; Clayton et al. 1942) were well aware of the dramatic hydrological changes reflected in figures 3.1 through 3.3. They were particularly concerned about permanent loss of the peat soil. Lowering of the originally above-ground water table to well below the ground surface had exposed the peat to oxidation:

> The outstanding problem of the Florida Everglades is concerned with prolonging the useful life of the cultivated land and conserving the virgin lands from the destructive effects of subsidence and fires. . . . A complete solution of this problem would require a return of the Everglades to their original condition, which is now impracticable. However if the available water is so distributed that a higher water table is maintained the losses from subsidence and fires can be much reduced. (Clayton et al. 1942, pp. 5–6)

The recommendations by Clayton et al. (1942) for reducing further soil loss, namely *raising* the water table as close as possible to 1.5 to 2.0 feet (0.5–0.6 m) below ground surface in the cropped agricultural areas, and higher in the virgin sawgrass areas, suggests that actual water tables in the Sawgrass Plains were generally lower than these values. This would be consistent with indications from the previous section: "The useful life of the cultivated lands can be increased by maintaining the highest water table compatible with good crop yields. A water table depth of 1.5 to 2.0 feet produces the best yield for most crops now grown in the Everglades. The water table in the virgin lands can be held somewhat higher by retarding the rate of run-off. The large diagonal canals [i.e., the muck canals] have increased this rate of run-off and their continuous flow bleeds the Everglades of much seepage water which if retained would result in a higher water table" (Clayton et al. 1942, p. 6). Widespread reduction of the water table to 3 or more feet (1+ m) below the peat surface was further suggested for the northern Sawgrass Plains by the 1937 agricultural recommendation that "[mole] drains in the peat soil [of the Everglades] should be placed 36 [inches] deep . . . [and that] as far as possible the moling should be done when the water table is below the depth of the mole [drains]" (Newell 1937, p. 21).

Hydrological and associated ecological effects were also noticed shortly after the muck canals opened in the southern and central Everglades. John King, civil engineer and surveyor, worked in the early 1900s for Miami developer Captain J. F. Jaudon (Writers Press Association 1926) and accumulated several years of field experience in the Everglades area west and southwest of Miami (King 1917a–g; Larned 1917, 1918a–h). By early 1917, just a few years after completion of the first major canals, King had already noticed definite changes in the Everglades:

> The drying up of the 'Glades, due to the various canals, is playing havoc with the birds here. The finer ones are fast disappearing. They lack feeding grounds. There are, occasionally, in the southern portion, a few green leg white herons as well as small blue and Louisiana blues, but five years has made a marked change. Of the food birds, the limpkin are found only occasionally. A guide told me his record was two in a season. . . . The best of [the migratory birds] is the big Florida mallard and a right prime favorite too. But his cupboard is no longer filled with food and, save in the surrounding country, even the mallard is migrating elsewhere. (King 1918, p. 150)

King's observations suggest that canal drainage caused marked, and apparently widespread, changes in the southern Everglades. An observation by Charles Torrey Simpson, a naturalist who began exploring the Everglades in the 1880s, suggests the same: "During 1915 and 1916 there was a considerable shortage of rainfall in the Everglade region and this loss, further increased by water taken from the lake by three canals, so lowered the level that perhaps a hundred square miles of its western and southern part were laid bare and no water at all could be found over the general surface of the great swamp" (Simpson 1920, p. 121).

Ornithologists also noticed changes. Harold Bailey commented in 1925: "Florida has often been referred to as the last frontier in the Union. From a natural history standpoint this was especially true until drainage, a destroyer of faunal life in all its branches, became rampant" (Bailey 1925, p. vii). Arthur H. Howell, for many years a student of Florida birds, noted reduced water levels and their effect on birds in the Everglades: "The digging of numerous drainage canals on the eastern side of the Everglades has resulted in lowering the water level several feet and has permitted the use of the land for agricultural purposes. With these profound changes in the character of the big marsh, many of the birds formerly living there have been driven out. Before the Everglades were drained, they furnished a home for thousands of Florida ducks, herons, egrets, limpkins, and Everglade kites, most of which have been compelled to leave the region entirely" (Howell 1932, p. 62).

Although initially planned, no major canals were built during the 1920s or 1930s in the portion of the Everglades south of Tamiami Trail. Tamiami Canal, a borrow pit dug to provide fill for Tamiami Trail and completed in 1928, does not appear to have contributed greatly to drainage of the Everglades; the gradient in stage was less than a third of typical Everglades gradients (Parker et al. 1955, pp. 487–491). Lowering of water levels in this area would, therefore, be primarily the result of reduced southward flows from the more northerly Everglades. A number of phenomena observed in the southern Everglades after Tamiami Canal construction began in 1916 are consistent with reduced southward flows. These include frequent fires in this area during the 1920s and 1930s (Robertson 1953, pp. 94–95), Captain Jaudon's walking trip across Shark River Slough in August 1921 (*Miami Herald* 1921; *Miami Daily Metropolis* 1921), and the observation by John H. Davis of being able to drive across Shark River Slough: "Colonel William S. Harney . . . explored the route from either Harney River or Shark River or both to Fort Dallas on the Miami River. This old route could be followed until recent years but it is seldom if at all used now because there is not sufficient water since the drainage canals were dug. In April 1943 a part of this route was so dry a truck 'glade buggy' could travel along it" (J. H. Davis 1943a, p. 26).

Pre–Tamiami Trail observations of extensive and frequent boat travel west and southwest of Fort Dallas (Miami) suggest that conditions southward were wetter prior to construction of Tamiami Trail. The Seminole Indians (Tebeau 1968, pp. 20–21), Colonel Harney (appendix E), Lieutenant Hugh Willoughby (appendix J), and the township surveyor Marcellus Williams (Williams 1874-T57 R39; see list of Township Survey References, page 321) all passed easily to and from the Miami River into the Everglades and from there to Shark River Slough or across to Big Cypress. Farther south, Glenn Simmons, an Everglades skiff builder and long-term resident of Florida City, reported that it was once possible to pass by flat-bottomed skiff from Florida City to Florida Bay (Simmons and Ogden 1998, p. 6).

Changes in coastal groundwater also suggest a lowering of water levels within the Everglades. Parker et al. (1955, pp. 585–587) discussed in detail the intrusion of salt water into the Biscayne aquifer in the Miami area as pressure heads were lowered in the Everglades and in the Everglades Keys by canal drainage. Leach et al. (1972, pp. 74–77) discuss similar intrusion in the Fort Lauderdale area. Figure 3.4 indicates that the intrusion began almost immediately after the start of drainage. This is consistent with the earlier conclusion that the four major canals were highly effective in rapidly lowering water levels in the Everglades.

H. Bestor, the drainage engineer for the U.S. Sugar Corporation from 1928 through at least 1943, wrote an insightful and comprehensive summary of the hydrologic condition of the Everglades by 1940:

> The zealousness of the past has finally accomplished, in a way, the drainage of the 'Glades. In fact, the insidious loss of water through the years has virtually over-drained much of the area to a disconcerting extent. The coastal settlements, which are dependent on the back country seepage for their water supplies, are already complaining about lack of water, and large areas of idle peat and muck land have been seriously burned and the soils have generally subsided. These effects are as serious as the soil erosion problems of other states. (Bestor 1942, p. 95)

Soil Subsidence

Prior to drainage, the peat soils (Fibric Histosols) of the Everglades were protected from rapid microbial decomposition by the anaerobic conditions created by the nearly continuous presence of surface water. As drainage lowered the water table below the ground surface, the soil initially subsided through physical compaction by dewatering. Additionally, the upper portion of the soil profile became unsaturated, exposing the profile to oxygen. As these soils were exposed to air for progressively longer periods each year, microbially catalyzed aerobic decomposition converted soil carbon to carbon dioxide, causing further subsidence of the ground surface (Clayton 1936, p. 2). Clayton's measurements of postdrainage subsidence for what he referred to as virgin land—that is, native sawgrass with little or no previous drainage—indicated that "a loss of one foot during the first five years; 1.5 foot during the first ten years; and a continuing loss of 1 inch per year thereafter should be expected" (Clayton 1943, p. 119). The spatial distribution of soil subsidence around the Everglades therefore provides a lasting record of the spatial distribution and degree of water table drawdown.

The engineer Lamar Johnson recognized soil subsidence as a widespread problem that began immediately after the opening of the first drainage canals:

Subsidence of organic soils is a natural result of drainage. Subsidence probably began in a relatively small area of the Everglades in about 1906. As the canals were extended into the heart of the Everglades, subsidence became more widespread.

One of the paradoxical observations of Everglades history is that the early gravity drainage system of canals, while insufficient to provide the flood protection needed [for agriculture], did provide sufficient drainage to set into motion the factors of continuous subsidence. (Johnson 1974, p. 165)

Although some contemporary individuals were aware of peat oxidation in drained wetlands, others either were not or possibly chose to ignore it. A 1913 assertion in an influential engineering report—that no subsidence would occur except for 8 inches (20 cm) from within the top foot (30 cm)—contrasts sharply with the actual losses of 5 to 7 feet (150–210 cm) by 1940: "the final plans [for the major drainage canals through the Everglades] herein presented are based on the assumption that the upper foot of muck will shrink vertically 7.8 inches, while the remainder of the [soil] column down to the water table will not shrink at all" (Florida Everglades Engineering Commission 1914, p. 63).

Forty years later a comment by Stephens and Johnson (1951a) gives a sense of the enormity of the early peat loss due to the lack of understanding of peat subsidence: "In making plans for the original drainage of the Everglades, apparently the main causes of subsidence were misunderstood. The original shrinkage of the peat due to drainage was considered, but the continuing losses by slow oxidation were not taken into account. Had the true nature and causes of subsidence losses been fully understood in the earlier days, the original plans might have been modified so as to have saved a large portion of the waste which has occurred since that time" (Stephens and Johnson 1951a, p. 191).

Quantification of Subsidence

Extensive information exists to quantify the losses of peat in and around the Everglades, both process-related time constants and spatial data. After the initial dewatering and compaction a strong, linear correspondence between the rate of subsidence and the depth of the water table was found: 0.054 feet of subsidence per year per foot of water table depression below the surface (0.054 cm/yr per cm, $R2=0.99$; data from Clayton 1943, p. 120) and 0.057 feet of subsidence per year per foot of water table depression below the surface (0.057 cm/yr per cm, $R2=0.81$; data from Stephens and Johnson 1951a, pp. 216–218). These rates, together with "the average rate of loss in the cultivated lands during recent years . . . [of] approximately one inch per year" (Clayton 1943, p. 118; see also Snyder and Davidson 1994, p. 107) suggest that average water tables were approximately 1.5 feet (45 cm) below the ground surface in the cultivated lands. Similar subsidence rates in unused lands suggest similarly depressed water tables there as well.

The spatial data include the following sources:

- maps of land surface elevation in the early 1940s (Jones et al. 1948d; plate 16) and bedrock elevation (Parker et al. 1955, p. 150)
- two different maps of peat thickness in the 1940s (Jones et al. 1948a; J. H. Davis 1946, p. 122)
- a set of 350 point measurements of peat thickness made in 1915 along the North New River Canal (Baldwin and Hawker 1915, pp. 781–795)
- time series measurements from various "subsidence lines" established in 1916 (see following discussion)
- transects of land surface elevation across the Everglades and along canal routes (Stewart 1907, pp. 59–74; Wright 1909c, pp. 158–159; Florida Everglades Engineering Commission 1914, pp. 120+)
- profiles of peat thickness along the proposed Tamiami Trail route (King 1917d; Southern Engineering & Construction Co. 1918)
- point measurements of peat depth from township surveys conducted from 1912 to 1918 (see following discussion)
- a contour map (U.S. Army Corps of Engineers 1960a) with topographic information for the areas along the edge of the Everglades where the peat soils were subject to subsidence and gave way to upland sand or rock of stable elevation

Estimation of soil subsidence in the Everglades after the onset of drainage up to the 1940s was based on a synthesis of observations of land surface elevations, peat thicknesses above the bedrock, and bedrock elevations. Subsidence estimates were combined with early measurements taken from 1900 to 1920 to estimate the predrainage topography. Several key conclusions were drawn from this analysis:

- The ground surface of the predrainage Everglades was only slightly lower than the adjoining "upland" landscapes; in other words, the basin was filled almost to the brim with organic (peat) soil.
- Significant soil subsidence and soil loss occurred wherever organic soils were originally present in the Everglades, including in the area south of Tamiami Trail (the present-day Everglades National Park).

- A substantial fraction of the total subsidence to date had already occurred by the 1940s.
- The most severe absolute losses in elevation (subsidence) occurred in the northern Everglades: a decrease of 5 to 7 feet (1.5–2 m).
- The most severe relative changes in elevation—fraction of the original peat profile lost—occurred along the eastern portion of the Everglades from near Fort Lauderdale south to the Miami area. Here, in places, losses approached 100 percent of the peat profile.

Evidence Used to Determine Changes in Topography

The range of information used to draw these conclusions and to estimate the predrainage topography is illustrated in figures 3.5 through 3.9, figures 3.11 through 3.13, and plate 16.

Figure 3.5 shows time courses for two "subsidence lines," one representative of the northern Everglades (formerly Sawgrass Plains landscape, now the Everglades Agricultural Area), the other of the Davie area near Fort Lauderdale (formerly Ridge and Slough landscape). A series of similar lines established by the U.S. Department of Agriculture documents the progress of subsidence from initiation of drainage (Clayton 1936; Clayton et al. 1942; Stephens and Johnson 1951a; Shih et al. 1979a, b, c; Shih et al. 1997). The lines consisted of established transects along which ground surface elevations were surveyed at multiple points. The lines were resurveyed at two- to ten-year intervals, and the point elevations were averaged and plotted versus time.

Exponential declines in soil surface elevation were typical. The rate constant was found to be a linear function of depth of the water table (Clayton 1936, p. 14; Stephens and Johnson 1951a, p. 218; Shih et al. 1978, p. 1154). Inflection points, such as the one seen for the Bolles canal line, reflect a change in the water table, often due to a switch from gravity-driven to pumped drainage. These lines indicate that between 1912 and 1940, as much as 6 or 7 feet (1.8 or 2 m) of the original 10 to 12 feet (3 to 3.6 m) of soil were lost near Lake Okeechobee (Stephens and Johnson 1951a, p. 198). Across the northern Everglades as a whole the average subsidence rate during this period was approximately 1 inch (2.5 cm) per year, with the initial subsidence occurring at a faster rate (Snyder and Davidson 1994, p. 107).

Subsidence stops only when the peat soil is resubmerged or all decomposable organic matter is completely oxidized. The original caption for Line "D" at Davie shown in figure 3.5 read: "This line was originally laid out on Everglades peat over sand. In some places along portions of the line, the peat has entirely disappeared with only the bare sand exposed" (Stephens and Johnson 1951a, fig. 11, p. 213).

Figures 3.6 and 3.7, generated from data collected by two different researchers, depict measured cross-sectional transects of bedrock and the original ground surface elevations of the organic soils of the northern Everglades basin as well as ground surface elevations measured in the early 1940s. Locations of the transects are shown in figures 3.8 and 3.9.

The original predrainage surface was flat or even slightly convex within the Sawgrass Plains landscape and only a few feet (about 1 m) lower than the surface of the bordering sand-based Eastern and Western Flatwoods. The Ridge and Slough landscape was slightly lower than the Sawgrass Plains and concave. By 1940 the surface along Transect A (fig. 3.6) had subsided approximately 4.5 feet (1.4 m) and, in some places, by as much as 7.5 feet (2.3 m). Farther south, along Transect B, subsidence was somewhat less, generally about 3.5 feet (1.1 m) with a maximum of 5.5 feet (1.7 m). Local "subsidence valleys," 2 to 4 feet (0.6 to 1.2 m) deep, formed along each canal due to depression of the water table being most severe closest to the canals (figs. 3.4, 3.9, plate 16). Transect C (see fig. 3.9 for location) shows a similar comparison of predrainage and 1940s elevations. The cross section of Transect C shown in figure 3.9 also shows the appearance of a new drainage-induced layer of "compact peat." Both figure 3.8 and figure 3.9 indicate that by the 1940s, soil subsidence resulting from regional drainage had caused the Everglades to become a sunken, concave basin (plate 16). Loss of this peat and the associated water-holding capacity may have had an influence on the hydrology of downstream portions of the Everglades.

Subsidence in the northern Everglades, as estimated from transects A, B, and C, was similar to that estimated from the individual subsidence lines. In the area directly south of Lake Okeechobee, a third to nearly half the depth of the original 10 to 12 feet (3 to 3.7 m) of peat soil was lost by 1940. The original topography of the Sawgrass Plains was profoundly altered. North of the Bolles Canal the original slope of the land was reversed by 1940—descending northward toward, rather than away from, Lake Okeechobee (Evans and Allison 1942, p. 38; J. C. Stephens, 1942, p. 24).

Figures 3.8 and 3.9 provide further support for development of a basin-wide view of the Everglades by showing two snapshots of the thickness of the peat soil present in the early 1940s. For comparison, the canals present in

1940, contours of bedrock elevation, and the location of the predrainage Everglades boundary have been superimposed on these soil thickness maps. The peat thickness contours shown in figure 3.8 are from a study of Everglades peats; those in figure 3.9 were measured during the course of the Jones et al. 1948a soil survey. The two studies have different strengths: the contours in figure 3.9 are probably based on a more systematic measurement grid, but the thickness resolution is quite coarse. The contours in figure 3.8 are based on unknown measurement locations but reflect finer depth resolution and give a clearer picture of the overall patterns, particularly the pattern of peat loss caused by the canals. Comparison of these two maps with the transects shown in figures 3.6 and 3.7 helps to illustrate the changes between predrainage topography and the topography that remained in the 1940s (fig. 3.10). Estimates of predrainage topography from this study are discussed further in chapter 11.

The spatial patterns of peat depth shown in the southern portions of both peat thickness maps help identify the predrainage location of Shark River Slough. The patterns clearly reflect the presence of a flow channel, with deepest peats in the center of the channel. A diagonal northeast-to-southwest orientation is visible, particularly in Townships 54 and 55. Prior to drainage the center of Shark River Slough flowed through what would become the future location of Tamiami Trail (Transect G) in Range 38, 9 miles (14.5 km) to the *east* of the future location of the L-67 extension canal. Water management subsequent to the 1940s has pushed the location of southward water flow progressively westward (Leach et al. 1972, pp. 36–42), such that it is presently restricted to Ranges 35 and 36, more than 12 miles (19 km) west of the original channel center.

One element of the pattern of 1940s peat thickness mapped in figure 3.9 may reflect an impounding influence of Tamiami Trail prior to the 1940s. The 5- to 9-foot (1.5–2.7 m) peat thickness contour ends abruptly just north of and parallel to the Trail. This would be consistent with lower peat subsidence north of Tamiami Trail, where water levels may have remained somewhat higher, and higher subsidence on the south, more drained side of the Trail. Outflows from Tamiami Canal to the ocean would have tended to reduce somewhat the water depth differential and associated peat accumulation differential north and south of Tamiami Trail, but this discharge was small, less than one third of the Miami Canal discharge (Parker et al. 1955, p. 493 compared to p. 427).

Figure 3.11 is based primarily on the unique dataset of more than 350 field measurements of peat thickness made in 1915, three years after opening of the completed North New River Canal and nine years after the opening of southeastern end of the canal. The measurements were made by two soil scientists, Baldwin and Hawker, along 6- or 8-mile-long (10 or 13 km) transects perpendicular to the North New River canal, one transect at each milepost. The measurements appear to have been carefully made, with attention to detail and to the substrate:

> The depth of the peat and the character of the underlying rock substratum are variable. From the second milepost below Lake Okeechobee to the tenth milepost the substratum is limestone, encountered at depths usually ranging from 110 to 130 inches below the surface. From this point toward the ocean the peat gradually becomes thinner until near the eastern boundary of the area it averages about 50 inches in depth. In this lower portion of the Everglades the character of the substratum becomes more variable. In many borings a thin stratum of sand, sandy marl, or marl is encountered lying between the peat and the limestone rock. Where the layer of sand is 10 inches or more in thickness the areas are indicated on the map as a separate phase. (Baldwin and Hawker 1915, pp. 781–782)

Figure 3.11 was created by combining the 1915 peat thickness data with data derived from two other sources: a set of land surface elevations taken along the canal in 1913 (Florida Everglades Engineering Commission 1914) and land surface elevations estimated for 1940 from contours drawn near the North New River Canal in the 1940s (Jones et al. 1948d). Polynomials were fitted to the 1913 and 1940 land surface data and to the peat thickness data (5th, 8th, and 8th order, respectively). The bedrock elevations shown were then estimated by subtracting the 1915 peat thickness polynomial from the 1913 land surface polynomial. The bedrock elevations shown in this figure compare favorably with an unrelated set of bedrock contours (Parker 1955, p. 150). As a result of the polynomial fitting process and subtraction, the variability present in the 1915 peat thickness measurements was arbitrarily and completely associated with the land surface. In reality, the bedrock surface is quite variable and the actual ground surface probably would have been more uniform than what is shown in figure 3.11.

The increased surface variability visible in figure 3.11 within the area of Ridge and Slough landscape relative to the Sawgrass Plains landscape may reflect actual point-to-point topographic variability once present in the pre-

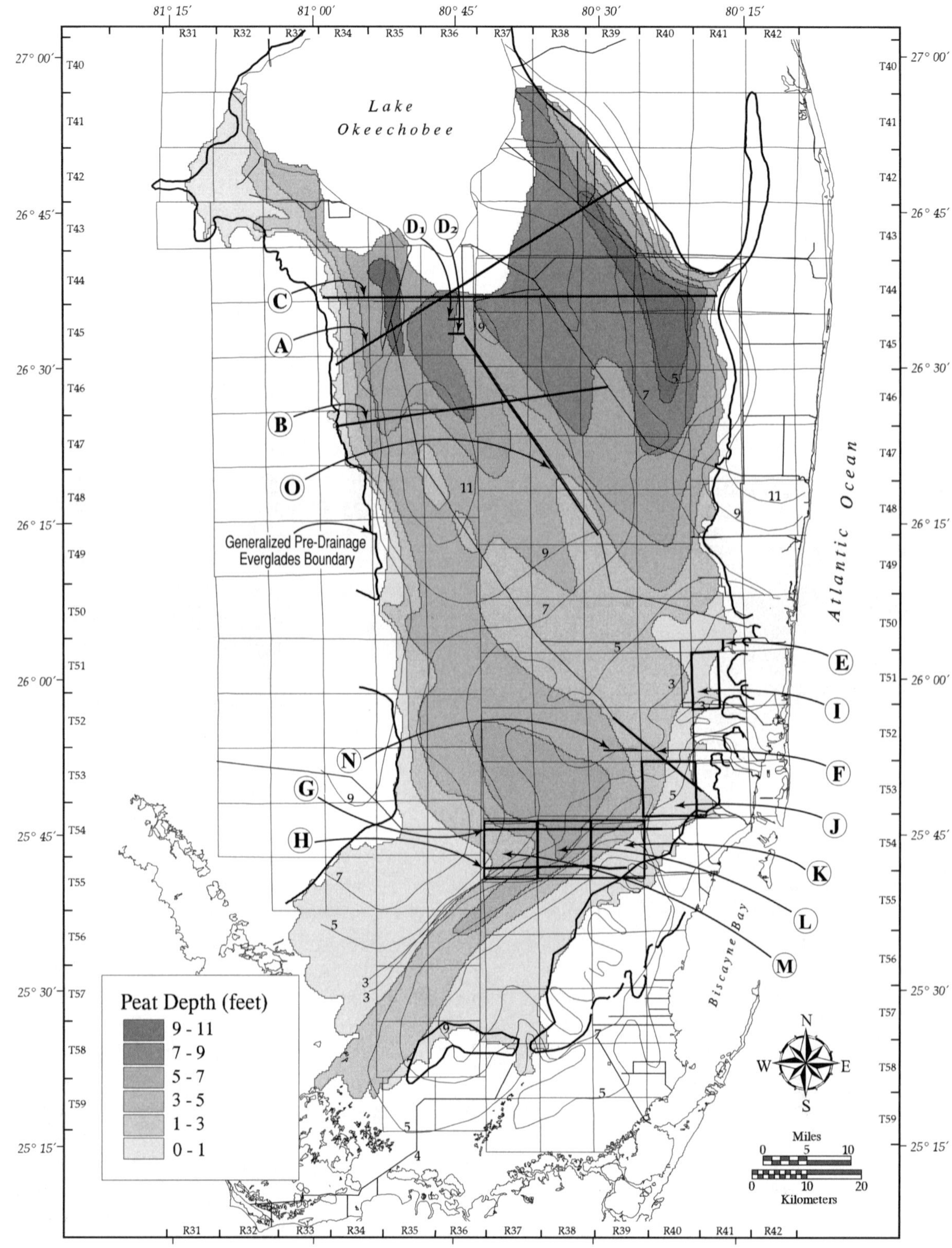

Figure 3.8. Map of approximate peat thickness, 1940–1945. The shading indicates peat depth, and contours indicate bedrock elevation. The canals present in 1940 are included as well as the location of the predrainage Everglades boundary (irregular black line). This is a location map for various transects referred to in this book (see also fig. 3.9). For transect profiles, see the following figures: A = fig. 3.6A, B = fig. 3.6B; C = 3.7; D_1 = 3.2 top, D_2 = 3.2 bottom; E = 3.5, line D; F = 3.12 top; G = 3.13 top, H = 3.13 bottom; I = 4.1; J = 4.5; K = 9.2; L = 8.8 bottom; M = 8.8 top; N = 3.12 bottom; O = 3.11. Sources: Peat thickness, J. H. Davis (1946); bedrock contours, Jones et al. (1948e), supplemented by Parker et al. (1955).

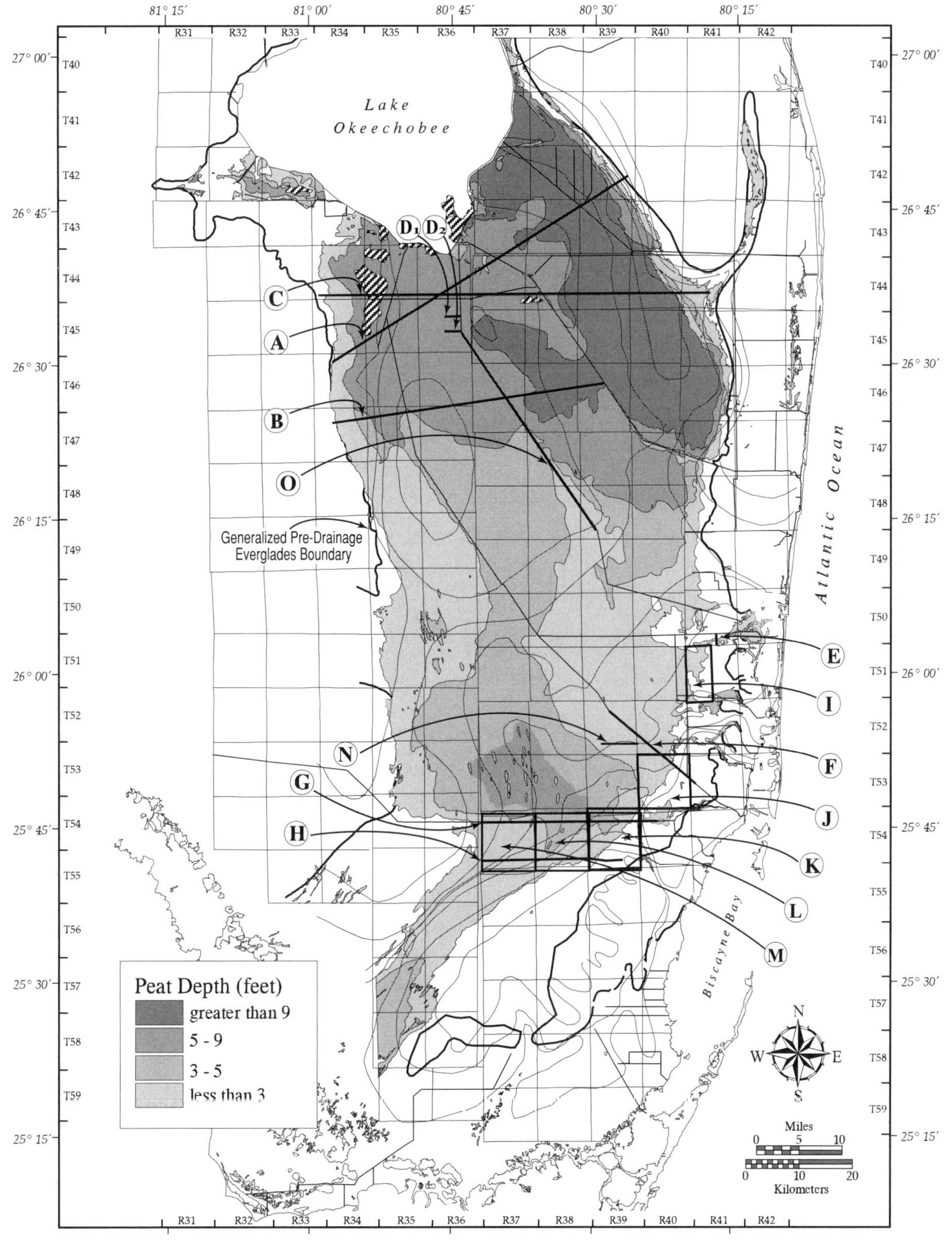

Figure 3.9. Map of surveyed peat thicknesses, 1939–1943. The shading indicates peat depth and contours indicate bedrock elevation. The canals present in 1940 are included as well as the location of the predrainage Everglades boundary. This is a location map for transects referred to in this book (see also fig. 3.8). For transect profiles, see the following figures: A = fig. 3.6A, B = fig. 3.6B; C = 3.7; D_1 = 3.2 top, D_2 = 3.2 bottom; E = 3.5, line 'D'; F = 3.12 top; G = 3.13 top, H = 3.13 bottom; I = 4.1; J = 4.5; K = 9.2; L = 8.8 bottom; M = 8.8 top; N = 3.12 bottom; O = 3.11. Sources: Peat thickness, Jones et al. (1948e); bedrock contours, Jones et al. 1948e, supplemented by Parker et al. 1955.

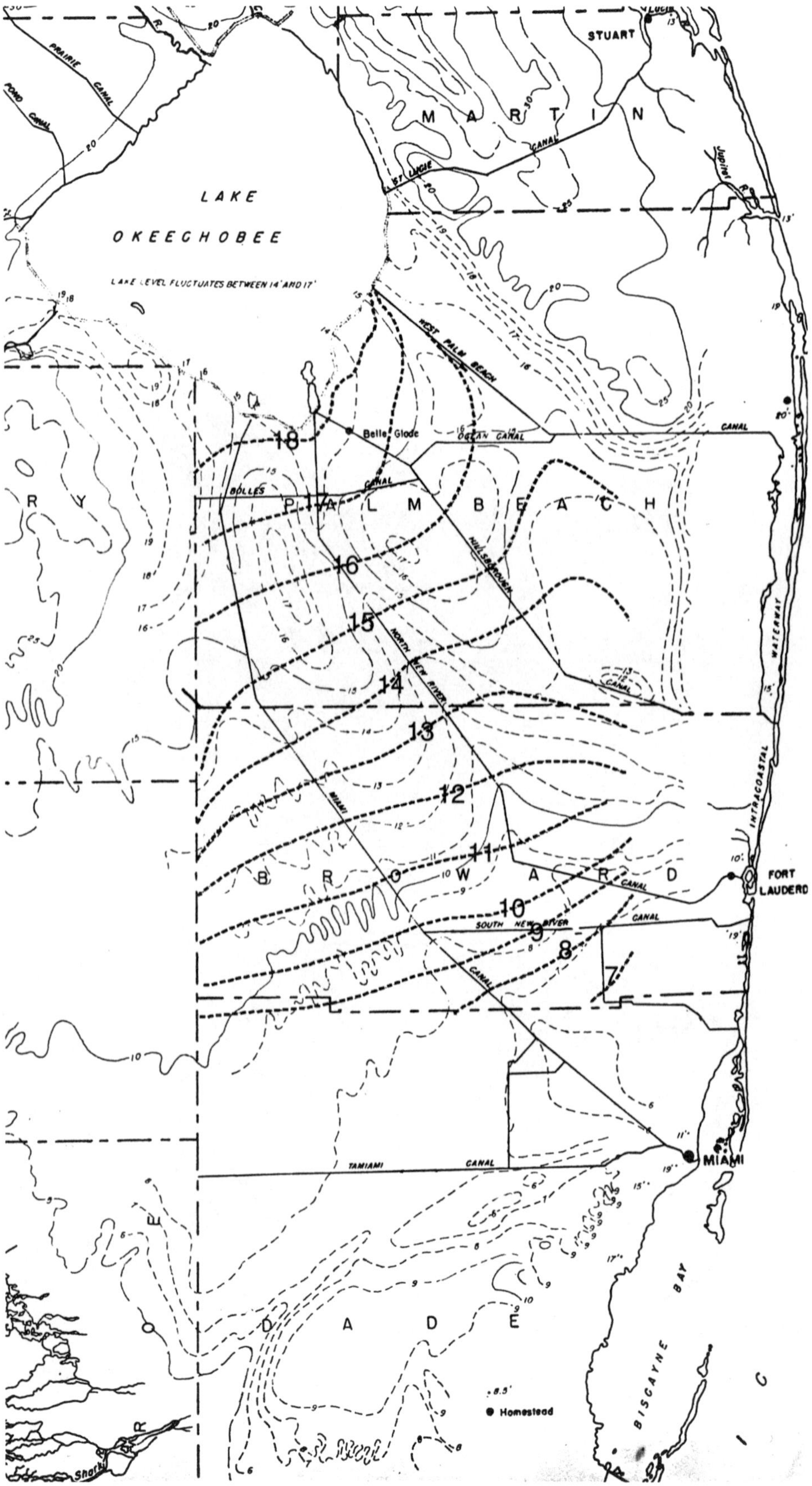

Figure 3.10. Predrainage land surface elevation contours estimated by J. H. Davis (dashed gray lines) are compared to actual 1940s contours (dashed black lines). Redrawn from figure 26 in J. H. Davis (1943a).

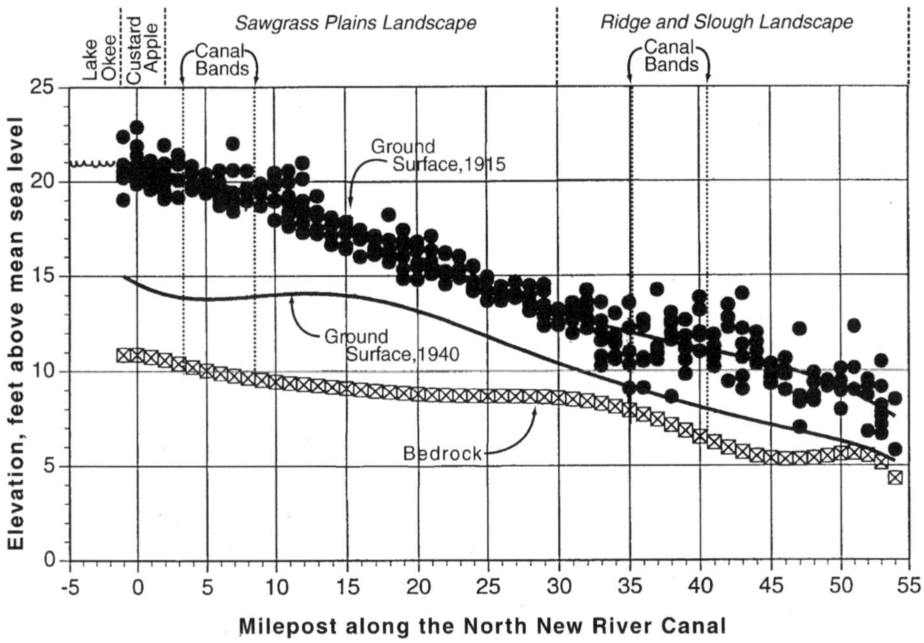

Figure 3.11. Ground surface elevations and peat thicknesses along the North New River Canal, ca. 1915 and 1940 (location shown as transect O on figures 3.8 and 3.9; see plate 10 left for milepost locations). Sources: Surface elevations, Florida Everglades Engineering Commission (1914); peat thicknesses, Baldwin and Hawker (1915); bedrock and 1940s surface elevations, Jones et al. (1948a, e).

drainage Ridge and Slough landscape. The profile of soil surface and underlying bedrock surface shown in this figure is probably representative of predrainage conditions as well as representative of profiles that would have been seen along the Miami and Hillsboro canals and the northern portion of the West Palm Beach Canal.

Figure 3.11 illustrates the Lake Okeechobee–Everglades system, in profile, as well as the dramatic loss of peat and elevation the system had experienced by the 1940s. Originally a thick wedge of peat soil tapered from 10 feet (3 m) thick at Lake Okeechobee to 4 or 5 feet (1.2 or 1.5 m) thick near the headwaters of the New River. The north end of the wedge formed the dam that allowed water in Lake Okeechobee to reach an elevation of about 21 feet (6.4 m) above sea level before overflowing into the Everglades. It is noteworthy that the slope of the Everglades that provided the driving force for southward water movement was not due to the slope of the underlying bedrock but, rather, formed by the peat itself.

Drainage-induced subsidence strongly altered the system. By 1929 the surface elevation at a point near Lake Okeechobee had dropped by 4 feet (1.2 m; Interbureau Committee 1930, p. 7). By the 1940s more than half of the original peat profile had been lost along much of the 55-mile (90 km) length of the North New River Canal. In the area of the Everglades that extends from the southern shore of Lake Okeechobee for about 15 miles (25 km) south, the land surface had become essentially flat and no longer sloped southward. By 1978 only a few feet (1 m) of the original peat remained at a point 4 miles (6 km) from the original lakeshore (Shih et al. 1979c, p. 70).

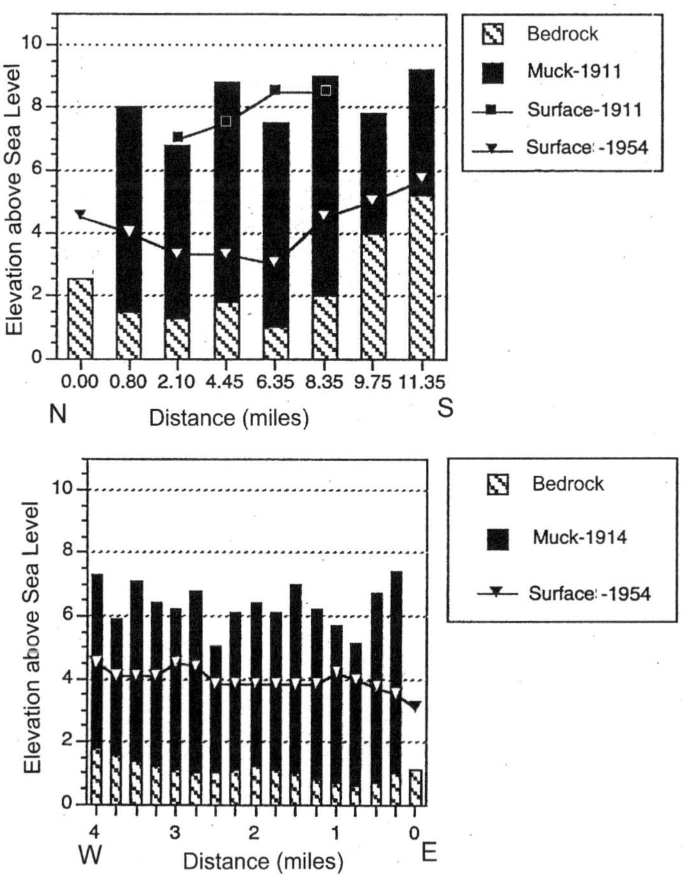

Figure 3.12. Reduction in peat thickness and ground surface elevation near the Miami Canal between 1911 (1914) and 1954. Top: Along the Miami Canal; Transect F in figures 3.8 and 3.9. Sources: Ensey and Elliot (1911); U.S. Army Corp of Engineers (1954). Bottom: Slightly west of the Miami Canal; Transect N in figures 3.8 and 3.9. Sources: Frederick (1914-T52 R39); U.S. Army Corp of Engineers (1954).

Figure 3.12 shows two peat profiles near the southern portion of the Miami Canal (transects F and N on figure 3.9). Measurements taken from a canal survey (Ensey and Elliot 1911), a township survey (Frederick 1914-T52 R39), and a postdrainage study (U.S. Army Corps of Engineers 1954, plate 14) indicate that early postdrainage (1911 and 1914) peat thicknesses of 5 to 6 feet (1.5 to 1.8 m) in this area are greater than those found farther north, near the southern portion of North New River Canal (fig. 3.11). Subsidence along these profiles was assessed by comparison with detailed local measurements made in 1954. Subsidence was especially severe beside the canal, with loss of more than 4 (1.2 m) of the original 6 feet (1.8 m) of peat. The depression or "subsidence crater" formed around the lower portion of the Miami Canal is clearly visible in plate 16 (see fig. 3.10 for identification of the canals).

Figure 3.13 shows peat thickness data for two parallel transects in the southern Everglades, one along the future route of the Tamiami Trail (transect G, fig. 3.9) and the second covering the same distance 4 miles (6 km) farther south (transect H, fig. 3.9). Subsidence losses along Tamiami Trail appear to have been about 2 feet (0.6 m), except to the east of the Dade-Broward Levee Canal, where essentially all of the original 3 to 4 feet (0.9 to 1.2 m) of peat had already been lost, and farther south along the Snapper Creek Extension, where about 3 feet (0.9 m) of peat were lost. Peat thickness remaining in the 1940s was estimated from figure 3.9 (Jones et al. 1948a); bedrock elevations were obtained from Parker et al. (1955, p. 150).

Figure 3.13 reinforces findings of the predrainage location and the diagonal northeast-to-southwest orientation of Shark River Slough that were previously discussed in relation to 1940s measurements of peat thicknesses (figs. 3.8, 3.9). Along the Tamiami Trail 1940s peat measurements indicated that the center of the slough was in the middle or western side of Range 38; whereas 4

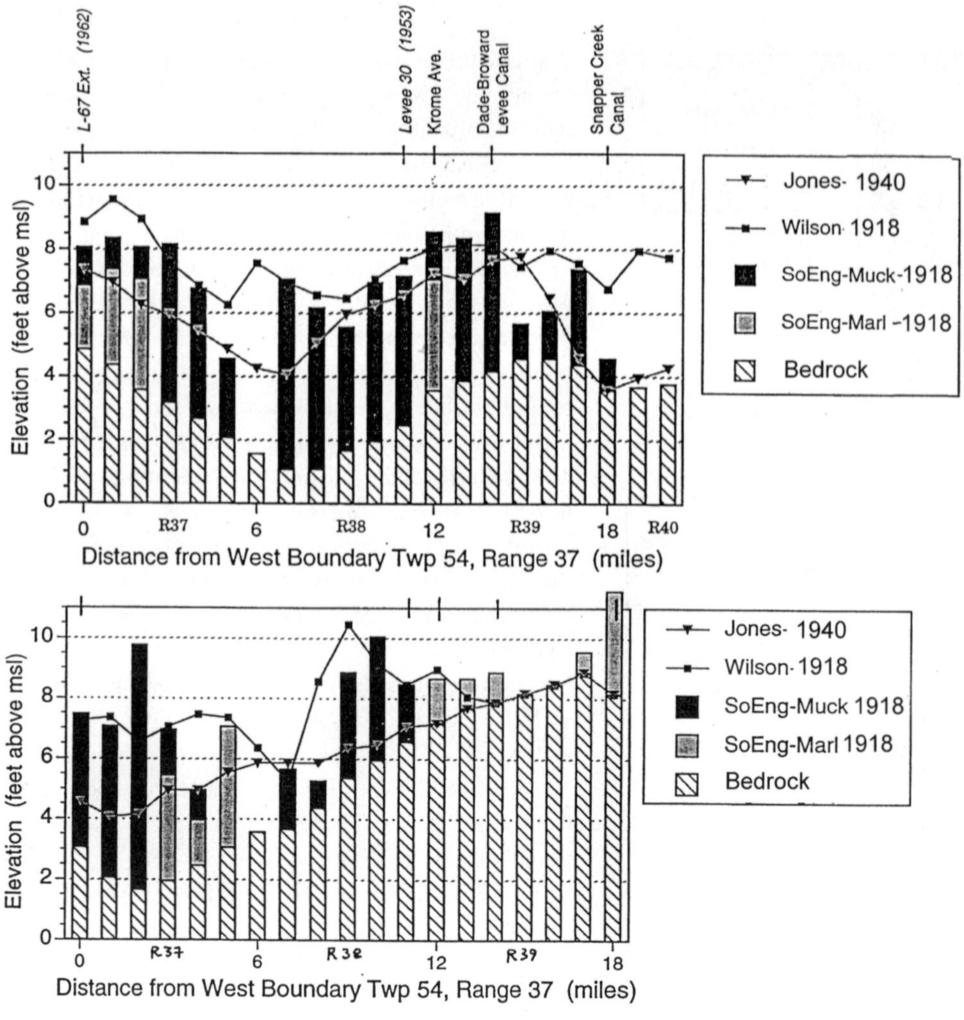

Figure 3.13. Reduction in peat thickness and ground surface elevation near Tamiami Trail between 1918 and 1940. Top: Tamiami Trail; Transect G in figures 3.8 and 3.9. Bottom: Along a proposed Snapper Creek Extension Canal; Transect H in figures 3.8 and 3.9. Sources: Southern Engineering & Construction Co. (1918; 1918-T54 R37; 1918-T54 R38; 1918-T54 R39); Jones et al. (1948a).

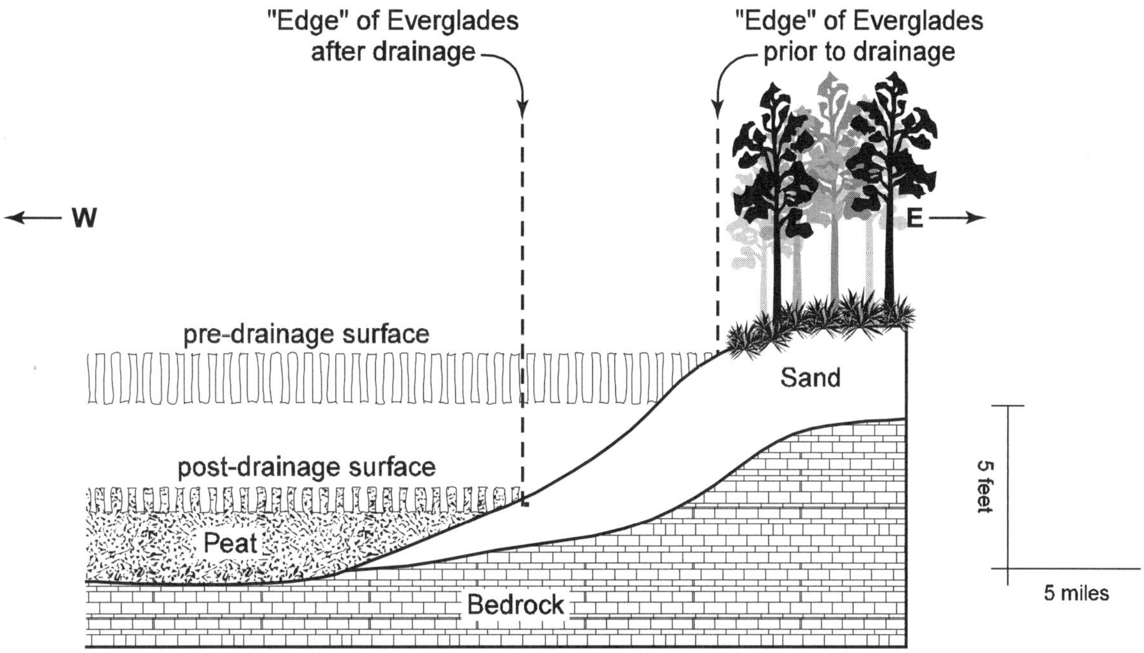

Figure 3.14. Postdrainage, anthropogenic formation of sandy areas along the eastern edge of Everglades. With the lowering of Everglades water levels, peat soils oxidized from the surface downward, exposing the underlying sand, and creating the so-called "peripheral short hydroperiod wet prairies." Sources: Chapter 4; appendix D.

miles (6 km) farther south, the predrainage center, as indicated by the thickest peat, is clearly farther west, in Range 37.

With a few exceptions that are probably the result of uncertainties in the estimates, peat thicknesses in the central and southern Everglades were less in the 1940s than in the period just after drainage (1911–1918). The most severe peat losses, up to 100 percent, occurred along the eastern edge of the Everglades, mostly between Townships 50 and 53 (transects E and F and townships labeled I and J on figs. 3.8 and 3.9). The severity of loss in this area is likely due to a combination of thinner (predrainage) peats, earlier and more intensive diking and drainage, and possibly the presence of underlying sand (fig. 3.14), which may have facilitated the lowering of water tables to levels below the peat horizon.

Peat Fires

In sawgrass and peat environments several different types of fires are possible, primarily distinguished by the depth of the water table: (1) leaf fires, which burn sawgrass (*Cladium jamaicense*) leaves without killing the submerged culms (Forthman 1973, pp. 48, 50); (2) hotter vegetation fires, which burn the exposed or partially exposed sawgrass culms and therefore kill the whole plant; and (3) peat fires, which burn not only the plants but the peat soil itself. Peat fires occur when water tables are sufficiently far below ground surface for the upper peat to dry. According to Cornwell and Hutchinson (1974, pp. 13, 15), water tables in the peat need to be deeper than the normal annual dry season minima of 4 to 6 inches (10–15 cm) for peat fires to occur. Frequent and extensive peat fires occurred shortly after drainage began lowering the water tables.

There is little doubt that leaf fires occurred in the Everglades even prior to drainage. Fires in sawgrass, one of the most ubiquitous species within the Everglades, are reported in the historical literature (e.g., Cooley 1851, p. 41). Modern scientific studies confirm that the periodic occurrence of leaf fires would not have destroyed the sawgrass landscape; thus leaf fires could well have been consistent with a continuous presence of sawgrass in the northern Everglades, as suggested by the deep soil profiles of sawgrass peat. Forthman (1973, pp. 50–52) found that sawgrass regenerates rapidly after fire, provided that the soil was sufficiently wet or inundated to prevent sawgrass culms from being killed. Hofstetter (1984, p. 472) suggests that periodic fires may even be beneficial, reducing the buildup of flammable leaf litter and the occurrence of "decadent sawgrass."

That scenario, based on long-term coexistence of sawgrass and fire, depends on water levels being sufficiently high to prevent sawgrass culms from burning. If water levels decline far enough below ground to dry out the surface of the organic soils, a different type of fire, the peat fire, can occur. From the postdrainage period it is known that peat fires not only kill sawgrass but also destroy a portion of the organic soil profile. The flammability of dry peat was recognized by early observers of the Everglades, such as J. T. Stewart (1907, p. 45), an agricultural scientist, and S. L. Lupfer, an engineer involved in surveying and dredging for the North New River Canal: "All of the marsh lands surrounding Lake Okeechobee are composed of 'peat,' very similar to the world-famed Irish fuel, and when these marshes become thoroughly dry they burn with little flame but intense heat. As long as the water level is within 4 or 5 ft. of the surface there is no danger of the land burning to any great extent; and, as this question of fire is a well understood one, care will be taken not to lower the waters of Lake Okeechobee sufficiently to invite this danger" (Lupfer 1906, p. 2).

Unfortunately, Lupfer's estimate that a water table within 4 to 5 feet (1.2 to 1.5 m) of the surface would prevent peat fires was off by almost a factor of 10; in addition, he was overly optimistic about the care that would be taken. R. M. Harper, who observed the eastern edge of the Everglades in 1909 and again sometime before 1927, related the peat fires to drainage: "in dry seasons the drained peat [of the Everglades] sometimes catches fire, and the soil then goes up in smoke" (Harper 1927, p. 126). Naturalist Charles Torrey Simpson also commented, in a 1930 letter to H. A. Kelly, on flammability of the peat soil after drainage:

> Most of the soil is composed of decayed or partly decayed vegetation, fairly rich in nitrogen but lacking in phosphoric acid and almost wanting in potash. A vast area is half decayed vegetation; in many cases the fallen plants seem as fresh as when they were buried thousands of years ago. Whenever the whole is dried out, as it becomes when it is drained, it is ready for the fire. Careless huntsmen, tourists, even residents here are constantly setting the Glades afire and when this is done, it is well-nigh impossible to put out such fires as they smoulder and burn away below the surface, much as they do in a coal mine. Year after year we here, at and near Miami, have been almost smothered in the dry season, whenever the wind got in the northwest, by clouds of acrid, black smoke from the burning Glades, and much money and labor have been expended in fighting these fires, which burn down to bed rock or the marl and leave vast ugly lakes when the rains come. Wherever they burn, the whole thing is destroyed forever. In a very informing article in the New York Times a few years ago, it was stated that over 200,000 acres had already been burned over. (Simpson 1930, pp. 40–42)

Prior to this letter, in 1920, Simpson specifically described the effect of only about five years of "partial drainage" on tree islands: "One of the results of partial drainage is that along this same east border [of the Everglades] numerous low, timbered 'islands,' which were formerly quite wet, have now been changed to dry land. A considerable part of the foundation of these groves is peat and in dry times it is very liable to fire, and once begun it is well-nigh impossible to extinguish it. These groves, despoiled of their only defense against fire, are often wholly destroyed" (Simpson 1920, pp. 126–127).

Extensive peat fires, with results similar to that shown in figure 3.15, covered tens to hundreds of square miles (30–300 square km; Bender 1943, p. 150) and spread during dry periods beginning in the 1920s (Robertson 1953, p. 94): "We [the then newly created Everglades Fire Control] were called upon to extinguish old muck fires of long duration, some reported to have been burning over a period of two years and some under old abandoned rock roads, making it necessary to wash through these rocks before reaching the smouldering muck underneath" (Bender 1937, p. 88). Similarly damaging peat fires continued into the 1950s (Florida Game and Freshwater Fish Commission 1956, p. 43; Wallace et al. 1960, p. 9). Such fires could burn for months and even through the wet seasons of multiple years (Bender 1943, p. 151). When they did, they could "destroy the upper, dry, compacted peat layers to a depth of three or four inches over large areas and up to a foot or more in localized situations" (Loveless 1959a, p. 8). Fires during the 1920s and early 1930s were of sufficient size and frequency to lead to the formation of a special fire control district in the Everglades. According to the district's director, "previous to the creation of a fire control district by the Legislature in 1935, vast areas of the Everglades were destroyed by fire. In one county alone one-third of the muck area was totally destroyed" (Bender 1943, p. 151). An earlier report of his gives a sense of the magnitude of the fires and their prevalence in previous years:

> If a fire is extinguished in a short time after starting very little damage to land will result and it is only when the fire is allowed to burn for several weeks and burn down two or three feet that the land is totally lost and becomes waste land on which no one cares to pay taxes.
>
> In my opinion, this is the first time in a number of

Figure 3.15. Photograph of drained, dried, and burnt peat of the Everglades, 1930s: "[The photograph] shows some of the white man's mistakes which are expensive to him and to the Indian. [It] reveals what a fire does to the muck land. The dust I am pouring from my hand is as pure ashes as ever taken from any fireplace. The foreground of the picture is a great bed of it. At my side are the banks of unburned muck. . . . These fires break out in the dry season of the year and rage for weeks, or until some rain puts them out. They usually burn down to wet muck, however deep that may be" (Glenn 1982, pp. 26–30). By permission of the Fort Lauderdale Historical Society, Inc. (James L. Glenn Collection).

> years that the east coast has been entirely free of the annoying and discomforting odor of burning muck and the residents of the Glades area have, for the first time, received from the State the long hoped for help and assistance in extinguishing burning muck. (Bender 1937, p. 90)

Effects of Peat Fires on Everglades Landscapes

Postdrainage peat fires often permanently altered Everglades landscapes, particularly in areas where the peat soils were naturally thin prior to drainage. Figure 3.16 shows a location originally within the Ridge and Slough landscape, close to its border with the Rockland Marl Marsh. At the time of the photograph this area had been strongly drained for more than twenty-five years by the Tamiami Trail Canal as well as the Snapper Creek Canal. By 1940 fire had burned off the remaining peat to reveal the bedrock, Miami oolite, below.

J. H. Davis (1943a, p. 81) described the effect of lowered water levels and of burning on the Rockland Marl Marsh area: "Excessive artificial drainage has recently created drier conditions promoting these fires and also causing the shallow organic soils to become oxidized and subside until now some areas once soil covered are returning to rockland conditions." In 1944 an Everglades Drainage District Advisory Committee noted the distinction between vegetation fires and destructive soil fires as well as noting lowered water tables as the cause of soil fires: "Such a state of saturation would prevent damage by burning to muck and peat soils swept by vegetation fires. The depth to which muck and peat soils burn by fire is conditional upon the depth of the water table below the surface. The frequent burning of muck and peat noted throughout the Everglades during the past two years reflects the low level to which the water table has fallen" (Allison et al. 1944, pp. 23–24).

W. B. Robertson also remarked on areas of exposed rockland in Everglades National Park that were created by peat fires: "In muckland areas [of the sawgrass glades] the soil destroying effects of fire are notable. Undoubtedly some of the present rockland areas of the glades have resulted from complete removal of the soil by repeated muck fires" (Robertson 1953, p. 81). These peat fires affected not only the sawgrass areas but the higher-

Figure 3.16. "Annual fires have burned off muck to reveal solution-pitted top of Miami oolite." Photo by Martha (Mrs. G. G.) Parker, July 30, 1940. Location: Coral Way, one half mile east of SW 102 Avenue, Miami (middle of the line between sections 8 and 17 of T54 R40). This was formerly Ridge and Slough landscape, near the border with Rockland Marl Marsh (see plate 5). Photo used by permission of Bernadette Parker.

lying tree islands as well. Bay heads, one of the higher elevation vegetation types, dried sufficiently in Everglades National Park to suffer repeated burns: "in some places the peat is burned below general glades level so that semi-permanent ponds are established" (Robertson 1953, p. 78).

Farther north C. M. Loveless noted similarly drastic effects of peat fires on tree islands in Water Conservation Area 3. His description is similar to the one given by Simpson (1920, pp. 126–127) almost forty years earlier: "These [early summer of 1956] fires completely destroyed many tree island communities by burning the peat substrate out from under the tree growth. Some of these areas are now open water ponds devoid of any type of emergent vegetation while others support sparse strands of sawgrass, water-lily, floating heart and other aquatic species" (Loveless 1959a, p. 8).

The postdrainage peat fires that became especially common and widespread between the 1920s and the 1950s had significant effects on Everglades landscapes. As noted by the State Commissioner of Agriculture, Nathan Mayo, "Already areas amounting in some instances to over two hundred square miles in a single body, have been denuded of all peat and muck by fires and have been rendered totally worthless for agriculture. The acreage of the areas thus ruined cannot be accurately determined but estimates have placed it as high as 20 percent to 25 percent of the 2,000,000 acres [suitable for 'intensive and profitable agriculture']" (Mayo 1940, p. 81). The numerous peat fires that became common in the first half of the twentieth century occurred as a result of removal of the overlying surface water and subsequent drying out of the peat soil itself. They highlight the efficacy of the early period of uncontrolled drainage.

Frequency of Peat Fires Prior to Drainage

While vegetation fires within the Everglades are not closely tied to water levels, and so do not shed much light on prevailing water level regimes, evidence of predrainage peat fire frequency certainly could be a potential source of information regarding the predrainage occurrence of below-ground water tables. The assumed occurrence of predrainage peat fires has sometimes been taken as an indication of dry conditions prior to man-made drainage. However, the available evidence regarding predrainage frequency, extremity, and especially the spatial extent of peat burns remains sketchy (Gunderson and Snyder 1994). Cohen (1984, p. 461) concluded that his soil core studies did not reveal evidence of widespread peat fires. At the same time, because of the intensive effort involved in microscopy studies, he was able to examine only a small number of cores. There has not yet been a charcoal study of soil cores with sufficient geographic scope to map the frequency and size of predrainage fires reliably.

In an often-referenced passage, J. H. Davis asserted: "There were sufficiently dry conditions over large areas and some periods of time when fires became extensive, as is evidenced by layers of ashes six or more inches thick that were found under some layers of peat in some areas when the canals were being excavated. Such deep layers of ashes signify the burning of thick peat layers and it is probable that the saw-grass peat has been built up and burned back a number of times in some areas" (J. H. Davis 1943a, p. 253). The observation of these layers during canal excavation was communicated to Davis by F. C. Elliott, former district drainage engineer. Garald Parker also observed ash layers in the field at two locations:

> Nevin Hoy and I hand-dug a number of deep pits both in the wet prairies and in the organic soils of the central glades. These pits were about 5' square and dug into bedrock. . . . It was in the walls of at least two of these pits, one dug in the vicinity of Miami Canal's Broken Dam [northwest corner of Section 10, T52 R39] and the other near the South New River confluence with Miami Canal, that we found ash layers up to about 2 inches thick. These buried ash layers indicate ancient, pre-historic times when there were severe droughts that dried the organic soils of the normally wet glades sufficiently to allow the soils themselves to burn as deep as the existing water table. (Parker 1974, p. 30)

Several points are important to note concerning the hydrologic implications of these two reports, both from well-qualified Everglades observers. First, the observations appear to be of limited spatial extent. This contrasts with the relative uniformity of the predrainage Everglades (see chapters 8 and 11), which would suggest that a drought sufficient to permit peat fires would have been widespread, leading to widespread and synchronous ash layers. These have not been found; in fact, reports of possible ash layers are rare (Peter Stone, pers. comm., 2004). Second, dark layers within peat may not always reflect peat ash; microscopic studies are required for unambiguous identification of charcoal (Patrick J. Gleason, pers. comm., 1997). Third, reported thicknesses combined with the very low mineral content of the peats of the Everglades (7–15 percent; Baldwin and Hawker 1915, p. 763) imply that implausibly thick layers of peat would have had to have been completely consumed by fire. A 6-inch-thick (15 cm) layer of presumed ash would imply that a layer of peat fully 60 inches thick (5 feet; 1.5 m) had been reduced to mineral ashes. Overall, the available data suggest that peat fires in the predrainage Everglades were exceptional rarities, suggesting that prior to man-made drainage, the wetland peats rarely became dry enough to burn.

Soil Changes

Soil oxidation caused by lowered water tables produced two major postdrainage changes in the soil types mapped by Jones et al. (1948a, e). The soil changes were paralleled by vegetation changes, resulting in a complete landscape change. As a consequence, the postdrainage landscape patterns shown on the 1940s maps of soils and vegetation were quite different from predrainage landscape patterns, as discussed further in chapter 4. Fortunately, enough post- and predrainage information is available for these areas to make it possible to reconstruct or "hindcast" the original landscape conditions reliably.

The first change was extensive and occurred in the middle of the Everglades. It transformed a large area of originally patterned peatland (Ridge and Slough landscape; chapter 8) into unpatterned or more weakly patterned peatland similar to that found in the Sawgrass Plains landscape (chapter 6). The second change affected the edges of the Everglades and transformed areas of patterned Ridge and Slough peatland into an unpatterned, primarily sand-based wetland, the so-called "peripheral wet prairies" (see chapter 4). Both changes were the result of long periods of strong anthropogenic drainage that caused lasting changes to the original vegetation, soils, and microtopography.

Ridge and Slough Soils to Sawgrass Plains Soils

(Loxahatchee Peats to Everglades Peats)

When the soils of the Everglades were mapped in the 1940s (plates 11, 15), a three-part spatial configuration was drawn: two disjunct areas of Loxahatchee peats separated by a central band of Everglades peat. The soil studies of the 1940s associated Loxahatchee peat with the Ridge and Slough landscape (Jones et al. 1948e, p. 149; see also chapter 8), while Everglades peat was associated with the Sawgrass Plains landscape (Jones et al. 1948e, p. 147; see also chapter 6). Because soils often retain predisturbance attributes for some time, the three-part soil configuration mapped in the 1940s (plate 15) has frequently been assumed also to depict accurately the predrainage configuration of the two principal landscapes of the Everglades. However, examination of earlier descriptions of these soils and landscapes indicates that by the 1940s, both landscape and soil had changed substantially, and the disjunct, three-part configuration mapped in the 1940s is in fact a postdrainage artifact.

One of the central, far-reaching conclusions of this study is that prior to drainage, the landscape configuration was not as mapped in the 1940s (plates 11, 14, 15) but instead was as shown in plate 5. According to our interpretation, discussed further in chapters 4 and 8, the Ridge and Slough landscape originally occurred as a single large entity—unbroken by a central section of Sawgrass Plains or Sawgrass Plains–like landscape (plate 5). An unbroken, non-three-part configuration implies several additional points, including that the two landscapes were arranged sequentially, rather than in parallel, with the Ridge and Slough landscape located downstream from the Sawgrass Plains. It also implies that in cross section the predrainage Everglades was likely close to level, rather than raised in the central portion, as might have been expected from a three-part configuration.

Soils form an important part of the evidence for our reconstructed predrainage landscape configuration shown in plate 5. Physical changes in the organic soils of the Everglades, together with changes in their classification, provided the background for accurately reconstructing the original configuration of the predrainage Everglades. Comparison of plate 5 (our reconstructed predrainage configuration) with plate 15 (the soils configuration mapped in the 1940s) indicates that the key area of difference is the central, northwest-to-southeast swath extending from Township 47 southward to Township 53 (see also fig. 4.9). Jones et al. (1948e, p. 14) labeled this area "1d hammock-sawgrass" and associated it closely with their physiographic region "1a sawgrass plains" (fig. 4.9). Significantly, area "1d" is aligned parallel to the Miami and North New River canals and encloses the area between them (plate 15, fig. 4.9). Plate 5 posits that area "1d" was originally part of the Ridge and Slough landscape, while plate 15 indicates that in the 1940s, only sixty years later, area "1d" was described as Everglades peat with a land cover of sawgrass (plate 14). By the definitions used by Jones et al. (1948e), area "1d" therefore became a southeastward extension of the Sawgrass Plains landscape. To reconcile the differences between plate 5 (this study) and plate 15 (i.e., Jones et al. 1948e) and to identify the predrainage condition correctly, it is necessary to track changes over time in the soils and landscape character that occurred in area "1d" (for this discussion referred to as the "central Everglades"), and to compare the changes with those that occurred in the "1a" area or "upper Everglades" also known in the present study as the Sawgrass Plains landscape. The two areas are compared at two time points, 1914–1915 and the 1940s. The 1914–1915 time point was chosen to include the earliest available soil descriptions; that is, as a proxy for predrainage descriptions. Fortunately the nature of the soils, as described in 1914–1915, suggests that they still retained predrainage characteristics.

Baldwin and Hawker created a soil map of a 6-mile-wide (10 km) swath along the full length of the North New River Canal in 1915. Their observations were made three years after opening of the full length of the canal. Observations along the upper half of the canal (milepost 3 down to mileposts 30–32; see fig. 3.17, top) correspond to the upper Everglades. Their description of the upper Everglades—that is, of the vegetation and the microtopography—is typical of the Sawgrass Plains landscape (chapter 6): "The surface of the upper part of the large area of this type ['Brown fibrous peat, deep over limestone'] is flat and nearly level. No natural drainage channels are apparent, and the flatness is broken only by infrequent alligator holes and runways" (Baldwin and Hawker 1915, p. 785).

Baldwin and Hawker mapped a single soil type for the upper Everglades, that is, for the "large area, extending from an irregular boundary within 2 or 3 miles of Lake Okeechobee to the thirty-second milepost below the lake on the North New River Canal" (Baldwin and Hawker 1915, p. 784), calling it "brown fibrous peat, deep over limestone." They described this soil as follows: "[Leaving Lake Okeechobee and] approaching the interior of the Everglades, the material becomes less decomposed [than the custard apple muck], and is nearly pure organic matter. Two and three miles south of the lake the surface material is a brown fibrous peat averaging 85 to 93 per cent of combustible organic matter. This area is uniform in composition and texture for many square miles" (Baldwin and Hawker 1915, p. 763). From borings and soil cores Alfred Dachnowski-Stokes reconstructed the peat history of the upper Everglades, or Sawgrass Plains, concluding that the result would be a brown, fibrous, and spongy peat, very similar to what Baldwin and Hawker observed in the field: "a fairly uniform saw-grass marsh, very much as today, with grasses, sedges, ferns, and herbaceous semi-aquatics having fleshy rootstocks, slowly encroached upon the exposed rocky floor. . . . Brown fibrous peat, spongy and water-soaked, formed a continuous layer over the limestone" (Dachnowski-Stokes 1930, pp. 99–100).

Strikingly, all other observations of the soil of the upper Everglades (area "1a") recorded between 1890 and

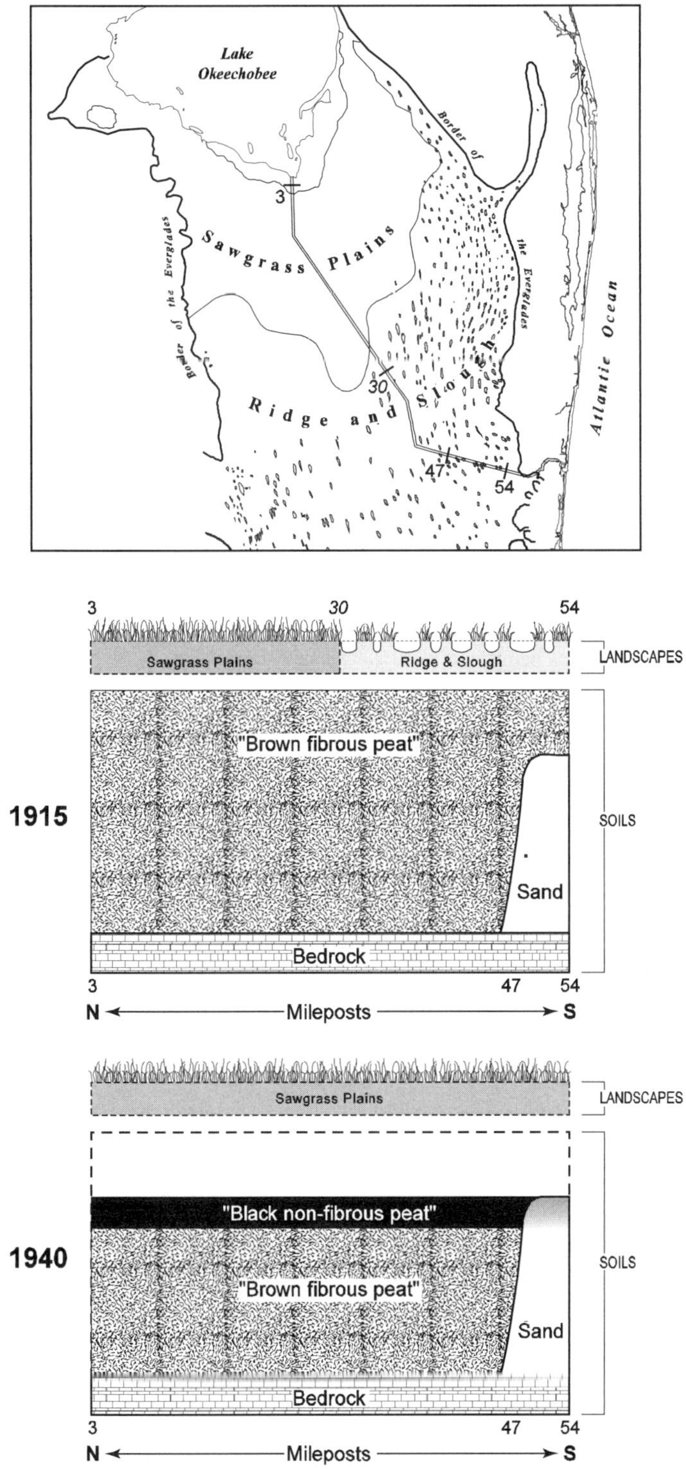

Figure 3.17. Landscape change in the upper Everglades during the first twenty-five years of drainage. Middle panel shows vegetation and generalized soil profiles along the North New River Canal in 1915; bottom panel, in 1940. Locations of mileposts along the canal and their relation to predrainage landscapes are shown in top panel. Note the disappearance of the Ridge and Slough landscape, subsidence of peat soils, and appearance of a surficial layer of decomposed, black, nonfibrous peat. Sources: Baldwin and Hawker (1915); Jones et al. (1948a).

about 1929 also described it as a minimally decomposed, brown, fibrous material (table 3.1). Carl Forsaith, who studied peat deposits from sawgrass and other marsh types in Florida, suggests that brown, fibrous, and spongy are all expected characteristics for peats that had formed under hydrologic conditions similar to those of the predrainage Sawgrass Plains; that is, year-round (or nearly year-round) surface water: "In those deposits which have been formed by the growth of Cladium, Sagittaria, Cyperaceae, etc., or, to be more general, where the peat is entirely herbaceous in texture, the samples appear as fibrous, light brown, spongy materials in which the individual parts of the plants can be distinctly seen and separated" (Forsaith 1916, p. 34). He further suggests that in "herbaceous peats there occurs little evidence of decay because of the presence of a constant water covering, as well as natural plant acids which prohibit the oxidation of the air and the destructive action of fungi, thus preserving its spongy, fibrous nature" (Forsaith 1916, p. 35).

In a slightly earlier study of Florida peat soils, Roland Harper similarly notes the contrast between fresh, undecomposed peat—brown and fibrous—versus more decomposed peat—black and more weakly structured, ultimately amorphous or plastic: "In texture, peat is more or less fibrous, being least so in the oldest and most thoroughly decomposed samples, which are rather plastic and look much like mud. The color is always some shade of brown, varying from light brown (this is usually in fresh or imperfectly decomposed peat) to nearly black" (Harper 1910a, p. 210). Overall it is clear from numerous sources that the soils of the upper Everglades were originally minimally decomposed, brown fibrous peats.

Baldwin and Hawker's description of the central Everglades area along the North New River Canal (mileposts 30 to 32 down to milepost 54) is obviously different from the upper Everglades and is typical of the Ridge and Slough landscape (chapter 8):

> The eastern half of this part of [the Everglades] is very gently undulating or hummocky, owing to a series of ridges and sloughs with local differences of 2 or 3 feet in elevation, having a general northwest-southeast direction. (Baldwin and Hawker 1915, p. 752)
>
> As the coast is approached [i.e., south of milepost 32] the topography changes somewhat [from the flat, nearly level Sawgrass Plains]. A series of ridges and sloughs trending in a northwest-southeast direction is apparent. (Baldwin and Hawker 1915, p.785)

A year earlier Harshberger (1914, pp. 166–172) had made similarly distinct observations of these two landscapes along the length of the same canal, with a similarly lo-

Table 3.1. Observations of upper Everglades (Sawgrass Plains) soils in predrainage or near-predrainage condition

Soil	Depth	Color	Structure	Decomposition	Organic Matter[a]	Comments
Soil of Sawgrass Plains[b]	Upper 7 feet (2.1 m)	—	"stringy rooty nature"	"did not cave after excavating"	—	"material excavated was all of one kind"
"Brown fibrous peat"[c]	Upper profile	"brown"	"fibrous"	undecomposed[d]	85–93	
"Peat"[e]	"near the surface"	"brown"	"loose, fibrous"	—	90–93	"very light when dry"
"Sawgrass peat"[f]	2 feet (60 cm)	—	"Coarsely fibrous"	Probably minimally decomposed[g]	93	
"Raw saw-grass soil"[h]	Profile (un-specified)	—	"fibrous"	"raw" = undecomposed (?)	—	
"Sawgrass soil"[i]	Profile (un-specified)	"brown"	"fibrous"	—	—	"sedge peat"
"Saw grass soil"[j]	0–5 feet (0–150 cm)	"brown"	"fibrous"; "light, loose, and soft"	"undecomposed"	82–93	"original [undecom-posed] plant material is rather abundant"
"Everglades Peat"[k]	0"–5" (0–13 cm)	"brown"	"fibrous"	—	88	"sawgrass peat"
"Everglades Peat"[k]	5"–15" (13–38 cm)	"yellowish-brown"	"coarsely fibrous"	—	93	"sawgrass peat"

Notes: a. Organic matter is either as given by authors cited or calculated as 100 percent minus the ash content given by the authors.
b. Observations by Fred Menge, dredge operator during 1890s dredging of Thirteen-Mile Canal (J. F. Menge 1907, pp. 40–41).
c. Baldwin and Hawker (1915, pp. 751–798).
d. Baldwin and Hawker (1915, pp. 762–63) distinguished "two distinct classes . . . in one case the surface material is black and decomposed and in the other it is brown and fibrous."
e. Miller (1918, p. 606). Ash content 7 to 10 percent.
f. Dachnowski (1924, p. 70). Ash content 7 percent. Sample from Okeelanta.
g. Decomposition is likely minimal because Dachnowski (1924, p. 70) described peats from other locations as "finely divided"; "fibrous, partly decomposed"; or "'Muck,' well decomposed."
h. Allison et al. (1927, pp. 38, 76). See text for extent.
i. Lloyd (1929, p. 55).
j. Hammar (1929, pp. 9, 11).
k. Evans and Allison (1942, p. 36). Soil profile actually taken and described in 1929.

cated transition at mileposts 28 to 30. Inspection of aerial photographs (USDA-SCS 1940) also shows a similar location of the boundary (mileposts 23 to 27) between the characteristic Ridge and Slough patterning (chapter 8) and the unpatterned Sawgrass Plains.

Baldwin and Hawker mapped the same peat soil type in the central Everglades ("1d") as in the upper Everglades ("1a"): "Brown fibrous peat, deep over limestone." The only difference they recorded was the additional presence in this area of irregular patches, approximately 4 by 6 miles (6 × 10 km) in size, of a sandy phase: "Brown fibrous peat, sandy phase, deep over limestone." The patches show a northwest-southeast orientation on the map, but it is not clear how well the patch shapes and orientations were known; that is, whether the patch outlines were explicitly surveyed or just interpolated from the regular 1-mile (1.6 km) grid of soil borings:

> As the coast country is approached areas are encountered in which the [soil] material contains sufficient quartz sand to raise the inorganic content to 16 per cent or more. These areas are indicated as sandy phases. (Baldwin and Hawker 1915, p. 763)

> [The] Brown fibrous peat, sandy phase, deep over limestone [soil type] . . . occurs in the lower glades below the thirtieth milepost from Lake Okechobee. . . . This portion of the Everglades is marked by low ridges and sloughs with a general northwest-southeast trend. There is a marked current of the water in the sloughs moving in a southeasterly direction, and it is probable that the sand [found in this phase] owes its distribution in part to the movement of water. (Baldwin and Hawker 1915, p. 788)

Robert Allison and Alfred Dachnowski-Stokes, in discussing the soils of the central Everglades, provide more detail and note the connection between landscape, plant species, and the same undecomposed brown fibrous peat soil:

> Lagoons and sloughs of open water are characterized by more open stands of this plant [sawgrass] though in these locations are to be found herbaceous plants such as arrow arum, *Peltandra virginica*; arrow-head, *Sagittaria lancifolia*; water-lily, *Nymphaea odorata* and pickerell weed, *Pontedaria cordata* which grows

> in nearly pure communities. More or less scattered as an admixture of less prominent members are found *Rhynchospora divergens, R. corniculata, Andropogon glomeratus, Cyperus speciosa*; and the ferns *Blechnum serrulatum, Acrostichum aureum* and *Osmunda regalis*. The remains of these plants are found throughout the larger part of the telmatic layers of brown fibrous peat. (Allison and Dachnowski-Stokes 1932, p. 227)

The mapping in 1915 of the same soil type, brown fibrous peat, in both the upper and central Everglades is in some ways surprising, in other ways not. Sawgrass would have been present in both areas, covering nearly 100 percent of the upper Everglades (the Sawgrass Plains), but also as much as 40 percent to 60 percent of the central Everglades (i.e., the sawgrass ridges; chapter 8). If the aquatic plant species found in the central Everglades sloughs described by Harshberger and by Baldwin and Hawker decompose more quickly and completely than sawgrass residues, then it is possible that the central Everglades peat might be almost as heavily dominated by sawgrass residues as the upper Everglades peat. Additionally, it is possible that non-sawgrass residues were present in the central Everglades peat, but that they did not materially alter the field appearance of the peat. Although Baldwin and Hawker (1915) did include laboratory analyses, their classification appears to have been based primarily on field characteristics.

It is recognized that the soils and landscape extents encountered in 1914 (Harshberger) and 1915 (Baldwin and Hawker) along the length of the North New River Canal, which persisted for perhaps another ten years (table 3.1), were observed after the onset of drainage. Nevertheless, the unoxidized nature of the soil in 1915 suggests that the soil and as well as the landscape extents still resembled their respective predrainage status. The predrainage condition of the upper (area "la") and central (area "ld") Everglades can therefore be summarized as: (1) two clearly distinct landscapes, with the transition between them along the North New River Canal at about the same location as the "la" to "ld" border (mileposts 28–32, fig. 4.9); (2) sawgrass stands present in both landscapes, covering almost 100 percent of the upper Everglades but only about 50 percent of the central Everglades, as it was confined to ridges distributed throughout the landscape; and (3) a single, minimally decomposed peat type—brown fibrous—present underneath both areas. Other early postdrainage authors also referred to the third point, the presence of only a single, sawgrass-derived peat soil type throughout most of the Everglades: "a vast plain (about 3,000,000 acres) of raw saw-grass peat" (Allison et al. 1927, p. 38). The only decomposed peat found within either landscape in 1915 was restricted to a narrow 4-mile-wide (6 km) area along the eastern edge of the central Everglades: "Near the rock rim two distinct classes are mapped. Both are high in organic content, with an average of about 90 per cent of organic matter, but in one case the surface material is *black* and *decomposed* and in the other it is brown and fibrous" (Baldwin and Hawker 1915, p. 763; emphasis added). Peat decomposition here may have been an anthropogenic effect of the longer period of drainage in this area.

As drainage continued during the ensuing decades and as the effects of drainage accumulated, the soils also changed. More specifically, the soils oxidized and decomposed, developing a surficial layer of black, amorphous or only finely fibrous peat over wide areas, including along the full length of the North New River Canal. Observations from agriculturalists during the 1920s, 1930s, and 1940s help clarify this transition from the predrainage condition to the conditions mapped in the 1940s (plate 15). Experience taught farmers in the upper Everglades that, in addition to drainage and removal of the original vegetation, the newly exposed dry land required a period of weathering or "breaking in" before it could become agriculturally productive: "Other large planters had the same experience with crop failures on the freshly broken peat soils. It was only by breaking the soils and allowing them to weather and oxidize for long periods of time that crops could be grown" (Lloyd 1929, p. 61). "The oxidation is naturally more rapid in the cultivated lands [than in the uncultivated raw soils]" (Lloyd 1929, p. 58).

An important descriptor of the soils, the degree of weathering, was recognizable as the soil changed, darkening from brown to black with a progressive loss of the original fibrous structure. The weathering and breaking in proceeded from the soil surface downward, suggesting that exposure to oxygen played a role: "Topsoil well broken: black, increasing reddish brown and undecomposed fibers with depth. 9 to 10 feet [to] rock" (Lloyd 1929, p. 78). "Typical sawgrass land. Little decomposition except in top 3 inches. High fiber content; little decomposition" (Lloyd 1929, p. 78). "Similar equipment [for measuring soil loss by oxidation] has been installed in Section 10 where the soil is still under native sawgrass cover and is more fibrous than that of the [already established] water table plots" (Newell 1937, p. 17). By the 1940s soil scientists had described the widespread appearance of a new, more decomposed surficial layer that was black and amorphous. They ascribed its origin to postdrainage oxidation, recognizing that the peat was originally minimally decomposed, and that it had previously been protected from decomposition by the covering layer of water:

> By far the greater portion of the peat soils of the Everglades is composed of the partially decomposed remains of sawgrass. The marshy condition of the Glades, during the period of formation, prevented a more complete decomposition of this material. In its original condition the sawgrass peat is a brown fibrous mass in which partially decayed sawgrass roots can be readily distinguished. These roots are approximately in a vertical position. After drainage, cultivation and weathering gradually transform the top soil into a condition approaching a true muck. The structure then changes into an amorphous mass, the density increases, the color becomes dark, and the rate of seepage through the soil is retarded. (Clayton et al. 1942, p. 10)

Early experimental measurements of seepage rates (soil hydraulic conductivity) further illustrate the oxidative transformation of the top soil. In samples of the original unoxidized brown fibrous peat, the unaltered, undecomposed soil structure provided a vertical conductivity much greater than the horizontal conductivity: "Comparisons of rate of seepage through the brown fibrous peat in a horizontal direction with that in a vertical direction showed the vertical seepage rate to be many times as great as the horizontal, due doubtless to the openings left in the vertical direction as the old sawgrass roots partially decomposed" (Newell 1937, p. 21). In contrast, after cultivation (and by implication, also after weathering and decomposition) the soil lost its original structure (became an "amorphous mass"), and the vertical conductivity understandably became much smaller than in the original undisturbed soil: "Experiments to determine rate of seepage through peat soil indicated that seepage through the top 18 [inches] of cultivated soil was much slower than that through the undisturbed soil at a depth of 18 to 36 [inches] below the surface" (Newell 1937, p. 21).

Observations of Everglades soils made by a committee of agricultural scientists in November 1929 appear to have captured a midpoint in the transition from the original brown fibrous peat to the ultimate black, amorphous state, as indicated by the intermediate description: "dark-brown, semifibrous, [and] slightly decomposed" (Interbureau Committee 1930, p. 14). At the same time, along canals, where both lowering of the water table and the degree of soil oxidation would have been more severe, the transition to the black, nonfibrous peat was already essentially complete:

> Peat is the typical saw-grass material of the Everglades. Perhaps there are one and one-half to two million acres of this material comprising the greater part of the so-called Everglades. It is a remarkably uniform body of material, consisting of brown fibrous to dark-brown, semifibrous, slightly decomposed vegetable matter, dominantly of saw-grass (*Cladium effusum*) composition. It varies in depth from about 4 to about 9 feet and is normally underlain by limestone. . . . In the southern portion of the Everglades, along the Tamiami Trail, are extensive areas of black nonfibrous peat with streaks of brown fibrous peat. This is also true in places along the Miami Canal. (Interbureau Committee 1930, p. 14)

The estimate of 1.5 to 2 million acres (0.6–0.8 million ha) of this peat soil, an extent that would have encompassed all of the upper Everglades (0.7 million acres or 0.3 million ha of Sawgrass Plains; table 4.2) as well as almost the entire Ridge and Slough landscape (1.5 million acres or 0.6 million ha; i.e., much more than area "1d") suggests that the Interbureau Committee, like Baldwin and Hawker (1915) and Allison et al. (1927) before them, felt that the soil underlying these two landscape areas was essentially the same.

Like the Interbureau Committee, soil scientists surveying Dade County in the 1950s described a decompositional transition and emphasized the role of soil oxidation under lowered water tables. The Dade County survey, which was made after the overall Everglades survey (Jones et al. 1948a e), included some of the same surveyors and shows almost identical mapping. The following observation may well reflect knowledge available during the Everglades survey: "Near the canals, where the water table has been lowered by artificial drainage, considerable oxidation of the organic material has taken place. As a result of this increased oxidation, the remaining peat has a higher mineral content, is less fibrous, and is more nearly black" (Gallatin et al. 1958, p. 18).

The modern soil classification system also recognizes the transition from unoxidized, undecomposed peats to well-decomposed ones. Most soils of the Everglades Agricultural Area (corresponding closely to the upper Everglades and the Sawgrass Plains extent shown on plate 5) would originally have been classified as Fibrists. After drainage, aerobic decomposition converted them into black, decomposed Saprists: "Histosols found in Florida are divided into Fibrist, Hemist, and Saprist suborders. Saprists are the most well decomposed and largely correspond to the term muck as defined in the 1951 Soil Survey Manual. Specifically, Saprists have a fiber content of less than one-sixth of the soil volume after rubbing and consist of almost completely decomposed plant remains. Their color is mostly black. Today, most of the soils of the EAA [Everglades Agricultural Area] are Saprists" (Snyder 1994, p. 30).

It is clear that by the 1940s, drainage-related oxidation and decomposition had altered organic soils in large portions of the Everglades relative to their first description in 1915. The upper horizon in the affected areas had developed a new, black, finely fibrous or even amorphous layer. It is also clear that the alterations leading to this new layer would have been most prevalent in the areas of organic soil most severely affected by drainage and decomposition. Likely areas would include those directly cut off from Lake Okeechobee—that is, the upper Everglades (area "1a")—and areas encompassed by drainage canals, such as the central Everglades (area "1d"). The new layer played an important role in the 1940s mapping of soil and landscape extents.

Jones et al. (1948a) mapped both the upper and the central Everglades as a single soil type, the "Everglades peats" (plate 15). More than 1 million acres (405,000 ha) were classified as this newly defined soil type (table 3.2). This mapping of the 1940s soil distribution thus echoes one aspect of the predrainage soil distribution (as estimated from Baldwin and Hawker 1915), namely the presence of only a single soil type along most of the North New River Canal. However, the soil profile (table 3.2) had become quite different from the original profile (table 3.1).

Jones et al. (1948a) considered both the upper and the central Everglades a part of the same sawgrass plains landscape category. They did so despite having at one point labeled area "1d" as "hammock-sawgrass" (fig. 4.9). Curiously, these surveyors never defined this label. Instead, they applied their description of the upper Everglades to the central Everglades as well—"broad, flat sawgrass plains" (table 3.2), mapping the vegetation of both areas as sawgrass (plate 14) and assuming this sawgrass to have been the source material for the Everglades peats (table 3.2).

This landscape configuration, i.e., the central Everglades apparently indistinguishable from the sawgrass plains of the upper Everglades, represents a dramatic change from the predrainage configuration estimated from observations made by Baldwin and Hawker (1915) and Harshberger (1914). The mapping by Jones et al. (1948a) implies that by the 1940s, the Ridge and Slough landscape characteristics documented in the central Everglades twenty-five years earlier—alternating ridges and sloughs, highly directional orientation, elevation differences of 1 to 3 feet (0.3–1 m), and aquatic plants such as white and yellow water lilies (*Nymphaea odorata* and *Nuphar advena*, respectively)—had all disappeared, presumably under the same influence of the strong drainage that had altered the soil profile.

Such a disappearance of the surficial aspects of a peat wetland is plausible, particularly under the drastic degree of drainage applied to the Everglades. The central Everglades (area "1d") would have been drained not only by the Miami and North New River canals that encompassed this area but also by the elimination of upstream inflows due to below-ground water levels in the upper Everglades (area "1a"). As water levels in the central Everglades were drawn below ground surface, water lilies and other aquatic plants would, of course, disappear, and sawgrass could be expected to spread into former sloughs. With peat oxidation, the elevation differences between ridges and sloughs would likely have diminished. The bottom panel of figure 3.17 illustrates these altered landscape and soil conditions, as seen by Jones et al. in the 1940s. For comparison, the middle panel illustrates the analogous near-predrainage situation (1915), as discussed earlier in this section. Mileposts 3 to approximately 30 correspond to the upper Everglades for this discussion; mileposts 30 to 54 correspond to the central Everglades.

Loss of the surficial aspects of the original ridge-and-slough landscape in the central Everglades (fig. 3.17, bottom) would have presented Jones et al. (1948a, e) with a mapping dilemma. Criteria used in soil mapping traditionally include both the soil appearance at the time of mapping and the parent material from which the soil was derived. Table 3.2 indicates that Jones et al. (1948e) were well aware of the ridge-and-slough landscape category, distinguishing it from the sawgrass plains landscape, albeit perhaps not as clearly as did Baldwin and Hawker (1915). They also distinguished a "Loxahatchee peat" soil, noting that it was derived, "for the most part," from the more aquatic species originally present in the sloughs of the ridge and slough landscape (table 3.2). Jones et al. (1948e) did not emphasize the contribution of sawgrass to this peat but presumably were aware of it, since sawgrass ridges covered approximately half the landscape (chapter 8).

Significantly, the description of the Loxahatchee peat as a "soft, felty, spongy brown fibrous peat" (table 3.2) is not strikingly different from the early (and presumably predrainage-like) descriptions of the peat soil associated with the sawgrass plains: "light, loose, and soft";

Table 3.2. The Everglades and Loxahatchee peat soils as defined in the 1940s

Source	Everglades Peat	Loxahatchee Peat
Landscapes associated with the soils (Jones et al. 1948e)	"The sawgrass plains include all the territory bounded by the Miami Canal, West Palm Beach Canal, and the coastal ridge except the Hillsboro Marsh section of ridge-and-slough land. The dominant native vegetation is sawgrass, a sedge which attains a height of 8 feet or more on the deep peat" (pp. 46–47).	"The ridge-and-slough area occurs in 2 sections—the Hillsboro Marsh section, mostly north of the Hillsboro Canal and immediately west of the coastal ridge, and a larger section west of Miami Canal and the rock rim in a body more than 100 miles long and from 5 to 20 miles wide, shaped somewhat like a flattened 'V' pointing eastward toward Miami. Each of these [two] sections contains thousands of small oval islands interspersed with sloughs and small ponds or lakes. The sloughs are filled with water most of the time and contain aquatic plants, especially the bladderworts [*Utricularia* spp.], coontail moss [possibly *Ceratophyllum demersum* L., although this species has not been reported by other historical sources], spider lily [*Hymenocallis* ssp.], bonnets [*Nuphar advena*], and several water or slough grasses that bend in the current" (pp. 47–48).
Extents (Figure 4.4) (Jones et al. 1948e, p. 14)	"1a, sawgrass plains" *plus* "1d, hammock-sawgrass"	"2, ridge-and-slough" (two separate areas)
Extents mapped (Jones et al. 1948e, p. 62)	1,050,000 acres (ha)	731,000 (ha)
Overall description (Evans and Allison 1942)	"*Everglades Peat*, which occupies the broad, flat sawgrass plains, is the most extensive soil type in the Everglades. The top six to eighteen inches usually is a black, finely fibrous organic material containing from 8 to 15 per cent mineral matter. This is underlain by brown fibrous material resting at varying depths on either rock or sand" (p. 44).	"*Loxahatchee Peat* is a very soft, felty, brown fibrous material, spongy in character, that has been formed, for the most part, from more succulent plants than the sawgrass that has given rise to the Everglades peats. Occasionally, lense of black, finely fibrous material occur within the profile, but, for the most part, this soil is a mass of raw, brown fiber from the surface down. Loxahatchee peat is usually found in more inaccessible sections of the Everglades" (pp. 44–45).
Source material (Jones et al. 1948e)	"It has been formed from sawgrass material" (p. 71).	"Loxahatchee peat is . . . composed of remains of [water] lilies, water grasses, and other water plants. . . . Ordinarily it is covered with water most of the year" (p. 72).
Surface soil (Jones et al. 1948e, p. 62)	"Black, finely fibrous, well-decomposed organic material, 6–18 inches."	"Black, finely fibrous, spongy peat, 1 or 2 inches."
Subsoil (Jones et al. 1948e, p. 62)	"Light, felty, fibrous brown peat."	"Soft, felty, spongy brown fibrous peat."
Shrinkage (Evans and Allison 1942)	"Shrinkage of this soil [upon drying] . . . is commonly found to be three or four volumes to one" (p. 38).	"Loxahatchee Peat, being light and fluffy, does not lend itself readily to diking, since it loses more than three-fourths of its volume upon drying" (p. 45).
Suitability for agriculture (Evans and Allison 1942)	"This type is suitable for cultivation . . . provided the depth to underlying limestone is sufficient" (p. 44).	"It is apparent that the land represented by this type should not be considered for agricultural purposes until all others in the glades are exhausted." (p. 44–45)
Suitability for agriculture (Jones et al. 1948e)	"If the [Everglades] peat is more than 5 feet deep, or is shallower but underlain by marl or sand, cultivation is possible and the land-capability class is III" (pp. 71–72).	"drainage [of the Loxahatchee peat] is not recommended because of the extreme shrinkage and settling that will occur. It is classified as class VIII land, not suitable for any cultivation" (p. 73).

"coarsely fibrous"; or "loose, fibrous" brown fibrous peat (table 3.1). The subsoils in particular seem to be nearly identical. The distinctions drawn between the Everglades and Loxahatchee peats by the 1940s surveyors appear to be twofold: (1) the field appearance; that is, the thick, black, less fibrous surficial layer in the Everglades peat; and (2) the parent plant material, essentially all sawgrass for Everglades peat and primarily non-sawgrass, aquatic species for Loxahatchee peat.

In mapping the upper Everglades (area "1a"), there was no conflict between these two mapping criteria. The upper Everglades was by all accounts dominated by sawgrass both before and after drainage, so the underlying peat soil would have been derived from sawgrass. By the 1940s drainage had affected essentially all of the upper Everglades, so the drainage-induced surficial black layer was likely widespread. Together, the sawgrass derivation and the surficial black layer lead directly to the designation "Everglades peat" with an overlying Sawgrass Plains landscape.

The central Everglades (area "1d") presented a more challenging situation. By the 1940s drainage apparently had produced a surficial black layer throughout this area as well (fig. 3.17, bottom), suggesting it should be classified as Everglades peats. However, regardless of the appearance of the landscape in the 1940s, according to the earlier accounts from 1914 and 1915, the central Everglades was originally part of the Ridge and Slough landscape. The many sloughs therein contained exactly the water lilies and other aquatic species that were considered indicative of Loxahatchee peats (fig. 3.17, middle). Thus, depending on whether the soil classifier chose to emphasize the presence in the soil of aquatic plant remains or the overall appearance of the soil, the peat soils of the central Everglades could legitimately have been classified either as Loxahatchee peat or as Everglades peat. Similarly, depending on emphasis on the original or then present appearance, the landscape apparently could have been classified either as sawgrass plains or as ridge-and-slough landscape.

Jones et al. (1948a) chose to map the central Everglades as Everglades peats and as sawgrass plains. To understand this choice, it is useful to consider the objectives of the time. First and foremost, their goal was practical. These surveyors were not charged with studying historical ecology, so mapping the natural, predrainage boundaries of Everglades landscapes as they might have looked more than sixty years earlier did not concern them. Instead, this soil conservation survey sought to "determine the various types of land in the Everglades Drainage District and their capabilities" (Jones et al. 1948e, p. 7). This practical focus was emphasized in the introductory remarks to their study: "The results of the research are presented herein only as they affect the use capabilities of the various land types and the water-control measures discussed" (Jones et al. 1948e, p. 7). These surveyors sought ways to identify soils that would be suitable for agriculture versus those that would not: "The soil-conservation surveyors . . . located and marked on maps the boundaries of the soil conditions that influence the use and management of the land" (Jones et al. 1948e, pp. 55–56).

As previously noted in the Impoundment Phase section of chapter 2, one of the practical concerns for the soil surveyors was whether the area being mapped had sufficient soil thickness for installation of drainage ditches. They were therefore mindful not only of the soil thickness at the time of mapping but especially of the expected future thickness after loss through dewatering, compaction, and the initial subsidence. This concern appears to have been felt most acutely for the less drained, more frequently water-covered portions of the ridge-and-slough areas, where it was feared that the soft fibrous peat would be subject to severe subsidence: "Loxahatchee peat [is] light and fluffy [and] loses more than three-fourths of its volume upon drying" (table 3.2). The discerning reader, however, will note from table 3.2 that the surveyors' fears regarding subsidence should have been equally strong for the Everglades peat: "Shrinkage of this soil [upon drying] . . . is commonly found to be three or four volumes to one."

The level of concern was in fact opposite. The Everglades peat was categorized as "suitable for agriculture" and the Loxahatchee peat as "not suitable for any cultivation" (table 3.2). This striking difference does not appear to have been truly based on any inherent physical differences between the two soils, nor on differing parent material. Instead, it appears much more likely to have been based simply on the degree of drainage and the resulting degree of changes to the soil, regardless of parent material. The Everglades peat soil of the upper Everglades, despite being subject to the same three-fourths shrinkage if fully dehydrated as the Loxahatchee peats, had proven itself excellent for agriculture. Subsidence and shrinkage of Everglades peats was not a future unknown but instead a previously, and successfully, overcome hurdle. In contrast, in areas of the Ridge and Slough landscape where water still remained on the surface for much of the year, the soil was "light and fluffy" and the degree of future subsidence a risky unknown. From a practical agricultural point of view, the new, postdrainage, well-decomposed black surface soil layer, when combined with greater than 5 feet (1.5 m) of remaining peat thickness, would logically have functioned as a de facto "seal of approval," indicating that surface water had already been largely removed, the soil had weathered, and sufficient thickness still remained. In short, this soil was an excellent predictor of agricultural soils with high potential.

For practical men of agriculture, not ecological historians of past soils, the "dilemma" mentioned—whether to map based on parent material instead of the likelihood of providing sufficient thickness for agricultural use—would probably not have given them pause. In their position it would have been irresponsible to map the large, drained central Everglades (area "1d") as Loxahatchee peat, and hence "not suitable for any cultivation," when it had already formed a surficial black layer and looked just as promising as the known soils of the upper Everglades. They had little choice but to map it as Everglades peat despite the area originally having been part of the Ridge and Slough landscape, as noted by Harshberger (1914) and Baldwin and Hawker (1915).

The mapping of the central Everglades as Everglades peat was only one of several examples of the drainage-induced, black, less fibrous surface layer being diagnostic. This is hinted at by the Dade County soil surveyors, several of whom also participated in the Jones et al. (1948a, e) effort: "[Everglades peat] is closely associated with the Loxahatchee peats but differs from them chiefly in having a black or very dark brown nonfibrous peat surface layer" (Gallatin et al. 1958, p. 18). Several other areas of unambiguous Ridge and Slough origin were also mapped as Everglades peat rather than Loxahatchee peat. Jones et al. (1948a) mapped a narrow strip of Everglades peat and sawgrass vegetation along the Hillsboro Canal in Township 47 Range 40 (plates 14, 15). At the same time, slightly farther away from the canal, and for large expanses to the north and south, the soils were mapped as Loxahatchee peat (plate 11) and the landscape as Ridge and Slough (plate 14). The narrow shape of this strip, exactly following the canal and canal bend, leave little doubt that prior to drainage by the canal, the entire area was Ridge and Slough landscape and Loxahatchee peat. Following the same process noted by the Interbureau Committee (1930, p. 14) and by Gallatin et al. (1958, p. 18), the Hillsboro Canal lowered the water table sufficiently to cause oxidation and apparently also formation of the diagnostic black surface horizon along this strip.

In a second example, the surveyors themselves make it clear that physical appearance—that is, the presence of the black, more decomposed, and less fibrous layer—took precedence in their mapping over the parent vegetation and landscape type. This example is a former "arm" of the Ridge and Slough landscape that once extended from the northeastern portion of the present Arthur R. Marshall Loxahatchee National Wildlife Refuge northwest almost to the shore of Lake Okeechobee. As Ridge and Slough landscape, this arm, parallel to what is now the L-8 canal (fig. I.1), was traversed twice in 1842 (plate 10 left; appendix G), with the expedition contrasting the last mile (1.6 km) or so of solid sawgrass with the much more passable sloughs of most of the arm. Figure 2 in Dachnowski-Stokes (1930) shows this slough in a peat profile cross section as a rounded, 2-foot (0.6 m) depression extending from West Palm Beach Canal to the edge of the Everglades (L-8 canal). The same slough is also visible as a similar, though wider, depression in a 1912 peat profile given by Stephens and Johnson (1951a, p. 234; see fig. 3.6, transect A, miles 35–39). Although there is no question that this arm or "old slough" was originally part of the Ridge and Slough landscape with Loxahatchee peat soil, by the 1940s it had been sufficiently altered by drainage that it was mapped as Everglades peats (plate 15). Curiously, the 1940s surveyors still considered the vegetation more similar to the Ridge and Slough landscape (plate 14): "Northeast of the West Palm Beach Canal from Lake Okeechobee to Twenty-Mile Bend is an old slough in which the peat was originally formed partly from water plants and was more loose than the Everglades peat. However, as a result of partial drainage for many years, it has become compacted and can scarcely be distinguished from the typical Everglades peat" (Jones et al. 1948e, p. 72).

It is clear that the mapping in the 1940s of the central Everglades (area "1d," fig. 4.9) as Sawgrass Plains landscape and as Everglades peat soil had little relation to the soils and landscape originally present in this area prior to drainage. Instead, it was a practical reflection of the drainage-induced changes that had already taken place and of the area's agricultural potential. It is also clear that the associations of Everglades peat, as defined by Jones et al. (1948e, pp. 71–72), with the Sawgrass Plains landscape, and of Loxahatchee peat with the Ridge and Slough landscape were only approximate, with drainage-induced changes potentially overriding parent material. The far-reaching division of the ridge-and-slough regions into two distinct areas separated by a swath of supposed sawgrass plains is not an accurate delineation of the predrainage Sawgrass Plains and Ridge and Slough landscapes. Instead, it is a depiction of the postdrainage situation present in the 1940s, primarily based on agricultural criteria.

Ridge and Slough Soils to Wet Prairie Soils

(Loxahatchee Peats to Davie Sands)

Drainage induced a different type of soil change along the eastern and western borders of the Everglades, north of Tamiami Trail (Township 54). A sloping sand layer, thicker toward the edges and outside the Everglades, thinning toward the interior, overlies the edge of the bedrock basin in these areas. Prior to canal drainage a layer of organic soil, thinner toward the edges and thicker toward the interior, covered the sand layer (figs. 3.6, 3.7). Predrainage water levels were sufficiently high and hydroperiods sufficiently long to ensure that the rate of peat accumulation exceeded or balanced the rate of peat decomposition, maintaining the surficial peat layer.

With the advent of drainage, both regional and local, average water levels dropped and hydroperiods shortened. Peats oxidized and, as noted in the earlier section on peat subsidence, were mostly or entirely lost along the eastern border of the Everglades by the 1940s. Two new soil types were defined to recognize the areas of newly exposed bare or nearly bare sand: "Davie mucky fine sand" and "Davie fine sand." The subsoils of the two soils were the same, so the soils were distinguished primarily by the amount of organic matter remaining in the surface soil (uppermost soil horizon). The surface soil of the Davie mucky fine sand was defined as a "dark gray mixture of well oxidized organic matter and fine sand" with medium to high organic matter content (Jones et al. 1948e, p. 66). The surface soil of the Davie fine sand was a "gray or light gray fine sand" with low to medium organic matter content (Jones et al. 1948e, p. 65).

The origin of the Davie soils as a drainage-induced phenomenon was explicitly recognized at the time. An article describing progress on development of the Jones et al. (1948a) map noted: "The Davie fine sand was originally covered with from one to three feet of peat. However, this peat has shrunk or has been oxidized or burned until only a shallow covering, if any, is left. This has been or will be incorporated into the top six to ten inches of the underlying sand" (Evans and Allison 1942, p. 45). A later "detailed reconnaissance" soil survey of Dade County published in the 1950s was equally explicit about the origin of the Davie soils:

Davie fine sands (generally closest to the bordering, non-Everglades higher ground, e.g., areas of Arzell fine sand):

> Davie fine sand has poor to very poor natural drainage, but large areas are now overdrained. . . . This soil occurs in fairly large areas in the sandy prairies that lie between Arzell fine sand and Everglades peat in the areas northwest, west and southwest of Miami. It was originally covered with a thin layer of peat or muck. Most of the organic matter in the surface soil has been destroyed by excessive drainage, natural oxidation and shrinkage, and fires. The sand is now at the surface. . . . The natural vegetation was mainly sawgrass, but following drainage of adjacent areas, myrtle, groundsel, primrose-willow, and wild grasses have spread over most of the area not cultivated. (Gallatin et al. 1958, p. 16)

Davie mucky fine sands (generally closest to the interior of the Everglades and to the remaining organic soils):

> This soil [the Davie mucky fine sand] occurs northwest and west of Miami on the edges of the sandy prairies that border the Everglades peats. The shallow peat layer of this soil has not yet been completely destroyed by fire or by slow oxidation following drainage. The soil is poorly to very poorly drained. It is closely associated with the Davie fine sand and differs from that soil mainly in having a thin layer of peat or mucky material over the sand layers. The vegetation is similar to that described for Davie fine sand. (Gallatin et al. 1958, p. 17)

The Davie sands are a clear example of drainage-induced change in the Everglades. By the 1940s the formation of the Davie sands had created a postdrainage sequence of four soils ordered by degree of organic matter content. In the direction from center to outside of the Everglades, and from very high to very low organic matter, the order generally was: Everglades peat, Davie mucky fine sand, Davie fine sand, and wet pineland sands such as the Arzell fine sand (e.g., plate 15, Townships 44–47 Range 34; Township 43 Range 40; Townships 48–53 Range 40 or 41). This ordered sequence (a "catena" in soil science) might logically suggest that the Everglades originally included, on the east and on the west sides, a broad transitional area distinguished by intermediate elevation, intermediate organic matter content, and intermediate hydrology. Ecologically, such a broad transitional zone, if present, would likely have been an important wading bird foraging area, with water depths and timing different from those present within the interior Everglades.

However, careful examination of the predrainage record indicates that such an intermediate, transitional area was not originally present. Instead, as discussed in more detail in chapter 4, the organic soils of the predrainage

Everglades and their associated vegetation patterns extended directly to the coastal ridge on the east and to the Western Flatwoods or cypress on the west, forming the sharp border shown on maps such as Ives (1856b; fig. 1.6) and Harshberger (1913; fig. 1.7); see also Steinberg (1980; plate 42). A soil catena does appear to have been present before drainage along the eastern and western borders of the Everglades basin, but the distinguishing characteristic was the subsoil, not the surficial organic material. From inside to outside of the Everglades, the predrainage sequence along the borders north of Township 54 appears to have been: peat, peat over shallow sand, peat over deep sand, and finally, outside the Everglades, wet sands such as Arzell fine sand (see for instance plate 11, the southeast corner of Township 49 Range 40). The increasing thickness of the underlying sand and decreasing thickness of the overlying peat did not prevent the peat from extending directly up to the edge of the Everglades where it gave way to pine flatwoods underlain by wet sands. Within the Everglades, the varying thickness of the sandy subsoil was likely not apparent, being masked by the overlying peat.

The cross section in figure 3.14 illustrates how the progressive loss of the surficial organic soil transformed the original relatively simple, predrainage, organic/nonorganic configuration into a postdrainage catena of intermediate soil types. The key feature was the shift from flat to sloped surface, as the flat peat (organic soil) was oxidized away, exposing a sloped sand layer as the new Everglades land surface. This newly exposed surface was in turn the origin of the spatial catena of successively drier soil types. Ecologically, the postdrainage catena and associated vegetation became the origin of the incorrect impression that the predrainage Everglades, Township 54 (north of Tamiami Trail), included peripheral, short hydroperiod wetlands.

A very similar pattern of progressive exposure of formerly peat-covered, sloped sands occurred in miniature within the Peat Transverse Glades (for locations see plate 5 and for detail, plates 41 and 43). Prior to drainage, outflows from the main Everglades apparently kept the peats in these transverse glades covered with water and protected them from oxidation. Predrainage township survey plats and field notes suggest that as a result, the peats filled the full width of these glades (fig. 3.18). With canal drainage of the Everglades, hydroperiods within the Peat Transverse Glades shortened and the peat began to disappear. Over time, strips of peat-free sand appeared along the edges of the transverse glades (fig. 3.18). As along the edges of the Everglades basin, these newly exposed sands were mapped in 1940 as Davie fine sands and, as such, likely also contributed to the inaccurate impression of peripheral, short hydroperiod wetlands having originally been present.

Both in the Peat Transverse Glades and in the overall Everglades basin, the Davie sands formed in strips approximately parallel to the borders of the glade or basin. An important exception to this configuration occurred in Township 50 Ranges 40 and 41, near Pine Island and west of Long Key (also called Sam Jones' Seven Islands; see plates 41, 42, and labels on 43). Here plate 15 shows a peculiar intrusion of Davie mucky fine sand extending almost perpendicularly 3 to 4 miles (5–6 km) into the Everglades. This intruding "tongue" of primarily sandy soil is conspicuously absent on all maps prior to 1939, including soil, vegetation, and township maps (U.S. Bureau of Topographical Engineers 1846, 1856; Florida Surveyor General's Office 1853; Ives 1856b; Fries 1898-T50 R40, 1898-T50 R41; Newman 1908-T50 R41; Ensey and Elliot 1911; Matson et al. 1913a; Harshberger 1913; Hassan 1915; Baldwin and Hawker 1915; Ashworth 1919b, c; Harper 1925; Marston et al. 1927; Board of Commissioners of the Everglades Drainage District 1935; Henderson 1939). The absence of this intrusion of sandy soil on the early maps is consistent with predrainage reports of this area having been part of the much wetter and organic soil-based Ridge and Slough landscape. Maps from about 1898 show multiple "Boat Landings" on Pine Island and Long Key (e.g., Fries 1898-T50 R40; Fries 1898-T50 R41), while other accounts describe boat travel from the New River out to these islands (Cooley 1851, pp. 42–43; Henshall 1883, pp. 52–53; appendices F and L). By the 1940s not only was the water around these islands long gone, but the organic soils were gone as well, leaving only the dry intrusion of Davie mucky fine sand. By 1932 the intrusion area included numerous citrus groves (Mitchell 1932), confirming the absence of waters of the Everglades and tending to confirm the absence of organic soils.

The intrusion first appears on the maps of J. H. Davis (1943b) and Jones et al. (1948a), and persisted on several later maps (e.g., Florida Agricultural Experiment Stations and Soil Conservation Service 1962; J. H. Davis 1967). It provides a stark example of the shrinking of the Everglades that occurred between 1880 and the 1940s, both vertically and horizontally. As overall water levels were lowered, and peat oxidized away to expose the underlying sloped sands, the edges of the Everglades were drawn toward the center. The extreme extent of this particular drawing into the center in the vicinity of Pine Island and Long Key likely reflects the geography and

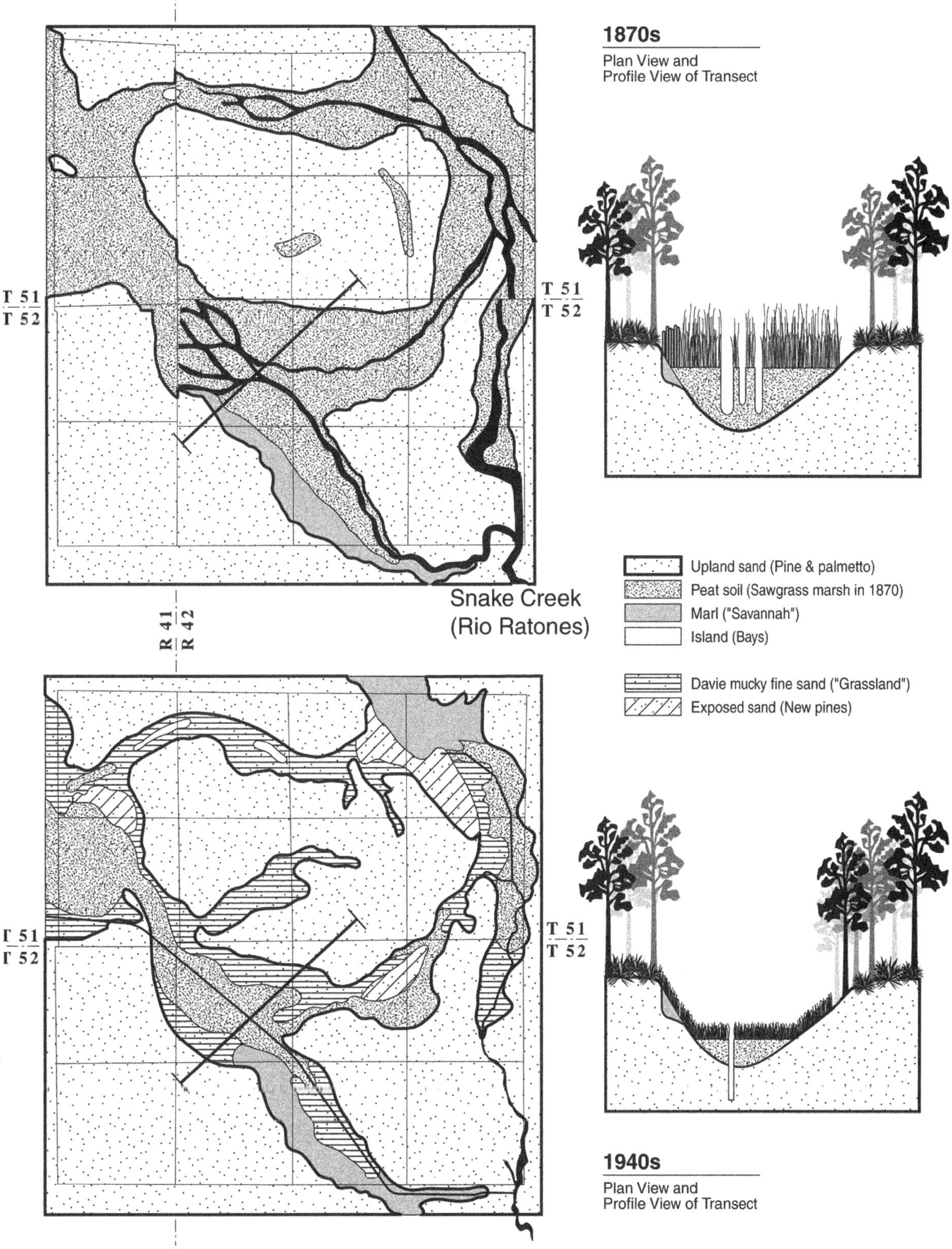

Figure 3.18. Postdrainage peat subsidence in the Peat Transverse Glades, Snake Creek area, 1870s to 1940s. Lowering of water levels in the Everglades reduced surface outflows through the transverse glades, diminishing creeks, eliminating the canoe route, and increasing oxidation of sawgrass peats. By 1940 peat remained only in the deeper centers of the transverse glades, and shorter hydroperiod grasses had replaced the sawgrass. Sources: For 1870s, Williams (1870-T51 R42; 1870-T52 R41; 1870-T52 R42); Newman 1908-T51 R41; for 1940s, Jones et al. (1948a).

early timing of the surrounding drainage. The North New River Canal just north of the intrusion likely reduced or eliminated the overland flow that swept southeastward through this area (plate 40) while the South New River canal drew off additional water. One of the first pump stations in the Everglades was placed here in 1913 by the Davie District (Stephens and Johnson 1951a, p. 211), and as shown on a development plat (Newman 1908-T50 R41), many secondary canals were installed even earlier.

Together, the various areas of Davie sands that had formed by the 1940s constitute a substantial area of postdrainage, apparently peripheral landscape. Their anthropogenic origin serves as a reminder that predrainage Everglades conditions and geography cannot be read directly from 1940s observations.

Vegetation Changes

With the onset of drainage, a number of observers noted widespread vegetation changes. These changes were generally thinning or weakening of the original predrainage communities, followed or accompanied by invasions of new species. Although some non-native species were recorded, most of the invading species had previously been present within the Everglades, either at higher elevations (drier conditions) or appearing ephemerally during drier years. Four main types of changes emerge from the historical observations: thinning of sawgrass stands, invasions of sawgrass areas by woody species, invasions of sloughs by shallower water species, and along the east and west margins of the Everglades north of about Township 54, outright replacement of areas that had originally been Ridge and Slough landscape by so-called peripheral wet prairies (see also previous section and chapter 4). The historical record regarding vegetation changes on tree islands of the Ridge and Slough landscape is lacking, so it is unclear to what extent they were affected by the initial phases of drainage. Later, as drainage further lowered water levels, soils on some tree islands became dry enough to burn (Craighead 1971, pp. 20, 126; Egler 1952, p. 228; Alexander and Crook 1974, p. 64), reducing the elevation of these (former) tree islands. Vegetation changes in the Custard Apple Swamp are not discussed here because that landscape was so quickly and completely cleared for agriculture (Will 1964, p. 36; J. H. Davis 1943a, p. 261).

Vegetation changes in the various landscapes throughout the Everglades are not all equally well documented. The earliest and most numerous observations were from the northern portions of the Sawgrass Plains landscape, closest to Lake Okeechobee, probably because these were among the first areas to be affected by drainage and because of the strong agricultural interest in these areas. Changes in the Ridge and Slough landscape were less documented, likely due to a combination of lower agricultural interest and difficult access. Documentation increased in the 1940s and 1950s as access increased with the availability of airboats. There seems to be little direct information available to document early changes in the Marl Marshes, but the excellent studies by Egler (1952) and Robertson (1953) provided some insight into previous conditions.

With drainage, sawgrass was observed to "become naturally weakened as the water recedes" (Allison et al. 1927, p. 39), allowing herbaceous and woody species to invade (Allison and Dachnowski-Stokes 1932, p. 228; J. H. Davis 1946, p. 89; Jones et al. 1948e, p. 47). The earliest reports of this were from the Sawgrass Plains, but slightly later observations suggest that as water depths decreased in the Ridge and Slough landscape, the sawgrass ridges also followed similar drainage-induced trajectories (chapter 8; appendix R). The herbs known as careless weed (*Amaranthus australis*) and dog fennel (*Eupatorium capillifolium*) were frequently noted invaders of the Sawgrass Plains (careless weed: *Miami Metropolis* 1907, p. 10; Wright 1912, p. 26; Small 1918, p. 695; Allison and Dachnowski-Stokes 1932, p. 228; J. H. Davis 1946, p. 89; dog fennel: Baldwin and Hawker 1915, p. 722; Harper 1927, p. 125; Allison et al. 1927, p. 39; J. H. Davis 1946, p. 89). The southward expansion (compare plates 5 and 13; see also chapter 6) of an initially narrow band of willow (*Salix caroliniana*) and elder[berry] (*Sambucus canadensis*) along the southern border of the Custard Apple Swamp landscape was a special case of woody invasion of the Sawgrass Plains (Baldwin and Hawker 1915, p. 722; Allison and Dachnowski-Stokes 1932, p. 228; see also chapters 6, Sawgrass Plains Landscape, and 7, Custard Apple Swamp Landscape).

As drainage continued, large portions of the Sawgrass Plains, as well as ridges of the Ridge and Slough landscape, were colonized by woody species. In the Sawgrass Plains landscape this seems to have occurred mostly as scattered clumps of wax myrtle (*Myrica cerifera*; Simpson 1920, p. 126). In the Ridge and Slough landscape, generally whole ridges were colonized by wax myrtle and also by willow (*Salix caroliniana*), dahoon holly (*Ilex cassine*), and/or bay (*Persea borbonia*; Simpson 1920, p. 126; Jones et al. 1947e, p. 47; Johnson 1958, pp. ii, 29). The drainage-induced process of shrub and small tree domination of sawgrass ridges was noticed early (Baldwin and Hawker 1915, p. 753) and eventually became quite widespread (Johnson 1958, pp. ii, 29). This colonization of former sawgrass ridges by trees may be what Sam Huff, a Mikasuki Seminole born about 1872 (Sturtevant 1956, p. 55–57) saw firsthand during the early 1910s:

> "Just as soon as [the dredges] hit the lake, the water is going to dry up in those Everglades, and as soon as the water dries up, they're going to start plantations"—that's what the white people said to the Indians. . . . But I didn't believe it, until they hit [Lake] Okeechobee. Then the water dried up, and even in Okeechobee it was dry too. The Everglades became small, and the trees grew very fast. (Sturtevant 1956, p. 65).

The resulting wooded ridges resembled tree islands, and should perhaps be referred to as "postdrainage pseudo tree islands." As such, they have the potential to complicate analysis of tree islands through confusion with "true" Everglades tree islands (i.e., those with a characteristic shape and thought to be hundreds to thousands of years old; Stone et al. 2002, p. 108). The early onset of the postdrainage invasion of sawgrass ridges by woody species (Baldwin and Hawker 1915, p. 753) might, for example, mean that detection of wooded areas on 1940s aerial photographs (USDA-SCS 1940) would indiscriminately identify both pseudo and true tree islands. The presence or absence of Gandy peats (i.e., tree island peats) mapped by Jones et al. (1948a, e) may be a useful tool for distinguishing between them.

Documentation of early observations of vegetation changes in the sloughs of the Ridge and Slough landscape is less common than documentation of changes in the Sawgrass Plains landscape. However, changes in sloughs were probably no less widespread than the better-documented vegetation changes in the Sawgrass Plains. A combination of factors is likely responsible for the scarcity of slough change observations: a possible lag in degree of drainage, with somewhat later drainage in the Ridge and Slough landscape; the subtlety of some of the changes—that is, increases in wet prairie species such as spikerush (*Eleocharis cellulosa*) and maidencane (*Panicum hemitomon*), which are not as strikingly obvious; the absence, for many observers, of clearly recognized, undrained reference sloughs for comparison; and finally the difficulty of access to the Ridge and Slough landscape after drainage began and before availability of airboats (Wallace 1955a, p. 2). Baldwin and Hawker were a notable exception to this. Their fieldwork brought them into direct contact with 100 square miles (260 square km) of Ridge and Slough landscape, which provided them a broad perspective, allowing them to compare, essentially simultaneously, at least two gradients in degree of drainage: longitudinally, with areas close to the ends of the North New River Canal (strongly drained) versus the middle canal reaches (weakly drained; see fig. 3.1), and laterally, with drainage effects decreasing with distance from the canal.

Historical documentation of vegetation changes in the Marl Marsh landscapes (plate 5) is also limited, but as in the case of sloughs, likely should not be interpreted as an absence of change. In the Marl Marshes the lack of well-recognized, undrained reference locations against which observers might compare other drainage-affected areas is even more severe. The slopes inherent in those landscapes in the cross-flow, transverse direction further complicate this, as it likely meant that the landscapes originally included a range of vegetation communities (e.g., Egler 1952, p. 229). Initial drainage likely shifted the communities along this elevational and hydrological gradient. Over time other species likely invaded, but as in the sloughs, the difficulties of access and the lack of unambiguous reference conditions apparently made it difficult to document changes in the Ochopee, Rockland, and Perrine marl marshes. That said, the displacement of sawgrass by woody species noted by Robertson (1953, p. 80) suggests a general drying as a result of drainage, and this in turn suggests that herbaceous shifts almost certainly occurred. Appendix T, comparing an 1880s land survey transect in the Perrine Marl Marsh with a 1950s transect, confirms that changes occurred but without providing detail regarding species or timing.

Firsthand Observations

The following quotations from firsthand observers, arranged chronologically, are representative of the range of postdrainage vegetation changes observed in the Everglades. The earliest, after only two years of lowered water levels in Lake Okeechobee and surrounding sawgrass marshes, are unfortunately not detailed but are still of interest for their earliness and for comparison with later, better-documented changes to the Sawgrass Plains:

> [The Okeechobee Marsh was] formerly covered with saw-grass, but [after two years of drainage it is] now partially clothed in new vegetation. (Duval 1884, p. 317)
>
> The extensive saw grasses, the dread of Government Surveyors along the [Kissimmee] valley and . . . the

> Northern borders of Lake Okeechobee are entirely disappearing from the prairies, and a different growth of succulent weeds, and prairie grasses, is rapidly taking its place. . . . The Western Shore of Lake Okeechobee [shows] the same results . . . prairie as far as the eye can reach is putting on a new growth of plants and grasses and grasses which do not grow in water. (Dancy 1884, pp. 333–334)

The drainage engineer James Wright was one of many early observers (e.g., *Miami Metropolis* 1907, p. 10; Small 1918, p. 695; Allison and Dachnowski-Stokes 1932, p. 228) to note the dramatic growth of careless weed, especially on any elevated spots but also within flatter, former sawgrass areas: "When the [sawgrass] land is once broken and this saw grass destroyed, it does not reproduce itself, but the ground is soon covered with a growth of coarse weeds, principally with what is called 'Careless Weed' [*Amaranthus australis*]. The writer has seen careless weeds along the canal banks, twenty inches in circumference at the ground and sixteen feet high—the growth of one season" (Wright 1912, p. 26).

The observations of the soil surveyors Mark Baldwin and Herman Hawker, made in 1915 after the Lake Okeechobee phase of drainage but only a few years into the muck canal phase, are particularly useful because of their known and extensive field exposure. Their description of sawgrass and *Sagittaria*, both native species, invading sloughs with drainage-reduced water depths is especially intriguing. The comment that these sloughs, when deeper, supported water lilies (*Nymphaea* and perhaps *Nuphar*) is consistent with numerous predrainage slough descriptions (chapter 8). "Edges of the sloughs" presumably refers to edges with the bordering sawgrass ridges:

> The drainage of the Everglades has proceeded sufficiently to induce noticeable changes in the character of the vegetation in certain places. In the interior of the glades, along edges of the sloughs which once supported a luxuriant growth of water lilies the lowering of the water table is accompanied by the invasion of saw grass and *Sagittaria* on the lower ground. Two or three miles south of the lake willows are gradually encroaching upon ground which under former conditions of poorer drainage supported a heavy growth of saw grass. In the "lower glades" saw grass is giving way to myrtle, maiden cane, and fennel. (Baldwin and Hawker 1915, pp. 752–753)

Given that Baldwin and Hawker's survey covered the length of the North New River Canal, "'lower glades' saw grass" here most likely refers to the sawgrass ridges of the Ridge and Slough landscape. This description of sawgrass ridges being invaded by wax myrtle parallels the shrubby invasions of the Sawgrass Plains. It also suggests that the drainage-induced process of sawgrass ridges being invaded by shrubs and small trees, and eventually coming to resemble tree islands, began at least as early as 1915.

Observations by the botanist John Kunkel Small of new species invading former areas of the Sawgrass Plains suggest that the changes there were rapid: "We collected on the prairie-like regions west of Moorehaven and also in the open places in and about the settlement, which probably a year or two before had been covered with saw grass. . . . One native weedlike plant was conspicuous, a pigweed, botanically known as *Amaranthus* and popularly as 'careless [weed]'" (Small 1918, p. 695).

The naturalist Charles T. Simpson lived near the Little River, close to Miami, so his observations were less tied to the progress of agriculture and the Sawgrass Plains and more focused on the Ridge and Slough landscape bordering the Atlantic Coastal Ridge. Simpson notes that even though the drainage works were not yet complete, the vegetation was already changing in this landscape. Drainage may have progressed further than he was aware by 1920, given the effects of reduced inflows from Lake Okeechobee (starting in the 1880s) and the effectiveness of the early muck canals: "Although only the preliminary work of drainage has been done yet it has had a marked effect on the vegetation. Along the banks of the canals and on all slightly elevated spots a variety of trees and shrubs are springing up, so that where formerly the eye swept over a monotonous even expanse of saw grass, the view now presents patches of incipient forests. . . . One of the results of partial drainage is that along this same eastern border numerous low, timbered 'islands,' which were formerly quite wet, have now been changed to dry land" (Simpson 1920, p. 126). Simpson's description of the highest points, the tree islands, becoming drier, is consistent with the next highest points, the sawgrass ridges, also becoming drier, in this case dry enough to support the "springing up" of trees and shrubs on the former ridges. The invasion of sawgrass ridges by shrubs and trees—that is, the creation of incipient drainage-induced tree islands—echoes an observation by Baldwin and Hawker made five years earlier.

The agricultural scientist Robert V. Allison and soil

scientist Alfred Dachnowski-Stokes studied and described changes in the increasingly drained Sawgrass Plains landscape of the 1920s and 1930s (e.g., Allison et al. 1927, p.39). They ascribe the changes to lowered water tables, with water tables notably referring to below-ground water surfaces instead of the above-ground water levels originally present in the sawgrass marsh:

> At present a displacement of the sawgrass vegetation . . . is going on due to the encroachment of broad-leaved and evergreen shrubs and trees as well as elder growth and miscellaneous weed growth. . . . On the Everglades side of this [custard apple] belt a transition zone of shrubs, bordering the Custard Apple belt and invading the sawgrass marsh, consists of willow, *Salix longipes* [*S. caroliniana*] (Fig. 6) and elder (*Sambucus intermedia*) [*S. canadensis*], as well as dense growths of weeds, especially pigweed, *Amaranthus* sp. (Fig. 7), *Eupatorium* sp., *Boehmeria* sp., and others. Among the coarser grasses, Bristly foxtail, *Chaetochloa magna*, *Panicum dichotomiflorum*, *Echinochloa walterii* and para grass, *Panicum baibinade*, are most common in the drained or partially drained sections. . . . The interesting manner in which these plant groups appear upon the higher land or in better drained positions, as along canal banks, is extremely interesting from the standpoint of ecology of the region as well as that of soil development. They compete with the sawgrass both directly and through its weakened condition brought about by the lowering of the water table that has followed drainage. (Allison and Dachnowski-Stokes 1932, pp. 228–229)

As noted in the previous section on soil changes and discussed in geographic detail in chapter 4, bordering strips sometimes referred to as "peripheral wet prairies" developed along the eastern and western borders of the Ridge and Slough landscape as drainage progressed. John H. Davis noted that as the original organic soil was lost from these areas, the Ridge and Slough vegetation disappeared and was replaced by various grasses able to grow on seasonally wet sands:

> Bordering wet prairies of low to tall grass vegetation on seasonally wet soils growing mainly on sand areas [are now seen along the eastern and some northwestern parts of the Everglades.] . . . The eastern prairie zone extends along the pineland border of the glades from a few miles north of the North New River Canal to South Miami. . . . Those [prairie areas] along the east border were formerly much more marsh-like with thin peat or muck soils over sand, but now since some of this border zone has been drained and some of it cultivated the top peat or muck layer has largely disappeared so that these formerly marsh areas are now essentially grassy prairies on the sand soils that have been left after the disappearance of the covering layer of organic soils. (J. H. Davis 1943a, p. 260)
>
> In some cases the original vegetation has been removed or altered and the drainage conditions so changed that instead of developing organic soils some areas are being depleted of them. . . . Along the eastern and some northwestern parts of the Everglades the mucks and peats over sand have been oxidized, burned, or washed off leaving what are now known as Davie and Clewiston fine sands which were formerly mucks or peats with sand subsoils. (J. H. Davis 1943a, p. 84)

In addition to the changes at the former edge of the Everglades, John H. Davis noted changes to the interior sawgrass, probably referring here to the Sawgrass Plains as well as the sawgrass ridges: "Large areas that were formerly dense saw-grass are now medium dense to sparse saw-grass because many such areas have been drained and burned over and weeds have become common. The wax-myrtle thickets and willow bushes are increasing in many saw-grass areas" (J. H. Davis 1943a, p. 261).

In a slightly later study of Florida peat soils, John H. Davis describes a general pattern of anthropogenic, drainage-induced plant succession within sawgrass marshes. The invading species closely match those noted by other authors:

> When saw-grass marshes are drained, as in the Everglades and a number of other large drainage districts, the saw-grass and other marsh plants decrease in number and size and usually weeds, shrubs, and some introduced cultivated plants invade the areas. A sparse growth of saw-grass with bushes, weeds, and in some places ferns usually indicates a drained marsh. The elderberry, *Sambucus simpsonii* Rehder [*S. canadensis*], sea myrtle or groundsel-bushes, *Baccharis* spp., primrose-willows, *Jussiaea* [*Ludwigia*] spp., and small willows, *Salix amphibea* Small, are some of the most abundant bushes. The careless weeds, *Amaranthus* [*A. australis*], fennel, *Eupatorium capillifolium* (Lam.) Small, and some broom grasses, *Andropogon* spp., are also usually abundant. (J. H. Davis 1946, p. 89)

Like Davis, the soil surveyors of the 1940s also noted a variety of herbs and ferns "coming in" as a result of drainage. Pigweed probably refers to careless weed, *Amaranthus australis*: "Since the land [the Sawgrass Plains and parts of the Ridge and Slough landscapes] has been drained or burned over many plants have come in, such as marsh ferns, royal fern, smartweed, pigweed, goldenrod, and castor bean. . . . Many clumps of wax myrtle have come in on the undeveloped sawgrass land since the canals were dug" (Jones et al. 1948e, p. 47). The "undeveloped sawgrass land," where clumps of wax myrtle had appeared, could refer to the Sawgrass Plains but more likely refers to area "1d" (fig. 4.9; see previous section on soil change), the former area of Ridge and Slough landscape that had changed sufficiently to resemble the Sawgrass Plains. If so, this would be a further progression of the changes to the Ridge and Slough landscape that had been reported twenty-five and thirty years earlier: sawgrass invading sloughs (Baldwin and Hawker 1915, p. 753) and wax myrtle shrubs invading sawgrass ridges (Baldwin and Hawker 1915, p. 753; Simpson 1920, p. 126). The reference to marsh fern (*Thelypteris palustris* var. *pubescens*) and royal fern (*Osmunda regalis* var. *spectabilis*) is interesting, as these species are associated with established (Wetzel 2002, p. 385) and incipient tree islands (Stone et al. 2002, p. 98).

The experienced drainage engineer Lamar Johnson felt that the vegetation changes in the Everglades were sufficiently major and influential to merit special comment in an engineering report. Significantly, Johnson's observations appear to confirm the widespread existence of "incipient tree islands"—that is, former sawgrass ridges—which under the influence of lowered water levels had been invaded by shrubs and trees:

> A prologue normally has no place in an engineering report and its only license for being is the writer's observations of changes in the Everglades over a period of 37 years. . . . While working on land surveys in the main floodway of the Everglades in the early 1920's, the eye could reach to the horizon without a single bush being seen projecting above the sawgrass. Today it is difficult to find a few hundred yards of glade that is not broken by brush growth. Marjorie Stoneman Douglas' "River of Grass" is rapidly becoming, in the vernacular of a native frogger, a "hells nest." (Johnson 1958, p. ii)

Several points indicate that Johnson is referring here specifically to the sawgrass of the Ridge and Slough landscape and not that of the Sawgrass Plains. First, Johnson's report is about the hydrology of Everglades National Park; that is, a part of the Ridge and Slough landscape far from the Sawgrass Plains. Second, the land surveys "in the main floodway" to which he refers were located within the Ridge and Slough landscape west of the Miami Canal (Johnson 1974, p. 97) and in Water Conservation Area 1 (Johnson 1974, p. 102). Third, the term "hell's nest" originates from Water Conservation Area 1 (chapter 8)—that is, a part of the Ridge and Slough landscape—and refers to an area of extensive, congested, shrubby growth that blocked the otherwise relatively easy travel through sloughs (J. H. Davis 1943a, p. 266; see also chapter 8).

Johnson's comment that in the early 1920s "the eye could reach to the horizon without a single bush being seen projecting above the sawgrass" is significant, all the more so because of its close similarity to Simpson's (1920, p. 126) comment, "where formerly the eye swept over a monotonous even expanse of saw grass." Both men were experienced observers of the Everglades, both were looking at the Ridge and Slough landscape, and this landscape naturally included tree islands. The tree island vegetation was clearly taller than the sawgrass ridges, and hence would have been easily visible above them. Why then would both observers have made a specific point of remarking on the vast vistas of sawgrass, without shrubs or bushes projecting above them? Plate 17 (top) and plate 18 suggest a possible answer and possible explanation for the drainage-induced changes that alarmed both observers. These plates, along with predrainage descriptions from expeditions (e.g., appendices E, H, and I) indicate that the tree islands were scattered and visually distinct, occupying a small fraction of the landscape surface area, less than 5 percent (chapter 8). Looking out, one saw predominantly sawgrass, with successive ridges merging behind each other at the low angle from which the landscape was usually seen. As obviously distinct and separate entities, the presence of scattered tree islands did not block the impression of endless sawgrass.

A photograph labeled "In the Heart of the Everglades—A Possible Lead" (Dix and MacGonigle 1905, p. 512), taken within the Ridge and Slough landscape before the invasion of woody species, shows exactly what Johnson and Simpson described—the eye sweeping over an even expanse of sawgrass. An isolated tree clearly rises above the sawgrass, and far in the background there is hint of the irregular line of a tree island along the horizon, but most of the photograph clearly shows a wide expanse of simply sawgrass, unbroken by any invading shrubs.

According to the observations of Baldwin and Hawker in 1915, Simpson in 1920, J. H. Davis in 1943, and Jones et al. in 1948, this situation or "viewscape" began to change, apparently quickly, under the influence of drainage until finally, by the time of Johnson's comment, it became difficult to find even "a few hundred yards of glade" where the view was not interrupted by shrubs extending above the sawgrass. Johnson was sufficiently concerned that he even used aerial photographs to map the shrub-invaded sawgrass ridges. Whether his concern was an ecological one is not known, but he definitely thought the change was sufficiently widespread to be a hydrological concern: "The widespread growth of myrtle, willow, holly and bay throughout the Everglades flood plain has not only changed its appearance but has influenced the flow of water" (Johnson 1958, p. 29). Interestingly, the shrub and small tree species that Johnson mentions, *Myrica cerifera*, *Salix caroliniana*, *Ilex cassine*, and *Persea borbonia*, are all characteristic of traditional tree islands as well. The overall impression suggests that drainage had reduced water depths sufficiently for the postdrainage elevation differential between sawgrass ridges and the water surface to resemble the predrainage elevation differential between tree islands and the water surface. Postdrainage ridges then developed hydrological regimes similar to predrainage tree islands and, as a result, became favorable environments for woody species. Interestingly, Stone et al. (2002, p. 99) posit a similar possible development sequence from sawgrass ridges to tree islands under natural, predrainage conditions.

Two botanists, Taylor Alexander and Alan Crook, working systematically in many different parts of the Everglades, also noted changes very similar to those reported by Johnson and earlier observers and noted that they could happen quickly:

> Sawgrass is being replaced by shrubs such as *Myrica cerifera* (wax myrtle), *Ilex cassine* (dahoon holly) and *Baccharis* sp. (saltbush). Invasion is common along canals especially where the hydroperiod has been shortened and the surface soil dried. Invasion occurs rapidly. This is dramatically seen along Alligator Alley [Highway I-75] where drastic changes have occurred in eight years since completion of the roadbed and canals. . . . Throughout the ecosystem, reduction of hydroperiod and depression of the water table has allowed shrub invasion of sawgrass along tree island margins. *Salix caroliniana* (willow) is a common invader of sawgrass in tree island margins and along with *Cephalanthus occidentalis* (buttonbush) seems to be replacing sawgrass in tree island margins and tails. . . . In some areas, as seen along Alligator Alley and Tamiami Trail, willow is becoming more common as sawgrass becomes less vigorous. (Alexander and Crook 1974, p. 64)

In Everglades National Park, south of Tamiami Trail, William Robertson also recorded displacement of sawgrass by woody or shrubby vegetation during the same period, the 1920s to 1950s. Robertson was most likely referring to the Marl Marsh landscapes rather than the Ridge and Slough landscape, but the type of change and underlying cause appear to have been similar:

> Local residents most familiar with the Everglades area [south of Tamiami Trail] agree that within the period of their observations (roughly 25 years) there has been a marked expansion of woody vegetation at the expense of the sawgrass area. It is also agreed that, in spite of fires which have destroyed many tree islands, this invasion has been sufficient to change the whole aspect of large areas particularly south of the Tamiami Trail. The species most involved in this thicket expansion is willow (Egler, 1952: 240; Poppenhager, Redding, Winte, pers. comm.). Some of the present willow areas mark sites of burned-out bayheads, but there seem to be many sites where willow scrub is independently invading sawgrass. It appears a reasonable hypothesis to attribute this development to the drying of the lower glades. It has, at least, occurred under conditions of lowered water levels, and cannot be interpreted as a fire effect because, as Egler points out (ibid.), fire acts to limit this thicket extension. (Robertson 1953, p. 80)

Spatial Patterns of Vegetation Change

In addition to the chronological examination of vegetation changes suggested by these observations, the changes can be considered spatially. Doing so suggests that there is a degree of both temporal and spatial order within them. In fact, early observers of postdrainage vegetation changes in the Everglades apparently recognized a sequential aspect (e.g., W. A. Graham 1951, p. 32). The ordering, if present, might have been simply a function of elapsed time since drainage, or of degree of drainage (drying), or of both together. Whatever the cause, spatial patterns of vegetation mapped in the 1940s reflect, in a number of places, a systematic, sequential component. This is particularly visible in John H. Davis's map of vegetation (plate 13). This map shows, in Townships 46 and 47 Range 40, a narrow strip of the "Saw-grass (with wax-myrtle thickets)" category extending eastward along the Hillsboro Canal almost 6 miles (10 km), as if drawn deep into the "Sloughs, Ponds, and Lakes (with aquatic plants)" category. Slightly farther west, it can be seen that this "drawing in" of a drier category into a wetter category (presumably by canal drainage) very likely converted a triangular portion of the "Sloughs, Ponds, and Lakes (with aquatic plants)" category into the much drier "Saw-grass (with wax-myrtle thickets)" category (southeast quarter of Township 46 Range 39). Still farther west and farther up the canal, drainage apparently drew a narrow strip of the even drier category, "Saw-grass Marshes (dense)" several miles (3–5 km) southeastward (downstream) into the "Saw-grass (with wax-myrtle thickets)" category. Analogously, drainage along the Miami Canal drew the same "Saw-grass Marshes (dense)" category 5 or 6 miles (8–10 km) down the canal and into the "Sloughs, Ponds, and Lakes (with aquatic plants)" category (Township 48 Range 36).

A much larger, but conceptually analogous, basin-scale sequential pattern is visible in plate 13 along the eastern edge of the Everglades in Townships 50 to 54 Ranges 40 and 41. In this area the pattern consists of three approximately north-south bands running parallel to the Atlantic Coastal Ridge (Ranges 41 and 42). These bands fall within the strongly drained central swath, oriented northwest to southeast, and surrounding the Miami and North New River canals. Other analyses (previous section on soil changes; chapter 4) indicate that this swath was originally all part of the Ridge and Slough landscape; that is, a landscape that included a significant percentage of sawgrass. From the Atlantic Coastal Ridge westward into the interior of the Everglades, the bands, as present and mapped in the 1940s, are ordered: (1) "Wet-Prairies," (2) "Saw-grass (with wax-myrtle thickets)," and (3) "Saw-grass Marshes (medium dense to sparse)." Farther away within the original Sawgrass Plains landscape area, the category "Saw-grass Marshes (dense)," which likely most closely resembles the original Sawgrass Plains landscape, makes a conceptual fourth "band" and an endpoint for the series.

The spatial ordering of these bands within the new sawgrass lands of the 1940s presents a picture similar to what might be expected as water levels progressively lowered within the basin; that is, a pattern of drier vegetation types encroaching inward from the edges of the basin. The conceptual fourth "band," dense sawgrass, is vegetatively closest to the original Sawgrass Plains and reflects the least drainage. Band (3), medium dense to sparse sawgrass, suggests the thinning and weakening associated with early drainage and mentioned by various observers (Allison et al. 1927, p. 39; Lloyd 1929, p. 58; Allison and Dachnowski-Stokes 1932, p. 228; J. H. Davis 1943a, pp. 261–262; Andrews 1957, p. B-30). Band (2), sawgrass with thickets of wax myrtle, suggests a stronger degree of drainage, sufficient not only to weaken the sawgrass but also to create a favorable environment for shrubby species (Baldwin and Hawker 1915, p. 753; Simpson 1920, p. 126; Jones et al. 1948e, p. 47; Johnson 1958, pp. ii, 29; Alexander and Crook 1974, p. 64). The last member, band (1), the "wet prairies," reflects soils that by the 1940s were essentially sands and did not support sawgrass. As such, it might not seem an obvious member of this drainage sequence. However, as discussed in the previous section on soil change, and later in chapter 4, the areas mapped as wet prairies in the 1940s were originally part of the Ridge and Slough landscape; they simply were the portion most drastically altered by nearly complete loss of the original peat soil (chapter 4). As such, the wet prairies of the 1940s are a logical end member of this sequence.

It is perhaps worth reemphasizing that not only the wet prairies seen in the 1940s, but also bands (2) and (3) were all originally part of the Ridge and Slough landscape. The tree islands that J. H. Davis shows within the central, strongly drained swath surrounding the Miami and North New River canals, as well as within bands (1), (2), and (3), are relicts carried over from the predrainage configuration of the Ridge and Slough landscape as a single unit. Similarly, the characteristic ridge-and-slough patterning (chapter 8) still visible in parts of the central swath area on aerial photographs (USDA-SCS 1940) and on later imagery (e.g., plates 24, 31, 32) is a relict of the original landscape condition.

Interestingly, the central, strongly drained swath of altered vegetation not only affected the Ridge and Slough landscape but extended continuously up into the Sawgrass Plains as well. The J. H. Davis (1943b) vegetation map (plate 13) shows that by the 1940s, this landscape too had been divided into two separated areas. The eastern and western areas of dense sawgrass (yellow) visible on plate 13 were the two least altered remnants of the landscape. The internal patch of "Saw-grass Marshes (with abundant ferns and cat-tails)" within the eastern area of dense sawgrass was likely an exception to this, with its location suggesting that it probably was a drainage effect. The area surrounding the North New River

Canal, and extending westward almost to the Miami Canal, shows the effects of drainage. Here the original dense sawgrass had been replaced by a swath of altered sawgrass, parallel to the canals, either thinned ("medium dense to sparse") or invaded by a woody species ("with wax-myrtle thickets").

Overall, as was the case for the soils, the vegetation mapped in the 1940s, rather than being an estimate of the predrainage landscape types and extents, was instead much more a cumulative record of the extensive alterations that had been wrought by drainage. Aspects of the original vegetation certainly remained, but the vegetation changes that had occurred by the 1940s were both extensive and intensive. Along the edges of the Everglades, new wet prairies had formed that bore essentially no resemblance to the predrainage vegetation. Through the center of the Everglades, a great swath of strongly altered vegetation had formed, splitting both the Sawgrass Plains and the Ridge and Slough landscape into two separated eastern and western areas. Sawgrass areas dried sufficiently to develop hydrologic regimes similar to predrainage tree islands. As a result, shrubs invaded both the Sawgrass Plains and the sawgrass ridges, eventually blurring the distinction between true predrainage tree islands and postdrainage "pseudo tree islands" formed on former predrainage ridges.

Summary

Many firsthand observations are available describing the drainage history of the Everglades that began in the 1880s. Similarly, many firsthand observations corroborate the severe and widespread ecological effects that drainage had caused by the 1940s. The rapidity of change varied. Hydrologic and vegetational changes were already visible within the first years; whereas changes in soil type, topography, and frequency of peat fires required five to twenty years to become obvious. By 1940 these changes were recognizable throughout the system.

The Everglades postdrainage history can be divided into three phases, a Lake Okeechobee drainage phase (1882–1896), a muck canal drainage phase (1906–1930), and an impoundment phase constructed between 1950 and 1975, the hydrologic effects of which continue to the present (2010). The first two phases greatly reduced water depths. The impoundment phase, corresponding primarily to the Central and Southern Florida Project for Flood Control and Other Purposes, actually raised average water levels in some parts of the areas north of Tamiami Trail (e.g., the downstream portions of the three Water Conservation Areas), but generally increased the annual and spatial variance in water level and greatly reduced sheet flow. South of Tamiami Trail the impoundment phase greatly decreased water flows and depths.

The drastic changes that occurred prior to 1940 underscore the irony of the wealth of excellent scientific data that became available in the 1940s, especially the comprehensive aerial photography and soil and vegetation maps. While all are accurate "snapshots" from that time, those studies may in no way be assumed to be accurate depictions of predrainage conditions. Nevertheless, when combined with historical evidence as presented in this chapter, they provide part of the basis for systematically reconstructing the characteristics and extents of the predrainage landscapes. This process of hindcasting (chapter 4) forms the core of the landscape descriptions presented in chapters 5 to 10.

4

Reconstructing the Landscapes of the Predrainage Everglades

According to Gordon Whitney (1994), a student of change in U.S. temperate forests, "the basic goal of the historical ecologist is to document the environmental changes of the past and to determine the factors responsible for their occurrence." He notes the difficulties of disentangling the web of natural and anthropogenic forces and tracing back the causes of environmental change over centuries. Although the subject of this chapter is certainly historical ecology, our goal was not to describe the full ecological past of the Everglades but rather its ecological condition for a short period just prior to drainage by the construction of canals. The primary focus was therefore on the predrainage Everglades of about 1840 to 1880. The environmental forces of the preceding five thousand years evidenced by the soil and pollen studies of others were of course kept in mind; for example, Dachnowski-Stokes (1930), Willard et al. (2001), Winkler (1995), Cohen and Spackman (1984), and Cohen et al. (1999). Indeed, review of these studies led to a sense that the mid-nineteenth-century conditions described in this study are likely representative of several previous centuries, perhaps longer. Nevertheless, the primary focus remained on characterizing the Everglades as they appeared just prior to the onset of mechanized drainage.

We simplified Whitney's goals for the historical ecologist by separating the important discussions of postdrainage changes (chapters 2, 3) from a description of the Everglades at the time point of interest (chapters 5–10). For readers with field familiarity concerning the postdrainage Everglades, the contrast between one's own experiences and the predrainage descriptions developed here may at times seem extreme. This is both understandable and accurate; the postdrainage changes *have* been extreme. A vast expanse of South Florida was once remarkably wet, but anthropogenic (i.e., mechanized) drainage has made that expanse both drier and smaller. If a predrainage landscape appears implausible, it may prove helpful to reread chapters 2 and 3. In writing the following chapters, the differences between predrainage and postdrainage conditions have been kept constantly in mind. A plausibility standard was applied throughout by repeatedly inquiring, "Are the qualitative and quantitative differences between postdrainage conditions and the predrainage ones presented in this study reasonably explained by the postdrainage influences catalogued in chapters 2 and 3?" The answer appears to be "yes," such that the equation holds well for the Everglades landscapes still remaining in the 1940s. Not surprisingly, the equation does not hold quite as well if the right-hand side is replaced with current (2010) postdrainage conditions, mainly because of the additional influence of post-1940s anthropogenic changes. That it would apply at all is a reflection of the extremity of the early (pre-1940) changes relative to the later ones. Thus chapters 2 and 3 go a long way toward reconciling the Everglades as mapped and described in the early to mid-twentieth century with what this study concludes was present in the early to mid-nineteenth century.

$$\begin{bmatrix} \textit{Original Conditions} \\ \textit{(Chapter 4)} \end{bmatrix} + \begin{bmatrix} \textit{Post-drainage Changes} \\ \textit{(Chapter 3)} \end{bmatrix} = \begin{bmatrix} \textit{Post-drainage} \\ \textit{Conditions of 1940s} \end{bmatrix}$$

The goals for the landscape descriptions (chapters 5–10) were both geographic and archival. Specifically, they were to map the boundaries of the Everglades as they existed in the nineteenth century, to describe the soils and vegetation of the Everglades at that time, and to publish the key observations used to establish these boundaries and descriptions—especially those extracted from unpublished manuscript sources. A similar though less intense effort was directed at the more elevated landscapes bordering the Everglades, with the goal of including sufficient detail to illustrate their connection to the Everglades.

Previous Landscape Studies of the Everglades

While the present study is the most extensive and detailed examination of predrainage Everglades landscapes to date, it builds, in part, on previous studies. This short review is restricted primarily to those investigations that treated the Everglades as a whole, rather than focusing only on specific subareas. Before examining the studies, however, it is helpful to listen to the words of several firsthand observers of the Everglades, before and shortly after the onset of mechanized drainage. The frequent references to "vastness" most likely reflect the wide horizons made possible by the extreme flatness of the accumulated peat surface and the absence of continuous tree cover.

Buckingham Smith's descriptions of the Ridge and Slough landscape, though familiar, bear repetition for their lyrical quality, their timing well prior to drainage (1848), and their consistency with what other observers described:

> The appearance of the interior of the Everglades is unlike that of any region of which I have ever heard, and certainly it is in some respects the most remarkable on this continent. Imagine a vast lake of fresh water extending in every direction from shore to shore beyond the reach of human vision, ordinarily unruffled by a ripple on its surface, studded with thousands of islands of various sizes, from one-fourth of an acre to hundreds of acres in area, and which are generally covered with dense thickets of shrubbery and vines. Occasionally an island is found with lofty pines and palmettos upon it, but oftener they are without any, and not unusually a solitary majestic palmetto is seen, the only tree upon an island, as if to guide in approaching it, or as a place of signal or lookout for its former denizens. The surrounding waters, except in places that at first seem like channel ways (but which are not), are covered with the tall sawgrass, shooting up its straight and slender stem from the shallow bottom of the lake to the height often of 10 feet above the surface and covering all but a few rods around from your view. The water is pure and limpid and almost imperceptibly moves, not in partial currents, but, as it seems, in a mass, silently and slowly to the southward. The bottom of the lake at the distance of from 3 to 6 feet is covered with a deposit of decayed vegetable substance, the accumulated product of ages, generally 2 or 3 feet in depth on the white sand and rock that underlies it over the entire surface of the basin. The flexible grass bending gently to the breeze protects the waters from its influence. Lilies and other aquatic flowers of every variety and hue are to be seen on every side, in pleasant contrast with the pale green of the saw grass, and as you draw near an island the beauty of the scene is increased by the rich foliage and blooming flowers of the wild myrtle and the honeysuckle and other shrubs and vines that generally adorn its shores. The profound and wild solitude of the place, the solemn silence which pervades it, unless broken by the splashing of a paddle of the canoe or light batteau, with which only can you traverse the Pahayokee, or by the voices of your "Campagnons du voyage," add to awakened and excited curiosity feelings bordering on awe. (B. Smith 1848a, p. 51)

Samuel Sanford, a field geologist exploring Florida for railroad interests and later for the U.S. Geological Survey, also speaks of looking out over vast distances. This was once a common description, particularly as explorers emerged from the enclosing woods of the Atlantic Coastal Ridge to gaze suddenly out across the open, apparently infinite extent of the Everglades. Experiencing this transition from closed to open expanse is now rare as almost all the original edges are obscured either by levees or by shrub thickets. Those fortunate to have spent time in the interior of the remaining Everglades can easily appreciate Sanford's description:

> It is difficult for a person who has not seen the Everglades to form even an approximate idea of that far-extending expanse of sedge, with its stretches of shallow water, its narrow winding channels of deeper water, its scattered clumps of bushes, and its many islands. Photographs fail to convey the impressions of distance, of remoteness, and of virgin wildness which strike the visitor who for the first time looks out across that vast expanse. (Matson and Sanford 1913b, p. 53)

Naturalist Charles Torrey Simpson refers to a wet, often inundated "prairie," a term often used by land surveyors in the 1840s and 1870s for expanses of sawgrass:

> If the stranger visits the lower part of the State he will find in the interior a vast extent of wet, often inundated prairie with wooded islets scattered along its borders. At the north of this great swamp, the Everglades, is Lake Okeechobee, which during the rainy season overflows the entire prairie. A low rocky ridge lies between the Everglades and the Atlantic coastal shore. It projects westward far into the swamp in southern Dade County, and finally disappears in the great prairie. This ridge is cut into numerous islands, and water from the Everglades passes through the channels between out to the sea. (Simpson 1920, p. 2)

Keeping these images in mind, one can focus on the maps and previous studies of Everglades landscapes. These can be categorized by time, approach, and spatial focus. Only one study, (i.e., Ives 1856a, b), was completed prior to the onset of mechanized drainage; all others recorded an environment already affected, to varying degrees, by drainage. Most of these studies are descriptive rather than reconstructive; that is, they simply describe and map what was present at the time of mapping. Only the last three studies fall into the purview of historical ecology, explicitly trying to (re)create a picture of the original, predrainage environment. Finally, these studies can be classified by their inclusion, to a relative degree, of spatial detail: the interior of the Everglades versus the more accessible exterior eastern border, the Atlantic Coastal Ridge. Recall that the Everglades was long considered one of the last American frontiers due to its inaccessibility. Prior to drainage, boat travel was possible, though arduous and carrying the risk of losing the way. As water depths dropped, overland access into the Everglades actually decreased, at least initially. Interior access awaited new technology: development of the airboat and later, as levels dropped even farther, the swamp buggy. Thus it was not until efforts by J. H. Davis (1943a, b) and Jones et al. (1948a–e), sixty years after the onset of drainage, that Everglades studies of the interior achieved a level of detail similar to that for the coastal uplands.

The *Military Map of the Peninsula of Florida South of Tampa Bay* by J. C. Ives (1856b) is justifiably well known and often reproduced (fig. 1.6). While Ives was neither a botanist nor a landscape ecologist, he was striving to provide a map with as much differentiation of the vegetation as possible, accurately located, to give soldiers and sailors a recognizable foothold for navigating the otherwise uncharted vastness of the Everglades. As a result his map distinguishes a remarkable number of ecologically useful vegetation types, including pine, cypress, and wet prairies, and does so quite accurately. His accompanying text (Ives 1856a) gives considerable insight into his sources and the understanding he developed through synthesizing them. Overlaying his map on later, more spatially accurate maps reveals the spatial distortions in Ives's map but also illustrates that the map provides a very good conceptual depiction of the features of the southern Florida peninsula. Especially well depicted is the cutting of the Atlantic Coastal Ridge into "numerous islands," as described by Simpson, with the waters of the Everglades passing to the sea through the channels between them.

John Harshberger, a landscape ecologist at the Wagner Free Institute of Science in Philadelphia, studied the vegetation of southern Florida during 1910–1914, making a number of excursions inland from the coasts and taking advantage of the newly opened North New River Canal to traverse the middle of the Everglades by boat. His investigations resulted in a monograph that provides numerous key observations of the early postdrainage system (Harshberger 1914) as well as a phytogeographic map (Harshberger 1913). The map provides little detail for the interior of the Everglades, but the considerable detail along the eastern edge of the Everglades is thorough and reliable. His observations aided formulation of landscape descriptions for the Atlantic Coastal Ridge, Eastern Flatwoods, Everglades Keys, Marl Transverse Glades, and the portion of the Ridge and Slough landscape at the headwaters of the Miami River. Additionally, Harshberger's observations along the North New River Canal, which he logged by milepost, were used to locate the transition between the Sawgrass Plains and the Ridge and Slough landscapes.

Roland Harper, a botanist, worked in South Florida from approximately 1909 to at least 1927, writing a number of articles and a monograph published by the Florida Geological Survey. While he did not attempt to map vegetation, his identification of plant communities and his detailed species lists provided useful background material. Unfortunately Harper's explorations did not extend into the Sawgrass Plains and reached only marginally into the Ridge and Slough landscape.

Like Harper, J. K. Small was a botanist, not a landscape ecologist, so he did not attempt to map the landscapes or plant communities of the Everglades. Nevertheless, he thoroughly documented his numerous collecting trips in South Florida. His narrative accounts and extensive collection of photographs from 1901 to 1938 provide important insights into both the predrainage Everglades and the rapid and dramatic ecological blows occasioned by the first decades of drainage.

John H. Davis Jr., working both as botanist and de facto landscape ecologist in the first part of the 1940s, provided the first synoptic overview of Everglades vegetation that included detailed mapping of the interior. His *Vegetation Map of Southern Florida* at 1:400,000 (1943b) and the accompanying monograph (1943a) have rightfully become classics. His work was based on extensive field observations and examination of the first comprehensive set of aerial photographs. Interestingly, Davis clearly did not intend to produce a spatially precise map of plant communities. Unlike that in a contemporaneous soils study, his drawing style indicates that he did not use the aerial photographs as a base map. His map is conceptual rather than spatially precise. Nevertheless, both map and monograph reflect keen observation and a comprehensive understanding of the vegetation communities. Particularly because his map is the earliest one to include interior detail and because it has become such a classic, it is important to emphasize that it is not, and was never intended to be, a map of predrainage landscapes. His goal and very substantial accomplishment were to map what was actually present in the early 1940s. His writings leave no doubt that he was aware of the great changes in the Everglades that had already occurred by the 1940s. The perception that J. H. Davis's map presents predrainage conditions probably arises in part because he did in fact hindcast two small areas: the already obliterated Custard Apple Swamp bordering Lake Okeechobee and some of the urbanized portions of the Atlantic Coastal Ridge.

Jones et al. (1948a–e) provided an interesting and extremely important suite of studies of the Everglades as a whole. The result of a massive multiagency, multidisciplinary field and analytical effort, it is not the typical county soil survey, such as was traditionally produced by the U.S. Department of Agriculture's Soil Conservation Service. It includes instead a detailed soil map of the multicounty area corresponding to the Everglades Drainage District, predecessor to the present South Florida Water Management District. The soil mapping is unusually coherent and logical because the soils were considered on a watershed basis, rather than county by county. The spatial precision has proven to be very good, probably due to the miles of surveying involved and especially to the use of aerial photography as a base map. Jones et al. (1948a; see also 1948e) did not map vegetation per se but did differentiate a small number of different land use categories, some of which correspond loosely to ecologically recognizable plant communities. Jones et al. aggregated these categories to produce a map of "physiographic divisions" of the Everglades. The reader will recognize some similarities (and important differences) between these physiographic divisions and the landscapes delineated in the present study. Like the contemporaneous J. H. Davis (1943b) map, the Jones et al. (1948a) soil map was not intended to illustrate predrainage conditions but is very accurate for the conditions present in 1940. The accompanying text indicates that the mapmakers were aware of changes from predrainage conditions. Their goal, however, was to map the suitability of the then present soils for agricultural and, to a lesser extent, urban uses. Knowledge of the predrainage condition of either the soils or the vegetation was of interest only to the extent that it shed light on soil suitability.

Robert Costanza (1975) undertook a first attempt at determining the historical ecology of the Everglades as a whole by producing an estimated map of the Everglades circa 1900. His map was a modification of the J. H. Davis (1943b) map, using 1950s aerial photographs to estimate vegetation changes between 1900 and the 1940s. Costanza's map is very similar to the Davis (1943b) map and is insufficiently hindcasted to be a good estimate of predrainage conditions.

A mapping study by S. M. Davis et al. (1994) is the most direct predecessor to the present study. These authors used a historical ecology approach to estimate a predrainage map of Everglades landscapes, based largely on the J. H. Davis (1943b) vegetation map and modified through comparisons with historical expeditions and maps. Change analysis using aerial photography and field inspection of sites previously studied by Alexander and Crook (1973, 1975) also contributed to estimation of this predrainage map. The pioneering use of historical sources to supplement scientific studies was a key precedent for the present study.

In summary, the predrainage, presettlement Atlantic Coastal Ridge and its predrainage relationship to the Everglades have been well documented, but the predrainage interior of the Everglades has not. Excellent synoptic studies of the Everglades were completed in the 1940s, but this was well past the onset of drainage, and their authors, by their own accounts, acknowledge that the studies do not depict predrainage conditions. The available historical ecology studies, especially S. M. Davis et al. (1994), began to fill the gap; however, substantially more time and detailed research were required to distinguish definitively the predrainage landscapes from the postdrainage conditions of the 1940s. Table 4.1 summarizes the previous mapping studies of the Everglades and highlights the need for the present study, which documents changes in the organic soils and vegetation (chapter 3), documents and characterizes the original landscapes (chapters 5–10), and develops a conceptual model of the predrainage topography and hydrology (chapter 11).

This Study: Reconstructing the Past

The exhortation to historical ecologists that "mastery of the written evidence should precede any ecologist's investigation of past events" (Whitney 1994, p. 10) was

Table 4.1. Characteristics of the principal landscape maps of the Everglades region

Time Period Described	Map Reference (and Location in Present Volume)	Comments
Predrainage (1850s)	Ives (1856b); (Figure 1.6)	Some spatial distortion of base map. Physiographically very accurate, especially for much of the eastern coastal uplands bordering the Everglades. Provides little detail on interior of Everglades.
Early postdrainage (1910)	Harshberger (1913); (Figure 1.7)	Some spatial distortion of base map. Physiographically very accurate, especially for the eastern coastal uplands bordering the Everglades. Provides little detail on interior of Everglades. Text differentiates Sawgrass Plains from Ridge and Slough landscape.
Postdrainage (1940s)	J. H. Davis (1943b); (Plate 13)	Physiographically very accurate overview. Spatial details are generalized. Interior mapped in detail. Does *not* depict predrainage conditions, except a few farmed and urban areas that were hindcasted to natural vegetation.
Postdrainage (1940s)	Jones et al. (1948a); (Plate 11)	Spatially very accurate. Interior mapped including shapes and orientation of large tree islands. No attempt was made to hindcast to predrainage conditions. Text recognizes many changes had occurred.
Predrainage (estimated)	Costanza (1975)	Modified from J. H. Davis (1943b) with only minor changes. Not fully hindcasted.
Predrainage (estimated)	S. M. Davis et al. (1994)	Modified from J. H. Davis (1943b). More extensive hindcasting to predrainage conditions, based on historical narratives. A number of 1940s postdrainage aspects persist.

taken to heart for this study. More than nine hundred different pre- and postdrainage sources were obtained. These were gleaned for observations of Everglades vegetation, soils, hydrology, topography, and/or fauna with relevant quotations recorded in an annotated database. The observation that North America possesses some of the best documentary evidence of environmental and land use change (Clawson and Stewart 1965, p. 81) appears to apply to South Florida as well, though with some limitations due to the difficulties of travel within the Everglades. Whitney's observation that "unearthing the appropriate records often involves sifting through a large amount of material of little use" (1994, p. 9) was experienced firsthand. John T. Sprague's history of the "Florida War" (First and Second Seminole Wars), for example, numbers more than 550 pages, but the gems of historical ecology it contains were often only a few sentences each and scattered throughout. Fortunately, many of the pitfalls typically associated with historical sources were less prevalent in South Florida sources. As this has bearing on the probable accuracy of the predrainage descriptions presented in this chapter, the strengths and weaknesses of these sources are reviewed here.

Whitney (1994, p. 10) classified the various types of sources useful for historical ecological research and discussed their pros and cons at some length. He found that early (1600–1700) travelers' accounts could be formulaic, replete with allegories, biblical imagery, and references to classical mythology, but short on reliable ecological information. In other cases, biases in travelers' accounts might be unintentional, such as a traveler concluding as he traveled only along the Ohio River, with its still wooded banks, that all of Ohio was unbroken forest. Promotional tracts obviously emphasized the most positive aspects of the landscape, sometimes distorting accounts of the soils, native vegetation, or climate. Pilgrims, for example, had on occasion written of woodlands with deep rich soil where more objective eyes found only scrub pine on sand. Early guides encouraging immigration to the American West sometimes published distorted information pumped from local tavern keepers. At the same time, sources written by settlers making a living on the land—surveyors, farmers, timber cruisers—often contributed objective and valuable information. Other sources of ecological information useful for quantifying the changing spatial relationships in the landscape included maps, aerial photographs and satellite imagery, with geographical information systems and scaled photocopies being useful tools for overlaying all these.

Federal land surveys made after passage of the Land Ordinance of 1785 using the thirty-six-square-mile township grid system were among the richest historical sources. The surveyors' field notes and accompanying plat maps usually included the species and diameter of the "bearing" or "witness" trees used to mark each half mile (0.8 km), general descriptions of the timber along the mile-long section lines, quality of the soil, general topography, and the width of any water bodies crossed. Despite or perhaps because of their wealth of information, Whitney points out several cautions, including the possibly of fraudulent surveys, surveys conducted after relevant environmental changes, misidentification or possible bias in the selection of bearing trees, and differences in the instructions given to the surveyors. He notes that most of these concerns are exceptions rather than the rule, and most can be overcome through awareness and cross-checking, and concludes that despite the cautions, the early land survey records are the best source of information on presettlement forests (1994, pp. 25–29).

By good fortune, most of the pitfalls identified by Whitney either do not occur in the Everglades narrative record or are easily recognized and avoided. Land booms, with the accompanying glowing promotional material, certainly occurred in South Florida; only one such source was used, and only the portion that was verifiable.

Travelers' accounts with allegories or mythology were not found, possibly because the rigors of travel into the Everglades selected against casual travelers. The available, and very useful, expedition logs tend to reflect observations from seasoned outdoorsmen knowledgeable of the flora and pragmatic in style. Most of the sources used were from individuals closely associated with the land, often with firsthand experience with the Everglades over periods of years. Nevertheless, the issue of unintentional bias in narrative accounts bears examination. The deliberate choices that were made as to which time of year to travel (i.e., wet season or dry season) and which routes to take across the Everglades might be expected sources of possible bias. However, neither seemed to be a serious problem. Time of year would certainly have affected the water depths observed, but the vast majority of the sources used in this study can be dated, usually to the individual day, making it possible to correct for season. It is true that most expedition routes were not *chosen* to provide a randomly distributed sample of observations of the Everglades, but it can reasonably be argued that they are not systematically biased.

With respect to the landscapes encountered, the routes chosen were effectively random as the choice was driven primarily by endpoints outside the Everglades, rather than by an effort to avoid or favor particular areas within. Most routes proceeded across rather than parallel to the orientation of ridges and sloughs; thus explorers were forced to experience both (as their records often ruefully note). Collectively, the various expeditions do cover much of the main landscape with some uniformity. Additionally, descriptive bias can be ruled out as the descriptions from different routes and from different points along the same route all tend to give a similar picture of the same landscape.

Land surveys, both plats and field notes, were extremely valuable sources for this study. While the cautions noted by Whitney could have applied, the way the survey information was used here avoided most problems. As the focus was on the interior of the Everglades where trees were scarce, misidentification or bias in the selection of bearing trees was not an issue. For this study, identification of the transition from forested upland to unforested wetland was sufficient; this was usually very clear and not dependent on exact species. Potentially fraudulent (i.e., not actually conducted) surveys were recognizable by overlay comparison with land surveys by other surveyors and/or on the Jones et al. (1948a) soil map. No fraudulent surveys were found; all plats containing landscape information were found to match remarkably well.

To the extent that problems with land surveys did arise, they were due to altered survey instructions and the occurrence of surveys after significant environmental change. Surveys used were from three main time periods: predrainage, early postdrainage through the 1940s, and 1950 to present. The Atlantic coastal townships, and often the first few miles of adjoining Everglades, were mostly surveyed for the federal General Land Office in the 1840s and 1870s (fig. 1.4). Later, after the onset of mechanized drainage, the Trustees of the Internal Improvement Fund (TIIF), a state entity, commissioned surveys of large parts of the actual Everglades. Since the TIIF-commissioned surveys took place after drainage began, water depths reported in those surveys may underestimate predrainage depths. However, the change in instructions (i.e., requiring a greater level of information) worked to advantage by providing many detailed transects and measurements, including the widths of sloughs and ridges and thicknesses (depths) of peat soil.

Overall, this exercise in historical ecology was greatly

aided, and to some degree buffered from the weaknesses of individual sources, by the very strong ecological links among the soils, vegetation, and hydrology. The strength of these links meant that only a narrow range of soil types, plant species, and water depths naturally occur together. Thus for example, if a historical source described peat soil, water 2 feet (60 cm) deep in October, and pine trees all at the same location, it would immediately be suspect: Why pine trees in such deep water? Even wax myrtle (*Myrica cerifera*) would raise questions, whereas water lilies (*Nymphaea* or *Nuphar*) or possibly even sawgrass (in a wet year) would not. Thus in addition to taking to heart immigrant Elias Pym Fordham's observation that "no dependence can be placed on any representation but that of an intelligent, honest man, long resident in the country" (1906, p. 71), each source was also screened for ecological credibility. The large majority of observations passed these credibility tests. Consistency among numerous sources was the rule rather than the exception.

The actual process of historical or forensic ecology followed to develop descriptions of each landscape and to delineate their extents was iterative, alternating back and forth between spatial and temporal focus and between description and delineation. As iterations progressed, the descriptions became more precise and better differentiated among landscapes, and the delineations more exact. We read sources and used the stated (or, in some cases, estimated) locations of the observations along with the maps of physiographic regions (landscapes) in Jones et al. (1948a, c, e) and in S. M. Davis et al. (1994) to assign the observations tentatively to the appropriate predrainage landscape. As observations associated with a particular landscape accumulated, they were ordered by date, permitting distinction of purely predrainage (pre-1884) from postdrainage observations. Depending on the location and history of drainage, it was possible to align some early postdrainage observations with the strictly predrainage observations. As observations accumulated, it sometimes became clear that certain boundaries between landscapes needed to be redrawn, or even that the observations needed to be shifted from one landscape to another. Overlays of satellite imagery, aerial photography, and maps, both early and present day, were extremely helpful in sharpening support for delineation of boundaries. As observations and iterations accumulated, the landscape descriptions that form the remainder of this chapter—soils, vegetation, and hydrology—to some degree "wrote themselves."

The landscape descriptions are generally devoid of citations. This is intentional; each description, and essentially each assertion within each description, is based on the full set of observations listed chronologically in the historical observations section for each landscape within the Supplementary Materials (DVD).

Development of the landscape descriptions was driven primarily by the accumulated and classified observations themselves. These were supplemented with two simple guiding assumptions, one temporal and one spatial. Temporally, the assembled observations can be loosely divided into three time periods: (1) predrainage, including some early postdrainage; (2) postdrainage through the 1940s; and (3) current (1970s through early 2000s). These divisions reflect the surge of scientific investigations in the 1940s, the influence of the impoundment phase that began under the C&SF Project, and direct observations of the presently remaining Everglades by the authors and others in the scientific community. Based on chapter 3, conditions in the Everglades were drier in the 1940s—shallower water depths, shorter hydroperiods—than in predrainage times. Comparisons of post–C&SF Project (post-1970s) hydrologic conditions with either 1940s or predrainage conditions are more complex. After the 1940s some areas actually became wetter as the impoundments of the Central and Southern Florida Project were completed in the 1950s and 1960s. Everglades National Park initially became drier before becoming somewhat rewetted as connections to upstream impoundments were managed differently. With the exception of the deeper, downstream ends of the impoundments, current water depths generally have not returned fully to predrainage depths. Recognizing that downstream ends of impoundments can constitute an exception, the following ordering of time periods by water depth was assumed:

$$[1920s\,to\,1940s] < [Current\;(1970s\,to\,2010)] < [Pre-drainage]$$

This ordering suggests similar ordering of hydrology-driven ecosystem aspects such as vegetation communities, peat oxidation/accumulation, and changes along the edges of the Everglades at the interface with forested uplands. This assumed hydrologic ordering was helpful in screening, evaluating, and interpreting predrainage and early postdrainage observations.

Similarly, spatial assumptions regarding the relative wetness of the different predrainage landscapes were also helpful. Where the peat-based landscapes (Sawgrass Plains and Ridge and Slough landscape) extended all the way to the mineral soil–based Eastern and Western Flatwoods, the flatwoods were assumed to have been slightly drier than the Everglades landscapes. In the southern Everglades an additional intermediate band was assumed to have been present; thus from wetter to drier, the landscapes were ordered:

[*Shark Slough* (*Ridge and Slough landscape, peat soil*)] > [*Marl Marshes* (*marl and peat*)] > [*Everglades Keys* (*sand or pinnacle rock*)]

As with the temporal assumptions, these spatial assumptions helped guide the evaluation and interpretation of the gathered observations.

A Sample Hindcasting: The Postdrainage "Peripheral Wet Prairies"

Space precludes individually documenting all the sources consulted and logic applied to hindcast each of the many different areas and boundaries that make up the predrainage Everglades. Instead, to give a flavor of the logic and range of sources used, one representative analysis is described here in detail. The changes hindcasted in this case were caused by oxidation of the organic soil of the Everglades. Similar soil oxidation occurred throughout the Everglades as water levels were lowered by drainage. While the details differed from place to place, the general phenomenon was widespread, so the sample of hindcasting—that is, exercise in historical ecology—described here is representative of the many others that produced this book.

Comparison of the postdrainage J. H. Davis (1943b) vegetation map (plate 13) with either the predrainage map by Ives (1856b; fig. 1.6) or the early postdrainage map by Harshberger (1913; fig. 12.7) reveals a transitional area that was mapped in the 1940s but not shown on the earlier two maps. The Davis map (1943b; plate 13) shows a 2- to 3-mile-wide (3–5 km) strip of "wet prairies" (light blue), just west of the coastal ridge pinelands, in Townships 50 to 54. A similar strip, with additional extension north to Township 47, is visible on the 1948a Jones et al. soil map (plate 15) as an area of "Davie fine sands" and "Davie mucky fine sands." Both Davie sands are mineral soils, lacking surficial layers of predominantly organic (peat) material. In contrast, the pre- and early postdrainage maps (figs. 1.6, 1.7) show no bordering strip—only undifferentiated Everglades extending directly east to the coastal pinelands. Despite the absence of the bordering strip from the early maps, it has been suggested that the strip mapped in the 1940s was in fact present prior to drainage as well, forming a transition between the Ridge and Slough landscape and the pinelands: the so-called "peripheral wet prairies."

The results of this study suggest otherwise. The analysis presented on the following pages leads to the conclusion that the "peripheral wet prairies" visible on plates 13 and 15 are strictly an artifact of drainage, and that the Ridge and Slough landscape originally extended continuously and fully to the coastal ridge. The transformation from peat-based Ridge and Slough landscape to sand-based wet prairies is the result of postdrainage oxidation: complete loss of the overlying organic peat to expose the underlying sands.

These conclusions were reached through spatial analysis of the area in question, particularly by overlaying different spatial sources such as surveys, maps, aerial photos, etc. For example, plate 40 shows the area under analysis: a composite of aerial photographs taken during the spring of 1940, overlain with a township range grid (yellow; 6 × 6 miles), the predrainage coastal rivers (blue), and the outlines of "islands" of upland sands and pine (red). Together this string of islands forms a discontinuous section of the Atlantic Coastal Ridge. The gaps between islands were the Peat Transverse Glades, low-lying areas of sawgrass growing on peat soil that extended continuously from the Everglades to the often multibranched headwaters of the coastal rivers farther east. In the western portion of the image, primarily in Township 50 Ranges 39–41 and in Townships 51–52 Ranges 39–40, a directional "grain" characteristic of the Ridge and Slough landscape is visible (cf. plate 24, left half), generally oriented southeast. (It is even more clearly visible on the individual aerial photographs than on the index sheets reproduced here). The dark color may indicate either surface water or the peat soil of the Ridge and Slough landscape.

However, closer to the coastal ridge, in Townships 50–52 Range 41 (southeast corner), the dark-colored directional grain of the Ridge and Slough landscape disappears on plate 40, replaced by a whitish area. The whitish area is also visible in the western half of Township 53 Range 41, though somewhat obscured by development. This 2- to 3-mile-wide (3–5 km) area just west of the islands of the coastal ridge corresponds to the bordering strip discussed earlier and mapped by J. H. Davis (1943b) as wet prairies. The white color may indicate light-colored sands, no longer covered by peat soil. In any case, the light color and general absence of directional grain suggest that at the time of the photographs, the Ridge and Slough landscape was absent; that is, it did not extend continuously eastward to the coastal ridge.

Turning now to the federal land surveys of the same area (plate 41; note slightly different scale and extent; Township 54 not shown), which were made prior to drainage, suggests a quite different picture. Comparison of plate 40 (1940s) and plate 41 (mostly 1870s) reveals a very close match between the shapes and arrangement of the "islands" of high ground. Recall that the red island outlines on plate 40 were mapped independently, seventy to one hundred years later, on the basis of soils. However, in contrast to the transitional strip configuration visible in the 1940s (plate 40), the predrainage surveys (plate 41) show no transition, instead marking the area directly west of the coastal ridge islands simply as "Everglades." This establishes that prior to drainage, some form of Everglades landscape bordered the islands directly. At the same time, simple absence of mention of a transitional feature in historic accounts does not prove that the feature was in fact absent. For instance, it is possible that a strip of wet prairie *was* originally present, and that the land surveyors saw no reason to distinguish wet prairie from Ridge and Slough landscape, and therefore simply labeled everything west as "Everglades." Additional information was still needed to establish what plant communities this label referred to, whether wet prairies, pure sawgrass, or Ridge and Slough landscape.

Fortunately the field notes from predrainage and early postdrainage land surveys of the Range 41 townships, encompassing the area of the purported transitional strip, are available and contain the needed plant information. Ten transects, each six miles (10 km) long, were compiled for this area (appendix D). Differing surveyors, time periods, and surveying instructions mean that the transects differ somewhat, but the overall picture is similar. For the portion of each transect west of the coastal ridge, the notes for Townships 50–52 repeatedly mention sloughs, lily pad sloughs, myrtle ridges, sawgrass, and hammocks (tree islands), all clearly reflective of Ridge and Slough landscape and not of the sand-based wet prairies observed by J. H. Davis in the 1940s. Comparison of eastern and western halves of the transects in Townships 50–52 confirms that the transition from Ridge and Slough landscape to upland vegetation was generally abrupt, occurring in less than a mile (1.6 km). (This blurred somewhat in the 1911–12 surveys, due to the rapid influence of the then recently excavated drainage canals.)

The hypothesis that the Ridge and Slough landscape originally extended directly to the coastal ridge, without an intervening wet prairie, is additionally supported by evidence of substantial predrainage thicknesses of peat soil close to the coastal ridge. Figure 4.1 shows measurements of multiple feet of peat present in 1912 in areas that by the 1940s were mapped as Davie mucky fine sands; that is, as wet prairie soils in which the originally distinct peat layer had oxidized away. From this it seems clear that the white areas visible in 1940 in Townships 50–52 Range 41 (plate 40), are very likely exposed sands that were originally covered with peat soil.

Determining the predrainage condition of the whitish, wet prairie area in the next township south, Township 53 Range 41, was somewhat more difficult. Transects 8, 9, and 10, which are all in this township and were all surveyed by George MacKay in 1845, mention "sawgrass," "water," "slues" [sloughs], and what appear to be small tree islands of cocoa plum. All of these are indicative of Ridge and Slough landscape. However, MacKay also mentions "wet prairies," raising the question of whether J. H. Davis's wet prairies of the 1940s might in fact already have been present in Township 53 Range 41 one hundred years earlier, well prior to drainage. On the other hand, early observers applied the term "prairie" to various Everglades landscapes (e.g., A. M. Brooks 1880, p. 253; Canova 1885, p. 65; Wright 1912, p. 17), so "wet prairie" in MacKay's survey could conceivably refer to elements of the Ridge and Slough landscape, but it equally well might not. Additional sources of information are needed to determine conclusively if this area was originally Ridge and Slough landscape or wet prairie.

Figure 4.2, an oblique view of the southeast quarter of Township 53 Range 41, clearly suggests tree islands, probably surrounded by a field of sawgrass. Unfortunately they are too far away and at too low an angle to assess the presence of sloughs or to distinguish sawgrass from wet prairie species. The tree islands do seem to approach the woods of the coastal ridge (foreground), but this too is difficult to assess precisely.

Figure 4.3 is unambiguous, clearly showing sawgrass and a tree island, both indicative of Ridge and Slough landscape. The location, "a few kilometers west" of the head of the Miami River and the coastal ridge, strongly suggests that navigable, open water sloughs must also have been present, as it is highly unlikely that Harshberger would have pressed several kilometers through solid sawgrass. (For other indications of sloughs in this area, see also Dix and MacGonigle, 1905, p. 526; Marchman 1947, p. 27; Willoughby 1898a, pp. 162–164.) This location would be in the purported wet prairie area and the whitish area on plate 40. The original presence of Ridge and Slough landscape in this area is further supported by figure 4.4, which shows the rapids present at the head of the Miami River, north fork; that is, at the transition from Everglades to coastal ridge. The narrative record suggests that these rapids flowed year-round in most years (e.g., Jones 1847, p. 61; Rodgers 1848, p. 61; Wright 1912, p. 24), which is consistent with a water source derived from Ridge and Slough landscape but would not be consistent with an area of seasonally drying wet prairies.

Photographs and the existence of the rapids favor the hypothesis of Ridge and Slough landscape rather than wet prairies in Township 53 Range 41, but additional evidence would be useful. Plate 40 suggests that

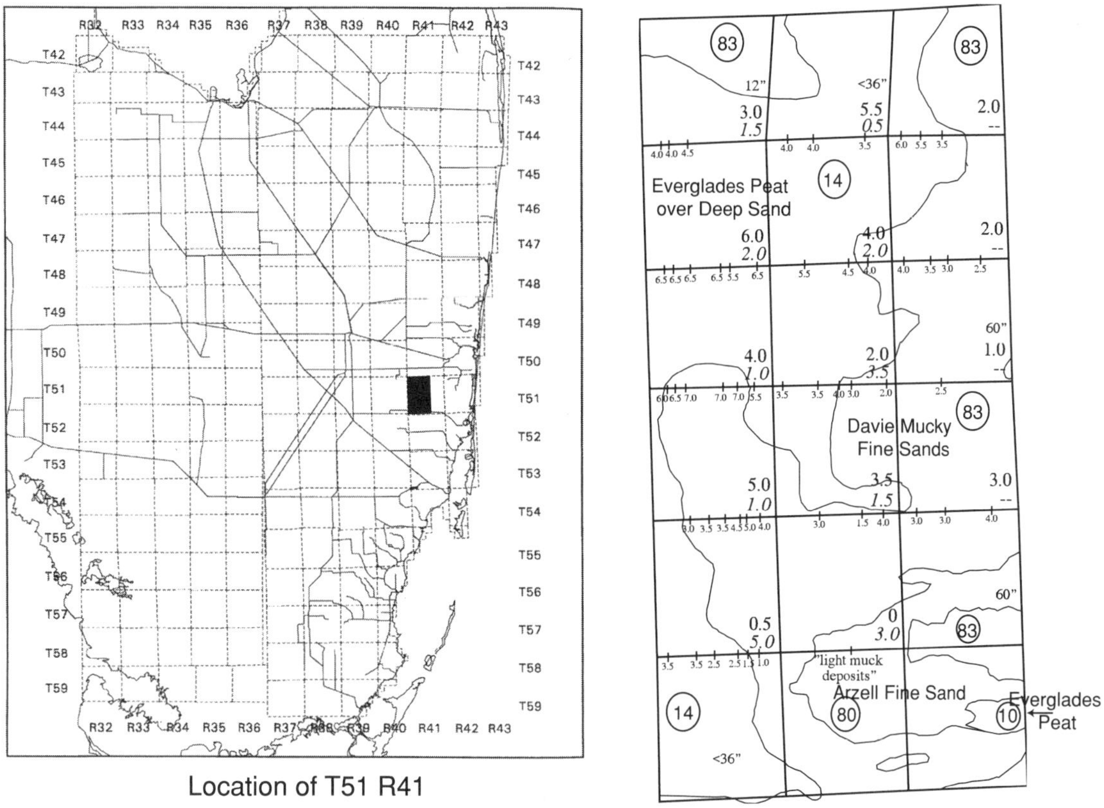

Figure 4.1. Peat loss in the western half of T51 R41, between 1912 and 1940. This half township was close to the eastern edge of the Ridge and Slough landscape (plate 41) and is presently bounded by Hiatus Rd. (W), University Dr. (E), Stirling Rd. (N), and NW 199 St. (S). The peat thicknesses of 3 to 6 or more feet (1–2 m) were measured during a township survey in 1912 (small numbers along tick marks). Irregular polygons marked with circled numbers indicate soil types mapped in the 1940s, each of which had an associated peat thickness: zero for types 83, Davie Mucky Fine Sand, and 80, Arzell Fine Sand, and 0 to 3 feet (0–0.9 m) for soil type 14, Everglades Peat over Deep Sand. Comparison with the 1912 peat thicknesses suggests soil subsidence of 1 to 5 feet (0.3–1.5 m) in this area, with 3 feet (1 m) being typical. Additional data from 1912 is indicated in the lower right section corners; from top to bottom: surface water depths when present (inches; smaller numbers); thickness of organic soil (feet); and thickness of underlying sand (feet; italics). Sources: Kackley (1912-T51 R41; see list of Township Survey References, page 321); Jones et al. (1948a, e).

Figure 4.2. View of the edge of the Everglades at the head of the Miami River, looking northwest from the former Cardale Tower, ca. 1913 (T53 R41, Sections 20, 28, 29; see also plates 33, 40). Foreground: North fork of the Miami River (left) and newly excavated Miami Canal (right), both crossing a wooded portion of Atlantic Coastal Ridge. The falls in the north fork (fig. 4.4) and nearby Burch Tower/Sallie Observatory, though not visible, were located at the edge of the Everglades (upper left corner of the photo). The road crossing the center is NW 27th Avenue (western border of section 34). Background: Ridge and Slough landscape, with sawgrass ridges (light gray) and tree islands (dark gray patches, more irregular elevation). In April 1845, U.S. Surveyor George MacKay described these sections (28, 29, and 20) as: "Everglades prairie[,] small islands [of] cocoa plum. . . . Open Everglades . . . post in water 6 inches. . . . Small clumps of myrtle + C[ocoa] plum all about" (MacKay 1845-T53 R41). By 1938 the Miami Springs Golf Course would have been visible in the center background. Photograph used with permission of the Historical Museum of Southern Florida, Miami.

Figure 4.3. "Tall saw-grass of Everglades and everglade hammock [tree island] at head of the Miami River, August 12, 1912. Original." In the text Harshberger (1914) gives the location as "a few kilometers west of the head of the Miami River." This photograph can be thought of as a close-up of one of the tree islands seen in the background of figure 4.2. It may also correspond to one of the "small islands of coco plum" reported by MacKay (1845-T53 R41). Photograph from Harshberger (1914). Used by permission of the Wagner Free Institute of Science, Philadelphia, Pennsylvania.

Figure 4.4. The famous rapids or falls on the north fork of the Miami River (see fig. 4.2 for location). People heading upstream "customarily did so on the south fork where there was less current and a shallow rapids falling less than a foot," whereas those heading downstream "normally used the north fork, [where] . . . the water fell about six feet while traversing some 450 feet of rapids" (Gaby 1993, p. 2). An 1896 account of paddling the smaller fork is instructive: "We struck the South Fork, the water becoming swifter and swifter. . . . I could hear Brewer panting behind me. . . . After about three-quarters of an hour of the hardest paddling I think I have ever done, the water slowed up a little and we could get some speed on the canoe. The trees opened up more, the stream becoming narrower and narrower, until we came to an opening where everything was clear ahead. This was the edge of the Everglades" (Willoughby 1898a, pp. 39–40). Top: The rapids in 1896, Florida Photographic Collection, used by permission of the Florida State Archives. Bottom: Photograph ca. 1890 by Ralph M. Munroe, Ralph M. Munroe Collection, used by permission of the Historical Museum of Southern Florida.

the whitish area seen in Township 53 in 1940, especially in the southwest quarter of Range 41 and the adjacent southeast quarter of Range 40 (i.e., in the wedge formed by the Miami and Tamiami canals) corresponds to the wet prairies mapped by J. H. Davis. It would be helpful to know whether organic soil, indicative of Ridge and Slough landscape, originally covered this area. The upper right corner of plate 33 shows this area two years earlier (1938) and in more detail. The whitish area between the canals appears to be mottled with small dark spots. Farther west, in Range 39, clear Ridge and Slough texture is visible, along with several oval tree islands. East of this, in Range 40 and approaching the whitish area, the Ridge and Slough texture seems to continue, here with a southeasterly orientation, but less clearly. Three oval tree islands are oriented northwest to southeast. The weakening texture with distance eastward, and the street grid present across the Range 40–41 border, south of the Miami Canal, are both suggestive of water levels lowered by drainage and soil oxidation, supporting the hypothesis that the whitish area might originally have been covered by peat. A comparison with aerial photographs taken thirteen years earlier in 1925 (not shown) is also suggestive; each of the dark spots forming the mottling in plate 33 appears to have been larger in 1925. The spotty pattern and the decreasing size of the spots after the onset of drainage suggest that they are low-lying, remnant patches of peat soil that originally formed a continuous layer over the whole area.

Measurements of peat soil thickness made in 1912 are available for Township 53 Range 40 (fig. 4.5), which includes the western half of the whitish area in question. As was the case for Township 51 Range 41 farther north, the measurements show the presence of one or more feet (30+ cm) of peat in 1912 in areas that were later mapped in the 1940s as peatless Davie sands. This confirms the postdrainage loss of peat suggested by the 1925, 1938, and 1940 aerial photographs, and lends further support to the hypothesis that peat soil, along with Ridge and Slough landscape, originally extended all the way to the coastal ridge.

Stepping back a moment for a more regional view of

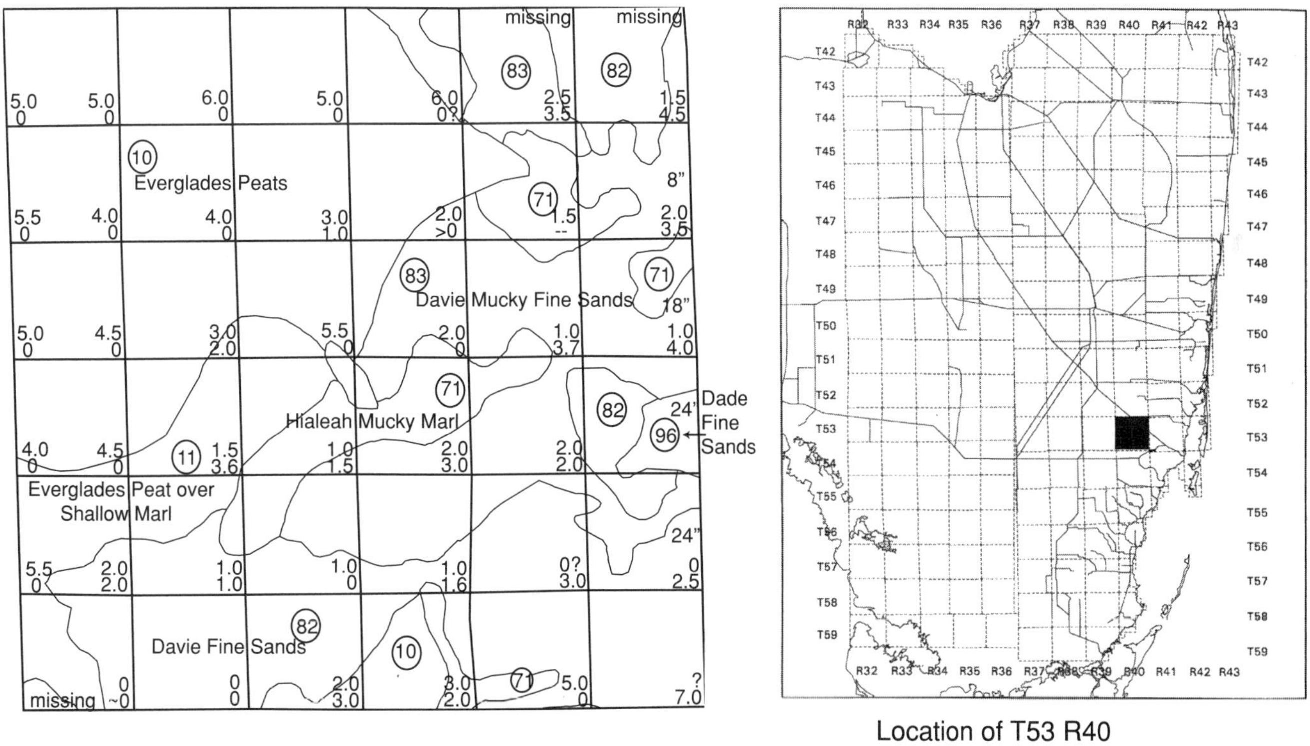

Figure 4.5. Peat loss between 1912 and 1940 along southeastern edge of the Ridge and Slough landscape, near the head of the Miami River (plates 40 and 41; box J in figs. 3.8 and 3.9). The southeast corner of T53 R40 presently bisects the Miami airport; the Homestead Extension of Florida Turnpike borders the western edge. Peat thickness in 1912 thinned to the southeast, from about 6 feet (1.8 m) down to about 1 foot (0.9 m); upper of the paired numbers in the section corners. Irregular polygons marked with circled numbers indicate soil types mapped in the 1940s, with associated peat thicknesses of zero for soil types 71, 82, and 83 and less than 3 feet (<0.9 m) for types 10 and 11. Comparison of 1912 with 1940 peat thicknesses suggests at least 1 to 3 feet (0.3–0.9 m) of soil subsidence toward the northwest and essentially complete loss of the peat toward the southeast. Additional data measured in 1912: thickness of sand underlying peat, increasing to the southeast (feet; lower of the paired numbers in section corners), and a few surface water depths (marked in inches). The large anomalous peat thickness of 5.5 feet (1.7 m) shown in Section 16 was likely measured on the tree island in the center of the township; see also plate 33. Sources: Frederick (1912-T53 R40) and Jones et al. (1948a, e).

the area mapped as wet prairies in the 1940s, one can make use of two maps of peat soil thickness measured in the 1940s, figures 3.8 and 3.9. Along the northern portion of the eastern border of the Everglades, from Township 41 south to about Township 48, both maps show that in the 1940s, peat still extended essentially right up to the border of the Everglades. The rapid thinning of the peat thickness near the border reflects the edge of the bedrock basin as well as westward thinning of a wedge of sand between bedrock and peat. However, farther south, exactly in the area mapped as wet prairie, from Townships 49–50 south to Townships 53–54, the situation is very different. Here figures 3.8 and 3.9 show that by the 1940s, the peat had shrunk 3 to 5 miles (5–8 km) westward from the border, just as one might expect if drainage and oxidation had lowered the surface of the peat. The contrast with the northern portion of the border is consistent with the differing drainage history. The northern portion was less effectively drained (e.g., Allison et al. 1944, p. 25; Lamb 1975), whereas the Township 50–54 portion west of the Fort Lauderdale to Miami area was actively, and apparently successfully, drained by a combination of coastal canals, the large North New River and Miami canals, and various north-south dikes installed to allow agriculture (e.g., Forman 1979, p. 6).

The loss of 2 to 4 feet (60–120 cm) of peat over the large areas suggested by figures 3.8 and 3.9 can be checked by comparison with 1912 measurements of peat thickness. Figure 3.9 shows that by the 1940s, Township 53 Range 40 (Box "J") retained essentially no peat in the eastern half. The maximum remaining peat in 1940 was in the northwest corner, mapped as "less than three feet [<90 cm]." For contrast, figure 4.5 indicates that thirty years earlier, in 1912, the northwest corner was covered with considerably more peat, 4 to 6 feet (120–180 cm) of peat, indicating a loss of 1 to 4 feet (30–120 cm). Similarly, in Township 51 Range 41 (fig. 4.1), comparison with figures 3.8 and 3.9 (Box "I") shows a similar loss of around 3 feet (90 cm) of peat, more than enough to eliminate the original Ridge and Slough landscape, to expose the underlying (white) sand, and to result in sand-based wet prairie plant communities. The words of a family who in 1910 pioneered along the North New River Canal within the area that later became transitional wet prairie help explain these changes: "Every spring muck fires would develop locally [T50 R41] and out in the dried Everglades. Most of the shallow muck that once covered the South Florida Education Center (T50 R41, Sections 22–23) was burned off long ago. The remainder of the muck was destroyed by bacterial oxidation when it became dry" (Forman, 1979, pp. 3–4).

At one location within what became the wet prairie area, the progressive postdrainage loss of peat was in fact systematically measured after the soil was exposed by drainage. Figure 3.5 (lower curve) shows average thickness of peat along a line near Davie (namesake of the newly exposed sandy soils), from 1916 to 1948. The original caption for the graph effectively captures the history. It is very likely representative of the whole bordering strip area: "This line was originally laid out on Everglades peat over sand. In some places along portions of the line, the peat has entirely disappeared with only the bare sand exposed" (Stephens and Johnson 1951b, fig. 11, between pp. 9 and 10).

Figure 3.14 summarizes, in schematic form, the hypotheses proposed at the beginning of this section. Originally a layer of peat several feet thick formed a higher elevation ground surface that extended fully eastward to the trees at the edge of the coastal ridge. With the initiation of drainage the peat began to dry and oxidize, the surface lowered, and the edge moved westward, exposing the underlying sand. This newly exposed area between the pre- and postdrainage edges of the Everglades corresponds to the whitish areas visible on 1940s aerial photographs (e.g., plates 33, 40) and eventually became the sand-based wet prairies. Similar losses of organic soil and shifts to wet prairie species occurred within the Peat Transverse Glades (fig. 3.18).

It is hoped that this section provides an impression of the approaches used to delineate the landscapes on the predrainage map of the Everglades (plate 5) and to develop the individual landscape descriptions in the chapters that follow.

The Predrainage Everglades: A Summary of the Landscapes

Plate 5 is a spatially accurate map of the predrainage Everglades and is one of the key results of this exercise in historical ecology. As illustrated in the preceding example of hindcasting, each line on this map—separating Everglades from uplands, landscape from landscape—is the culmination of considerable research aided by geographical information systems: overlays of land surveys, explorers' routes, soil maps, vegetation maps, aerial photographs, and satellite imagery.

Several features of the predrainage Everglades visible on plate 5 are noteworthy. First, nearly the entire length of the eastern uplands (Eastern Flatwoods or cypress) borders directly on a single Everglades landscape, the Ridge and Slough landscape. Thus all explorers emerging from the coastal woods anywhere between the Jupiter Inlet (Townships 40–41) and the Miami River (Township 54) would have gazed out upon alternating sawgrass

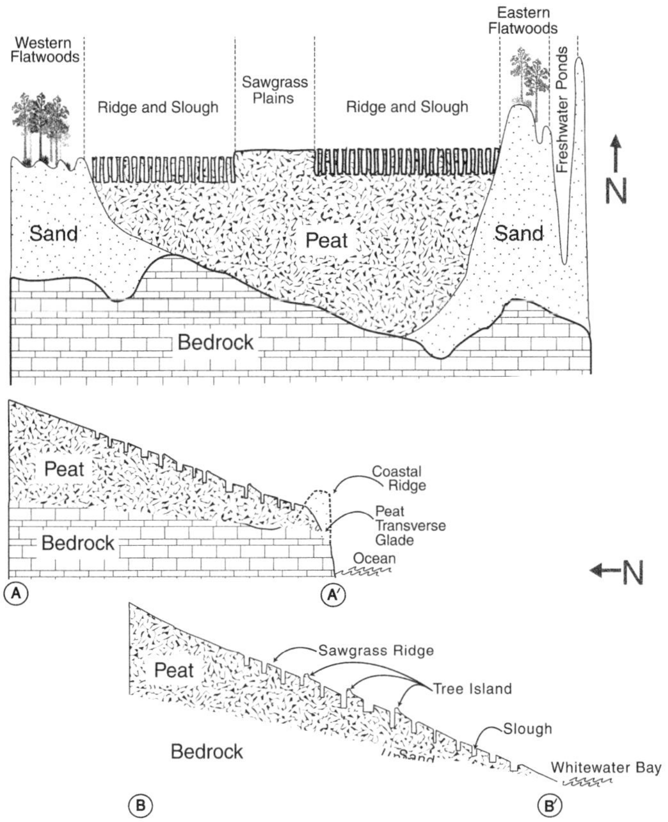

Figure 4.6. Estimated predrainage soil cross sections through the peat soil portions of the Everglades. Top: Transverse direction, approximately perpendicular to flow direction, vicinity of Townships 47 and 48. Bottom: Two downstream cross sections, along the eastern and western flow paths (see chapter 11, Drainage Patterns section). The western cross section B–B' includes Shark Slough and terminates simply by declining steadily to sea level. The eastern cross section A–A' has a more complex termination, with the coastal ridge rising higher than the surface of the adjacent Everglades, while the Peat Transverse Glades pierced the ridge and extended the downward slope from the Everglades to the sea.

ridges and open sloughs, with scattered tree islands rising between them; that is, a landscape like that described by Buckingham Smith at the beginning of this chapter.

Second, the landscape configuration differs significantly north and south of about Township 54 (parallel with Miami). North of Township 54, the peat soil substrate filled the bedrock basin to within a few feet of the rim elevation (Wright 1912, p. 17), allowing the two peat-based landscapes, the Sawgrass Plains and the Ridge and Slough landscape, to cover the full width of the Everglades, thus bordering directly on the adjacent pine or cypress forests on either side (fig. 4.6, top). The topography of the resultant peat surface was essentially independent of the underlying bedrock topography. South of Township 54, the soils and geology were quite different; the bedrock made a steeper sided valley and the shallower soils masked less of the bedrock. This combination meant that the Ochopee and Rockland marl marshes were slightly elevated above Shark River Slough and that they sloped slightly upward with distance away from it (fig. 4.7).

Third, the uplands forming the eastern boundary of the Everglades north of Township 54 are configured in two hydrologically distinct ways. A long, wide section from Townships 40 to 49 had only one major connection to tidewater, the Loxahatchee Slough, which does not seem to have been either a large inflow or outflow (Parker et al. 1955, p. 143). Along a shorter section farther south, from Townships 50 to 54, the uplands narrow and break into a series of separated "islands." The low spots between these islands, the Peat Transverse Glades, formed eastward extensions of Everglades soils and vegetation (Buckingham Smith's "covers [coves] and inden-

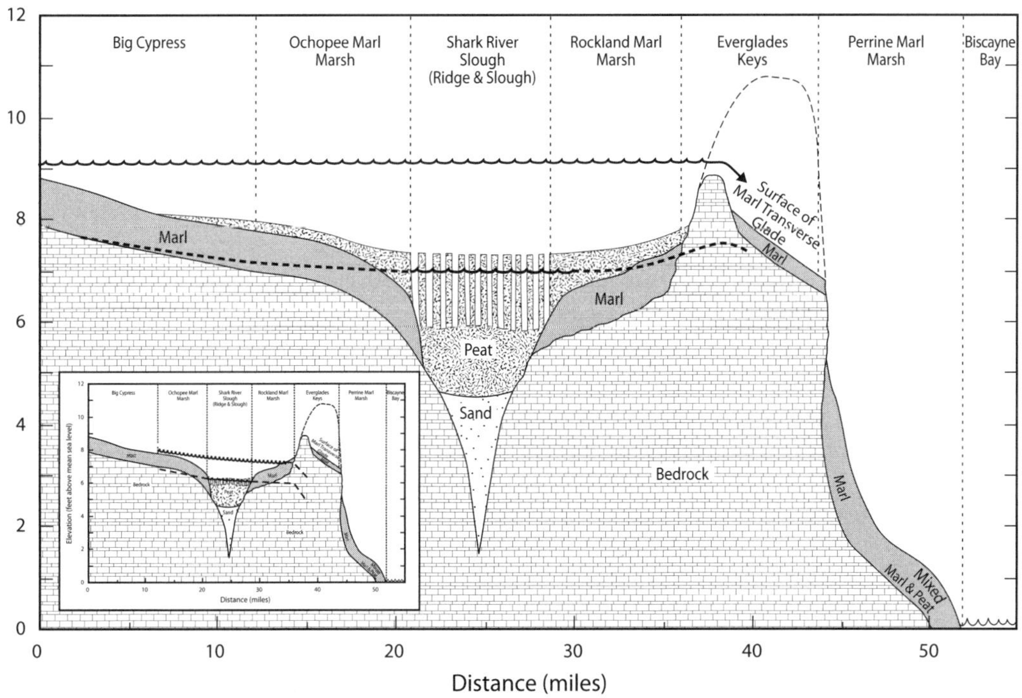

Figure 4.7. Estimated predrainage cross section through the southern Everglades and the Everglades Keys. Location: see plate 35, dotted transverse line. A thin, variable layer of peat extended across much of the Rockland and Ochopee marl marshes. Not shown are the known intermixing and interbedding of marl and peat (e.g., fig. 1.9). Water depths (wavy lines) and hydroperiods on the Marl Marshes were sufficient to support sawgrass and slough-like areas. Contrast with postdrainage conditions shown in plate 36 and fig. 11.5 (reproduced here as inset). Sources: See text; chapters 9 and 11.

tations"; Smith 1848a, p. 49) and conducted water to the various short coastal rivers.

Fourth, along a significant length of the Lake Okeechobee southern shore, sawgrass grew right up to, and formed, the shoreline. The Custard Apple Swamp was a narrow and relatively localized feature forming only a portion of the outflowing interface between Lake Okeechobee and Everglades.

Plate 5 can be used to examine landscape sequences along several different longitudinal and transverse axes. A southeasterly longitudinal axis, stretching from Lake Okeechobee to the Peat Transverse Glades (and ultimately to Biscayne Bay), was frequently mentioned in early accounts, and is suggested by the overall shape of the Everglades, north of Township 54. South of Township 54 the configuration changes, with landscape features oriented along a southwesterly longitudinal axis within Shark River Slough. Considering the Everglades as a whole, both north and south of Township 54, it appears most helpful to think in terms of two parallel axes, an eastern and a western one. Both of these longitudinal axes originated from the southern shores of Lake Okeechobee, and both initially followed an approximately parallel, generally southeasterly course.

Ultimately however, the two axes diverged. The eastern axis is oriented generally southeast with some slight bends paralleling the shape of the eastern border and ending at the Peat Transverse Glades and coastal rivers. The western axis is longer and more noticeably curved: initially the orientation is southeasterly, then south, then curving back southwest into Shark River Slough. While the alignment of both axes is suggested by the overall shape of the Everglades, it is delineated in even more detail by the pattern of the tree islands (cf. plates 5, 18 [eastern axis], and 31 [western axis]). Axes and directionality are addressed in more detail in chapter 11.

Topography is also discussed in detail in chapter 11; suffice it here to note that the western axis sloped from approximately 22 feet (6.6 m) above mean sea level at the south shore of Lake Okeechobee to approximately zero at Whitewater Bay (fig. 4.6, bottom; B-B' transect). The eastern axis sloped from approximately 22 feet (6.6 m) above mean sea level to approximately 7 feet (2.1 m) along the western edge of the Atlantic Coastal Ridge (fig. 4.6, bottom; A-A' transect). Eastward of this point the discontinuous configuration of the coastal ridge between the New and Miami rivers meant that two separate slopes emerged: on one hand, "islands" of the higher coastal ridge, well above the Everglades peat; on the other hand, low points—the Peat Transverse Glades and eventually the coastal rivers—which continued the downward slope toward the ocean.

Interestingly, much of the slope present in the predrainage Everglades north of about Township 49 and along both of the longitudinal axes was due not to the slope of the underlying bedrock but rather to the slope of the accumulated peat itself. Figure 4.8 (top) indicates that going northward along the route of the North New River Canal from milepost 30 to milepost 5, the peat thickness increased by 5 feet (1.5 m), whereas figures I.2 and 3.11 indicate that the bedrock along the same route was approximately level, all at elevation about 9 feet.

In the downslope direction, the principal landscape sequence was:

$$[\textit{Lake Okeechobee}] \Rightarrow \left\{\begin{matrix}[\textit{Sawgrass Plains}]\\ [\textit{Cust. Apple Swamp}]\end{matrix}\right\} \Rightarrow [\textit{Sawgrass Plains}] \Rightarrow \begin{bmatrix}\textit{Ridge \&}\\ \textit{Slough}\end{bmatrix},$$

where the curly brackets indicate that the sequence could transition either directly from Lake Okeechobee to the Sawgrass Plains or first through the Custard Apple Swamp. The direct sequence applied more to the western axis, the Custard Apple Swamp sequence more to the eastern axis. Figure 4.8 illustrates some of the differences between the landscapes present along the course of the eastern axis.

In the transverse direction, and north of Township 54, the landscape sequence was generally simple:

$$[\textit{Western Uplands}] \Rightarrow [\textit{Everglades Peatlands}] \Leftarrow [\textit{Eastern Uplands}],$$

where the uplands were pinelands, cypress, or a mix, and the peatlands were either the Sawgrass Plains or the Ridge and Slough landscape. The extent and flatness of the central component, the peat-based portion of the Everglades, was one of the truly unusual, if not unique, aspects of this ecosystem. With less than one or two feet (30–60 cm) of vertical variation across a width of more than thirty miles (48+ km)—a ratio of 1:100,000—this was the source of the "vastness" so often mentioned by early explorers of the Everglades.

South of Township 54, where the peatlands did not fill the width and depth of the bedrock valley, landscape sequences along the transverse axis across Shark River Slough were more complex than farther north. Wetlands formed both on the level peat surface in the center of the southern transverse axis and also on the higher and sloped floodplains of intermixed pinnacle rock, marl, and thin layers of peat (fig. 4.7). The varied ground surface elevation in turn varied the hydrology, soils, and vegetation, yielding the following sequence from west to east:

$$\begin{bmatrix}\textit{Big}\\ \textit{Cypress}\end{bmatrix} \Rightarrow \begin{bmatrix}\textit{Ochopee M. Marsh}\\ \textit{(marl \& peat)}\end{bmatrix} \Rightarrow \begin{bmatrix}\textit{Ridge \& Slough}\\ \textit{(peat)}\end{bmatrix} \Leftarrow \begin{bmatrix}\textit{Rockland M. Marsh}\\ \textit{(marl \& peat)}\end{bmatrix} \Leftarrow \begin{bmatrix}\textit{Everglades}\\ \textit{Keys}\end{bmatrix}$$

Each of the marl marsh areas varied somewhat along the axis, from wet conditions closer to Shark River Slough to drier conditions closer to the bordering uplands.

Figures 4.6 and 4.7 and plate 16 help explain the soil and hydrological differences between the Peat Transverse Glades and the Marl Transverse Glades. The downward slope of the Everglades into the Peat Transverse Glades (fig. 4.6, lower half, top profile) meant that the Peat Transverse Glades were nearly always covered by water. In contrast, water reached and covered the Marl Transverse Glades only after water levels within Shark River Slough had risen high enough to cover the full width of the Rockland Marl Marsh (fig. 4.7; plate 16). Thus these transverse glades were only seasonally inundated. At least as early as 1912, prior to the onset of local, mechanical drainage, tomatoes were grown here in the dry season (Waldin 1912, p. 15).

Sawgrass (*Cladium jamaicense*) helps contrast the different landscapes analyzed in this study. Sawgrass was present in all, with the exception of the Custard Apple Swamp. In the Sawgrass Plains it was by far the dominant species, covering the landscape with a tall canopy so dense as to be invariably described as "impenetrable." The Ridge and Slough landscape also contained dense and relatively tall sawgrass, but it was confined to the long linear ridges and to the margins of some tree islands (Williams et al. 1911, p. 64; Marchman 1947, p. 20; Forthman 1973, p. 13). Between these ridges, the deeper water sloughs with water lilies allowed relatively easy wet season passage by dugout canoes and even, at least once, by "deep-hulled ships' cutters" (Buker 1975, p. 56). Numerous slightly higher tree islands, apparently absent from the Sawgrass Plains, dotted the landscape. Both Sawgrass Plains and the Ridge and Slough landscape occurred on a substrate of accumulated plant remains—peat soil—but the former landscape was unpatterned while the latter formed a distinctly patterned peatland.

Sawgrass also grew in the marl marshes, fairly ubiquitously, but may have been somewhat lower in stature and possibly less dense. Both the Peat Transverse Glades and the Marl Transverse Glades were covered by sawgrass, though likely the sawgrass was considerably denser and taller in the Peat Transverse Glades.

Table 4.2 tabulates the predrainage and current extents of the various landscapes, showing numerically what can be seen qualitatively in plates 2 and 4: the almost complete loss of the original Sawgrass Plains land-

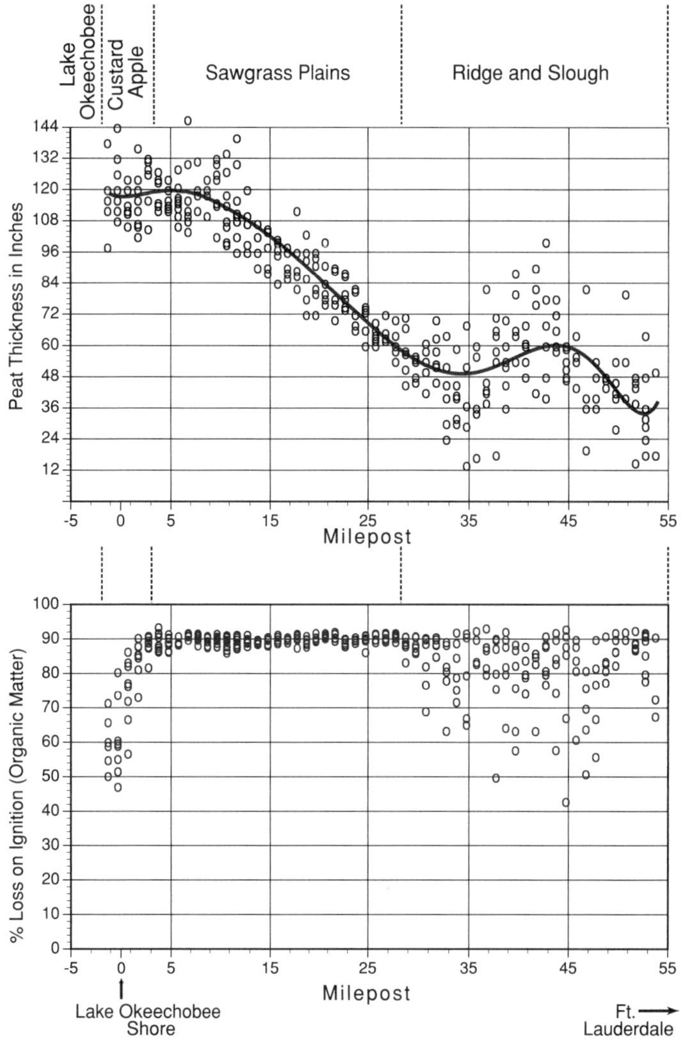

Figure 4.8. Soil thickness (top) and soil organic matter content (percent weight loss on ignition; bottom) measured in 1915 along both sides of the North New River Canal (Baldwin and Hawker 1915; see fig. 2.2 for view of canal and fig. 3.1 for locations of sample points). Near Lake Okeechobee lower organic matter values reflect higher mineral content, likely from silt washed in from the lake. In the Sawgrass Plains note the very low variance in organic matter content, and generally low variance in peat thickness, which likely reflect surface flatness. Greater variability in thickness within the Ridge and Slough landscape likely reflects the differing elevations of ridges, sloughs, and tree islands. Baldwin and Hawker (1915, p. 788) ascribed the reduced organic matter content within the Ridge and Slough landscape to increased presence of fine quartz sand. Noting that there was a "marked current of the water in the sloughs moving in a southeasterly direction," they thought it "probable that the sand owes its distribution in part to the movement of water." Data from Baldwin and Hawker (1915).

Table 4.2. Everglades landscape extents prior to development and drainage (ca. 1850) and at two postdrainage time points, 1940 and 2005

Landscape	ca. 1850 (acres)[a]	% of total	ca. 1940[b] (acres)	% of 1850	ca. 2005[b] (acres)	% of 1850
Custard Apple Swamp	36,000	1%	0	0%	0	0%
Sawgrass Plains	660,000	24%	440,000	66%	43,000	7%
Ridge and Slough	1,500,000	55%	880,000	59%	670,000	45%
Degraded Ridge & Slough[c]	—	—	430,000	29%[d]	420,000	28%[d]
Peat Trans. Glades	15,000	1%	0	0%	0	0%
Rockland Marl Marsh	160,000	6%	150,000	93%	80,000	51%
Degraded Rockland MM[c]	—	—	—	—	22,000	14%[d]
Marl Trans. Glades	17,000	1%	3,000	20%	1,000	5%
Perrine Marl Marsh	220,000	8%	180,000	84%	160,000	74%
Ochopee Marl Marsh	140,000	5%	140,000	100%	140,000	100%
Total	2,730,000	100%	2,220,000	82%	1,540,000	56%

Sources: 1850 extents—this study (plate 5); 1940 extents—intersection of this study with Jones et al (1948a) soil and vegetation map and with U.S. Department of Agriculture–Soil Conservation Service (1940) aerial photographs; 2005 extents—intersection of this study with satellite imagery.

Notes: *a*. 1 acre = 0.405 hectares = 0.00156 square miles.

b. Extents of the original (1850) landscapes remaining in 1940 or 2005 were calculated by deducting areas that had been drained and converted to other land uses. See also note c.

c. Areas where drainage had substantially altered the vegetation and/or the soils, but the land use had not been changed, were classified as "degraded."

d. Area of degraded landscape is referenced to original (1850) landscape extent.

scape to agriculture, loss of a substantial eastern block of the Ridge and Slough landscape to agriculture and urbanization, loss of the northern and eastern portions of the Rockland Marl Marsh, and the nearly complete urbanization of the Eastern Flatwoods and Atlantic Coastal Ridge including the Everglades Keys.

Comparison with Previous Studies

The landscape map produced in this study (plate 5) and the detailed landscape descriptions following this chapter provide a basis for comparison with earlier estimations of the predrainage Everglades—Ives (1856b), Harshberger (1913), and S. M. Davis et al. (1994)—and also with the postdrainage maps of Jones et al. (1948a) and J. H. Davis (1943b).

However, neither Ives (fig. 1.6) nor Harshberger (fig. 1.7) differentiates many landscapes within the Everglades, somewhat limiting the possibilities for comparison. The eastern and western boundaries of the Everglades shown in plate 5 compare very closely to Harshberger's map and also to Ives's map. Ives's map is less spatially exact but does indicate the presence of the Rockland and Perrine marl marshes. Both maps show the Marl Transverse Glades, and Harshberger very clearly indicates the Peat Transverse Glades. Harshberger (1913) overestimated the Custard Apple Swamp, which is understandable, since the detailed land surveys used for the delineation in plate 5 were not available to Harshberger, having been done in 1916, three years *after* completion of his map.

Significant differences are evident between the mapping results of this study (plate 5) and the map of predrainage landscapes estimated by S. M. Davis et al. (1994). This study indicates that the S. M. Davis et al. map reproduces a number of aspects of the Everglades that had developed by the 1940s, but that do not appear to have been present prior to drainage. These include the eastern and western areas of "peripheral wet prairies," an overestimated extent of the (custard apple) "swamp forest," and a central band of sawgrass that in the 1940s appeared to partition the Ridge and Slough landscape into two wetter areas separated by a drier "sawgrass dominated mosaic." The partitioning of the Ridge and Slough landscape largely followed the postdrainage maps of Jones et al. (1948a) and J. H. Davis (1943b), which distinguished an eastern "Hillsboro Lakes" area of Ridge and Slough landscape (Townships 44–49 Ranges 39–41, on plates 13 and 15) and a western "Shark River Slough" area (approximately Townships 48–58 Ranges 35–37; note that this name is now applied only to the area south of Township 53), separated by a drier area between the Miami and North New River canals. As part of the present study, reexamination of the characteristic Ridge and Slough patterning seen in early aerial photographs from these three areas indicated no basis for distinguishing the three, instead finding similar spacing and size of ridges as well as qualitatively similar patterning in all three areas. Additionally, careful reexamination of 1940s hydrology and soil descriptions and mapping objectives used by Jones et al. (1948e) and J. H. Davis (1943a, 1946) indicates that the separate, drier, more sawgrass-dominated central area was an artifact of drainage, combined with the agricultural rather than ecological focus of the 1940s studies.

The differences between the results of this study and earlier historical ecological studies often can be traced to insufficient hindcasting of postdrainage changes in the Everglades. The overall effect of these changes has been to create a spatially more complex landscape; that is, a patchwork of anthropogenically differentiated areas. This man-made increase in spatial complexity is not the same as an increase in ecological complexity; rather it is simply a recognition that water management has increased the number and differentiation of jigsaw puzzle pieces beyond the original five main landscapes. The increased spatial complexity is the combined result of two effects: (1) uneven application and influence of mechanized drainage on the organic soils of the Everglades, and (2) uneven properties—prior to drainage—of these organic soils, particularly in the depth of the organic layer and variations in the underlying substrate. As drainage progressed, the combination of these two

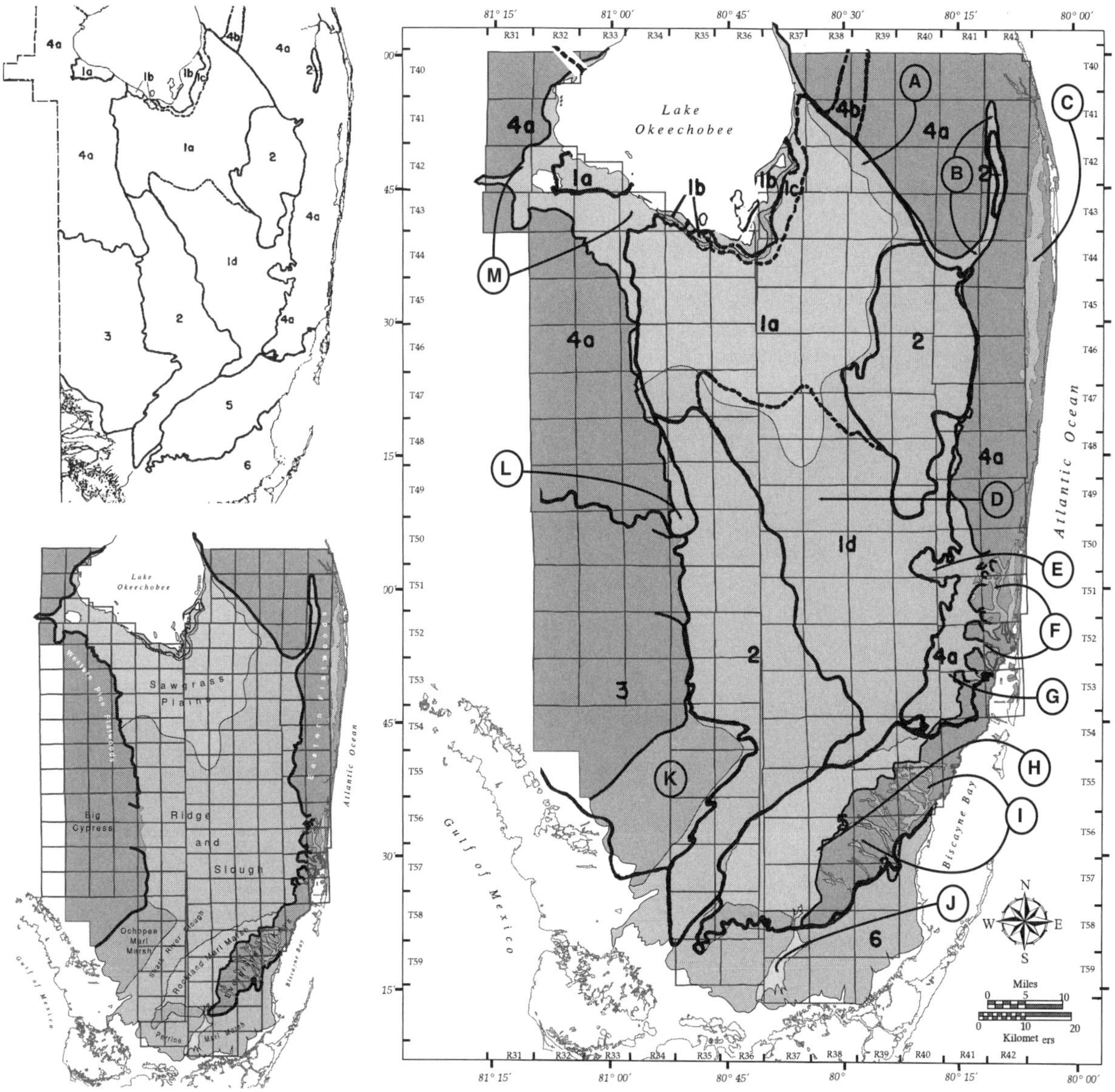

Figure 4.9. Comparison of the predrainage Everglades landscape extents mapped in this study for approximately 1850 (lower left) with physiographic divisions mapped by soil scientists in the 1940s (upper left; Jones et al. 1948a, e). The physiographic divisions mapped in the 1940s reflect a combination of the original predrainage landscapes and the altering influence of up to sixty years of drainage. See table 4.3 for a detailed discussion of the individual areas of difference (circled letters on overlay diagram; right). Physiographic divisions from Jones et al. (1948e) were labeled: 1a = sawgrass plains, 1b = custard-apple, 1c = willow-and-elder, 1d = hammock-sawgrass, 2 = ridge-and-slough, 3 = hammock-and-glades, 4a = coastal ridge and sand prairies, 5 = Miami rock rim, and 6 = coastal marsh.

Table 4.3. **Landscape changes in the Everglades between predrainage conditions (pre-1880s) and the 1940s (letters = circled labels on figure 4.9)**

	Comments	Vegetation/Landscape		Soils	
		Pre-1880	1940	Pre-1880	1940s
A	"An old slough." This arm of the Ridge and Slough landscape originally extended from the present Loxahatchee National Wildlife Refuge north to a cypress stand bordering Lake Okeechobee. Appendix D3 documents an expedition in this area. Jones et al. (1948a) included it in the Sawgrass Plains because, due to drainage, the soil was indistinguishable from Everglades Peat (Jones et al. (1948e, p. 72).	ridge and slough[a]	sawgrass	peat	peat
B	Loxahatchee Slough. Another arm of the Everglades. At different times drained both into and out of the Everglades.	ridge and slough	weedy shrub-land	peat	sands
C	"Freshwater Ponds." 20-mile-long, 2-mile-wide sawgrass marsh surrounded by pine flatwoods and scrub. This area is unnamed in historic literature. It corresponds approximately to the location of Congress Avenue in northern Palm Beach County.	sawgrass	wet prairie	peat	sands
D	Central Ridge and Slough landscape. Drainage by the bordering North New River and Miami canals eliminated sloughs in this large area, replacing them with sawgrass. See chapter 3 for detailed analysis of the changes. Hind-cast ing reestablishes the original configuration of the Ridge and Slough land-scape as a single, continuous unit, rather than two artificially isolated areas.	ridge and slough	sawgrass and tree islands	peat	peat
E	"Sand Protrusion." Appears on maps after 1930. Earlier maps and narratives indicate that this westward extension of sandy grasslands was a postdrainage phenomenon. Effective drainage in the wedge between the North New River and South New River canals oxidized the peat and exposed the underlying sand.	ridge and slough	weedy/ grassland	peat	sands
F	Peat Transverse Glades. The Everglades originally ex tended eastward in arms of peat soil and sawgrass along the borders of the Atlantic coastal rivers (cf. fig. 3.18). Although the peat was already oxidized by the 1940s, the areas are still recognizable on maps of flood risk.	sawgrass	weedy/ agriculture	peat	sands
G	"Peripheral Wet Prairies." J. H. Davis (1943a, pp. 84, 260) notes that drainage-induced oxidation of overlying peat converted the original Ridge and Slough landscape into sand prairies in this area.	ridge and slough	weedy/ grassland	peat	sands
H	Everglades Keys, Marl Transverse Glades, and Rockland Marl Marsh area. Originally a series of pine islands separated by the seasonally wet transverse glades and bordered by marl marsh to the east and west. Drainage eliminated the transverse glades and blurred distinction with marl marsh to west.	pinelands; sawgrass in the Marl Transverse Glades; marl marsh	pinelands; agriculture and overdrained marl marshes. (Distinctions blurred, leaving a continuous "Miami Rock Ridge")	marl and/ or pinnacle rock	marl and/ or pinnacle rock
I	Marl Transverse Glades. Prior to drainage were seasonally wet.	sawgrass	agriculture	marl	marl
J	Taylor Slough. Originally an arm of the Ridge and Slough landscape. Affected by regional drainage.	ridge and slough	sawgrass	peat + marl	peat + marl
K	Ochopee Marl Marsh. Jones et al. (1948a, e) did not distinguish from Big Cypress, perhaps because the area was not of agricultural interest.	marl marsh	hammock and glades	marl	marl
L	Pseudo-wet prairies (western). Drainage-induced wet prairies caused by oxidation of peat down to underlying sand (J. H. Davis, 1943a, p. 84).	ridge and slough	weedy shrub land	peat	sands
M	Okeechobee Marsh and Northwest Sawgrass Plains. Chapter 5 establishes that the southwest side of Lake Okeechobee was bordered by an extensive sawgrass marsh on deep peat. Drainage began in 1882. By the 1940s, extensive areas of peat had oxidized down to the underlying sand.	sawgrass	grassland, mostly non-sawgrass	peat	sands

Note: a. "Ridge and slough" refers to the slightly elevated sawgrass ridges and the lower, aquatic sloughs characteristic of the Ridge and Slough landscape (chapter 8).

spatial non-uniformities transformed an originally vast, extremely flat, and regionally very uniform surface of accumulated peat into a patchwork of areas that now vary discontinuously in substrate, topography, hydrology, and vegetation.

To illustrate this transformation, one can imagine a broad, shallow bowl, the bottom of which is covered with a topographically irregular surface of gravel. A sufficient volume of an opaque slurry (say of flocculent organic peat) poured into the bowl will obscure the irregularities of the gravel bottom and also will settle to form a flat surface. This corresponds to the predrainage Everglades condition. If the slurry of flocculent peat is then allowed to oxidize, simulating the effects of drainage, the surface would lower, reexposing some of the higher irregularities in the underlying gravel. Within the postdrainage Everglades, the decline of the original peat surface was not uniform. Differences in the intensity of drainage and, within impoundments, differing water regimes at the downslope and upslope ends, resulted in differing rates of subsidence and/or loss to fire, further accentuating the spatial differences across the postdrainage Everglades.

Table 4.3 and figure 4.9 provide additional details about landscape differences between predrainage and 1940s postdrainage conditions. Together, this table and figure explain many of the differences resulting from the soil and vegetation changes induced by drainage. The processes used for hindcasting are, in each case, similar to those described earlier in the sample hindcasting section.

Landscape Descriptions

The following chapters consist of detailed descriptions of the individual landscapes of the predrainage Everglades—their vegetation, soils, topography, predrainage spatial extent, and hydrology—and of the adjoining areas: Lake Okeechobee and the slightly higher "uplands" bordering the Everglades to the east and west. The focus is on one landscape at a time; chapter 11 addresses the topographic and hydrologic relationships of all the landscapes together; that is, of the predrainage Everglades as a whole. Each landscape description includes three sections: (1) general description—a synthesis of all available historical observations or descriptions, (2) hydrology—principally water depths and hydroperiods, and (3) hindcasting—including a comparison of 1940s and predrainage extents and some discussion of the reasoning used to hindcast each landscape. The observations forming the basis for the syntheses presented in each chapter are included in the Historical Observations section of the Supplementary Materials (DVD). Longer source materials or specialized compilations of materials are also found in the Appendices section of the Supplementary Materials (DVD).

5

Lake Okeechobee and the Okeechobee Marsh

Prior to drainage efforts, which began in 1882, the hydrology of the Everglades was directly connected to that of Lake Okeechobee. It was clear to all observers, as early as the 1840s and continuing into the early 1900s, that water flowing from Lake Okeechobee into the Everglades formed a significant part of the predrainage Everglades water budget. Accordingly, all plans for drainage of the Everglades, from Buckingham Smith onward, emphasized the necessity of first lowering Lake Okeechobee to eliminate the lake outflows (e.g., B. Smith 1848a, p. 49; Meigs 1879, p. 864; Hamilton Disston in 1881, as recorded in Trustees of the Internal Improvement Fund 1881, pp. 463–467; Broward 1905, p. 100; Wright 1909a, p. 132; F. C. Elliot 1927, p. 30). Wright, a drainage engineer, stated that the lake needed to be "permanently lowered 4 or 5 feet [120–150 cm]" (1909c, p. 160); other engineers suggested similar reductions of 6 feet (180 cm; e.g., J. A. Williams et al. 1911, p. 197).

This chapter focuses on a portion of the predrainage interface between Lake Okeechobee and the Everglades, namely the approximately 25 miles (40 km) of western and southern shoreline where sawgrass marshes directly bordered the lake. (The border between the Custard Apple Swamp and Lake Okeechobee is discussed separately in chapter 7). The Okeechobee Marsh, an important subregion of the Sawgrass Plains, is discussed here in detail because of its particular relevance to the overall lake-Everglades interface and to the changes in that interface that accompanied early (1880–1900) lowering of Lake Okeechobee water levels. The chapter focus is on the geographic, vegetative, and dimensional aspects of the interface; the flows themselves are discussed in more detail in chapter 11, under Inflows from Lake Okeechobee.

Okeechobee Marsh

Unlike the mineral sediments forming the eastern shore of Lake Okeechobee, the western and southern shoreline, from north of Fisheating Creek on the west to Pelican Lake on the east, was formed by the accumulation of organic soils, the partially decomposed remains of aquatic plants. Gleason et al. (1984, p. 301) ascribe the very existence of Lake Okeechobee to the impounding effect of this organic soil. This organic shoreline can be subdivided into three sequential sections, from west to east (for locations, see plate 4): (1) the great sawgrass or "Okeechobee Marsh" surrounding Lake Hicpochee and bordering the southwestern shoreline of Lake Okeechobee (from Fisheating Creek south and southeast to the middle of Range 33); (2) a second, shorter section of sawgrass shoreline, where the main portion of the Sawgrass Plains landscape directly bordered Lake Okeechobee (from the middle of Range 33 eastward to about the eastern border of Range 34); and (3) a third section still farther east, where the Custard Apple Swamp bordered the lake (from Range 35 east to Pelican Lake in Township 42). A hydrologically and chemically significant aspect of this shoreline arrangement is that under predrainage conditions, water flowed from Lake Okeechobee into the Everglades not only in the better known sequence {Lake Okeechobee} => {Custard Apple Swamp} => {Sawgrass Plains} but also in a parallel sequence that passed from Lake Okeechobee directly into the Sawgrass Plains.

The distinction between the herbaceous sawgrass shoreline and the wooded custard apple shoreline was obvious. In contrast, the distinction between the two sawgrass components, the Okeechobee Marsh and the rest of the Sawgrass Plains, was probably invisible, both at the shoreline and farther inland. Both sections were extensive, very flat regions of uninterrupted dense, tall sawgrass growing on deep sawgrass peat soils. Both were frequently described as "impenetrable." Prior to drainage, water depths were similar and similarly connected to water levels in Lake Okeechobee. The strong similarities explain why many maps (including plate 5 of this study) did not distinguish the two areas, showing the Okeechobee Marsh simply as an extension of the Sawgrass Plains.

The differences between the Okeechobee Marsh and the Sawgrass Plains were geographic, hydrologic, and historical. The Okeechobee Marsh, at the southwest corner of Lake Okeechobee, drained westward, forming the headwaters of the Caloosahatchee River and, significantly, the hydrologic link between that river and Lake Okeechobee (fig. 5.1). In contrast, the Sawgrass Plains drained southward into the Ridge and Slough landscape. Thus both sawgrass marshes formed outflows from Lake Okeechobee, but they sloped and drained toward separate outlets. The easier access to the Okeechobee Marsh (from the Caloosahatchee River, Lake Okeechobee, and from the nearby pinelands) meant that the Okeechobee Marsh was explored earlier. It also was the earliest target of large-scale efforts to drain Lake Okeechobee and the Everglades.

The combination of easier access and earlier drainage (1881–1883) created a considerable collection of predrainage and early postdrainage observations of the Okeechobee Marsh area. These help clarify the predrainage hydrologic connection between Lake Okeechobee and the Everglades and also help estimate predrainage conditions within the initially less accessible and hence less documented Sawgrass Plains.

The area referred to here as the "Okeechobee Marsh" —also known variously as the Great Okeechobee Marsh, the Hicpochee and Okeechobee sawgrass, and the Great Sawgrass Marsh—was well known to early drainage organizers. James Kreamer, drainage engineer for Hamilton Disston's company, described it as follows (see also fig. 5.1): "The water along the west margin of Lake Okeechobee, extending for a distance of over thirty miles [along the border of lake], naturally finds an outlet through the Caloosahatchie valley. The great [sawgrass] marsh region west of Lake Okeechobee is triangular in form, the long base bordering the lake, the sides converging towards Fort Thompson, near which point the [Caloosahatchee] river affords egress for the water. In the center of this great marsh is Lake Hickpochee" (Kreamer 1890, p. 95).

J. W. Sackett, assistant engineer with the U.S. (Army) Engineer Office, noted the great density of the sawgrass within the Okeechobee Marsh as well as the presence of a bordering strip of wet prairie: "The greater portion of the [Caloosahatchee] river above Fort Thompson is bordered by an almost impenetrable saw-grass marsh, as is also Lake Okeechobee to a still greater degree. This

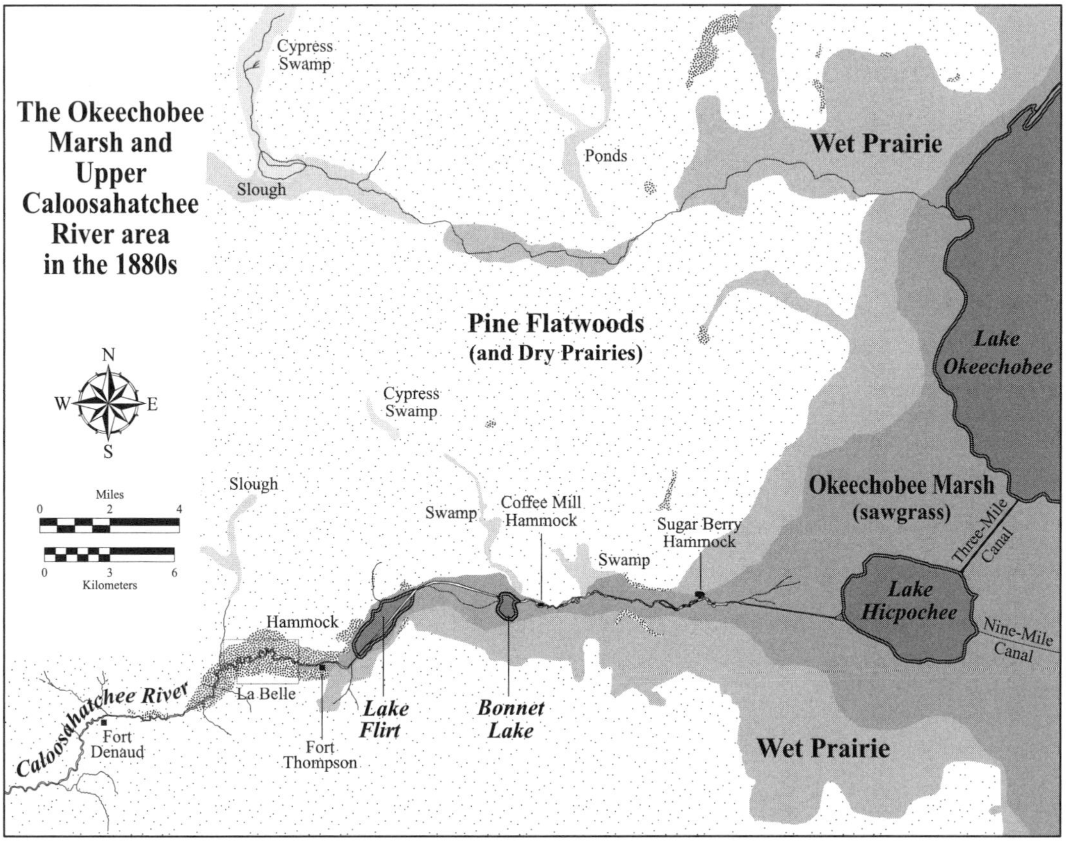

Figure 5.1. Map of the Okeechobee Marsh portion of the Sawgrass Plains, showing principal landmarks, the natural upper limit of the Caloosahatchee River (west of Lake Hicpochee), and location of canals dredged in 1882–1883 to connect the Caloosahatchee River to Lake Okeechobee. Sources: Tannehill (1871-T42 R30, 1871-T42 R31), Meigs (1879), Sackett (1888), Kreamer (1892), U.S. Army Corps of Engineers (1929), and Antonini et al. (2002, pp. 133–135).

marsh is in turn bordered by a [slightly higher lying] prairie which, with a few exceptions, intervenes between it and the [still higher] timbered land" (Sackett 1888).

The bordering wet prairie and the impenetrable sawgrass were also noticed and mapped by Government Land Office surveyors, especially those working between 1860 and 1873 (table 1.3, fig. 1.4). The sequenced elevation of the lower wetland (peat-based sawgrass marsh), higher wetland (sand-based wet prairie), and the still higher timber (sand-based pinelands) reflects the origin of the Okeechobee Marsh through accumulation of sawgrass peat within a sandy valley. Several peat depth transects recorded by Kreamer (1892; see inset to plate 9) illustrate this origin as well: V-shaped profiles with peat thickness tapering to zero toward the edges of the sawgrass and reaching 10 feet and occasionally 12 feet (3–4 m) deep in the middle of the valley. The high uniformity of the sawgrass vegetation throughout the Okeechobee Marsh suggests that the peat surface was very level. By the 1880s this accumulated peat had not quite fully filled the valley, leaving room for the bordering seasonally flooded wet prairies.

As in the rest of the Sawgrass Plains (chapter 6), the process of peat accumulation and the uniformity of the sawgrass stand within the Okeechobee Marsh both suggest that water depths there were quite uniform over long distances. Early descriptions and leveling by surveyors confirm this, indicating that that the level of the water in the marsh was the same as that in Lake Hicpochee and in Lake Okeechobee:

> The water of the saw-grass country [Okeechobee Marsh] is in large part an oozing out product derived from the lake [Hicpochee], just as the waters of the more southern Everglades represent a similar outflow from Lake Okeechobee. Indeed, in view of the very nearly uniform level occupied by the two lakes, and the swamp character [sawgrass marsh] of the intervening territory, it is more than probable . . . that the last named lake [Lake Okeechobee] is itself . . . the most important contributor to the river's basin, largely regulating the height of its waters, and of those of the smaller sister lake [Hicpochee] lying to the west. (Heilprin 1887, p. 406)

Eight years earlier, prior to any drainage efforts, U.S. Assistant Engineer J. L. Meigs (1879) came to the same conclusion after an extensive effort to push through the sawgrass marsh to Lake Okeechobee. With the help of "water and mire for the most part 2 feet deep" (Meigs 1879, p. 865) and by burning the dense sawgrass stand, they were able to force a skiff to within ¼ mile (400 m) of the western shore of Lake Okeechobee. Despite being stopped there by "vast beds of water-lilies [probably *Nymphaea odorata*, possibly also spatterdock, *Nuphar advena*], careless [*Amaranthus australis*] and frog weeds, and wild lettuce [*Pistia stratiotes*]," (p. 865) they concluded that with the water surface in the marsh "being manifestly on the same level with the water of the lakes, it was evident that the leveling might properly be commenced on the south shore of Lake Hikpochee" (p. 865).

Survey measurements made almost thirty years later found the same level connection between marsh and lake. The survey party of Lawrence Brett, E. W. Chadwick, and Paul Funderhide, under direction of John T. Stewart, drainage engineer with the U.S. Department of Agriculture, ran a line of levels from "Travers House" (southwest corner of Township 43 Range 32) through the Okeechobee Marsh, northeast to Lake Okeechobee. Despite the drainage efforts of the 1880s, water was present on the Okeechobee Marsh at the time of their 1907 survey, with the water level being the same as that in Lake Okeechobee: "The men waded the swamp and continued the line [of levels] in the direction of Lake Okeechobee until a point was reached where there was no slope in the surface of the water for several miles. This condition extended to the open water of Lake Okeechobee" (Wright 1909c, p. 153).

Figure 5.2, a profile through the length of the Okeechobee Marsh drawn from survey measurements (Sackett 1888), adds important detail to the preceding narratives of Heilprin; Meigs; and the party of Brett, Chadwick, and Funderhide. Like those observers, at the shore of Lake Okeechobee Sackett found the water level in the Okeechobee Marsh to be one and the same as the water level in Lake Okeechobee. Like the other observers, Sackett found the water depths within the sawgrass of the Okeechobee Marsh to be quite similar at all points. Sackett's key additional result is that both the peat soil surface and the water surface of the marsh sloped downward away from the shoreline. Thus, where the Okeechobee Marsh bordered Lake Okeechobee (southwest shore), the lake shoreline formed a very subtle high point (actually a "high line"), with the lake bottom sloping downward toward the center of the lake (roughly eastward), and the Okeechobee Marsh land surface sloping downward toward the west (fig. 5.3A). Assuming that any peat subsidence that might have occurred between 1882 and 1888 had occurred evenly across the Okeechobee Marsh, then figure 5.2 suggests that the downward slope was about 8×10^{-5}, somewhat greater than the overall slope of the Everglades, 5×10^{-5}.

Several observers described the thickness of the layer

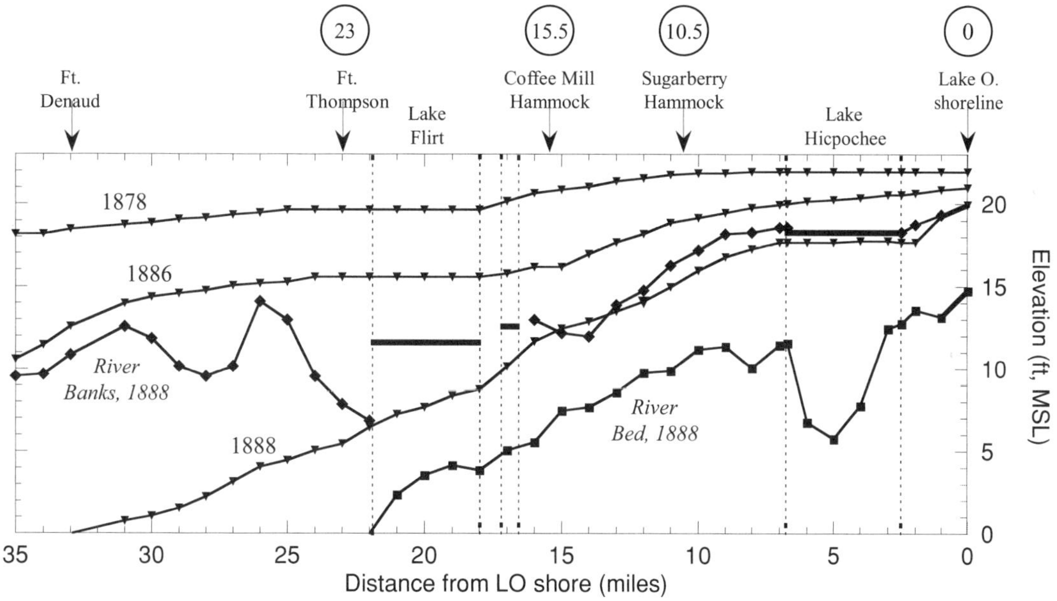

Figure 5.2. Profile of early ground and water surfaces of the upper Caloosahatchee River and Okeechobee Marsh: high water of 1878, prior to existence of the canals; high water of 1886, after canal completion; river (or canal) bed and banks in 1888, five years after dredging of canals; and dry season water level in 1888. Circled numbers are "canal miles" downstream from Lake Okeechobee. From mile 0 to about mile 15, the river bank elevation corresponds to the elevation of the Okeechobee Marsh surface and will reflect any subsidence that may have occurred between 1883 and 1888. In 1878, note the level water surface within the full width of the Okeechobee Marsh (miles 0 to 10.5); this would have been continuous with the water surface of Lake Okeechobee. In contrast, by 1888, drainage had drawn the water table below ground surface throughout the width of Okeechobee Marsh. Source: Sackett (1888).

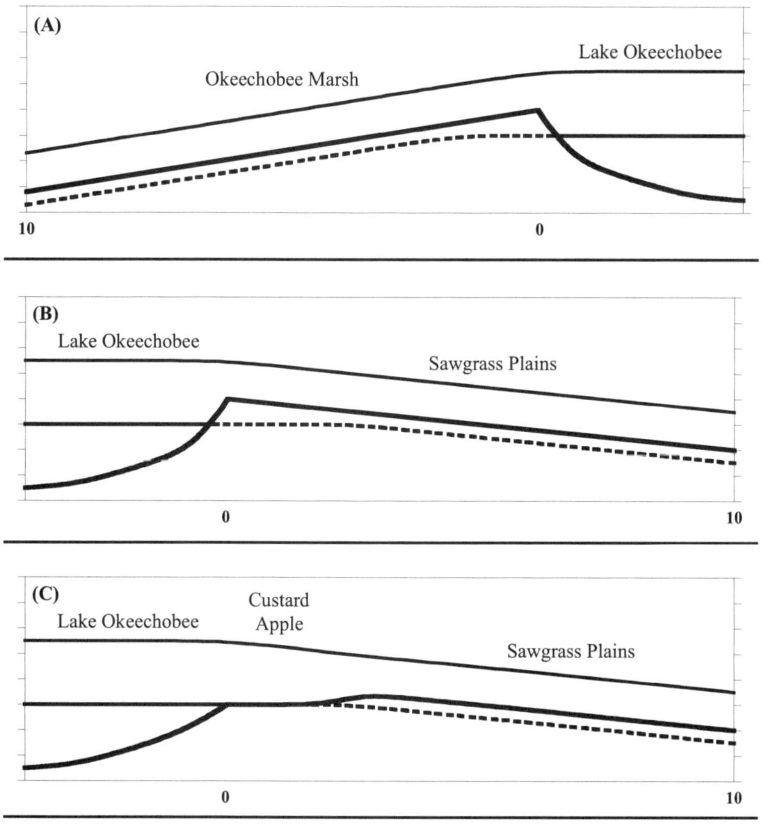

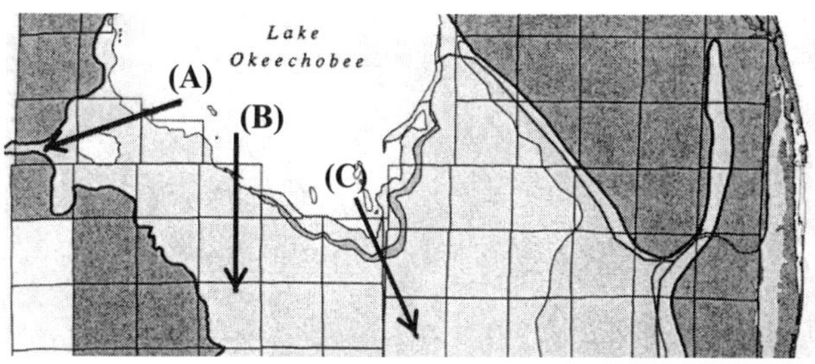

Figure 5.3. Cross sections of predrainage water and ground surfaces, extending 10 miles (16 km) from Lake Okeechobee into the Everglades in three directions: (A) southwestward into the Okeechobee Marsh (Lake Hicpochee not shown), (B) southward into the Sawgrass Plains, and (C) southeastward through the Custard Apple Swamp. Upper line illustrates typical end-of-wet-season conditions; lower line (dashed to indicate below ground) illustrates typical end-of-dry-season conditions. Vertical axis tick marks equal 1 vertical foot (30 cm). Sources: Okeechobee Marsh, Sackett (1888); Sawgrass Plains, chapters 6 and 11; Custard Apple Swamp, chapters 7 and 11.

of sheet flow passing over the Okeechobee Marsh. James Kreamer, engineer for Hamilton Disston's drainage company, indicates that "prior to beginning our [drainage] operations [in 1882], the entire marsh country was permanently covered with water. At high stages this sawgrass was inundated to a depth of 24 to 30 inches" (1890, p. 95). Meigs's observation of fairly deep water, 10 to 20 inches (25–50 cm), toward the end of a dry season during a year not known for exceptional rainfall (1879) suggests that Kreamer's observations of 24 to 30 inches (60–75 cm) may in fact be representative of normal wet season depths, rather than exceptional "high stage" events. Tannehill's description further suggests that 24 or more inches (60 cm) would have been closer to typical: "the saw grass extends SE-E-NE + NW in an unbroken and dense marsh with two feet [60 cm] or more of water at all times" (Tannehill 1871-T42 R31; see list of Township Survey References, pages 321).

Under exceptional circumstances the normal condition shown in Sackett's data (fig. 5.2)—that is, marsh water sloping downward to the west, away from the lake—could alter dramatically if westward outflows from Lake Okeechobee exceeded the outflow capacity at the rapids near Fort Thompson (canal mile 23). At such times the backed-up water surface could extend as a flat surface from Lake Okeechobee to the western edge of the Okeechobee Marsh, and from there with only a two-foot drop over the next 12 miles (0.6 m over 19 km) to Fort Thompson (fig. 5.2, High Water of 1878): "at low stages . . . the water very slowly passed from Lake Okeechobee [through the sawgrass of the Okeechobee Marsh] into the river below; while a rise of only a foot in the lake permitted the water to pass with such freedom that the entire Caloosahatchee valley would be very quickly flooded" (Sackett 1888).

The predrainage vegetation at the shoreline appears to have been consistent with the subtlety of the high point; the dense sawgrass stand of the interior of the Okeechobee Marsh simply extended, without change, and largely without trees, right up to the shoreline: "The chief difficulty was to connect them [the line of levels from Caloosahatchee River] with Lake Okeechobee, as the saw-grass margin surrounding the lake is very wide and presents no breaks in the vicinity of the Upper Caloosahatchee" (Sackett 1888). "The marsh here is intersected by many creeks and bayous, but very few trees appear. . . . In a long stretch of coast, but one tree is in sight" (Ober 1874, p. 593). Despite the smooth continuity of the water surface between Lake Okeechobee and the Okeechobee Marsh, the transition between the two areas was distinct and easily recognized from the vegetation: "The border line of the lake is in most places not absolutely defined, owing to a continuous passage of the open waters into those of the Everglades; on the whole, however, the delimitation of the latter region is fairly well marked, the growth of saw-grass or flag terminating rather abruptly" (Heilprin 1887, p. 413).

The abrupt edge of the sawgrass, somewhat surprising given the continuous water level between marsh and lake, may reflect a possibly sharp change in the substrate material and elevation. Predrainage nearshore lake data is lacking, but lake soundings in the southwest portion of Lake Okeechobee a few miles (3–5 km) from the shore indicate water depths of 4 to 10 feet (1.3–3 m), as reported by Heilprin in 1887—7 to 10 feet (2–3 m, p. 411); by Bullock in 1891—6 feet (2 m, p. 27); and by Wright in 1909—4 to 7 feet (1.3–2 m, 1909b). Heilprin (1887, p. 412) indicates a "firm bed of sand" as substrate for the "lake proper." The shoreline edge of sawgrass may, therefore, have coincided with a fairly sharp transition from approximately 2 feet (60 cm) of water depth on top of deep peat within the Okeechobee Marsh to 4 or more feet (>1.3 m) of water depth and a mineral substrate within Lake Okeechobee.

The trees or shrubs one might have expected at this well-marked shoreline were conspicuous in their absence: "The [southwestern] shore of the lake [Okeechobee] was marked by only 3 groups of trees, 2 cypress and 1 of willows, being everywhere else, as far as could be seen, bare of trees" Meigs (1879, p. 865). Navigators approaching the shoreline from within Lake Okeechobee noticed the same absence of anything but sawgrass: "[In July of 1857] we . . . began looking for an outlet into the Everglades, but without success. We were bounded by a high, almost impenetrable wall of saw-grass, through which the water from the lake oozed its way into the Everglades" (Canova 1885, pp. 29–30). "We soon find ourselves sailing along a shore destitute of trees, the saw grass marsh reaching to the waters edge" (A. P. Williams 1882). "[In 1914] it was all sawgrass for miles and miles around the lake" (Storter 2000, p. 81). The general absence of trees along the sawgrass shoreline probably explains the effectiveness of a once well-known shoreline landmark, the Lone or Sentinel Cypress (fig. 5.4). If trees had been common along or near the shoreline, the Lone Cypress would not have stood out as a landmark.

The ubiquity of sawgrass along the southwestern shoreline under predrainage conditions together with the absence of trees or shrubs constitute key information for understanding the predrainage interface between Lake Okeechobee and the Okeechobee Marsh. The vegetation record both clarifies the physical configuration of the outflowing shoreline and constrains the likely

Figure 5.4. The Sentinel or Lone Cypress, a well-known landmark on the predrainage shoreline of Lake Okeechobee, now within the city of Moorehaven. Views are from about 1883 (top) and the 1920s or 1930s, both looking to the north. When the Three-Mile Canal was dredged between Lake Okeechobee and Lake Hicpochee, the Sentinel Cypress was on the west side of the canal where it joined Lake Okeechobee. Referred to at least as early as 1888 (Sackett 1888), the Sentinel Cypress still stands in 2010, although lake levels are now far below the base of the tree and the lake shoreline is more than 0.5 mile (0.8 km) away.

range of predrainage lake stages. In addition, the presence of sawgrass, exposed along miles of shoreline to nearly year-long outflows from Lake Okeechobee, carries strong implications regarding the phosphorus concentrations (i.e., low concentrations) in predevelopment lake water.

The continuity of the sawgrass right up to the shoreline and the almost complete absence of shrubs or trees indicate that there was not a distinct, elevated rim at the shoreline. Although the peat surface of the Okeechobee Marsh sloped downward away from the lake, the same slope, without local distortion, apparently continued all the way to the shoreline. If a nearshore distortion in slope—that is, an elevated rim—had been present, it would have experienced a shallower, shorter hydroperiod environment and would have been visibly colonized by woody species.

This continuous, rimless connection with Lake Okeechobee also suggests that variations in lake water levels usually occurred within the 2- to 3-foot (60–90 cm) annual range that supports the growth of tall, dense sawgrass (fig. 5.3A, B). If in this area lake levels had frequently decreased and remained below ground elevation for long periods of time, the sawgrass would have been invaded and replaced by species such as dog fennel (*Eupatorium capillifolium*), careless weed (*Amaranthus* spp.), and shrub thickets that prefer drier conditions. Such an invasion pattern was in fact reported after drainage by Disston's canals and later after the early state canals had lowered lake levels (Baldwin and Hawker 1915, pp. 752–753; Wright 1912, p. 26; Allison et al. 1927, p. 39; Graham 1951 p. 32; Allison and Dachnowski-Stokes 1932, p. 228). Similarly, if lake stages had regularly persisted at high levels corresponding to very deep marsh depths—that is, greater than 2 or 3 feet (60–90 cm)—the sawgrass marsh would have been replaced by sloughlike conditions of open water and water lilies. The presence of deep sawgrass peats (Meigs 1879, pp. 866–867; Kreamer 1892; Jones et al. 1948a) suggests that in fact predrainage lake levels generally remained in a range that was optimum for the adjoining sawgrass: an approximately 2-foot (60 cm) annual rise and fall, with the water table in the marsh dropping slightly below the peat soil surface during the last few months of the dry season.

Finally, the existence of a long stretch of sawgrass marsh, directly and continuously flushed by lake water, suggests that predevelopment phosphorus concentrations in Lake Okeechobee were quite low. Recent studies indicate that when water column concentrations in sawgrass marshes of the Everglades exceed background levels (about 10 ppb), a vegetation shift typically occurs, with cattails (*Typha* spp.) replacing sawgrass (McCormick et al. 2002, pp. 109–114; Sklar et al. 2005, p. 162). No reports of cattails along the predevelopment, predrainage shoreline were found.

The unusual rimless connection between Lake Okeechobee and the Okeechobee Marsh has sometimes led to the impression that part of the Everglades, especially the Okeechobee Marsh and other parts of the Sawgrass Plains nearest the lake, formed a predrainage littoral zone for Lake Okeechobee (Aumen 1995, p. 3; Aumen and Gray 1995, p. 344; Steinman et al. 2002, p. 23; Havens and Gawlik 2005, pp. 908–909). Similarly, some authors have suggested that Lake Okeechobee was significantly larger prior to construction of levees in the 1920s and 1930s. The latter impression appears to have grown out of a related concept of considering the Okeechobee Marsh, at least seasonally, as part of the waters of the lake. For instance, Lawrence Will states: "Here [on the 'southwestern arc'] the lake bottom was so flat, grown over with flags, bonnets, and high grass, that when the lake rose after the summer rains, the shore line could move back for half a dozen miles or so" (Will 1964, p. 8).

Whether the Okeechobee Marsh should be considered as a littoral zone for Lake Okeechobee is in part a semantic question. The two waters were certainly connected, and in such a way that water was more likely to be present within the Okeechobee Marsh when lake levels were high, and less likely when lake levels were sufficiently low, thus meeting one of the criteria for a littoral zone. At the same time littoral zones are normally the result of rising lake levels inundating successively higher elevation zones, whereas for the Okeechobee Marsh (as well as the other portions of the Sawgrass Plains bordering the lake) the relation was the opposite; one of outflow into a lower-lying marsh (fig. 5.3A, B).

Lake Okeechobee Areal Extent

There may be less uncertainty regarding the postdrainage size of Lake Okeechobee relative to its predrainage size. The lake shoreline was generally taken as the edge of the sawgrass or custard apple vegetation. This is reflected in general maps of southern Florida (e.g., Florida Surveyor General's Office 1849, 1853; Ives 1856b; D. H. Elliott 1890; Trustees of the Internal Improvement Fund 1908b) and in plat maps by the U.S. General Land Office surveyors (cf., Hancock's account of meandering the western shoreline, 1907-T40 R32). The fact that these mapmakers were aware of and drew the sawgrass areas (Okeechobee Marsh and Sawgrass Plains) but did not include them as part of Lake Okeechobee suggests that the sawgrass areas were considered as separate, downstream areas. Accepting the mapmakers' shoreline definition makes it relatively straightforward to compare current and predrainage lake areas. The present levee-defined shoreline is fairly similar to the original, generally being not more than a mile or at most 2 miles (1–3 km) distant from the original. A 1921 map (Trustees of the Internal Improvement Fund 1921) shows an even smaller distance between an approximate predrainage shoreline (20 feet above Punta Rasa datum) and a (drained) shoreline at 16 feet above Punta Rasa datum. The average distance inward was about 0.4 mile (0.6 km); the maximum about 0.9 mile (1.4 km). Thus the current lake area is slightly but not dramatically less than the original. Overall, the fact that the Okeechobee Marsh and the Sawgrass Plains sloped downward away from Lake Okeechobee suggests that the sawgrass areas were not simply extensions of the lake but instead formed hydrologically and ecologically distinct units.

Prior to 1883, when Disston's company completed several connecting canals, the Caloosahatchee River did not connect directly as a river to Lake Okeechobee, or even to Lake Hicpochee: "Before the Drainage Company's canal was open there was no clearly defined channel [in the Okeechobee Marsh] . . . within about 8 miles of Lake Okeechobee" (Sackett 1888). Instead, as shown on a plat map by Tannehill (1871-T42 R31) and described by J. Meigs, an assistant engineer with the U.S. (Army) Engineer Office (1879), the river drained from the Okeechobee Marsh, which in turn received water from Lake Okeechobee. During his examination Meigs found that the Caloosahatchee River divided into many prongs, west of Lake Hicpochee (fig. 5.1), and that these became ever narrower and shallower "until the canoe could be forced no farther through the dense [saw]grass. . . . No one of these prongs of the Caloosahatchee furnished an entrance into Lake Hikpochee, though it was manifest that the water in them was supplied by the lake" (1879, p. 865).

Meigs's observations from 1879, made prior to the dredging of any canals, provide a good sense of the flow regime within the triangular-shaped Okeechobee Marsh: a sheet of water with an even depth over the marsh, with velocity highest toward the west (narrowest part of marsh) and decreasing eastward toward Lake Okeechobee, where the marsh was widest. Although no canal or river connection was present at the time of Meigs's observations, it is convenient to organize them by "canal mile," measured along the route of the future canals, with zero at the Lake Okeechobee shoreline (figs. 5.1, 5.2). At about canal mile 10, near the narrowest, westernmost portion of the Okeechobee Marsh, Meigs found the river channel to be "about 15 feet wide and from 5 to 8 feet deep, the water flowing westward with a strong current" (1879, pp. 864–865). One and a half miles (2.4 km) farther upstream (canal mile 8.5), at the head of the river and west of Lake Hicpochee, the water in the prongs of the river was slower moving, "running from the east with a slow but very perceptible current" (p. 865). At about the same distance from Lake Okeechobee, but in the surrounding marsh rather than in the river prongs or rills, the velocity was lower: "everywhere the water in the saw-grass marsh at the head of these rills [prongs] was from 10 to 20 inches [25–50 cm] deep, and was observed to flow slowly westward" (p. 865). Still farther upstream, on the east side of Lake Hicpochee (canal mile 1.5), Meigs found the same depth of sheet flow, but the marsh had widened sufficiently that the flow (presumably still occurring) was no longer perceptible: "The water of the [sawgrass] marsh [between Lake Hicpochee and Lake Okeechobee] had no current whatever, and in depth varied generally from 10 to 20 inches [25–50 cm]" (p. 865).

The funneling effect of the westward narrowing of the Okeechobee Marsh is also reflected in the widths and flows of the canals constructed between 1881 and 1883. The Three-Mile Canal was cut through the widest portion of the Okeechobee Marsh (fig. 5.1), from Lake Okeechobee to Lake Hicpochee, and was cut one dredge-width wide, 24 feet (7 m), and about 6 feet (2 m) deep (Duval 1884, p. 318; Sackett 1888). In contrast, the next segment of canal, from the western shore of Lake Hicpochee to Lake Flirt (canal mile 7 to 18), was initially cut to the same dimensions, 24 by 6 feet (7 × 1.8 m; Duval 1884, p. 318), but prior to 1885 a second cut was made, for a total width of 48 feet (15 m; Menge 1947, p. 2; Daniel et al. 1886, p. 22). The same widening was continued another 9 miles (14 km) along and westward of Lake Flirt (Daniel et al. 1885, p. 22). Despite the factor of two difference in width, Daniel et al. (1885, p. 22) observed the same current in both the single-width Three-Mile Canal and the double-width canal west of Lake Hicpochee: 1.5 miles per hour (0.7 m s^{-1}) in both locations. Apparently sheet flow from the marsh around Lake Hicpochee was contributing significantly, on the same order as the Three-Mile Canal discharge. A similar twofold downstream increase was calculated by Fennema and Bhatt (2003, p. 17), who used Sackett's (1888) field survey results to estimate canal flows at the time of Sackett's 1887 dry season observations. They estimated an average of 175 cubic feet per second (cfs) (5 m^3 s^{-1}) for the Three-Mile Canal flowing into Lake Hicpochee and 428 cfs (12 $m^3 s^{-1}$) for the canal flowing out of Lake Hicpochee (canal miles 7–10).

Initial Impacts of Canals on Lake Okeechobee

The preceding sections of this chapter suggest that a quite detailed picture of the predrainage nature of the Okeechobee Marsh and its hydrologic connection to Lake Okeechobee can be synthesized from numerous first-hand observations. A key additional question is whether the canals dredged through the Okeechobee Marsh in 1882–1883 by the Disston company significantly lowered Lake Okeechobee stages and, in so doing, significantly altered flows from Lake Okeechobee into the Everglades.

The timing of anthropogenic lowering of Lake Okeechobee stages is, on the one hand, important for understanding the environmental history of the Everglades and, on the other hand, difficult to know from the historical sources. The issue would be simple if regular, continuous stage gage records were available from about 1860 forward, providing two decades of precanal stage information to compare with the early postcanal construction period. Unfortunately, the earliest stage records begin much later, in 1915, with a partial record for 1912–1913. Therefore the ecologically important question of when man-made canals first lowered Lake Okeechobee water levels must be answered through analysis of the historical record.

The ecological importance of the timing of lake lowering arises from the nature of the predrainage connection between Lake Okeechobee and the Everglades. As clarified in the earlier sections of this chapter, under predrainage conditions the water level of Lake Okeechobee was typically higher than the ground surface of the bordering sawgrass marsh during most of the year. As these marshes of the Everglades sloped downward, away from Lake Okeechobee, the presence of lake water levels higher than ground surface meant that water flowed from Lake Okeechobee into the Everglades during most of the year.

Lowering Lake Okeechobee stages relative to the marsh ground surface would obviously reduce these flows. Thus if the Disston company's efforts to lower lake levels were successful, then the Everglades experienced reduced inflows and lowered water depths beginning not in the 1910s with State of Florida's drainage efforts, as suggested by H. K. Brooks (1984, p. 39), but instead thirty years earlier, in 1883. If Disston's efforts successfully lowered Lake Okeechobee, then J. H. Davis's observations of the vegetation of the Everglades and the U.S. Department of Agriculture–Soil Conservation Service (1940) aerial photographs would both be observations of an Everglades that had been subjected to sixty years of reduced hydrology, not just thirty years. Similarly, if Disston's drainage efforts were successful, then only observations of the Everglades prior to 1883 can be considered representative of "natural," predrainage conditions.

Answering the question of whether Disston's efforts succeeded in lowering Lake Okeechobee, well prior to the state canals of the 1910s, is not straightforward. Surprisingly, not only are there no stage records for the period of Disston's work; no evidence has surfaced indicating that any measurements were ever made. Table 5.1 and figure 2.1 together help illustrate the problem. Table 5.1 lists various estimates related to the elevation of predrainage water levels in Lake Okeechobee. While there clearly is uncertainty associated with these values, there does appear to be some consistency. The peat (muck) ground surface appears to have been about 21 feet (6.4 m) above mean sea level, with water levels reaching up to about 23 feet (7 m) above msl and a mean at or slightly less than 22 feet (6.7 m) above msl. These estimates of predrainage conditions would apply up to 1882. The gap between 1882 and 1915 (fig. 2.1) is the challenge, along with the drop from around 22 feet (6.7 m; predrainage) to 15–17 ft (4.5–5 m; 1915) above msl. The shape of the trajectory is the key issue, particularly as there is some indication that by 1900–1910 the Disston company's canals may have become less effective over time than they were originally. If this is true, then the early period, about 1885 to 1900, may have experienced even lower Lake Okeechobee levels, prior to a gradual rise back up to the levels recorded in 1915.

Determining the validity of the narrative record, summarized in the historical observations found in the Supplementary Materials (DVD), is also challenging as this aspect of Everglades history was subject to substantial pressure. Disston's contract with the state Board of Trustees of the Internal Improvement Fund specified that he was to be paid as a function of the number of acres of land successfully reclaimed (drained). Obviously both Disston and the trustees had vested interests in any declarations of successful drainage. The narratives must therefore be sifted for physical observations that are consistent across observers as well as being hard to falsify. Table 5.2 provides a timeline of the main drainage activities, and observations of drainage activities, undertaken by the Disston company between 1878 and 1900.

The narrative record is also clouded by several confounding factors: the financial panic of 1893–1894, Hamilton Disston's death in 1896, and repeated complaints of flooding in the upper Caloosahatchee valley by settlers

Table 5.1. Estimates of predrainage stages of Lake Okeechobee

Description	Elevation (feet)[a]	Citation
Predrainage Observations		
"High-water of Lake Okeechobee"	25.08	Meigs (1879, p. 870)
"Surface of water in Lakes Hikpochee & Okeechobee" (Dry season; March 1879)	21.94	Meigs (1879, p. 870)
"Lake Okeechobee"	25	W. W. Harney (1884, p. 604)
"Lake Hickpochee"	20.24	W. W. Harney (1884, p. 604)
Shore of Lake Okeechobee, "High water 1878"	23.4	Sackett (1888)
Lake Hicpochee, "High water 1878"	23.4	Sackett (1888)
Postdrainage Observations[b]		
Shore of Lake Okeechobee, "ordinary [postdrainage] High water 1886"	22.4	Sackett (1888)
"These [surveys by Gen. Gillmore in 1880–1882] and other surveys by Col. Charles Hopkins, Maj. Wirts, V. P. Keller, J. W. Newman, and others established the altitude of Lake Okeechobee, 'the head of the Everglades,' at 21–23 feet above tide level."	21–23	Jennings (1903, p. 86)
"A subsequent observation and examination, made by W. H. Caldwell, assistant United States engineer, December, 1901, shows the level of the lake to be 20.42 feet above mean low-water level of the Gulf of Mexico."	20.42	Jennings (1903, p. 86)
"present elevation [about Nov. 1, 1912] is 21.6 feet. . . . The lake continues to overflow all along its south shore."	21.6	Elliot (1912, p. 558)
"Lake Okeechobee, Extreme High Water"	24.4	Slattery (1913)
"Lake Hicpochee, Extreme High Water"	24.4	Slattery (1913)
"When the lake rises to about 22 feet above mean sea level it is said to overflow into the Everglades."	22	Matson and Sanford (1913b, p. 54)
"Lake did not rise to a higher stage than 22½ feet because the lake overflowed its banks along the entire south shore."	22.5	Dachnowski-Stokes (1930, pp. 93–94)
"Muck soil about the lake had . . . a surface elevation about 21 feet above sea level."	21	Hills (1931, p. 3)
"Lake at times of flood"	21–22	Herr (1943, p. 48)

Notes: a. Elevations are as originally stated in the references indicated. Earlier references may reflect Punta Rassa datum, later determined to be 1.44 feet below mean sea level (U.S. Army Corps of Engineers 1978, pp. 37A–38A).
b. Jennings (1903) and all subsequent sources are postdrainage observers; their estimates may reflect a combination of predrainage and postdrainage conditions, thus underestimating predrainage lake levels.

there. Particularly as Disston's death often has been described, likely incorrectly, as a suicide, these factors have sometimes merged into a popular myth of suicide due to perceived failure of his Florida drainage efforts (Knetsch 2006, pp. 203–205). More careful examination suggests otherwise.

Knetsch notes that out of the many newspaper obituaries for Disston, only one, from a newspaper of opposing political persuasion, indicated suicide. The others echoed the coroner's report: death by natural causes, probably of a weakened heart; that is, heart attack or stroke. According to Knetsch, Disston and his drainage company were in good financial condition. There seems to be no evidence of discouragement about the performance of his company's drainage efforts. Neither the circumstances of Disston's death nor his finances lend any support to the argument that his company failed to lower water levels in Lake Okeechobee.

The repeated complaints about flooding made by settlers of the upper Caloosahatchee Valley, and their associated requests for the closing of the Disston company's canals (Lilly et al. 1890, p. 87; McIntosh 1895, p. 313; Hendry 1897, p. 404; McIntosh 1898, p. 465), might at first glance constitute a different line of evidence suggesting that the company had in some way failed, and in particular that it had failed to lower water levels in Lake Okeechobee. Further investigation actually suggests the opposite. The fact that the settlers consistently asked for the canals (from Lake Okeechobee to Lake Hicpochee and from Lake Hicpochee to the Caloosahatchee River) to be

Table 5.2. Timeline of early drainage history and explorations of the Okeechobee Marsh

Year	Event[a]
1860–71	U.S. Government Land Office township surveys of areas bordering the Okeechobee Marsh; Marsh considered "impracticable" to survey due to water depths and dense sawgrass.
1879	Survey of Caloosahatchee River and Okeechobee Marsh to within 0.5 mile (0.8 km) of Lake Okeechobee shoreline by Colonel Meigs, U.S. Engineer Office.
1882	Disston's company begins dredging canals from Lake Okeechobee to Lake Hicpochee, through Bonnet Lake and Lake Flirt, and into upper Caloosahatchee River.
1883	Canals to Caloosahatchee River finished; steamboat *Bertha Lee* arrives in Kissimmee from Fort Myers.
1884	Report by James Dancy on Disston company drainage, finding western shore of Lake Okeechobee with "defined margin"; Okeechobee Marsh "at least three feet above the water level."
1884	State Engineer H. S. Duval reports that "the rich muck prairies, at the foot of the lake, normally submerged . . . [are] now four feet above water."
1886	Exploration of Lake Okeechobee, Calooshatchee River, and Disston canals by Angelo Heilprin.
1886	Report by Legislative Committee appointed by governor November 17, 1885, finding Lake Okeechobee "not materially lowered" but canals below Lake Hicpochee "evidently producing some results."
1888	Survey of Caloosahatchee River and Disston canals by J. W. Sackett, U.S. Engineer Office.
1888	Disston's company dredges the Thirteen-Mile Canal south from Ritta River.
1889	End of Disston's company's dredging efforts south and west of Lake Okeechobee.
1890	Petition by settlers of upper Caloosahatchee valley to Trustees of the Internal Improvement Fund (TIIF), complaining that Disston's canals were draining not only the Okeechobee Marsh but also "a large volume of water from Lake Okeechobee," thus "tending to endanger the bordering lands along the river," and asking TIIF to close canal(s) from Lake Okeechobee.
1893	(Overly?) favorable report of the TIIF on Disston's drainage, finding that the canals were "discharging more water than is flowing into Lake Okeechobee, as a consequence the water level of this great lake has been lowered about four and one half feet below its normal level," and further finding that the lowering was not due to a period of "droughty years."
1894	Dredge Captain Fred Menge writes to Disston that "surface of the [Okeechobee] marsh is from 4 to 5½ feet above the water in the canals from Lake Flirt up to Lake Okeechobee. . . . Lake Okeechobee must be 7 feet lower than when we tapped it in 1882." Two weeks later, after a very heavy September storm, he wrote, "the water ran up stream by Ft. Thom[p]son," yet the "banks of the lake [Okeechobee were still] from 2½ to 3 feet above the surface of the lake."
1895	Second request, transmitted by F. A. Hendry, from settlers of upper Caloosahatchee valley requesting TIIF to close Disston canals from Lake Okeechobee.
1896	Disston dies at age fifty-one of natural causes.
1897	F. A. Hendry again requests that TIIF close canal from Lake Hicpochee into the Caloosahatchee River to protect the upper Caloosahatchee valley from flooding.
1897	Chief Engineer James Kreamer of Disston's drainage company again recommends against closing canal. He states to TIIF that a "simple closing of the Canal will not prevent the waters from Lakes Okeechobee and Hicpochee [from flooding the upper valley], noting that closing canal would again "fully charge [the subsoil,] and [that] the surface of the great lakes would again raise to the old level." The three components that Kreamer indicates would simultaneous be needed to prevent flooding in the upper valley suggest that Lake Okeechobee had in fact been lowered below marsh surface: (1) closing canal; (2) erecting levee fully separating the former Okeechobee Marsh from Lake Okeechobee; and (3) extension of Thirteen-Mile Canal to convey water southward into Everglades.
1897	TIIF apparently agrees with Engineer Kreamer, as they took no action on Hendry's 1897 request to close the canals, ordering it "filed for future reference."

Note: a. Additional detail and citations can be found on the accompanying DVD in the chronologically arranged, annotated Historical Observations for the Okeechobee Marsh and Lake Okeechobee (Chapter 5).

closed clearly indicates that the canals were performing their function: conveying water out of Lake Okeechobee. These outflows were above and beyond predrainage outflows, so by definition the canals were contributing to the lowering of Lake Okeechobee stages. Since the canals were dredged down 7 feet or more (2 m) into the marsh peat, they did, as Kreamer noted, "effectually reduce the general water level of the region, by steadily carrying off the impounded water" (1892, p. 94).

If the Disston company's canals were successfully lowering Lake Okeechobee water levels, why would the settlers petition for closing them? The answer appears to lie in the geography of the Caloosahatchee Valley. The settlers arrived prior to the first drainage efforts, so they naturally did not settle and cultivate the Okeechobee Marsh, which was inundated much of the year. Instead, they settled farther west, around and below Fort Thompson (canal mile 23), in the "rich hummock lands bordering the river" (Meigs 1879, p. 864). Unfortunately, as noted by Sackett (1888), the portion of the river between canal mile 23 and 32 (i.e., about Fort Thompson to Fort Denaud) was narrow and "very much obstructed with snags and drift" that greatly impeded flows during high water times. As shown in figure 5.2, prior to the canal dredging, the difference between normal (14 feet [4 m]) and high (18 feet [5 m]) water stage along this portion of the river was enormous, reflecting the restrictions present, especially during high outflows. While the Disston company's canals were not the original cause of the floods, as an obvious and additional source of floodwater, they were naturally blamed.

The answer the Disston company's chief engineer James Kreamer supplied to the settlers' petitions seems physically accurate. Kreamer noted that simply closing the canals as requested would not solve the problem; that it would instead cause a reversion to precanal conditions. Stages in Lake Okeechobee would again rise, flooding the Okeechobee Marsh, which would again drain into and occasionally flood the upper Caloosahatchee River. Truly to prevent flooding of the river valley, Kreamer indicated that it would be necessary to complement closing of the canals with construction of a full-length, north-south levee separating the Okeechobee Marsh from Lake Okeechobee. Additionally, he recommended construction of a canal southward through the Everglades to provide a replacement outflow for the former outflows into the Caloosahatchee River that would be cut off by these levees (Kreamer 1890, p. 96).

In summary, the petitioners' complaints of flooding in the Caloosahatchee Valley and the company's proposed solutions suggest that the canals through the Okeechobee Marsh were, in fact, effective at lowering Lake Okeechobee water levels. The settlers were essentially arguing that the canals needed to be closed because they were *too* effective at conveying water from Lake Okeechobee into the upper Caloosahatchee River.

Historical observations of the canals themselves and of early landscape changes further reinforce the impression that the Disston canals succeeded in lowering Lake Okeechobee stages. Multiple observers agreed on the actual widths and depths of the canals (Dancy 1884, pp. 333–334; Daniel et al. 1886, p. 22; Sackett 1888; Conrad Menge 1947, pp. 1–4) and also on the observed flow velocities: 1.5 to 6 miles per hour (0.7–2.7 m s^{-1}; Hopkins 1884b, p. 102; Dancy 1884, p. 334; Heilprin 1887, p. 404; Daniel et al. 1886, p. 22). Those observations can be translated into estimates of annual lake lowering using the following assumptions: for example, midrange velocities of 2–4 miles per hour (0.9–1.8 m s^{-1}), year-round flow at these velocities, canal side slopes of 1:1, and an estimated predrainage area of the lake corresponding to the present-day (2010) area plus an additional 1-mile-wide (1.6 km) concentric strip. Using these assumptions, a simple velocity-times-area calculation suggests that Disston's canals contributed about 1 to 2 feet (30–60 cm) of lake lowering per year. While a portion of this lowering would have occurred naturally from flow through the Okeechobee Marsh, Fennema and Bhatt's (2003, p. 17) calculations for east and west of Lake Hicpochee suggest that the natural fraction would have been substantially less than half. The magnitude of the purely canal-induced lowering appears to be consistent with the predominant indication of the historical record; that is, a progressive lowering of Lake Okeechobee over somewhere between five and ten years.

Historical observations include descriptions of relative elevations, such as newly visible "defined margins" on the lakes (Dancy 1884, p. 333; J. F. Menge 1894c, p. 334; Disston 1895, p. 333), creeks being "entirely within their banks" (Dancy 1884, p. 333; J. F. Menge 1894a, pp. 333–334), and land surfaces becoming higher than adjacent water surfaces, such as, "the rich muck prairies, at the foot of the lake [Okeechobee Marsh], normally submerged, [are] now four feet above water and in a state of cultivation" (Duval 1884, p. 316). See also Hopkins 1884a, p. 324; Dancy 1884, p. 333; Kreamer 1890, p. 95; Kreamer 1891, p. 166; McIntosh 1893, p. 8; J. F. Menge 1894a, pp. 333–334; J. F. Menge 1894c, p. 334; and Disston 1895, p. 333 for other similar observations. These observations of water levels lower than adjacent ground surfaces may be doubly indicative of successful lake lowering, as many of the ground surfaces formed of peat (e.g., all of the sawgrass marshes) also may have been lowered during the course of drainage. Soil subsidence would have begun as soon as water tables were lowered below the soil surface, both slowly due to microbial oxidation of the peat and more rapidly due to shrinkage and compaction as the soil was dewatered (Clayton et al. 1942; Stephens and Johnson 1951a). Thus an observation of a creek "being entirely within its banks 2 feet" may be indicative of a water level drop even greater than 2 feet (0.6 m), measured relative to the original predrainage water elevation. It is possible that this phenomenon provides an explanation for the apparent contradiction in the Daniel et al. report (1886), which found that Lake Okeechobee had not been "materially lowered," (p. 22), yet at the same time found that canals below Lake Hicpochee "were evidently producing some results" (p. 23). Subsidence of peat soils may also have contributed to Conrad Menge's impression that the canals "did not succeed in lowering the lake level by 4 feet" (1947, p. 2). In both cases the accompanying decline in ground elevation may have led to underestimates of the absolute decline in water level. Similarly, the various historical observations describing dry land vegetation invading the former sawgrass marshes suggest substantial lowering of Lake Okeechobee—that is, lowering that must have been in excess of the subsidence of the ground surface.

Historical observations also include an additional, implicit type of information concerning relative Lake Okeechobee water levels—flow reversals in the canals. During some high water events, instead of flowing from Lake Okeechobee westward into the Caloosahatchee River, the flow direction reversed, from Fort Thompson back into Lake Okeechobee (C. Menge 1947, pp. 2–3; J. F. Menge 1894b, p. 334; Disston 1895, pp. 334–335). Sackett's observations plotted in figure 5.2 provide a likely estimate of the water elevation at Fort Thompson during such reversals. The clearing and snag removal activities carried out by Disston's company suggest it is unlikely that river stages would have returned to the precanal, exceptionally high stages seen in 1878. More likely the levels would resemble the postcanal high levels seen in 1886, which Sackett described as "average high water." This suggests that at the times of reversal, Lake Okeechobee levels would necessarily have been lower than the 1886 level of 15.5 feet (4.7 m) above mean sea level at Fort Thompson. Lake Okeechobee levels of less than 15.5 feet above msl indicate a substantial lowering, more than 6 feet (2 m), below the original predrainage elevation of about 21 feet (6.4 m) above msl. Similar flow reversals occurred in the Thirteen-Mile Canal as well; after the storm of September 25–26, 1894, both this "south canal" and the Caloosahatchee canal flowed backward into Lake Okeechobee (J. F. Menge 1894c, p. 334).

Summary

The previous pages have presented a number of different lines of evidence. All indicate that by the mid- to late 1880s the drainage efforts of Disston's Atlantic Gulf Coast Canal and Okeechobee Drainage Company had lowered water levels within Lake Okeechobee sufficiently to draw them below the ground surface of the adjoining sawgrass marshes (i.e., the Okeechobee Marsh and Sawgrass Plains) during all or most of the year. Since these sawgrass marshes formed the controlling elevation regulating outflows from Lake Okeechobee into the Everglades, this lowering created a highly significant hydrological and ecological change. In likely less than ten years, the hydrological link between Lake Okeechobee and the Everglades shifted from one where outflows connected the two areas throughout most of the year during most years to one where outflows became the exception. The last year of undisturbed, predrainage connection would have been 1881; from about 1890 onward the Everglades would have become primarily rain-driven, as outflows from Lake Okeechobee into the Everglades would mostly have ceased. F. C. Elliot (1912, p. 558) notes such an exceptional outflow toward the end of the 1912 wet season, stating that Lake Okeechobee was then 1.5 feet (45 cm) above "normal," and that the lake "continued to overflow all along its southern shore."

The Okeechobee Marsh formed the westernmost segment of outflowing shoreline of Lake Okeechobee. The second outflowing segment was formed by the remainder of the bordering Sawgrass Plains. The nature of this landscape and of its interface with Lake Okeechobee, both discussed in more detail in chapter 6, were very similar to the Okeechobee Marsh. The primary differences were that the Sawgrass Plains did not narrow and terminate as a river and that the slope was slightly less. The same presence of sawgrass on the shoreline of the Sawgrass Plains as described for the Okeechobee Marsh suggests the same shoreline elevation. This means that any lowering of Lake Okeechobee water levels achieved by Disston's canals would similarly have reduced outflows into the Sawgrass Plains.

The third and last segment of outflowing shoreline was formed by the Custard Apple and Cypress Swamp (chapter 7). This segment was different in that the predrainage elevation of the Custard Apple and Cypress Swamp appears to have been lower than that of Sawgrass Plains landscape immediately downstream. As a result, outflows from Lake Okeechobee along this portion of the shoreline would have been controlled primarily by the elevation of the Sawgrass Plains. Thus here too, lowering of Lake Okeechobee by Disston's drainage efforts would have reduced outflows through the Custard Apple portion of the southern shoreline.

Overall, despite the absence of early stage records, the narrative historical record suggests that for a period of time, the canals dredged by Disston's Atlantic Gulf Coast Canal and Okeechobee Canal Company were successful in lowering Lake Okeechobee, very likely by several feet (0.6–1 m) and perhaps at times by as much 7 feet (2 m), measured relative to the original elevation. While lake levels would have continued to rise and fall seasonally, the fraction of the year when levels were above shoreline ground surface would have been reduced, perhaps greatly. These reductions would have applied to all landscapes along the originally outflowing shoreline, and would have altered the Everglades water budget substantially for some portion of the period 1882 to the late 1890s.

The Atlantic Gulf Coast Canal and Okeechobee Canal Company did not dissolve with Disston's death in 1896—James Kreamer, longtime chief engineer and superintendent, continued to be active—but the company's involvement in dredging and in maintaining the canals appears to have diminished considerably, if not stopped altogether. Information concerning canal flows and Lake Okeechobee stages for the late 1890s and early 1900s appears to be scarce. There are indications of sand bars forming in the canals, and it seems plausible that without maintenance, snags and drift might again have accumulated in the narrows between Fort Thompson and Fort Denaud, but the timing and influence of such changes are not clear. Nevertheless, taken together, the decline in Disston company activity as well as examinations in relation to the State of Florida's drainage activities from about 1905 to 1915 suggest that after the initial lowering of Lake Okeechobee levels by Disston's canals, lake levels began to rise again, perhaps returning almost to predrainage levels. Then, as reflected in the historical observations for this chapter and for chapter 3, around 1910–1920 the new state canals became very effective at lowering water levels in Lake Okeechobee.

6

Sawgrass Plains

At 1,025 square miles (265,000 ha), the Sawgrass Plains landscape was the second largest within the Everglades (table 4.2, plate 5), once comprising 25 percent. It originally stretched southward from the shores of Lake Okeechobee about 20 miles (30 km) to the Ridge and Slough landscape. The predrainage extent of the Sawgrass Plains corresponded closely to the present Everglades Agricultural Area (fig. I.1), with the addition of a short extension southeastward between the present Miami and North New River canals. The Okeechobee Marsh, discussed in detail in chapter 5, is considered here to be a part of the Sawgrass Plains.

Essentially all observers described the Sawgrass Plains as "impenetrable," a characterization easily understood in light of the density, height, and sharpness of the vegetation (fig. 6.1) as well as the presence of surface water much of the year. This might have left a dearth of information regarding the "Big Sawgrass," but a fortunate combination of predrainage and early postdrainage sources makes it possible to reconstruct the landscape with considerable confidence.

Sources used include surveyors' township survey notes from the U.S. General Land Office (predrainage), state township surveys for the Trustees of the Internal Improvement Fund (early postdrainage), accounts of navigation on Lake Okeechobee, drainage reports, botanical accounts, soil surveys, and studies of soil cores. During the 1800s U.S. surveyors reaching the Sawgrass Plains from the east or west universally declared the landscape "impracticable" and relinquished their lines. Some noted water depths at the edge; all noted the location of the edge, so the eastern and western predrainage boundaries can be mapped with considerable (surveyed) accuracy. The north boundary extended either directly to Lake Okeechobee or, along the southeast shore, to a thin intervening band of custard apple or cypress swamp (chapter 7). The section of the Sawgrass Plains directly bordering the lake (i.e., the Okeechobee Marsh) is documented in chapter 5 by navigational accounts, by a detailed agricultural map (Kreamer 1892; see plate 9), and by point observations. The southern border with the Ridge and Slough landscape is somewhat less precisely known, having been mapped by the authors from postdrainage aerial photographs (U.S. Department of Agriculture–Soil Conservation Service 1940), with supporting evidence from a predrainage expedition sponsored by the *Times-Democrat* newspaper (appendix H), and two early postdrainage expeditions by John Harshberger (1914; appendix M) and J. K. Small (1914).

All reports indicate that sawgrass (*Cladium jamai-*

Figure 6.1. Typical tall, dense sawgrass of the Sawgrass Plains. Note the height of the canopy above the surveyor. Baldwin and Hawker (1915, fig. 2 of plate 24).

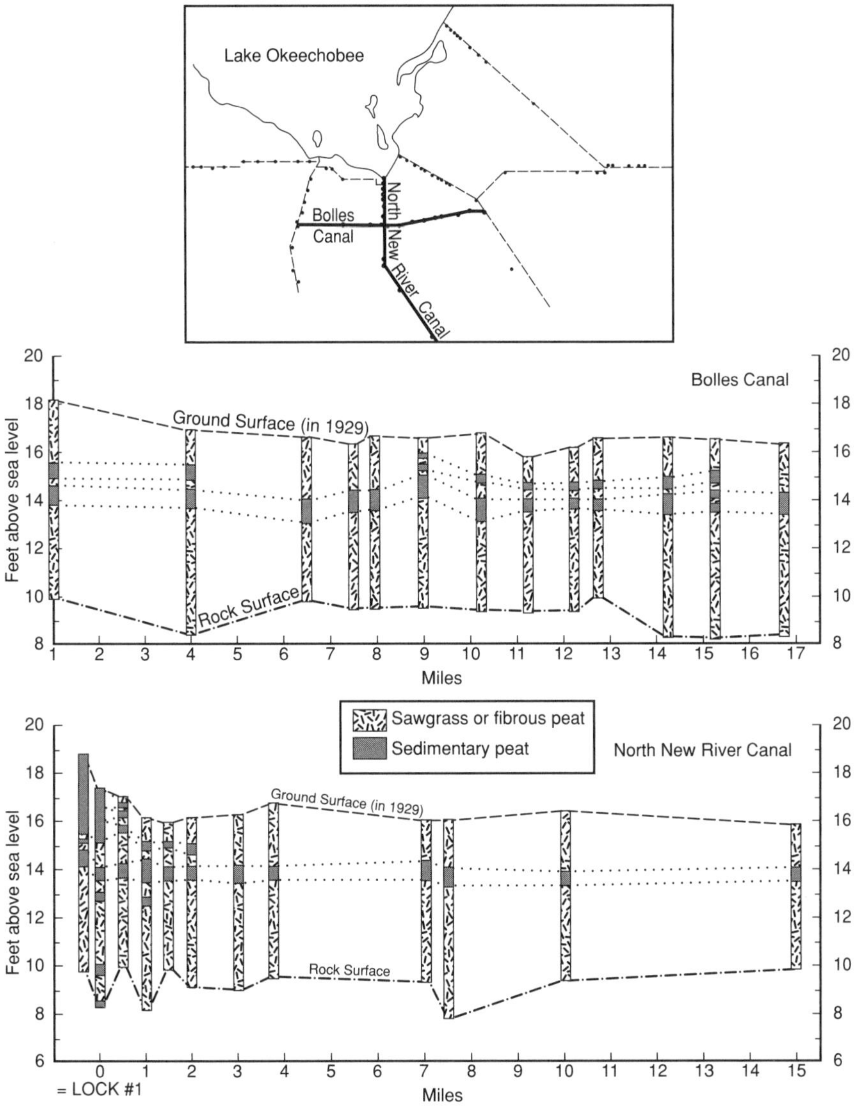

Figure 6.2. Soil profiles and elevations in the Sawgrass Plains, 1928–1929. Dachnowski-Stokes (1930) described most of the length of all of these cores as peat derived from sawgrass. He ascribed the thin but widespread layer of blackish-brown sedimentary peat (ca. 14 feet elevation) seen in these soil cores to a period of persistent, above-average Everglades-wide inundation. This dark layer is likely the same as black layers reported from canal dredging in the 1910s (J. H. Davis 1943a, p. 253) and from soil pits (Parker 1974, p. 30). Davis and Parker ascribed these to peat fires, but charcoal in the Everglades is difficult to identify reliably in the field without microscopic examination (P. J. Gleason, pers. comm., 1997). Dachnowski-Stokes's attribution of the dark layers to an unusual period of widespread, persistent inundation seems more likely. Note that ground surface had already subsided 2–5 feet by 1929. Redrawn from original figures in Dachnowski-Stokes (1930).

cense) was the dominant plant in this landscape with occasional patches of marsh vegetation such as "flags" (*Pontedaria cordata* and *Sagittaria* sp.; see appendix H). The presence of uniform sawgrass within the interior of the landscape was established from two predrainage expeditions (Meigs 1879; Hopkins 1884b); from the twelve state township surveys available for the area comprising the Sawgrass Plains (early postdrainage; cf. fig. 1.4), covering 430 square miles (111,000 hectares); and by inspection of 1940s aerial photographs (U.S. Department of Agriculture–Soil Conservation Service 1940). Examination of the field notes from the twelve township surveys, representing most portions of the landscape, indicated that of a total of 187 recorded observations of vegetation, 183 (98 percent) were "sawgrass." Photointerpretation of the 1940s aerial photographs suggests that the sawgrass reported in the township survey notes was present throughout the landscape. Soil cores analyzed by Dachnowski-Stokes (1930) further support the ubiquitous presence of uniform sawgrass. Even more significant, his analysis suggests that the Sawgrass Plains had been dominated by sawgrass throughout the history of the Everglades (fig. 6.2), with the exception of one or two periods corresponding, perhaps, to differing climate during or following the Medieval Warm Period (Willard et al. 2001).

As discussed in detail in chapter 5, prior to the drainage of the Sawgrass Plains and the subsequent lowering of Lake Okeechobee water levels, the Sawgrass Plains extended to Lake Okeechobee, forming more than 30 miles (50 km) of shoreline. The intersection of the landscape with the lake was direct; that is, without an intervening elevated soil rim and without an intervening strip of differing vegetation.

Chemically, the presence of sawgrass directly adjoining the lakeshore indicates that sawgrass was able to persist under nearly year-round exposure to outflowing predrainage lake water. The presence of an intervening transitional or filtering zone with differing soil type or differing vegetation apparently was not necessary for sawgrass persistence. No evidence was found of predrainage cattail (*Typha* spp.) stands along or near the shore. Together these points suggest that predevelopment nutrient concentrations in Lake Okeechobee were below the thresholds that presently cause sawgrass to be replaced by cattail stands.

The uniformity of the Sawgrass Plains was remarkable. All aspects—soils, microtopography, vegetation, and hydrology—appear to have varied little across this landscape 35 miles wide and 20–25 miles long (55 km wide, 30–40 km long): "The surface of the upper part of the large area of this type is flat and nearly level. No natural drainage channels are apparent, and the flatness is broken only by infrequent alligator holes and runways" (Baldwin and Hawker 1915, p. 785). Baldwin and Hawker's description echoes that made during an expedition thirty years earlier (see appendix H): "The upper or northern Glades being a solid body of dense saw grass extending 30 miles East and West and 40 miles North and South, and which has hitherto been considered impenetrable" (Hopkins 1884b, p. 100).

Conditions were apparently ideal for sawgrass. It formed a dense stand, 6 to 10 or even 12 feet (2–4 m) high, and was invariably described as impenetrable. The sawgrass stand was continuous—unbroken by sloughs, channels, or tree islands. Although the Seminole Indians regularly crossed the more open Ridge and Slough landscape, they understandably avoided the wall formed by the Sawgrass Plains (Simpson 1920, p. 120). The only exploring expedition to cross this landscape resorted to burning the sawgrass ahead of their passage (appendix H; see also plates 10 left, 10 right).

The slight trend that was present, a downward slope to the southeast, does not appear to have altered the landscape's vegetation. The physical and vegetative uniformity of this landscape likely results from its aquatic origin; that is, the process of peat soil (fibric Histosols) accumulating under a water surface tending to be leveled by gravity. The peat, formed from partially decomposed sawgrass remains (Miller 1918), consisted of 85–90 percent organic material (Baldwin and Hawker 1915; Miller 1918; Forsaith 1916, 1917; Hammar 1929). Further decomposition (and loss) of the peat was limited by anaerobic conditions in the soil profile. The profile remained anaerobic much of the time because the landscape was covered with water during the greater part of most years.

The soil depth (thickness) measurements shown in figure 4.8, 150 of which are from the Sawgrass Plains, were made by soil scientists in 1915, three years after the opening of the North New River Canal. While water levels at each end of the canal were already lowered by that time (see fig. 3.1), soil conditions along the central reaches still resembled predrainage ones. Figure 4.8 indicates that in 1915 the peat of the Sawgrass Plains formed a wedge, ranging from about 10 feet (3 m) near the northerly, upstream end of the landscape down to about 5 feet (1.5 m) toward the downstream end. Early drainage and any subsidence that may have occurred prior to 1915 could have affected the results shown in figure 4.8, but not enough to greatly alter the basic shape; that is, a peat ground surface that sloped southward more steeply than did the underlying bedrock surface. Thirty years earlier, an engineer leading the New Orleans *Times-Democrat* expedition during November and December 1883 described the same peat wedge, although apparently with a greater thickness of peat: "We enter the saw grass ocean in which we find but a few inches of water not sufficient to float our light boats, but flowing perceptibly in a southerly direction. I sounded regularly with a 10-foot rod and for 30 miles . . . I found no evidence of rock or sand . . . 30 miles south of lake I found rock at 8½ feet below surface" (Hopkins 1884b, p. 100).

The slope of the peat surface of the Sawgrass Plains, rather than reflecting the slope of the underlying bedrock, appears to have reflected more closely the slope of the surrounding uplands forming the rim of the Everglades basin (cf. U.S. Army Corps of Engineers 1960a). It seems likely that by the 1800s (or before), accumulation of peat had come to equilibrium with the surrounding uplands; in other words, the basin had filled sufficiently with peat so that typical water depths on top of the peat brought the water surface to within a foot or two (30–60 cm) of the surrounding upland surface (Wright 1912, p. 17; Lodge 1994, pp. 15–16).

The transition between the Sawgrass Plains and the Ridge and Slough landscape was progressive, with the unbroken plain giving way along the southern edge to breaks made by increasing numbers of open water sloughs and channels. The sawgrass of the Sawgrass Plains continued southward in the form of sawgrass ridges.

Several authors have commented on the annual weedy species that invaded the landscape after the canals began lowering water levels and decreasing the hydroperiod (Wright 1912, p. 26; Harper 1927, pp. 124–125; J. H. Davis 1943a, pp. 261–262; Jones et al. 1948e, p. 47; Loveless 1959a, p. 2; Cornwell and Hutchinson 1974, p. 9; see also chapter 3). These include dog fennel (*Eupatorium capillifolium*), pigweed or "careless weed" (*Amaranthus* or *Acnida australis*), and goldenrod (*Solidago* spp.). Historical accounts suggest that some of these species may occasionally have been present prior to drainage as well, during exceptionally dry years or during multiple dry years in a row.

Hydrology

A number of information types contributed to the estimates made here of long-term average water depths and hydroperiod: (1) direct, recorded observations of predrainage depths (appendix A); (2) determination of hydrologic conditions optimal for sawgrass growth (J. H. Davis 1943a, p. 278; R. Andrews 1957, p. B-30; Forthman 1973, pp. 5, 14); (3) determination of the hydrologic conditions necessary to foster the accumulation of peat soil; (4) measurement of water depths associated with conversion of sawgrass tracts into open water sloughs; (5) observations of postdrainage changes in vegetation; and (6) determination of the direct connection of this landscape to water levels in Lake Okeechobee (chapter 5). Recorded observations ranged from low values sufficiently dry to ride a horse over the landscape to high values of 2 feet or more (60+ cm) of surface water. Consideration of all of the informational sources together yielded a somewhat narrower range for the long-term average predrainage water depths estimated here for the Sawgrass Plains: 0.5 feet below land surface (-15 cm) at the end of the dry season to 1.5 feet (45 cm) above land surface at the end of the wet season, with a hydroperiod from nine to ten months.

Hindcasting

The Sawgrass Plains landscape, as defined and mapped in this study, differs from the mapping and classification of previous authors. Relative to studies from the early 1900s, the differences tend to reflect a lack of a distinction between the Ridge and Slough and the Sawgrass Plains landscapes. Relative to more recent studies, differences mostly reflect underestimation of the substantial drainage-induced changes that occurred between the 1840s and the 1940s (i.e., incomplete hindcasting).

As defined here, the Sawgrass Plains landscape resembles, to varying degrees, the following categories from other studies (see also chapter 4; fig. 4.9): Harshberger's "Everglade Formation" (1914), Harper's "Sawgrass marshes" (1927), Small's "Saw-Grass Prairie" (1929b), and J. H. Davis's "Saw-grass Marshes (dense)" (1943b). Harshberger (1914) did not draw the Sawgrass Plains as distinct from the Ridge and Slough landscape, although he did note differences characteristic of the two landscapes along the course of the North New River Canal. Harper (1927) also did not distinguish between the two landscapes. Mapping by J. H. Davis (1943a, b) distinguishes between the Ridge and Slough and the Sawgrass Plains landscapes, but by the 1940s, postdrainage vegetation changes (chapter 3) had greatly altered the Sawgrass Plains. As a result, J. H. Davis subdivided the originally uniform landscape into four categories of sawgrass, "medium dense to sparse," "dense," "with abundant ferns and cat-tails," and "with wax-myrtle thickets." Additionally, the total area of uniform sawgrass mapped by Davis in the 1940s was substantially larger than the predrainage extent estimated in this study, reflecting conversion of areas that were originally Ridge and Slough landscape into continuous sawgrass. The Sawgrass Plains category of S. M. Davis et al. (1994) and of this study are similar in extent; however, the composition of the "sawgrass dominated mosaic" of S. M. Davis is very different.

To hindcast the extent of the Sawgrass Plains (plates 13, 14) and the Sawgrass Plains–related Everglades peat soils (plate 15) back to predrainage conditions, we had to compensate for the substantial increase in apparent area of the Sawgrass Plains that had occurred after drainage (cf. tables 4.2, 4.3), particularly the expansion of sawgrass into the Ridge and Slough landscape. To the northeast, the long thin slough bordering the Eastern Flatwoods (the current L-8 canal area) was drained sufficiently by 1930 to be classified as a Sawgrass Plains–like landscape (Dachnowski-Stokes 1930, pp. 99–100). Similarly, sawgrass expanded into small areas along the western sides of the present Water Conservation Areas 1 and 2. The main, and major, sawgrass expansion occurred to the southeast, in the area between and surrounding the Miami and North New River canals and extending almost to the Atlantic Coastal Ridge (see plates 13 and 14 for 1940s extent; cf. plate 4 for hindcasted predrainage landscape boundaries). Expansion would have continued all the way to the ridge except that by the 1940s this area had already lost essentially all of its original peat profile, forming Davie mucky fine sand and Davie fine sand soils (fig. 3.14) and supporting wet prairie vegetation. These specific areas of increase are discussed in more detail in the Ridge and Slough chapter (chapter 8) as they were hindcasted back to their original ridge-and-slough composition.

The major postdrainage changes to the Sawgrass Plains were increases in extent; however, some small decreases also occurred in the northwestern, western, and northern portions of the landscape. To the northwest, in Townships 41–42 Range 32 and Township 43 Ranges 33–34, Everglades peat was lost to oxidation, creating new areas of Davie mucky fine sand and Davie fine sand (Jones et al. 1948a). J. H. Davis (1943a, b) mapped the same areas as wet prairie. For this study those areas were hindcasted to Sawgrass Plains on peat, based on descriptions of sawgrass in predrainage township surveys, on Kreamer's (1892) detailed vegetation and land suitability map, and on peat soil thicknesses recorded in early postdrainage soil surveys. Kreamer's map distinguished four main categories: (1) "First Class Sugar Land," also labeled as "A Very Fine Piece of Saw Grass Marsh" and "First Class Marsh"; (2) "Slough Land Bordering on Saw Grass," also labeled as "Good Slough Land for Rice, Pard Grass, and Vegetables"; (3) "2nd and 3rd Class Prairie for Pasture"; and (4) several grades of pinelands. Comparison with land classifications used in the township surveys suggests that Kreamer's (1892) "First Class Sugar Land" (saw grass) would have been mapped by Jones et al. (1948a, e) as "Everglades Peats"; the sloughs as either "Everglades Peats" or "Loxahatchee Peats"; the prairies as "Wet Sands"; and the pinelands as "Upland Sands." Peat soil cores taken by Dachnowski-Stokes in about 1928, within areas later mapped by Jones et al. (1948a) as Davie fine sands, indicated that there were originally several feet (ca. 0.5–1 m) of peat on top of the sand (1930), confirming the hindcasting of these areas to Everglades peat and Sawgrass Plains.

A very narrow band of Davie mucky fine sand along the western border also was hindcasted to Everglades peat, based on association with the northwestern areas and on a government township map (Stearns 1874-T47 R34). Near the Custard Apple Swamp, the Sawgrass Plains decreased slightly after drainage due to southward expansion of willows and elders. This was corrected by this study when the custard apple and willow/elder areas were redrawn using early postdrainage township surveys (chapter 7).

7

Custard Apple and Cypress Swamp

Descriptions of the predrainage Custard Apple Swamp landscape have often included a number of misconceptions. This is not surprising as most of the descriptions were made after, or were based on observations made after, one or both of the periods of lake lowering. The Custard Apple Swamp, and the adjacent willow and elder area immediately parallel to it, were almost completely cleared for agricultural use early in the development of the Everglades (Will 1964, p. 36). Additionally, as an area directly bordering Lake Okeechobee, it would have been influenced by any lowering of lake water levels, such as Disston's efforts begun in 1882 and the State of Florida's efforts begun in 1906. While sufficient predrainage reports of custard apple exist to confirm that the Custard Apple Swamp was a natural part of the Everglades (i.e., not an artifact of drainage-induced alterations), those reports are not numerous, and they do not provide much detail regarding predrainage swamp conditions. Postdrainage sources and descriptions are more numerous and more widely accessible.

Postdrainage accounts described the Custard Apple Swamp as the sole gateway for Lake Okeechobee water to enter the Everglades. This gateway was thought to be in the form of a noticeable rim, over which the lake would occasionally overflow. In addition, the Custard Apple Swamp was characterized as being up to 4 miles (6.4 km) wide, 45 miles (72 km) long, and extending as far west as the Three-Mile Canal at Moorehaven.

As discussed in the Sawgrass Plains chapter (chapter 6), sawgrass directly bordered Lake Okeechobee along 25 miles (40 km) of the southwestern and southern shoreline. No elevated rim was present; the waters of the lake simply extended as a level surface, flowing directly into the sawgrass (chapter 5). Rather than being an exceptional high-water occurrence, this extension of lake water into the sawgrass persisted for most of the year during most years.

A similarly direct, "level water surface" connection from Lake Okeechobee into the Custard Apple Swamp appears to have been present prior to drainage. The consistent description of the area as "swamp"; the presence of tree bases buttressed and enlarged, presumably in response to inundation; the natural habits of custard apple trees; and the descriptions of the Custard Apple Swamp as a location of lake outflows together all suggest that surface water from the lake was present throughout much of most years.

The predrainage Custard Apple Swamp was a narrow 1-mile-wide (1.6 km) band of nearly monotypic swamp forest, dominated by what was called custard apple (*Annona glabra*), and extending approximately 30 miles (48 km) along the southern and eastern shores of Lake Okeechobee (plate 5, fig. 7.1). Note that the common name "custard apple" now refers to *Annona reticulata*, a cultivated plant, while the native *Annona glabra* that once populated the Custard Apple Swamp is now generally referred to as "pond apple." In keeping with historical usage, we use "custard apple."

Observations of custard apple in 1857 (Canova 1885, p. 26, 29), in 1874 (Ober 1874, p. 593), and in 1883 (Hopkins 1884b, p. 104; Williams 1884, pp. 70–73) firmly establish that the Custard Apple Swamp was not a postdrainage or drainage-induced phenomenon. The forest was dense, with intertwined branches and additional vines that together formed a closed canopy and almost impenetrable thicket. The trees were variously reported as 15, 20, or 30 feet (4–9 m) tall. At times the trees were covered by a high-climbing vine, the moonflower (*Ipomoea alba*). Individual cypress trees were occasionally found among the custard apples. A similar feature still exists at the south border of Lake Myakka in Sarasota County, Florida (William Loftus, pers. comm., 2009).

On the downstream, southern side of the Custard Apple Swamp was a transitional marsh, described in 1883 as "yma [yuma] grass, wampu or warmpea [pickerel weed, *Pontedaria cordata*], mixed with scrub willow [*Salix caroliniana*]." This transitional marsh formed a narrow strip between the Custard Apple Swamp and the Sawgrass Plains. Township surveyors working in the transitional marsh in 1915–1920 mention "open glades," "flags,"

Figure 7.1. Custard apple forest: the transition between Lake Okeechobee and the Everglades. Top: "Onions on Muck. Custard Apple in Background. South Shore of Lake Okeechobee [in 1915]" (Baldwin and Hawker 1915, fig. 1 in plate 25). Bottom: "Custard-apple forest, south shore of Lake Okeechobee, June 22, 1912. Original" (Harshberger 1914); from the Wagner Free Institute of Science, Philadelphia, Pennsylvania.

and "[water] lillies," all suggesting that the transitional marsh was wetter (deeper) than the sawgrass area: "from willow into open glade, wild millet and lillies" and "from open glade of wild millet and flags into willow." "Flag" likely refers to fire or alligator flag (*Thalia geniculata*) as well as pickerelweed or blue flag (*Pontederia cordata*) while "lillies" indicates *Nymphaea odorata* and possibly *Nuphar advena* and *Nymphoides aquatica*. Many postdrainage descriptions refer to thickets of willow (*Salix caroliniana*) and elderberry (*Sambucus canadensis*), either interspersed between patches of custard apple or in a band paralleling the southern boundary of the Custard Apple Swamp. However, a number of postdrainage observers also state that willow and elders invaded areas of sawgrass after water levels were lowered by drainage. The width of the bordering band of willow and elderberry mapped any time after the onset of drainage in 1882 is therefore almost certainly greater than the predrainage width. The width of the Custard Apple Swamp alone, when surveyed in 1915–1916, was 1 to 1.5 miles (1.5–2.5 km). At that time, the elderberry-willow band was 0.5 to 0.75 miles (0.8–1.2 km) wide.

Predrainage descriptions of the actual custard apple shoreline were not found, but postdrainage accounts, written after Lake Okeechobee levels were lowered, indicate "no banks" (W. A. Roberts 1907, p. 39) and "low muddy and irregular shore with its dense growth of pond-apple" (Small 1914, p. 74). The reference to no banks is probably an explicit differentiation from the peat banks that occurred along the sawgrass areas found farther west on the shoreline. "Muddy" is likely a reference to the highly (50 percent) inorganic, fine-grained fraction in the custard apple soils or to fine-grained inorganic silts that apparently washed in from the lake.

At least eight short rivers exited from Lake Okeechobee into the Custard Apple Swamp (Stewart 1907). Those rivers, sometimes referred to as "dead rivers," flowed southerly, out of the lake rather than into it, reflecting the southward slope of the Everglades (fig. 7.3). While the rivers were as much as 100–200 feet (30–60 m) wide and up to 6 or 8 feet (1.8–2.4 m) deep, all seemed to narrow quickly, all were short, and most included several forks. A number were short enough, 0.5–2 miles (1–3 km) long, to disappear while still within the Custard Apple Swamp; a few fully crossed the swamp, but did not continue into the Sawgrass Plains, apparently disappearing within the transitional willow/elderberry area. The forks and the main reaches of the rivers themselves were oriented in widely varying angles, ranging from west to east. The rivers were confined to the custard apple portion of the Lake Okeechobee southern shoreline; with high certainty, none were present farther west in the areas where sawgrass and sawgrass peat soil directly bordered the lake. This distribution along the southern shoreline might be due to differing erodibility of the two substrates; the sawgrass peat was known to be fibrous and strong, while the custard apple peats, with higher mineral and lower organic content, likely had lower tensile strength.

It is difficult to know the hydrologic importance of these short rivers. The widths and depths at their mouths as well as reported velocities suggest significant flows, but the short lengths, apparently random orientations, and especially the multiple branches suggest only minor contribution at the scale of the Everglades or even of the Sawgrass Plains landscape. The shortness of these rivers suggests that surface flow properties of the Sawgrass Plains landscape probably were the primary factors controlling inflows from Lake Okeechobee, rather than the rivers. However, after drainage lowered Lake Okeechobee water levels sufficiently to reduce overland flow through the Custard Apple Swamp, it is likely that for a time, the rivers became the only remaining outflows. As lake levels continued to drop under drainage, even these rivers dried out. The name "dead rivers" may reflect this postdrainage drying out or possibly their short, dead-end character. Already in 1915 and 1916 state surveyors working found dry river beds and referred to some as "dead rivers."

Ecological factors that determined the predrainage extent of the Custard Apple Swamp are of particular interest. Chapters 5, 6, and this one indicate that the outflowing southern shoreline of Lake Okeechobee formed from organic soils, included significant lengths of both sawgrass shoreline and custard apple shoreline, raising the question of why the full shoreline length was not Custard Apple Swamp. Harshberger (1914) suggested that the custard apple extent might be related to frost and temperature effects due to the moderating influence of the lake. It is not clear how temperature buffering would explain the presence of custard apple east of Sand Point (Township 43 Range 34) but not west of it.

Several authors noted that the much higher mineral content of the custard apple soils (fig. 7.2), in comparison with the sawgrass soils (50 percent instead of 10 percent), likely reflected the washing in of inorganic silts from the southeastern corner of Lake Okeechobee. Comparison of the Custard Apple Swamp extent, as mapped in this study (plate 5) using state township surveys, with the extent of custard apple soils called "Okeechobee mucks" (plate 15; Jones et al. 1948a) shows that the custard apple extent falls almost entirely within the extent of Okeechobee mucks. The spatial relation is particularly intriguing because the two mapping approaches were independent. The vegetation was mapped by land surveyors around 1916, while the soils were mapped almost thirty years later when the custard apple vegetation had been almost completely removed. Probably because of their high agricultural value (and perhaps partly because of the absence of native vegetation normally used in delineation), Jones et al. (1948a) mapped these soils in detail: "in the areas of peat and muck soils, the land now cultivated was investigated in every field where necessary, to locate the significant boundaries" (Jones et al. 1948e, p. 59). The similar spatial extents suggest a possible linkage between the predrainage custard apple distribution and the lake-derived minerals in the soil.

Many descriptions of the predrainage Everglades mention a "custard apple rim." Such a feature, if present as a nearshore band of elevated soil, would have been an important aspect of the predrainage relationship between Lake Okeechobee and the Everglades, increasing the separation between these water bodies. However, it seems likely that no such nearshore elevation was present in the original, predrainage landscape. It is more likely that the concept of a rim arose either (1) from the visual effect of the custard apple canopy rising above the lake to the north and the sawgrass to the south; or (2) as an actual but postdrainage rise in the soil elevation that developed as a result of drainage. It is probably significant that references to a "rim" begin to appear in the historical record only after the onset of drainage.

Prior to the onset of drainage in 1882, the soils of the Custard Apple Swamp do not appear to have been higher than the lake water level. Before drainage, surface wa-

Figure 7.2. Soil of the custard apple zone, 1915: "Muck. The non-fibrous character and relatively firm mineral texture of the muck, due to the presence of fine sand, silt, and clay, characteristic of the 'custard-apple' region of the shores of Lake Okeechobee, is indicated in the illustration" (Baldwin and Hawker 1915, plate 26).

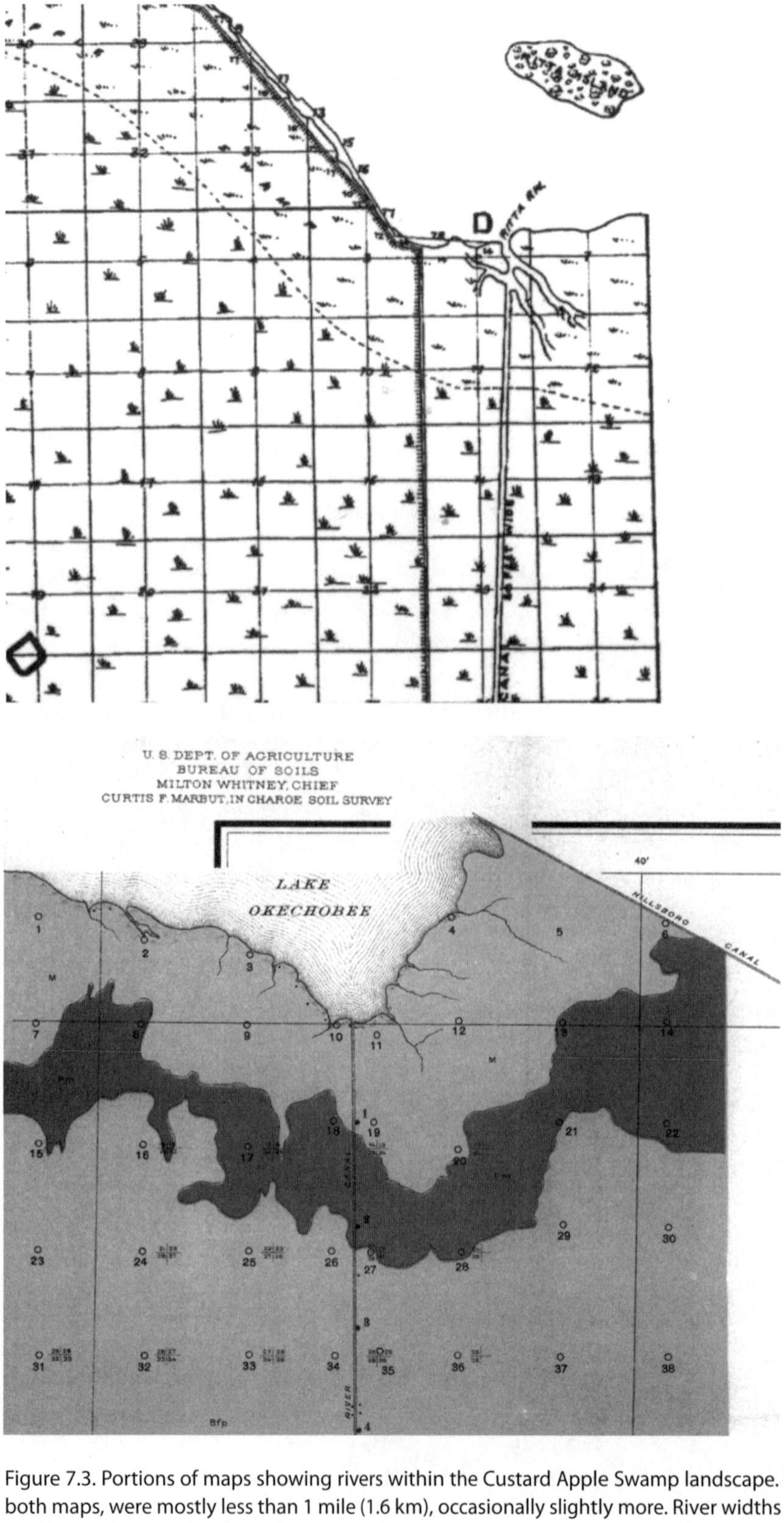

Figure 7.3. Portions of maps showing rivers within the Custard Apple Swamp landscape. River lengths, likely drawn to scale on both maps, were mostly less than 1 mile (1.6 km), occasionally slightly more. River widths shown by Kreamer (1892; upper map) are likely to be to scale; those shown by Baldwin and Hawker (1915; lower map) are likely not. Dashed line in upper map is the south edge of the Custard Apple Swamp present in 1892; the north-south canal is the Miami Canal. In the lower map, the three bands of soils shown from north to south are: Muck, Peaty muck (dark band), and Brown fibrous peat; the north-south canal is the North New River Canal. On the lower map, the south edge of the Custard Apple Swamp fell within the Peaty muck soil band. All the rivers terminated within the Custard Apple Swamp; that is, before reaching the Sawgrass Plains. Note the branching and apparently random, nonsystematic orientation. Sources: Kreamer (1892), T44 R35; Baldwin and Hawker (1915), T43–44 R35–36.

ter was present in the Custard Apple Swamp, indicating that the soil surface was lower than the lake level. This relationship—lake level above or even with the soil surface—persisted even after drainage. As the lake was progressively lowered, subsidence of the custard apple soil appears to have roughly kept pace (fig. 2.1), with both lowering by about 4 or 5 feet (1.2–1.5 m) by the 1940s: "The elevation of the land is now from 16 to 17 feet above sea level which is about level with the normal lake level" (Herr 1943, p. 12). Therefore it appears unlikely that the concept of a custard apple rim arose through comparison of soil and lake levels.

Although it can only be hypothesized, it seems likely that the appearance of a rim arose from differing soil properties to the south and that it was, in fact, a man-made artifact of uneven, drainage-induced soil subsidence. The custard apple soils differed from the adjacent sawgrass soils by having much lower organic matter content (50 percent organic material in the custard apple soil versus 90 percent in the sawgrass soil). Almost all of the subsidence was due to the organic fraction, so for equal lowering of water tables, the subsidence in the sawgrass soil would be almost twice that in the custard apple soil. After a time this would leave the custard apple soil higher, forming an elevated "rim." Indeed a drainage engineer noted exactly such differential subsidence prior to 1936: "In the heavier muck soils, such as the 'custard apple' belt adjacent to Lake Okeechobee, the subsidence appears to be less than that in the sawgrass peat farther out from the lake. This is due to the higher mineral content and greater density of the custard apple soil" (Clayton 1936, p. 14).

The Custard Apple Swamp extended up the eastern shore of Lake Okeechobee as far as the border between Townships 41 and 42 (Reyes 1855-T41 R37; Hardin 1916-T42 R37; Livingston and Heister 1919-T41 R37), which is approximately the point where the West Palm Beach Canal enters Lake Okeechobee. North of this, the narrow strip of swamp forest separating Lake Okeechobee from the sawgrass plains changed completely. Cypress swamps replaced custard apple, the shoreline changed from low and swampy to distinct beaches of white sand, and a sand ridge several feet (60+ cm) high paralleled the shore. While these sandy features might have been an artifact of postdrainage lowering of lake levels, or of oxidation of a surficial layer of organic material, their mention by predrainage observers (Ober 1887; Canova 1885, p. 25; Preble 1842, p. 35) suggests that they were original and nature features. The cypress forest extended north along the shore approximately to the present location of Port Mayaca.

Hydrology

Estimation of predrainage water depths for the Custard Apple Swamp is complicated by the almost complete absence of predrainage water depth point observations and by the severity of early drainage (chapter 5). Lowering of Lake Okeechobee levels during the early 1880s caused water tables to drop several feet (60+ cm) below ground in the sawgrass areas bordering the lake, presumably lowering water tables in the Custard Apple Swamp as well. The scarcity of observed surface water depths in postdrainage accounts may in part simply reflect the absence of surface water.

Estimates of the predrainage depths and hydroperiods—periods with water above the average soil surface—in this landscape are based on the characteristics of the vegetation (buttressed, enlarged bases of the custard apple trees) and the repeated identification of the area as a "swamp," which suggested that surface water was present much of the time. The misconception that this swamp was higher than the adjacent sawgrass derives primarily from the term "custard apple rim" applied to this area. As noted previously, the rim appears to have been a postdrainage artifact of subsidence. Additionally, the descriptions of a water depth sequence—deepest in custard apple, somewhat shallower in the transitional area (willow, elderberry, pickerelweed band), and shallowest in the Sawgrass Plains—and the requirement of deeper water depths for custard apple compared to sawgrass, confirm that predrainage water depths were deeper in the Custard Apple Swamp than in the Sawgrass Plains, and hydroperiods were longer. The average long-term water depths were estimated to range from zero feet (0 cm) above land surface at the end of the dry season to 2 feet (60 cm) above at the end of the wet season, with a hydroperiod of 11–12 months.

Hindcasting

Four mapped (or mappable) sources, all postdrainage, are available for estimating the predrainage extent of the Custard Apple Swamp: Kreamer's map (1892; see plate 9), Harshberger's vegetation map (1914; see fig. 1.7), field notes from the state township surveys conducted in 1915–1916, and the J. H. Davis vegetation map (1943b; see plate 13). Kreamer's map has been found to be detailed and accurate in areas where it could be compared with earlier U.S. surveyor township surveys. Unfortunately, it extends only as far east as Range 35. Much of the Custard Apple Swamp is therefore not included. As discussed earlier, Harshberger's sketch is known to have errors and is too general to be useful for quantifying predrainage extents. The J. H. Davis map of the vegetation of southern Florida (1943b) has frequently been used as an estimate of the predrainage extent of the Custard Apple Swamp. The Custard Apple Swamp was almost completely obliterated by that time, but Davis considered it sufficiently important that he explicitly deviated from his practice of mapping the landscapes he actually saw and instead estimated a predrainage extent. Unfortunately, though his estimates of the total east-west length along the shoreline match well with other sources, he seems to have overestimated the width of the landscape substantially. As the original custard apple forest was largely cleared and removed for cultivation by the time of Davis's work, it is probable that he simply estimated the original extent from the extent of the nearshore soils, such as the Okeechobee muck, Okeelanta peaty muck, and Everglades peaty muck (cf. plates 11 and 13).

The most detailed information on the width of the landscape is available from the field notes of state township surveys carried out in 1915 and 1916, mostly by civil engineer and surveyor Otis Hardin. Those surveys mapped not only the township lines but, more important, the subdivision lines, providing a network of transects 1 mile (1.6 km) apart, with vegetation described and transitions mapped to the nearest hundredth of a chain (approximately 0.5 foot; 15 cm). The notes mention custard apple, willow, elderberry ("alder"), and sawgrass. Mapping out these notes results in a band of custard apple generally 1–1.5 miles (1.6–2.4 km) wide and a willow and elderberry band 0.5–0.75 miles (0.8–1.2 km) wide. As noted previously, there are indications that the width of the willow and elderberry band may have increased after drainage. It is unlikely that the band of custard apples, generally preferring wetter rather than drier conditions, would have expanded after water levels lowered.

The mapping derived from the state surveys can be compared with the Kreamer (1892) map where they overlap in Townships 34 and 35. In Township 35, where the Custard Apple Swamp is at about typical width, the combined width of the custard apple and willow/elderberry bands is very similar to the single band labeled "Willow & Custard Apple" by Kreamer. However, farther west in Township 34, where the landscape narrowed to zero, the state survey notes showed a wider extent than Kreamer's (1892) map. The reason for the difference is not known.

The predrainage extents shown on plate 5, green for the Custard Apple Swamp and yellow for the willow/elderberry thicket, were mapped directly from the state survey notes, with modification at the west end to reflect the earlier source, Kreamer (1892). The specific township surveys used are shown on figure 1.4.

8

Ridge and Slough Landscape

The Ridge and Slough landscape was the heart of the Everglades. At 1.5 million acres (600,000 ha), it was the largest landscape, originally covering 55 percent of the Everglades (table 4.2). The landscape once stretched across the full width of the basin, almost 40 miles (60 km) from the Eastern Flatwoods on the east to Big Cypress on the west, and two-thirds of the length of the basin, 80 miles (130 km) from the Sawgrass Plains south to the mangroves at the foot of Shark River Slough (plate 5). The well-known name "River of Grass" (Douglas 1947) captures a portion, but only a portion, of the predrainage aspects of this landscape: "river" accurately reflects the flow, directionality, and absence of stagnation noted by many early observers, and "grass," of course, reflects the vast quantities of tall, dense sawgrass (*Cladium jamaicense*). However, "River of Grass" calls to mind a lawnlike uniformity that belies the observations of early explorers or even, in certain locations, of current observers. While the smaller upstream Sawgrass Plains landscape (chapter 6; plate 5) was indeed highly uniform, we show in this chapter that the Ridge and Slough landscape was much more varied and that the arrangement of this landscape's varied subcomponents reflected an organized, systematic structure. This organization, developed on a peat soil substrate, classifies the predrainage Ridge and Slough landscape as a "patterned peatland" (H. E. Wright et al. 1992). Jones et al. (1948e, p. 15) referred to the area as the "ridge-and-slough country," likely taking the name from earlier soil scientists who described the area as a "series of ridges and sloughs" (Baldwin and Hawker 1915, p. 752). As a reflection of the original strongly directional patterning, we have adopted this name throughout the present work to describe this landscape.

The Ridge and Slough landscape was horizontally extensive, and at the regional scale of the Everglades basin, it was also remarkably flat. From an elevation of about 15 feet (4.5 m) above sea level at the north end, the landscape sloped very slightly downward over the course of 80 miles (130 km) to a few feet (0.5–1 m) above sea level where it merged with the mangrove communities. The transverse direction (crosswise to the downstream direction) was even flatter. Viewed from a low angle, the many individual sawgrass ridges collectively took on the appearance of a continuous, endless prairie stretching off into the distance.

However, at a smaller scale, say that of a canoeist, or when seen from a steeper angle, the landscape was anything but uniform or flat (plates 17, 20, 22). The characteristic patterning included elevation differences of as much as 4 feet (1.2 m) due to differences in soil thickness. These differences in elevation were, and to some extent remain, a key feature of the Ridge and Slough landscape. The elevation differences between the landscape subcomponents: (a) supported distinct vegetation types; (b) provided important upland, aquatic, and seasonally aquatic habitats; (c) created enormous lengths of interfaces between these habitats; and (d) made navigation possible.

The main subcomponents of the predrainage Ridge and Slough landscape—aquatic sloughs, sawgrass ridges, and semi-terrestrial tree islands—were differentiated by elevation, soils, hydrology, and vegetation. Sloughs were mostly free of vegetation extending above the water surface, being limited primarily to water lilies floating on the surface (*Nymphaea odorata* and *Nuphar advena*), so they appeared lower and much more open than the surrounding sawgrass ridges or tree islands. Ridges were elongated, slightly elevated areas covered by dense sawgrass stands rising 6 to 8 or more feet (2–2.5+ m) above the water surface. Sloughs and ridges together created an alternating, corrugated effect, especially when crossed in the transverse direction (plates 23, 17 top, 20, 24 left). Tree islands occurred as large lenticular ("teardrop-shaped") strand islands and also as smaller circular or oval islands sometimes called "clumps" in the early literature. Various tree and shrub species colonized the tree islands, as is reflected in the correspondingly diverse names used to describe them; for example, "bayheads," "willowheads," and "tropical hammocks."

Perhaps one of the most remarkable features of this

large landscape of 1.5 million acres was its similarity over extremely long distances. In the course of traveling 10 to 20 miles (15–30 km) in the transverse direction, the early explorer might have passed roughly 30 to 60 sawgrass ridges, a similar number of sloughs, and perhaps 5 to 10 large tree islands. There would have been slight differences from one slough to the next and from one ridge to the next, but overall the sloughs were all clearly sloughs, the ridges all clearly ridges. The similarities, ridge to ridge or slough to slough, were much more striking than the differences. This remarkable degree of predrainage uniformity appears to be due to formation of the landscape on a peat soil substrate, leveled by the process of accumulation under a level water surface.

The structure and composition of the landscape changed considerably as a result of canal construction and drainage activities (chapters 2 and 3), a point that underlines the importance of predrainage observations. Information from historical surveys and descriptions of the area by early explorers were combined with information from recent detailed studies of existing hydrology, soils, and vegetation patterns to develop descriptions of the landscape conditions that existed in the early nineteenth century. As in the other landscape chapters, the reconstruction of the predrainage Ridge and Slough landscape presented here was developed from the full set of chronologically arranged, annotated historical observations included in the Historical Observations sections for chapter 8 on the DVD. Because development of each aspect drew on the collective historical information, we have generally not included individual citations.

Early explorers ascending the swiftly running coastal rivers left vivid descriptions of their sudden emergence from the closed canopies of the coastal forests into the wide open expanse of the Everglades. The vista they gazed out upon was the Ridge and Slough landscape, as this landscape extended directly to the bordering woodlands (chapters 3 and 4). On the eastern and western borders, these woodlands were cypress or pine (plates 5, 41–43); to the south the border was formed by mangroves (e.g., Willoughby 1898a, pp. 38–41; Dimock 1911, pp. 116–119). They frequently described the landscape as a vast lake, with the water surface partially obscured by tall, dense sawgrass and numerous tree islands (fig. 4.2). As they poled, paddled, or even sailed canoes through the landscape, these explorers found a network of waterways or sloughs between the ridges of sawgrass and noted tree islands that ranged in size from one-quarter acre to hundreds of acres. In the sloughs, which in places expanded into wider ponds ("lagoons" in the early literature), the water was sufficiently deep to remain open or be dotted on the surface with water lilies (*Nymphaea odorata* and *Nuphar advena*; plates 17, 19, 20; fig. 8.1). When the explorers' course was parallel to the alignment of the landscape, the early explorers could easily pole or even paddle canoes through the sloughs, covering as much as 15 miles in a day (plates 10 left, 10 right). However, when traveling crosswise to the landscape, the same explorers reported significantly slower going, being forced to alternate slough travel with pushing their boats through the dense sawgrass and boggy soil of the sawgrass ridges (plate 18). Numerous tree islands were encountered, and some were used as campsites, although predrainage explorers frequently remarked on the scarcity of dry locations on the islands and often awoke to find that they had sunk into wet ground during the night. Interestingly, explorers often, though not always, reported dispensing with their mosquito bars, finding these unnecessary. This might be a reflection of active mosquito fish (*Gambusia holbrooki*) populations being present throughout landscape, including within sawgrass ridges during the wet season (chapter 11).

A description from 1841 by one of Colonel Harney's engineers mentions the three main subcomponents forming the landscape—sawgrass ridges, sloughs (here referred to as "channels," elsewhere sometimes called "leads"), and tree islands—and gives an impression of its vastness: "The appearance presented upon entering the Everglades is that of an immense prairie, stretching farther than the eye can reach, covered by thick saw-grass, rising six feet above the surface of the water, which it conceals—the monotony varied by numerous snake-like channels and verdant islands, scattered few and far between—the average depth of water over the whole extent being from two to four feet" (Anonymous 1841, p. 253).

The same engineer gave a clear impression of the open and apparently vegetation-free nature of the aquatic sloughs—deep enough even in January (dry season) to move swiftly and to allow the use of paddles—as well as the contrast between the open sloughs and the dense sawgrass ridges: "We then moved forward swiftly and noiselessly, at one time following the course of serpentine channels opening out occasionally into beautiful lagoons, at another forcing our way through barriers of saw-grass. After several hours of hard paddling, we came in sight of Chitto's Island" (Anonymous 1841, p. 247).

The descriptions of navigable, open water alternating with much less passable sawgrass ridges are consistent with early photographs and maps of the Ridge and Slough landscape (plates 17–19, 20; figs. 8.1, 8.2). Narrative descriptions of this landscape indicated that the contrast between sloughs and ridges was generally sharp and dis-

Figure 8.1. Ridge and Slough landscape west of Fort Lauderdale, ca. 1911. Sawgrass ridges (dark gray) with distinct edges rise above the open water of the sloughs (light gray). The location is likely Township 50 Range 40 or 41. The landscape pattern is similar to plates 17–20, farther south. Photograph by H. Gunther taken from the dredge *Okeechobee*. Reproduced from Sellards (1912).

tinct. Field notes from township surveys also suggested that this was the case. Numerous State of Florida surveys conducted for the Trustees of the Internal Improvement Fund around 1911–1912 recorded each transition between slough and ridge (and vice versa) along the survey lines (e.g., fig. 1.3). The extra work involved in measuring and recording the location of these transitions—as many as 10 in a mile—would only have been plausible and possible if the boundaries were distinct, and the vegetation and/or elevations were significantly different.

James O. Wright, at one time chief drainage engineer in charge of Everglades reclamation, provided a detailed and early description specifically of the Ridge and Slough landscape, particularly of the relationship between the peat ("muck") thickness and the relative elevations of the ridge and slough landscape subcomponents. However, before his description is examined, Wright's observations merit special comment, given his notoriety (e.g., Hanna and Hanna 1948, pp. 152–162; Purcell 2001). Wright was intimately involved with Everglades drainage from about 1906 to 1912, arranging for instance for John Stewart's (1907) very useful studies. In the early 1910s Wright was much vilified for his possible self-interest in drainage efforts in Florida and especially in North Carolina (Hanna and Hanna 1948, pp. 158–159). Particularly, he has been accused of presenting overly optimistic engineering calculations regarding the potential for successful drainage. That said, examination of Wright's descriptions of the Everglades themselves reveal considerable detail, suggesting repeated firsthand observation of extensive areas. Wright's descriptions do not shy away from aspects that would have made the job of drainage harder or its outcome more uncertain; instead they give every appearance of being straightforward, empirical accounts of what he saw. Comparison with descriptions of other observers reinforces the impression that Wright, despite other likely shortcomings, was in fact a trained and careful observer of the natural world:

> Beyond this ["practically level" area of the Sawgrass Plains, the soil material] becomes uneven, being marked with irregular depressions and ridges. As the surface is level, this unevenness of the sub-stratum causes a varying depth of muck within a limited area.
>
> The surface of the Everglades, taken as a whole, appears to be a level plain, having a slope of three inches per mile toward the south and east, yet this statement is subject to slight modification. Throughout the entire area there are numerous winding shallow depressions or channels 100 to 500 feet wide, and one to three feet lower than the land through which they pass. These depressions, locally called "Strands," [i.e., sloughs] wind through the 'Glades in all directions, though their general trend is from north to south. In other places there are slight depressions, like ponds or lagoons, covering one to forty acres. These are most numerous along the eastern margin and throughout the southern portion. The muck in these depressions

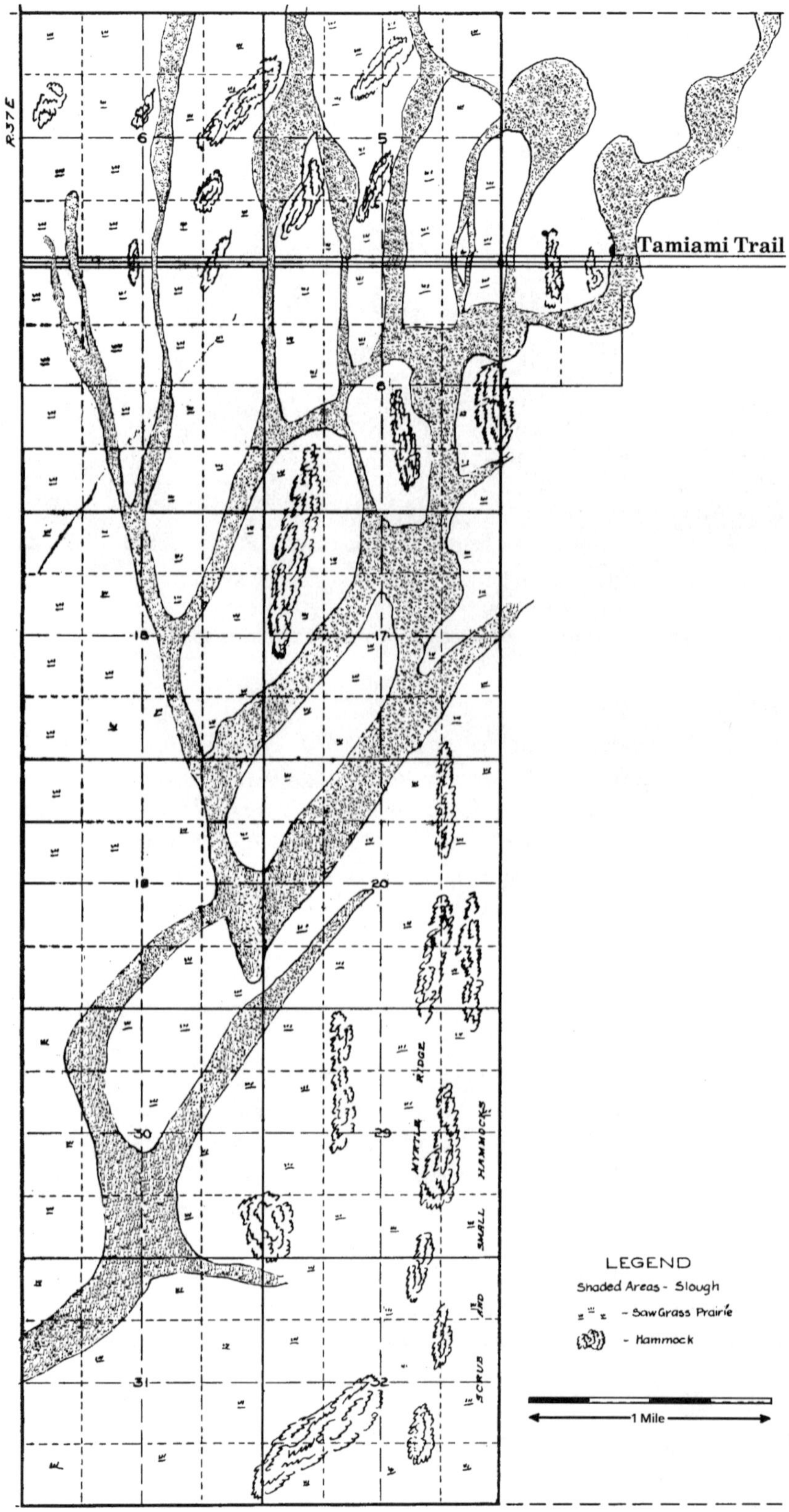

Figure 8.2. Plan view of a 12-square-mile portion of the Ridge and Slough landscape, prepared by John King in 1917, covering the northeast Shark River Slough area (T53–54 R37). Map elements correspond to sloughs, sawgrass ridges, and strand tree islands ("hammocks"). Dimensions are approximate. The overall orientation is most likely inaccurate (see fig. 1.8 for more accurate rendering), but the pattern is representative. Drawn by John King (1917e) for Miami developer James F. Jaudon.

> is usually less firm than the land on either side [i.e., than the peat of the adjacent sawgrass ridges]. These irregularities of the surface also affect the depth of muck at these points, as the underlying hard material [i.e., bedrock] is no lower in these surface depressions than under the adjacent land. These low places are usually filled with water, while the general surface of the Everglades is comparatively dry. (J. O. Wright 1912, pp. 18–19)

Wright's detailed Ridge and Slough descriptions are consistent with those of other observers (see Historical Observations for chapter 8 on the DVD). His observations also emphasize the importance of quantifying the predrainage dimensions of the Ridge and Slough landscape.

Wright's last comment, that the "low places" (i.e., the sloughs) usually contain water at the same time that the "general surface" (i.e., the sawgrass ridge surface) is dry, illustrates a critical difference between the patterned peatland of the Ridge and Slough landscape and the unpatterned Sawgrass Plains. In the Sawgrass Plains there was only one "general surface," so when water levels dropped below ground during the course of the dry season, surface water disappeared from the entire landscape.

Subcomponents of the Ridge and Slough Landscape

This patterned landscape differed from other Everglades landscapes by being composed of distinct subcomponents—ridges, sloughs, and tree islands—each of which was reproduced many times across the landscape. The following section provides predrainage estimates of their physical dimensions, after which each of the subcomponents is discussed, along with some variants.

Physical Dimensions

Appendix C tabulates observations of the lengths, widths, and heights of the subcomponents of the Ridge and Slough landscape, which were gleaned from predrainage and early postdrainage observers. Postdrainage aerial photographs (e.g., plates 24, 26) are an additional source of length, width, and spacing information, provided the measurements are taken from areas where vegetative or other aspects suggest that their horizontal dimensions may still closely resemble predrainage conditions.

Present-day field measurements from such areas might also be a source for vertical dimensions, particularly for the height of sawgrass ridges and of tree islands relative to surrounding sloughs. However, the reconstructions of past hydrology, and especially of peat subsidence discussed in chapter 3, suggest that relative vertical dimensions are less likely to have remained constant since predrainage times. Those changes are discussed further in chapter 11. In this chapter we note only that an apparent postdrainage "flattening" of the landscape appears to have occurred as early as the 1950s. As the elevation difference between ridges and sloughs decreased, water depths in sloughs became shallower, allowing emergent wet prairie species, and sometimes even sawgrass, to invade the sloughs, severely restricting traditional forms of navigation:

> General Considerations: Prior to intensive drainage much of the Glades area was dissected by relatively deep open water "sloughs." These natural drainage channels permitted traverse into the interior of the Glades by "pole-boat" [likely dugout cypress canoes, flat-bottomed "John boats," or "Glades skiffs" (Simmons and Ogden 1998, pp. 6, 165)]. Journey by "pole-boat" for example from Dania on the east coast to the Big Cypress, forty miles to the west, or from Ft. Pierce south through the Glades to Ft. Lauderdale were not uncommon [e.g., see similar reports in Lamb 1975, p. 1, and Padgett 1998, p. 2]. A "pole-boat" trip such as this would be a remarkable feat indeed at the present time even during periods of high water. The major portion of these open water "sloughs" have now reverted to *Rhynchospora* sp. or maidencane [i.e., wet prairie species; might also have included spikerush (*Eleocharis cellulosa*)] or have been invaded by sawgrass, making navigability possible only by airboat and then only during the wet season. (Wallace 1955a, p. 2)

Similarly altered conditions can be seen at present, both in portions of Water Conservation Area 3A to the north of Alligator Alley (Highway I-75) and in Shark River Slough to the south. Interestingly, the slough shapes and dimensions north of Alligator Alley are still very similar to ones seen elsewhere in remaining open water sloughs. As in Wallace's description, the sloughs north of Alligator Alley have been completely filled by dense stands, usually spikerush (*Eleocharis cellulosa*; plate 28 bottom). This appears to support the hypothesis that postdrainage changes to the Ridge and Slough landscape may have been more rapid in the vertical than in the horizontal dimensions.

Table 8.1 estimates the predrainage vertical and horizontal dimensions of the Ridge and Slough landscape subcomponents. The vertical estimates are based on pre-

Table 8.1. Physical dimensions of subcomponents of the predrainage Ridge and Slough landscape

Landscape Element	Dimension	Estimate
Sawgrass Ridge	Width	100–2,500 feet
	Length	1–5 miles
	Height	1–2 feet[a]
Strand Tree Island	Width	400–2,500 feet
	Length	0.5–1.5 miles
	Height	2–4 feet[b]
Bayhead	Diameter	50–100 feet
	Height	2–4 feet[c]
Slough	Width	300–1,300 feet
	Length	> 6 miles[d]
	Depth	1 foot; 3 feet[e]
Channel[f]	Width	3–300 feet
	Length	> 1 mile ?
	Depth	3–5 feet[g]

Source: Based in part on historical observations compiled in appendix C.

Notes: a. Refers here to the height of the average ground surface of the ridge above the average bottom surface of the surrounding sloughs.

b. Height of the center of the elevated north end of the island above average bottom surface of the surrounding sloughs.

c. Height of the center of the island above average bottom surface of the surrounding sloughs.

d. As the sloughs, together with the channels, formed a continuous flow network through this landscape, the length of sloughs and channels are somewhat ill-defined.

e. Depth refers to the average annual low and high water depths in the sloughs.

f. Also referred to as "creeks," "runs," "streams," "gullys," or "leads."

g. Note that the authors cited in appendix C give channel depths as the depth from water surface to bottom of the channel. We estimated the elevation of the bottom of the channels by subtracting the water depth from the average slough water depth (2 feet); thus the channels extended 1 to 3 feet below the bottom of the slough.

drainage and early postdrainage sources (appendix C), while the horizontal dimensions are based primarily on measurements from selected portions of the remaining present-day landscape in combination with historical observations.

One of the estimates in table 8.1, the height of sawgrass ridges above sloughs, is sufficiently central to this landscape to merit special comment. The estimate is based on Wright's previously mentioned description of "one to three feet" (30–90 cm; Wright 1912, p. 17) and on another description by soil scientists of "two to three feet" (60–90 cm; Baldwin and Hawker 1915, p. 752). Wright's writings suggest that his observations were based on extensive field experience over large areas, but the details of exactly where and how were not encountered in the course of this study. In contrast, the field basis for Baldwin and Hawker's observations is well documented. In 1915 the two soil surveyors made more than three hundred soil borings, down to bedrock, at 1-mile intervals along the full length of the newly opened North New River Canal (fig. 2.2). To make these borings, at each mile marker along the canal the two surveyors proceeded laterally from the canal, taking cores close to the canal and again at 1, 2, and in some cases 3 miles (1.5, 3, 4.5 km) away from the canal, repeating this same procedure for the opposite side of the canal. In the course of their sampling within the Ridge and Slough landscape Baldwin and Hawker passed through more than 80 miles (130 km) of the landscape. Their estimate of ridge height, therefore, seems very strongly grounded in extensive empirical observations. The estimate of the sawgrass ridge height given in table 8.1 of 1 to 2 feet (30–60 cm) above the surrounding sloughs is a conservative interpretation of Wright's and Baldwin and Hawker's well-grounded observations.

The estimates included in table 8.1 and illustrated in figure 8.3 and plate 23 suggest that much of the ecological importance of the Ridge and Slough landscape is due to the elevation differences associated with the original predrainage ridge-slough microtopography and to the interaction of that microtopography with the predrainage hydrology. Plate 23 illustrates this using a "unit cell" approach to characterizing the landscape, based on the assumption that the overall landscape can be captured by replication of many of these unit cells.

The Ridge and Slough landscape can also be conceptualized by imagining a set of three planes, each resting on one of the classes of subcomponents. Thus one imaginary plane would rest on the peat surface of all the sawgrass ridges—this plane would correspond to the "general surface" indicated by Wright (1912, p. 17). A second plane would rest on the bottoms of all the sloughs and a third on the heads of the tree islands. The distances between these imaginary planes would correspond to the vertical dimensions given in table 8.1.

An additional fourth plane was formed by the surface water present in the landscape. This plane oscillated over the course of the wet and dry seasons. Interestingly, despite the slight regional slope, under predrainage conditions all four planes appear to have been parallel to each other, a topic discussed further in chapter 11.

The relative topography of the ridges, sloughs, and tree islands—that is, the separation of the three imaginary planes—was different prior to drainage than it is at present. Peat oxidation and fires occurring between the 1920s and 1950s, made possible by water levels lowered by drainage, diminished the elevations of tree islands and sawgrass ridges relative to the sloughs. In addition, reduced water flow in the impounded, postdrainage Everglades may have allowed sloughs and channels to fill in partially, through increased rates of accumulation of organic material. If so, this would have diminished further the elevational differences between sloughs and the tree islands and ridges. Overall, the microtopography of the Ridge and Slough landscape has tended to flatten since the initiation of drainage. As discussed in chapter 11, this flattening complicates efforts to estimate (and also to restore) predrainage water depths in sloughs.

Sloughs

Sloughs were the truly aquatic subcomponent, being covered by surface water year-round during most years. This created a network of elongated waterways winding through the landscape (fig. 8.2; plates 23, 31 upper left, 24 left). Borders of the sloughs were sharply defined by the edges of the sawgrass ridges (above water; plates 17, 20, 21) and by the corresponding edges of the elevated peat of the sawgrass ridges (mostly below water, except toward the end of dry season). The appearance of sloughs varied somewhat, from open water ("open sloughs") to open water partially covered with water lilies ("lily pad sloughs") to even open water with a scattered growth of graminoid species ("grassy sloughs"; see also appendices B and C). Most sloughs appear to have been sufficiently open to reflect blue sky strongly from the water surface, creating a vivid contrast with the light gray-green of the sawgrass (plates 21, 22) and the darker vegetation of the tree islands (plates 17 top, 18 top). Although generally forming a continuous network in the downstream direction, the sloughs could narrow and widen, varying from wider "ponds" or "lagoons" to narrower and shorter "runs" or "leads" (e.g., fig. 8.2).

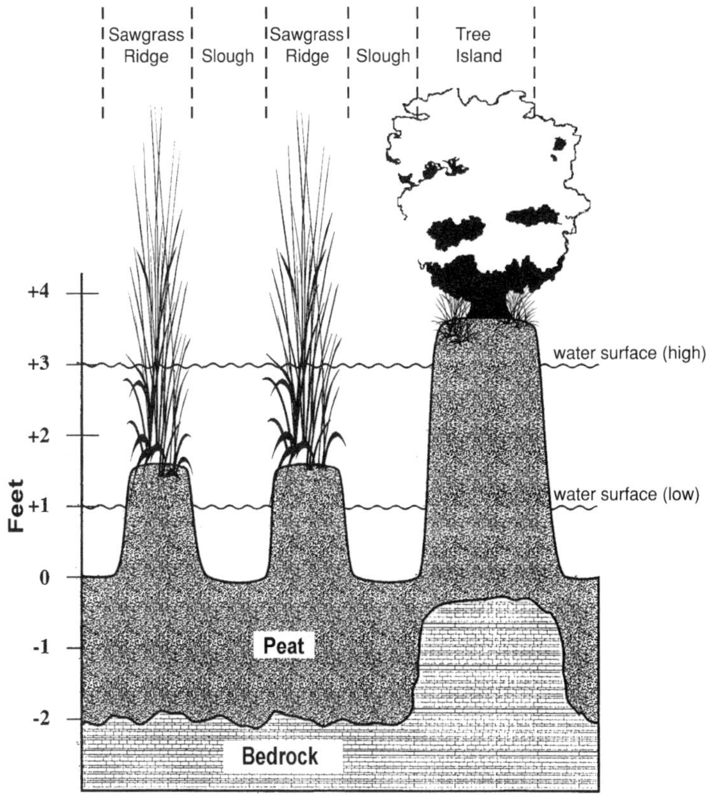

Figure 8.3. Estimated predrainage cross section of the Ridge and Slough landscape. Dimensions are based on table 8.1. Vertical scale is exaggerated for display; vegetation is not to scale.

Open sloughs with a partial cover of water lilies appear to have been the most common. Probably for this reason, the term "slough" has been applied both to indicate areas with the characteristic elongated shape and bordering sawgrass ridges and also to indicate areas of open, water lily marsh; originally, those were usually one and the same. Of the lilies, the most consistently found was the white water lily (*Nymphaea odorata*), suggesting this as an indicator species for predrainage sloughs. Floating hearts (*Nymphoides aquaticum)* are not mentioned by name but may also have been present; *Nymphaea* and *Nymphoides* can be difficult to distinguish by leaves alone. Spatterdock, also called "bonnets" (yellow water lily, *Nuphar advena*), is rarely seen now except in alligator holes, canals, and the upper portions of tidal creeks (southern Everglades National Park), but it was often mentioned in predrainage and some postdrainage accounts. Figure 8.4 illustrates a slough densely covered with *Nuphar*, photographed in the dry season of 1916. The density of *Nuphar* seen here at the headwaters of

Figure 8.4. An unusual slough filled with yellow water lilies (*Nuphar advena*, also called spatterdock or bonnets) was photographed in February 1916 by J. K. Small in the upper reaches of Taylor Slough (T58 R37, probably Section 14 or 15), close to the Rockland Marl Marsh. More typical predrainage sloughs contained an open cover of white water lilies (*Nymphaea odorata*) and open water, with occasional *Nuphar advena* (see fig. 8.1; plates 17, 19, 20, 22). Small's caption reads: "Deep slough in the Everglades east of Royal Palm Hammock. One of several sloughs that have preserved the hammock from the prairie fires of the Everglades. This channel is seldom perfectly dry. It, like similar places in the vicinity, is a rendezvous for the native amphibians in the dry season. Its edges are fringed with saw-grass, common-reed [*Phragmites australis*(?)], and small hammocks of low trees and shrubs" (1929, p. 84). Photograph from the J. K. Small Collection, Florida Photographic Collection, used by permission of the Florida State Archives.

Taylor Slough was not typical of the Ridge and Slough landscape, suggesting locally deeper water. There is some indication from the historical record that yellow water lily prefers deeper water than white water lily (see also chapter 11).

"Flag" sloughs were sometimes mentioned by land surveyors working for the Trustees of the Internal Improvement Fund of the State of Florida during the 1910s. Flags apparently referred to duck potato or arrowhead (*Sagittaria lancifolia*), possibly to pickerelweed (*Pontederia cordata*), and, much less likely, to alligator or fire flag (*Thalia geniculata*). Flag sloughs were probably so named because these emergent species visibly stand out in sloughs, either as isolated individuals or as small patches.

Predrainage slough descriptions and numerous accounts of rapid canoe travel of 10–15 miles (15–25 km) per day together suggest that predrainage plant densities were generally low enough to present little or no resistance to canoe travel—provided of course that travel was parallel to the prevailing ridge and slough direction; transverse travel that required pushing through sawgrass ridges was much slower.

Predrainage sources occasionally mention "grassy sloughs," sometimes even mentioning a typical wet prairie species, maidencane (*Panicum hemitomon*), by name. This naturally draws a comparison to the various "whitegrass flats" (wet prairies) widely observed in former Ridge and Slough landscape areas from the 1920s or 1930s up until impoundment of the Water Conservation Areas beginning in the 1960s. Those wet prairies developed a dense, emergent canopy (plate 30, panel C) of one or more characteristic "wet prairie species": spikerush (*Eleocharis* spp.), beakrush (*Rhynchospora* spp.), or maidencane (*Panicum hemitomon*; Loveless 1959b, p. 3; Craighead 1971, pp. 144–145; Goodrick 1974, pp. 185–189; Gunderson 1994, p. 332). Analysis of observations from the 1950s through 1970s, the periods just before and after impoundment of the Water Conservation Areas (see Historical Observations 8.1–8.5, 8.7, 8.11, 8.14, 8.16; see also tables HO8.1–HO8.5 on the DVD), strongly suggest that the wet prairies of the 1950s were an anthropogenic phenomenon, arising from water depths lowered by the strong drainage of the 1920s through 1940s (chapters 2 and 3). In fact, early water managers actively promoted this phenomenon: "The wet prairie community is of primary importance in the management of the Loxahatchee Refuge. . . . Water depth necessary for its maintenance will vary from 12 to 18 inches during the rainy season to a few inches below the ground surface during the dry period" (U.S. Department of Interior, Fish and Wildlife Service 1972, p. 8; see also Hagenbuck et al. 1974, pp. 14–15). After impoundment, when wedge-shaped water depths developed within the Water Conservation Areas (shallow/dry at upstream end; deep/wet downstream), wet prairies developed most strongly in former sloughs with depth regimes intermediate between slough hydrology and sawgrass hydrology. These postdrainage observations suggest that predrainage sloughs would have been "grassy" either at times when water depths had been less than typical or in shallower locations.

Predrainage sloughs described as "grassy" probably would not have resembled postdrainage wet prairies. In at least some cases it apparently was possible to canoe through the grassy sloughs at a rate similar to canoeing through sloughs (e.g., *Times-Democrat* expedition; appendix H), suggesting that the density of wet prairie species was low. Water lilies were mentioned as being present along with the grasses (e.g., Ingraham expedition; appendix I), suggesting that the water depths were not strongly different from the more common water lily sloughs. Predrainage water depths in most sloughs were usually too deep to encourage wet prairie species. Nevertheless, at least maidencane, and possibly spikerush and beakrush as well, may have been present at low densities, especially during dry years. Development of dense postdrainage wet prairies within former sloughs, as suggested by Wallace (1955a, p. 2), corresponds to paleoecological analysis. Pollen in soil cores is suggestive of a postdrainage increase in *Poacaeae* and *Cyperaceae* pollen (Debra Willard, pers. comm., 2007) around 1900, but it has not yet been possible to narrow this specifically to the characteristic wet prairie species.

Typical predrainage sloughs (i.e., the open or water lily type) are likely to have included some degree of the following three constituents: (a) the submerged aquatic plant bladderwort (*Utricularia* spp.), (b) microalgal periphyton, and (c) flocculent organic material ("floc"). In each case, examination of a combination of present-day observations, historical predrainage observations, and palynological studies of soil cores is likely to produce the most accurate information regarding predrainage slough conditions.

Bladderwort is ubiquitous in present-day sloughs that resemble predrainage conditions; that is, sloughs with low levels of phosphorus (P) and calcium (Ca); water depths similar to predrainage depths; and an open, partial cover of water lilies. Vegetation studies frequently consider *Nymphaea* and *Utricularia* together to be the indicator species for sloughs. Although very widely distributed in sloughs, bladderwort does not usually fully fill or "clog" the water column, either vertically or horizontally. No predrainage reports specifically referring to bladderwort (*Utricularia* spp.) were encountered, but the current distribution suggests that this species would have been common. Loveless (1956, pp. 39–40) felt that *Utricularia foliosa* plays a "major role in the formation of organic soils in the slough communities," perhaps referring to flocculent detritus. The long distances that early explorers were able to travel per day through predrainage sloughs suggests that then, as now, *Utricularia* must not have completely filled the water column.

Periphyton—that is, assemblages of microalgae covering the submerged portions of macrophytes (Browder et al. 1994, p. 380)—was not reported by name in historical documents. The absence of the modern name periphyton is of course not conclusive, but the scarcity of references to periphyton-like materials within predrainage sloughs may be suggestive. Periphyton is not evident in Willoughby's photos from early 1897, yet is now common in the same areas, perhaps because they are typically now much drier. Captain Dawson's two expeditions in 1855 mention "fungi," which is assumed to refer to periphyton (see Historical Observations 8.11 and 8.12 on the DVD). The scarcity of references to periphyton, "fungi," or other descriptions likely to refer to periphyton may reflect a comparative rarity of at least the highly visible, whitish "sweaters" that surround plant stems in some present-day sloughs. Although the darker-colored periphyton sometimes found on bladderwort might have been overlooked, it seems less likely that the present-day quantities and widespread distribution of the whitish "sweaters" would have gone unnoticed or unremarked upon by predrainage observers.

The current system, particularly Water Conservation Area 3A, provides an additional indication of the likely presence or absence of periphyton sweaters prior to man-made drainage. Under the present impounded conditions, a depth gradient is present within the sloughs of Water Conservation Area 3A, with the shallowest water depths at the northern, upstream end and the deepest water at the southern end (McVoy 2006; U.S. Geological Survey 2006). Visual inspection of the full length of this depth gradient suggests a parallel but opposite periphyton pattern. In the shallow water sloughs at the north end, periphyton sweaters are very obvious on the stems of wet prairie species, but they become less common and almost completely disappear in the deeper sloughs at the southern end. The southern sloughs more closely resemble, in water depth and vegetation, the descriptions of predrainage sloughs. While the historical record is not definitive, the scarcity of references to periphyton, along with its present-day absence from deeper sloughs, suggest that at least the sweater type of periphyton may not have been common in predrainage sloughs.

"Floc," an opaque brown layer of loose, noncohesive, flocculent plant detrital material of near-neutral buoyancy is often present within a large portion of the presently remaining Ridge and Slough landscape. Specifically, this layer at the bottom of the water column, typically from 0 to 12 inches (0–30 cm) thick, is found in areas where water depths and water chemistry resemble predrainage conditions. The near-neutral buoyancy and noncohesive nature mean that the layer is easily stirred or dispersed upward into the water column by walking or paddling, gradually resettling afterward. Floc layers are typically seen in Water Conservation Area 3A to the south of Alligator Alley (Highway I-75), in Shark River Slough, and in the Loxahatchee National Wildlife Refuge. In deeper water, tannins in the water column can extinguish enough light that the floc layer may not be noticed unless a white object is lowered down into the layer. Floc may also be present in areas with elevated P or Ca, but its properties become quite different. Elevated Ca especially seems to result in a much more cohesive, stabilized flocculent layer, notably in Water Conservation Area 2A.

The current presence of floc in portions of the Ridge and Slough landscape that resemble predrainage conditions is suggestive that unconsolidated floc was also present naturally, prior to drainage. Assuming that tannin coloration of the waters was similar, the deeper water typically present would have made it easy to overlook the presence of a floc layer. The historical record mentions the presence of material softer than the fibrous peat, sometimes referred to as "mud." When these references are made in areas of peat soil where no marl soil was present, it is probable that they refer to floc, especially when sinking into the material is mentioned: "Here [in sloughs] the vegetation is more completely decayed and is so loose when saturated with water that one sinks to the bottom sand, or rock" (Stewart and Brett 1907, p. 159). The "mud banks" referred to by Captain Dawson late in the 1855 dry season likely also refer to floc, in this case to a floc layer that has completely filled the vertical profile, even becoming compressed and more viscous, as the water level dropped (compare Historical Observations 8.11 and 8-WP-13 on the DVD).

If, as seems likely, floc was present prior to drainage, in sloughs it may have been derived in part from partially decayed bladderwort (*Utricularia*), with or without a periphyton covering, and on ridges from partially decayed sawgrass leaf litter. The differing hydrology of ridges and sloughs would likely have influenced floc-peat interactions. On ridges, the approximately annual dry down would have compacted the floc layer onto the peat each year, probably irreversibly. In contrast, in sloughs, where predrainage dry downs were rare, removal of accumulated flocculent material by compaction or oxidation would have been rare. Experience with the managed, presently impounded portions of the Ridge and Slough landscape (the Water Conservation Areas, especially 2A), indicates that significant quantities of flocculent material can accumulate, even to the point of raising concerns about slough loss (see Historical Observations 8.10, 8.11, 8.14, and 8.15 on the DVD). The long-term persistence of sloughs under predrainage conditions suggests that, at that time, they did not become filled in by flocculent plant detritus, which in turn suggests the existence of a predrainage export mechanism that counterbalanced floc production (see also appendix R). As discussed further in chapter 11, net downstream sediment transport via sheetflow might be such a mechanism. While predrainage water velocities were low, perhaps too low to resuspend floc, the near-neutral buoyancy of the floc means that other processes might provide resuspension, allowing downstream transport to occur. Likely processes include animal movements (fish, grass shrimp, alligators, etc.), wind blowing along the long axis of the sloughs, and gas production by photosynthesis and/or decomposition.

Further studies and different approaches may yield additional conclusions, but at this point, the available historical and current system information suggests that bladderwort (*Utricularia*) and flocculent organic material were likely both present in predrainage sloughs, while periphyton "sweaters" probably were not common.

An important aspect of sloughs not explicitly addressed in table 8.1 is their predrainage cross-sectional shape. It would be reasonable to expect that sloughs might have been concave in cross section. For instance, they could have formed part of a very low amplitude, approximately sinusoidal cross section: deeper in the middle and shallowing out toward the bordering sawgrass ridges, with no sharp discontinuity in slope at the interface between ridge sawgrass and slough vegetation (fig. 8.5A). Sinusoidal and linearized sinusoidal ridge-and-slough cross sections are in fact hypothesized and drawn in Everglades literature; for example, Spackman et al. (1976, p. 207); Bernhardt et al. (2004, figs. 4–6 be-

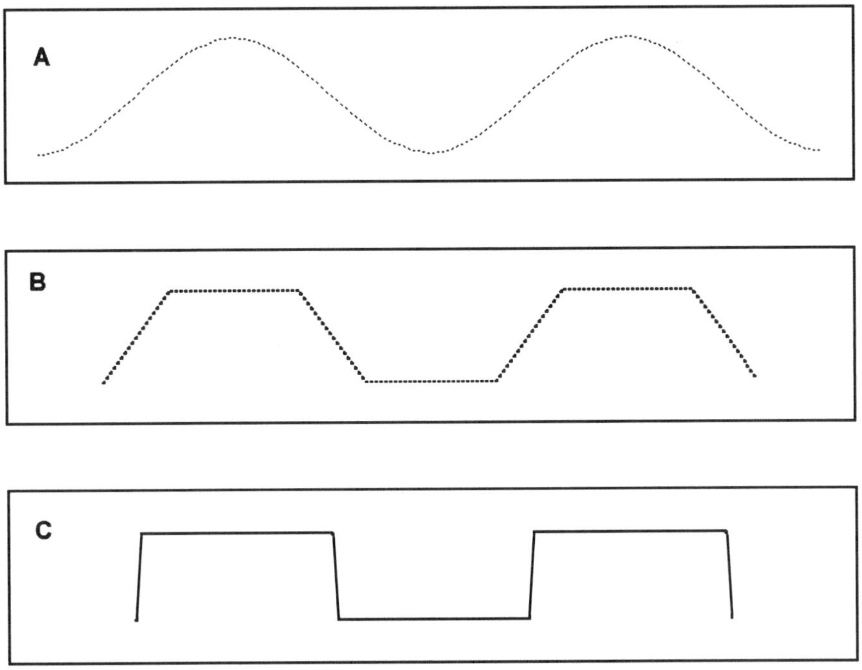

Figure 8.5. Three possible shapes of the transverse cross section through predrainage ridges and sloughs (see text). Centerline spacing is 1,000 feet (300 m). Shapes: (A) sinusoidal; (B) linearized sinusoidal; and (C) near-square wave. Widths and approximate percentages of the landscape that would be occupied by the transitions shown in panels B and C (no definable transitions in panel A) are: (B) = 150 feet (50 m), ca. 30 percent; (C) = 15 feet (3 m), ca. 3 percent. Early photographs (e.g., plates 17 and 19 bottom) and descriptions suggest that (C), the near-square wave, would most closely resemble actual predrainage cross sections. The transitions may well have been even shorter and sharper, occurring over only one to two feet (< 1 m).

tween pp. 22–24); and Bernhardt and Willard (2009, p. 1727). A smooth, continuous slope from slough to ridge is intuitively appealing as a simple, stable configuration that would have resisted erosion of ridges into sloughs. However, the available historical evidence suggests that such a configuration was not originally present.

Figure 8.5B illustrates a linearization of the sinusoidal cross section shown in figure 8.5A, created explicitly to compare the extents of ridges, sloughs, and the interfaces between them. The linearization was chosen to (1) approximate a sinusoidal shape, (2) yield equal widths of ridge and of slough, and (3) preserve the smooth, continuous slope between them. The ridge-to-ridge (centerline) spacing was assumed to be about 1,000 feet (300 m). It can be seen that attempting to preserve the smooth slope results in a very wide transition area between each ridge and slough, on the order of 150 feet (50 m) wide. In this hypothetical scenario, fully one-third of the landscape would have been transition area. Given the extreme sensitivity of Everglades vegetation to elevation and hydrology (e.g., Harshberger 1914, pp. 164–165), such transitional areas very likely would have resulted in transitional vegetation, probably wet prairie species closer to the sloughs and perhaps *Pontederia* and/or *Sagittaria* spp. closer to the ridges. The combination of wide width, one-third of the landscape cross section, and differing vegetation together suggest that such transitional areas, if present, would have been very obvious.

Neither historical nor current observations give any indication of such wide transitional areas. Instead, photographs and descriptions suggest a sharp transition in the vegetation, shifting from aquatic slough to sawgrass ridge within a distance of only a few feet (1 m). Neither photographs nor narratives suggest the presence of peripheral vegetation within sloughs that would indicate shallower water. Instead, water lily leaves are often seen right next to the ridge sawgrass, with water lily leaf densities not strikingly different from those close to the center of the sloughs.

Overall, predrainage observations suggest a steep elevation change at the interface and an essentially flat cross section within the slough. Vegetative evidence from ridges also suggests a nearly flat cross section, further emphasizing the steepness of the elevation change at the slough-ridge interface. Figure 8.5C represents this configuration, here drawn with the conservative assumption that the transitions were 15 feet (5 m) wide. An intriguing aspect of the resulting square wavelike peat configuration is the implication that some process or mechanism must have acted under predrainage conditions to maintain these sharp transitions. If the effectiveness of such a process decreased under postdrainage conditions of managed water, as it seems to have, the transitions likely would have changed, becoming more diffuse, like those shown in figures 8.5A and 8.5B. In any areas where the postdrainage transitions did become wider and more diffuse, appearance of a new, correspondingly wide postdrainage ecotone would be expected. The ecotonal vegetation would be intermediate in hydrological requirements, falling between slough plants and sawgrass.

In fact, exactly such a configuration was reported by postdrainage observers: ecotonal wet prairie vegetation on the edges of sloughs, forming elongated bands parallel to and between the sloughs and the sawgrass ridges (Loveless 1959b, p. 2; Goodrick 1974, p. 185).

The mechanisms that form and maintain sloughs are still unknown. Alligator trails, wind, fire, and periodic flushing by high water have all been suggested as possibilities. The strongly directional organization of ridges and sloughs parallel to the direction of regional water flow suggests that water flow was an important influence, whether directly or indirectly. The near-neutral buoyancy of flocculent organic material suggests the possibility that water transport of floc might have contributed to slough and possibly ridge maintenance through net export of carbon. Both Baldwin and Hawker (1915, p. 788) and Parker et al. (1955, p. 151) remarked on flows sufficient for sediment transport within the Ridge and Slough landscape. Further discussion of forces that might have shaped the Ridge and Slough landscape appears in chapter 11.

Under postdrainage conditions, sloughs have, in places, "grown shut" with wet prairie species (see plate 28), or even with sawgrass, as can be seen today in the northern portions of Water Conservation Areas 2A and 3A (fig. 8.6) and much of Water Conservation Area 3B (Rutchey et al. 2005). Lateral expansion of sawgrass into sloughs may also be identified in aerial views (plate 24 right). Wet prairies, as the name implies, can form dense, lawnlike stands, sometimes to the point of hiding the surface of the water from view. Annotated observations from the 1950s (postdrainage, but pre-impoundment of the Water Conservation Areas) and 1970s (post-impoundment) clearly document the close relationship between sloughs, wet prairies, and water depths (see Historical Observations 8.1–8.5, 8.7, 8.11, 8.14, and 8.16 on the DVD). Those observations make it particularly clear that postdrainage wet prairies are simply sloughs with lowered water depths and annual drying and colonized, as a result, by some combination of the wet prairie species: beakrush (*Rhynchospora tracyi*), spikerush (*Eleocharis cellulosa* or *elongata*), and/or maidencane (*Panicum hemitomon*). As water depths increased after impoundment of the Water Conservation Areas in the early 1960s, sloughs that were described in the 1950s as "whitegrass flats" (wet prairies) again became sloughs, with wet prairie species replaced by the original white water lilies and bladderwort.

The labile postdrainage transitions back and forth between sloughs and wet prairies during the course of changing water depth management were mirrored by parallel back and forth transitions between sawgrass ridges and sawgrass ridges invaded by woody species (see Historical Observations 8.2–8.5, 8.7, 8.11, and 8.14 on the DVD). This suggests that both ridges and sloughs were responding to the same raising and lowering of the regional water level.

The postdrainage invasion of formerly open water sloughs by dense wet prairies or by sawgrass may simply be the result of shallower water depths due to regional lowering, or it may have been additionally influenced by accumulation of sediments within the sloughs. In either case, vegetative shifts from slough to wet prairie have implications for water flow and sediment transport: the high stem density of wet prairies would be expected to have higher hydraulic resistance than the more open water column associated with sloughs. This suggests a possible feedback loop. If the loop proves to be an accurate characterization of predrainage conditions, it suggests an instability that has direct bearing on a mechanistic understanding of the long-term persistence of predrainage sloughs. A likely positive feedback loop that would promote slough to wet prairie shifts might include:

(1) increased stem density of wet prairie species;
(2) increased resistance to floc/sediment transport;
(3) decreased floc transport, causing increased floc/sediment accumulation;
(4) increased slough bottom elevation, causing decreased water depths;
(5) increased wet prairie species growth; reduced slough species.

We note that if this feedback loop had been the only mechanism controlling slough vegetation, wet prairies would have become the predominant form. The fact that sloughs, rather than wet prairies, appear to have been the predominant predrainage form suggests the existence of a separate predrainage mechanism that effectively counteracted the tendency toward wet prairies. The nature of this geomorphological mechanism has not yet been established. Possibilities include export of organic material (floc) from sloughs, plant growth–based mechanisms, and/or nutrient-based mechanisms.

Channels

A special class of sloughs was sometimes mentioned in the historical literature. Variously called "channels," "creeks," "runs," "streams," "gullies," or "leads" by some of the early explorers and land surveyors from the 1910s, these sloughs appear to have been narrower and deeper. They were probably located at points where the network of sloughs constricted and appear to have been associ-

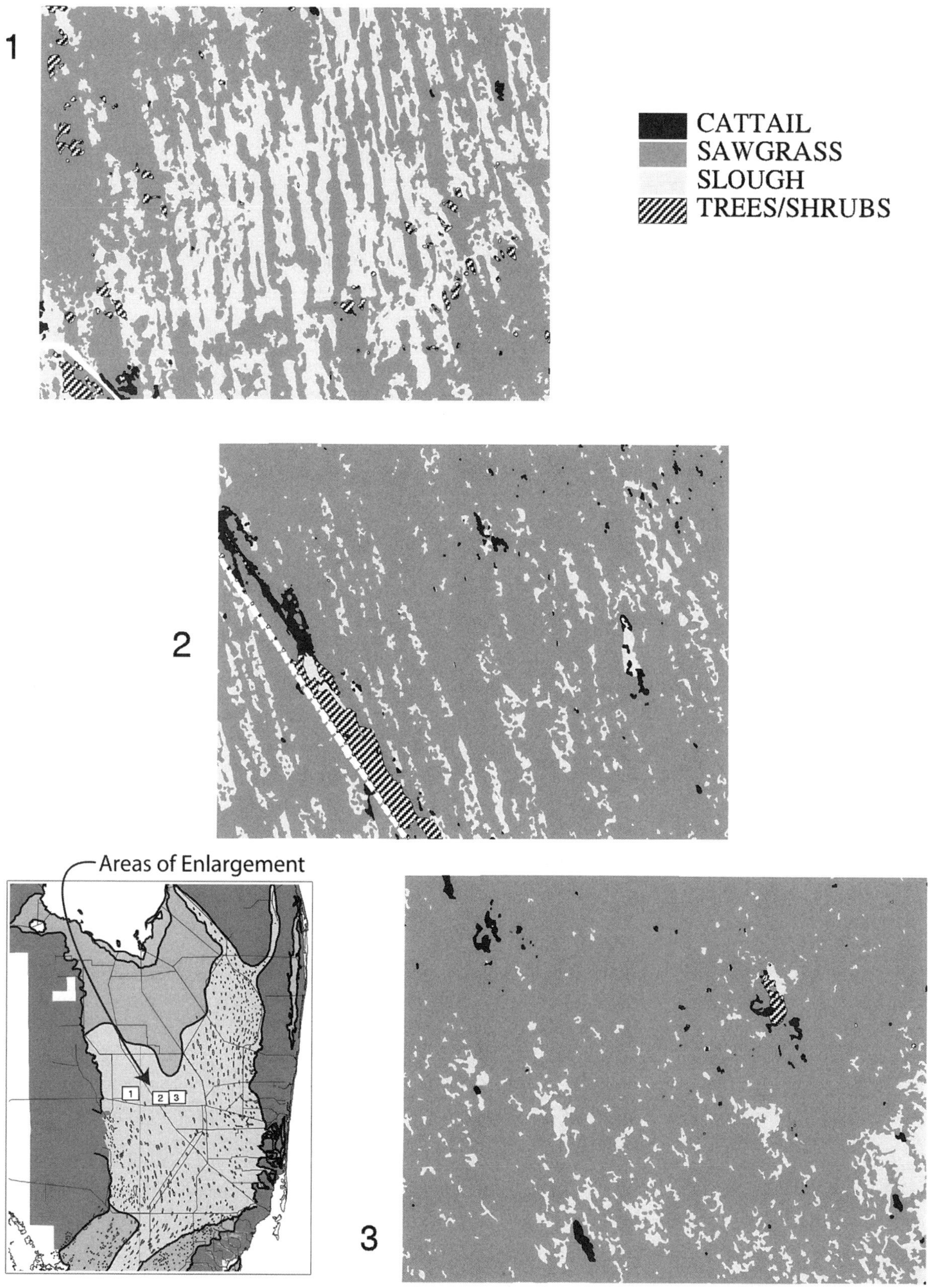

Figure 8.6. Detailed vegetation mapping of portions of three areas within the Ridge and Slough landscape, based on 1995 infrared photography. Panel 1 retains much of the original patterning; panel 2 retains some directionality but sloughs have substantially disappeared and been replaced by sawgrass; and in panel 3 almost all directionality in patterning has been lost and the area is now dominated by sawgrass.

ated with higher flow rates. There is some indication that less vegetation was present, perhaps a result of the deeper water and higher flow rates. Some of the predrainage and early postdrainage observations of visible currents (appendix S) appear to have been made in those channels. Surveys of a number of Everglades townships include references to 5 feet (150 cm) of water in what were almost certainly channels (see appendix A). Predrainage accounts frequently refer to those channels as having rock or sand bottoms. This may be an indication of lower rates of accumulation or of more effective scouring of loose, flocculent organic material. Scouring is seen in some tidal creeks at the "foot" of the Everglades when wet season currents are strong (William Loftus, pers. comm., 2009).

References to channels become much less frequent after about 1910–1920, perhaps because drainage and later impoundment reduced flow and water depths so much that channels filled in, becoming indistinguishable from other sloughs.

Sawgrass Ridges

Sawgrass ridges were long, narrow areas of slightly elevated peat covered with sawgrass (plates 23, 17 top, 20 top). The elevation of the ridges above the sloughs created a hydropattern similar to that of the Sawgrass Plains: inundated most of the year with the water table dropping below the peat surface of the ridge at the end of the dry season (fig. 8.3). As a result, the vegetation resembled that of the Sawgrass Plains: dense, tall, and nearly monospecific stands of sawgrass. Based on the uniformity of the sawgrass, the elevation along the long axis of the ridges appears to have been quite uniform. Present-day observations, if also representative of predrainage conditions, suggest that the crosswise profile may have been very slightly convex. New airboat trails across a sawgrass ridge will sometimes reveal a few scattered myrtle (*Myrica cerifera*) or buttonbush (*Cephalanthus occidentalis*) shrubs toward the center of the cross section, suggesting slightly higher and dryer conditions there. In addition to occurring in distinct elongated ridges, sawgrass is also sometimes found in the current system as much smaller, approximately circular patches, less than 150 feet (50 m) across. These may have been present in the predrainage system as well.

Predrainage accounts give the impression that sawgrass ridges were almost pure sawgrass stands. Certainly the density of the canopy severely restricted the light for other species, especially at the water surface. Present-day observations indicate that dense sawgrass stands in areas with water depths similar to predrainage ones generally have only a limited number of scattered individuals of other species; sawgrass is clearly dominant (plate 22). Areas with more altered hydrology tend to have higher proportions of nonsawgrass species.

Maximum elevations of the sawgrass ridges were likely determined by a feedback mechanism balancing peat accumulation and decomposition, controlled by prevailing water levels. As partially decayed sawgrass biomass from leaf litter and/or roots accumulated, it would have added to the peat soil, increasing the ridge elevation. As the ridge elevation rose relative to the water level in surrounding sloughs, the upper peat would have become subject to progressively shorter hydroperiods, exposing the peat to increased oxidative subsidence, thus creating a counter force that would tend to diminish the ridge elevation. It seems likely that the competing forces of accumulation and oxidation would have resulted in an equilibrium point. In this simple model, the equilibrium would have resulted in a ridge height generally stable relative to the long-term average water surface elevation.

Under drier, postdrainage conditions, sawgrass ridges clearly have been colonized by woody species, especially wax myrtle (*Myrica cerifera*) and holly (*Ilex cassine*; Historical Observations 8.2–8.4 on the DVD; see also chapter 3), sometimes blurring the distinction between ridges and true tree islands. However, historical descriptions suggest that under truly predrainage conditions—that is, before 1880—woody species were largely absent. Given the number of expeditions that pushed boats across sawgrass ridges, it seems likely that woody species, if present, would not have gone unremarked. A further and compelling indication that predrainage ridges probably had little woody vegetation is provided by the observation that sawgrass ridges that had been invaded by woody species during the pre-impoundment, low water depth period of the 1950s actually reverted to dense sawgrass with much less woody vegetation when water levels were raised by impoundment in the 1960s (Historical Observations 8.5, 8.7, 8.11, and 8.14 on the DVD). These observations suggest parallel phenomena occurring in sloughs and on sawgrass ridges; as water depths were lowered by unrestricted drainage, predrainage sloughs were invaded by wet prairie species while sawgrass ridges were invaded by woody species such as wax myrtle and holly. When water depths were increased by impoundment, both sets of invading species retreated, returning sloughs and ridges closer to their predrainage conditions.

Township surveys from the 1910s describe ridges with shrubs, particularly myrtle (*Myrica cerifera*), which seems likely to have been an early postdrainage change, particularly given later indications that myrtle readily expands into sawgrass areas during drainage-induced dry conditions (J. H. Davis 1943a, p. 257, 261–262; Johnson 1958, pp. 29–32; Johnson 1974, p. 211). Pollen profiles for ridge, slough, and transitional area soil cores from Water Conservation Area 3A show recent increases in *Myrica* pollen percent abundance (figs. 7–13 in Bernhardt et al. 2004), but it is not clear how closely the onset of this increase matches the onset of Everglades drainage. Similarly, *Myrica* pollen abundance increased somewhere around 1900 in a core from the south end of a strand tree island in the Loxahatchee National Wildlife Refuge (Willard et al. 2006, p. 575) and in a "near tail" core from Nuthouse Island in Water Conservation Area 3B (Willard et al. 2002, p. 136).

Tree Islands

Tree islands were conspicuous as being the only places that trees occurred within the Ridge and Slough landscape. Almost all explorers mentioned tree islands, usually noting "hundreds," "thousands," or "innumerable" numbers of them and noting a wide range of sizes, from "clumps" of a few to 50 feet across (1 to 15 m) to large islands encompassing hundreds of acres (50–100+ ha). Several different types currently exist and likely were present predrainage (Loveless 1959a, pp. 4–5; Craighead 1971; Gleason and Stone 1994, p. 182; Brandt 1997). We discuss only what appear to have been the two main types: large, elongated strand tree islands and smaller, approximately circular tree islands, often called bayheads.

Tree islands were elevated above the surrounding sloughs, usually slightly higher than the sawgrass ridges (appendix C). The elevation appears to have been due to varying combinations of anomalies in the bedrock elevation, accumulated peats, and in some cases the presence of tussocks. Chapter 11 discusses in more detail the many predrainage observations of tree islands as being just slightly higher than the surrounding water levels and damp or boggy. Willoughby provides a representative summary: "I expected always to find dry islands; such were conspicuous in their absence" (Willoughby 1898a, p. 149).

Postdrainage investigations reviewed in Stone et al. (2002, pp. 78–79) indicate that Everglades tree islands are variously associated with (1) peat on top of mounds in the mineral substrate, (2) peat on top of depressions in the mineral substrate, or (3) simply accumulated forest peat independent of any special feature in the underlying mineral substrate. The third type, tree islands unrelated to mineral substrate anomalies, appears to be associated with the portion of the Ridge and Slough landscape with the deepest peats, namely the present Loxahatchee National Wildlife Refuge and extending some distance southward into Water Conservation Area 2A. These tree islands may have originated from peat "pop-ups" or "batteries" that break loose from the slough peat substrate, float away, and create instant elevation increases when they come to rest (Stone et al. 2002, p. 100). In the other tree island localities, Water Conservation Areas 3A and 3B and Shark River Slough within Everglades National Park, the islands appear to be associated with localized high or low spots in the bedrock.

The tree islands for which elevation was due to peat or to a combination of bedrock and peat were the most numerous in the Everglades. A much smaller fraction was sand based. In the now suburbanized eastern Everglades, Stone et al. (2002, p. 85; see also Carr 2002) note that at least some of the former tree islands there appear to have been associated with slight sand mounds, supplemented by pre-Columbian human settlement–related material. Additionally, two groups of former islands located west of Fort Lauderdale in Township 50 Ranges 40–41 (plate 43; see also plates 40, 41) deserve separate mention. Prior to drainage, Pine Island and Sam Jones' Seven Islands (also called "Long Key") were both tree-covered islands but were neither peat-based strand tree islands nor bayheads. Instead, much older Pleistocene sand provided the elevation for these two important island groups. Parker et al. noted that "Pine Island, so called from the fact that prior to drainage of the glades this large, pine-covered dune was entirely surrounded by water, is the best developed of these ice-age dunes" (1955, p. 145). Although all of these islands are now surrounded by houses rather than water, they are still distinct features. Pine Island is now the Broward County Pine Island Ridge Natural Area; the small island south of Pine Island (plate 41, T50 R41 Sec 29) is the Broward County Tree Tops Park; the western portion of Sam Jones' Seven Islands is part of the Broward County Long Key Natural Area, while the eastern portion is part of the Flamingo Gardens botanical garden.

Soil cores from peat-based tree islands suggest origins ranging from 3200 years before present (BP; Willard et al. 2002, pp. 143–144) to about 1300 BP (Stone et al. 2002, p. 108), or even as recently as 300–500 BP (Willard et al. 2006, p. 577). These ages are substantially younger than the surrounding marsh peat. Gleason and Stone (1994, pp. 150, 182–186) suggest a possible trend within the Everglades; younger tree islands farther north and older ones farther south. The exact mechanisms responsible for the formation of strand and bayhead tree islands are not fully understood, but floating "batteries" of peat, succession from sawgrass ridges, and downstream transport of nutrients have all been suggested (Sklar and Van der Valk 2002).

Historical accounts provide no quantitative assessments of tree species encountered on predrainage tree islands and only somewhat limited qualitative information, generally not differentiating between bayheads and strand tree islands. Historical accounts do consistently report low trees or shrubs (cf. plates 18, 19) with the occasional lone cabbage palm (*Sabal palmetto*) standing out as an exception. Wax myrtle (*Myrica cerifera*) is reported, as are cocoplum (*Chrysobalanus icaco*) and "whitewood," which probably refers to dahoon holly (*Ilex cassine*; Daniel Austin, pers. comm., 2004). The frequency of references to cocoplum in early reports, especially in township surveys, may reflect somewhat greater predrainage frequency. Other bays, such as the red bay (*Persea borbonia*) and some sweet bay (*Magnolia virginiana*), were likely present and certainly were widely reported in the 1940s. Custard apple (*Annona glabra*) was occasionally mentioned, as was wild fig, (*Ficus* spp.). *Ficus* may have tended to occur as isolated and very large individuals. Willow (*Salix caroliniana*), is mentioned and appears to have increased after drainage. Buttonbush (*Cephalanthus occidentalis*) does not seem to have been specifically mentioned but likely was present. We did not find references to hackberry or sugarberry (*Celtis laevigata*) in the predrainage literature. Cypress (*Taxodium distichum*) does not appear to have been common, but when it was present on tree islands, it appears to have been the dominant tree. Postdrainage studies discussed in Armentano et al. (2002, pp. 239–241) record a number of tropical species on the more southerly tree islands, increasing south of about Township 53. Several of Willoughby's photographs looking outward from tree islands (Willoughby 1898)—"Filling the Camp Bucket," "On Willoughby Key," and "Outlook from Station 9"—are interesting in suggesting an open understory, although this may simply reflect man-made openings in the vicinity of boat landing points.

It was noted previously in this chapter that analysis of observations of the Ridge and Slough landscape from the 1950s through the 1970s suggested a parallel trend for ridges and sloughs; within both, the deviations from predrainage conditions that appeared during the period of uncontrolled drainage (ca. 1910s to ca. 1960) were to a noticeable degree reversed as water depths increased with impoundment of the Water Conservation Areas. A similar reversal to more predrainage-like conditions on tree islands did *not* occur (see Historical Observations 8.5, 8.9, 8.11, and 8.12 on the DVD). The explanation for the difference appears to lie with loss of tree island elevation during the period of uncontrolled drainage. As the highest elevation landscape subcomponent, tree islands would logically have been the most severely affected by soil oxidation or even peat fires when water levels were lowered. As this elevation loss could not be quickly recovered, the rise of water depths upon impoundment created unnaturally deep water on top of the post-impoundment islands. Rather than reverting tree islands closer to predrainage conditions, conditions for trees were moved further from predrainage ones. When this occurred, after previous lethal burning of trees, as in Water Conservation Area 2A, the results were catastrophic for tree islands (see Historical Observation 8-WP-12 on the DVD).

Strand Tree Islands

Strand tree islands usually have a characteristic shape in plan view, either teardrop or lenticular (plates 31, 32, 36). Heads of strand tree islands, when present, are typically small, loosely circular elevated areas located at or near the upstream end. The heads are typically a very small percentage of the overall tree island area (Heisler et al. 2002, p. 286). Currently, strand tree islands typically also include an elongated oval "near-tail" area of shrubs trailing downstream, and a similarly elongated oval "tail," often of sawgrass, concentric around the near-tail (plate 26). In areas with lower managed water depths, the near-tail shrubs may expand partially or fully into the tail, losing the distinction between the two. Historical sources do not provide much information on the predrainage structure of strand tree islands as a whole but do record the presence of surrounding sawgrass fringes and of slightly elevated heads. Visual inspection, the multicentury age of tree islands, and the strong association between tree islands and their underlying peat mounds (Stone et al. 2002, p. 83) together suggest that the predrainage outlines of a number of tree islands were probably similar to those presently seen. Within the present Loxahatchee National Wildlife Refuge the elevations and vegetation of strand tree islands appear to be fairly uniform throughout, so distinct heads and near-tails are not visible (Stone et al. 2002, p. 84). Although not yet confirmed, it seems likely that this undifferentiated, uniform configuration was present in this area under predrainage conditions as well.

The named tree islands reached during Colonel Harney's expeditions (appendices E, F) seem to have been mostly strand tree islands, in some cases cultivated by the Seminole Indians (Sturtevant 1953, pp. 60–62; Carr 2002, pp. 198–199).

Strand tree islands were consistently aligned, reflecting the direction of predrainage regional water flow. This alignment and more particularly the shape in plan view—streamlined and often differentiated into a rounded upstream head and a long, tapered tail (plates 36, 24-left side)—have led to a widely accepted impression that the flow of water is involved in shaping strand tree islands (Stone et al. 2002, pp. 87–88).

Strand tree islands share some similarities with sawgrass ridges, leading Stone et al. (2002, pp. 98–99) to propose a successional route from ridges to tree islands. The similarities include the sawgrass fringe around many strand tree islands; the ability of sawgrass ridges to be colonized by shrubs such as wax myrtle or buttonbush when water depths are low; and the partial overlap in the elevations of ridges and tree islands, relative to their surrounding sloughs.

At the same time, several other observations appear to argue against a simple successional relationship between ridges and strand tree islands. First, strand tree islands and sawgrass ridges coexist and have done so for centuries, begging the question of why some ridges would have succeeded to strand tree islands while the majority did not. Second, the spacing of tree islands across the landscape is fairly regular (plates 31, 32), raising the question of why only a systematic subset of the ridges would have succeeded. Third, close inspection of sawgrass ridges and strand tree islands visible on aerial photographs from 1940 to 2004 suggests that although both are elongated and oriented in the same direction, there may be consistent differences in shape.

The differences between sawgrass ridges and strand tree islands relate both to overall shape and to the perimeter itself (for this discussion see plate 25, especially left half; plate 31 upper left; plate 26; plate 24 left half). Sawgrass ridges are typically longer and narrower than strand tree islands. Sawgrass ridge perimeters tend to be rougher with more irregular protrusions or indentations; that is, more fractal in outline. In contrast, strand tree island perimeters tend to be better defined, smooth, and streamlined, with few or no protrusions or indentations. The overall width of sawgrass ridges, perimeter irregularities aside, tends to be fairly constant along the full length of each ridge. In contrast, strand tree islands appear quite different in plan view, tending to have a tapering width, widest near mid-island or near the upstream end. While all of these differences are qualitative, they appear to be sufficiently widespread to merit being addressed in any successional hypotheses of tree island and ridge formation.

Bayheads

Bayheads are small round or ovoid tree islands (plate 17 top: dark clumps at mid-distance; plate 18 top, especially upper left). They were dominated, if present conditions are an accurate indication, by various bays: *Persea borbonia*, *Myrica cerifera*, *Ilex cassine*, and *Magnolia virginiana*. Systematic measurements of predrainage bayhead elevations relative to the surrounding sloughs are not available. However, postdrainage measurements suggest a range of about 2 to 4 feet (60–120 cm) above the bottom of the surrounding sloughs (Gleason et al. 1975; Stone et al. 2002, pp. 80–81; Heisler et al. 2002, pp. 295–297); that is, greater than the predrainage relative elevations of sawgrass ridges, 1 to 2 feet (30–60 cm). Postdrainage peat losses by oxidation or fire (chapter 3) suggest that predrainage relative elevations would have been greater than or equal to the current measurements of 2 to 4 feet (60 to 120 cm). In predrainage accounts, bayheads (often referred to simply as "clumps") are frequently described as being wet or overflowed, suggesting that water depths in sloughs could attain similar depths of 2 to 4 feet (60 to 120 cm). At the same time, the average annual range of 2 feet (60 cm) between the ends of the wet and dry seasons would have allowed an annual period of drier soil, consistent with the intolerance of bays for long periods of inundation and with the dry conditions required for germination.

Predrainage elevations of the tree islands were likely controlled similarly to elevations of the sawgrass ridges—by a feedback mechanism based on water levels and peat oxidation rates. The presence of woody (higher lignin content) peats might have made tree island peats more resistant to oxidation than the sawgrass peats, allowing higher equilibrium elevations on islands. Differences in nutrient concentrations may also have played a role in elevation differences between tree islands and sawgrass peats. Willard et al. (2002, pp. 118, 143) found strongly elevated phosphorus contents on island soils relative to the surrounding marshes during the entire period of island formation, apparently due to higher avian fecal deposition rates.

Soils

The entire Ridge and Slough landscape was underlain by a blanket of organic soil (peat). North of about Township 54 the peat blanket filled the full Everglades basin, out to the bordering flat sand ledges of the Eastern and Western Flatwoods (plate 5; fig. 4.6). South of about Township 53, within Shark River Slough, the peat blanket was not thick enough to fill the basin fully; in that area the Ochopee and Rockland marl marshes, landscapes with limited, probably patchy predrainage peat, extended up the mildly sloping sides of the basin (plate 36; fig. 4.7). Over the Everglades as a whole the thickness of the peat blanket decreased with distance downstream, meaning that the slope of the peat surface was steeper than that of the underlying bedrock surface (fig. 3.11). This effect was strongest within the Sawgrass Plains but also continued through the Ridge and Slough landscape (figs. 3.8, 3.9). In some places the peat blanket thinned enough for some of the slough bottoms actually to be on bedrock; that is, with no floor of peat at all. An early explorer stated that here "the oar or pole sounds ringingly, and rebounds" upon hitting the limestone bottom (Mallory 1847, p. 62). Possibly similar areas of exposed bedrock are presently seen in the middle of Shark River Slough where sunfish and bass repeatedly nest on them (William Loftus, pers. comm., 2009).

Prior to drainage the peat slope formed two main axes, northwest to southeast in the area from Lake Okeechobee south to about Township 53 and northeast to southwest in the area south of about Township 53. As noted by Stone et al. (2002, p. 74), and as visible from measurements along the North New River Canal (Mitchell 1932), infilling of the Everglades basin by the peat blanket obscured the local relief of the underlying bedrock. Additionally, in many places a layer of marl is present at the interface between peat and bedrock and apparently formed early in the history of the Everglades. Longtime Everglades observer and National Park Service ecologist W. B. (Bill) Robertson suggested that this marl layer may have functioned as "caulking," sealing the bedrock (William Loftus, pers. comm., 2009).

The various elevations of ridges, sloughs, and tree islands—that is, the microtopography that distinguished the Ridge and Slough landscape—were formed almost entirely by differing thicknesses of accumulated organic materials (fibric Histosols or peat). Whether the nature of the peat under each of the ridge, slough, and tree island subcomponents of the landscape originally differed is less clear. In a hypothetical reconstruction, Spackman et al. (1976, p. 207) distinguished three separate peats corresponding to these subcomponents. They illustrate sawgrass ridges and tree islands as local mounds of sawgrass or tree island peat on top of a thick substrate of water lily peat underlying the entire landscape. However, Spackman et al. (1976, p. 238) also state that "Saw grass peat is a very black material that is homogeneous in appearance when examined with the naked eye. Because of the nature of the source material, it is devoid of any woody, fibrous, or coarse granular texture and is grossly amorphous." Based on the changes documented in chapter 3, this description is clearly *not* accurate for the predrainage condition of sawgrass peat; instead it describes postdrainage, oxidized sawgrass peat, in fact the "black nonfibrous peat" already noted by soil scientists Baldwin and Hawker (1915, plate 28). Prior to drainage, sawgrass peat was originally brown and fibrous (Clayton et al. 1942, p. 10) and, therefore, quite similar to the field description given by Spackman et al. for water lily peat: "in hand sample this [water lily] peat type is light brown to reddish-brown in color and finely-fibrous to finely granular in texture" (1976, p. 109).

Soil scientists working in the Everglades in 1915 also found a strong similarity between sawgrass peat and peat from the Ridge and Slough landscape. Baldwin and Hawker (1915, pp. 781–786) mapped a 50-mile-long swath along the North New River Canal (figs. 3.17 and 2.2) all as a single peat type, despite the fact that the upper half of the swath was within the Sawgrass Plains landscape and the lower half within the Ridge and Slough landscape (fig. 3.17). Their mapping approach apparently did not find enough difference between the peats of these two landscapes to warrant classifying them into two separate types.

Interestingly, in addition to not finding a map-worthy difference between the two landscapes, Baldwin and Hawker apparently also did not find map-worthy distinctions between the subcomponents of the Ridge and Slough landscape. The regular 1-mile grid they used would almost certainly have intersected both sawgrass ridges and water lily sloughs, yet only a single soil type was mapped. Baldwin and Hawker describe the soil as "typically consisting of brown fibrous to dark-brown semi-fibrous, slightly decomposed organic matter . . . varying from 36 to 140 inches. In many borings a poorly defined stratification is observed, layers being encountered in the vertical section which consist of black, well-decomposed peat or peaty muck. Generally the material below 3 or 4 feet is darker brown, more thoroughly decomposed, and of a higher ash content than the surface peaty material" (Baldwin and Hawker 1915, p. 781). Baldwin and Hawker's plate 27, between pages 784 and 785, reproduced here as figure 8.7, mentions a "light, fibrous or felty character and corky texture" for the brown fibrous peat.

Overall, although peat differences were apparently not visible to the naked eye, it seems likely that microscopic palynological techniques would have detected

Figure 8.7. Brown fibrous peat: "The light, fibrous or felty character and corky texture of the brown fibrous peat, representative of hundreds of square miles of the Everglades, can be seen in this illustration" (Baldwin and Hawker 1915, plate 27). In 1915 brown fibrous peat was mapped across both the Ridge and Slough and the Sawgrass Plains landscapes, from within a few miles south of Lake Okeechobee to a few miles west of the head of the New River. The separation into two different soils, Everglades and Loxahatchee peats (Jones et al. 1948a, e), did not occur until 1930–1940, when drainage had already oxidized the brown fibrous peat into a "black nonfibrous peat" (cf. fig. 3.7; Baldwin and Hawker 1915, plate 28).

differences between water lily (slough) peats and sawgrass peats, whether the latter came from the Sawgrass Plains or from ridges within the Ridge and Slough landscape. At this point it is not clear whether it is appropriate to speak of a single Ridge and Slough soil—such as, for example, the Loxahatchee peat mapped by Jones et al. (1948a, 1948e)—or whether it would be more appropriate to define two separate peats, one associated with sawgrass ridges, the other with sloughs. There appears to have been a fairly distinct peat associated with tree islands; for example, the Gandy peat mapped by Jones et al. (1948a, 1948e).

In the southern part of the Ridge and Slough landscape, south of about Township 52 (plate 11), the peat soil profiles were further complicated by the presence of scattered patches of marl. Soil surveys, township surveys, and road profiles indicate areas with alternating horizontal layers of peat and marl, such as the "Hialeah Mucky Marl" (plate 11), as well as areas where vertical "columns" of marl were mapped within the peat profiles (fig. 1.9; King 1917d). The sloped sides of some of the columns of marl drawn in figure 1.9 suggest that King probed at multiple depths within the profile and that the vertical columns were actually present. Similar vertical columns of marl within peat areas have been seen in soil cores in Shark River Slough and Taylor Slough (William Loftus, pers. comm., 2009). The two township surveys shown in figure 8.8 (Township 54 Ranges 37–38) further confirm the co-location of marl and peat ("muck") at multiple points in the vicinity of Tamiami Trail. The average total thickness (marl plus peat) of the soil profiles in the two townships seems to have been similar, but the easternmost township (Range 38) included more peat and less marl. The greater fraction of peat in Range 38 compared to Range 37 is consistent with greater peat thickness mapped for Range 38 in figures 3.8 and 3.9. Both phenomena may be reflective of Range 38 being approximately centered, east to west, within the Ridge and Slough landscape, at least in Township 53 and the next several townships north (plate 4). Overall, the mixed presence of marl and peat within the southerly portion of the Ridge and Slough landscape suggests that the factors controlling the distribution of marl are complex and determined by more than hydroperiod alone.

Landscape Patterns

Landscape pattern here refers to the size, spacing, and arrangement of the Ridge and Slough landscape subcomponents described in the preceding sections. A striking aspect of the predrainage Ridge and Slough landscape is the prevalence of organization or structure, rather than randomness. This structure is apparent in shape,

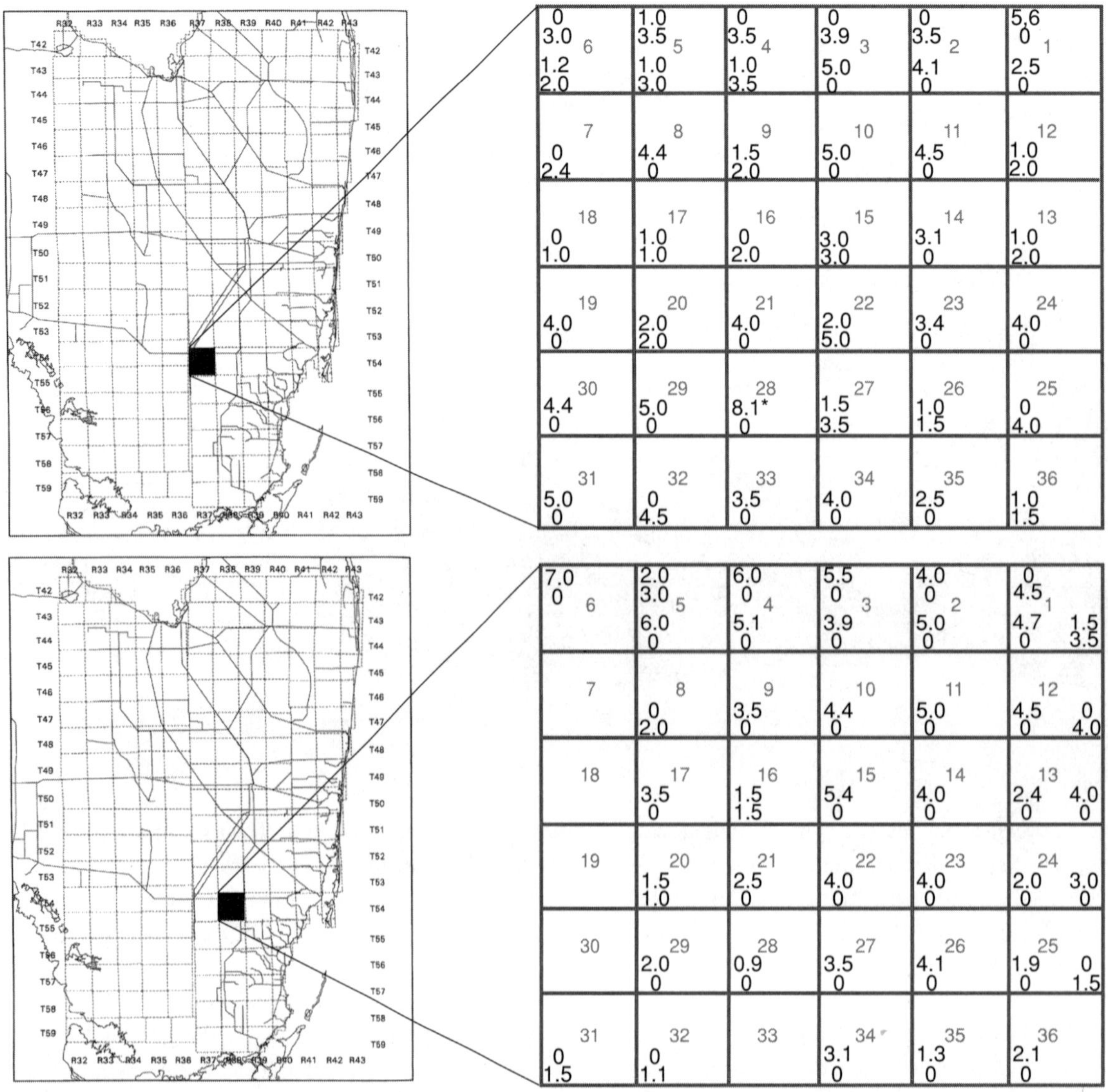

Figure 8.8. Peat thicknesses measured in 1918 in the center of Shark River Slough, Ridge and Slough landscape. The western boundary of Township 54 Range 37 (top) coincides with the L-67 Extension Canal (fig. I.1), while T54 R38 (bottom) extends one mile east of the L-31N levee. Measurements indicate feet of peat soil (upper of the paired numbers in the section corners) and of marl soil (lower of the paired numbers). See text for a discussion of marl and peat balance. Comparison of these peat thicknesses with those in figures 3.8 and 3.9 suggests subsidence of 1 to 2 feet between 1918 and 1940. Note the very thick peat soil (8.1 feet) in southwest corner of Section 28, T54 R37, measured on a tree island labeled in the 1918 field notes as "dense hammock." This tree island is still visible in 2010 at 25.703, -80.639. Sources: Southern Engineering & Construction Co. (1918-T54 R37; 1918-T54 R38).

in orientation (i.e., directionality), and in distribution of the landscape subcomponents. The structure also takes the form of a high degree of similarity of landscape subcomponents across the full breadth of the landscape (i.e., uniformity). Interestingly, the patterning is, and probably was, not particularly obvious at the scale of an individual slough and ridge (e.g., plates 17, 20, 22), yet it was the (micro)topographic and vegetative differences between these ridges and sloughs that formed the pattern visible at the larger scale. Taken as a whole, this structure, formed largely from accumulated peat, classifies the Ridge and Slough landscape as a patterned peatland. Similarities, as well as important differences, between this patterned peatland and the patterned string bogs of boreal areas are discussed further in chapter 11.

This discussion relies heavily on visual inspection of aerial photographs, ranging from the earliest available (1940) to 2004, and covering, in 1940, essentially the full extent of the Ridge and Slough landscape. The assumption, based on this spatially and temporally wide-ranging inspection, is that photographs from the area corresponding to approximately Townships 51–52

Range 36 and Township 51 Range 37, presently west of the Miami Canal and south of Alligator Alley (Highway I-75), closely resemble predrainage spatial patterns (see plate 4 for townships; for detail see plates 31 upper left, 24 left). This important assumption is based on synthesis of three elements: (1) inspection of numerous aerial photographs (admittedly all postdrainage) taken throughout the original Ridge and Slough landscape covering six decades, (2) comparison of the photographed patterns with all predrainage descriptions (see Historical Observations for chapter 8 on the DVD), and (3) extensive field experience in these areas by the lead author of this study. The field experience provided an important bridge between current conditions and the firsthand accounts of predrainage explorers as well as forming the basis for interpretation of the aerial photographs.

A number of aspects of the Ridge and Slough landscape suggest self-organization, in both horizontal and vertical dimensions. In the horizontal plane the widths of ridges and tree islands and the spacing (centerline to centerline) of ridges were all narrowly distributed. Narrow distribution here means, for example, that ridges were almost all about the same width, with the deviations being small relative to the width. Interestingly, while some tree islands do seem to be positioned on local high points in the bedrock, examinations to date suggest no relation between sawgrass ridge locations and the bedrock. If further study confirms that sawgrass ridges are always independent of the bedrock, this makes their fairly regular spacing all the more remarkable. In the vertical dimension, ridge and tree island heights and slough water depths were also narrowly distributed. Self-organization in two dimensions is suggested as well: the strand tree islands and most sawgrass ridges were much longer than they were wide, suggestive of an elongating or shaping process. Qualitatively, there appears to be somewhat more variability in the length (downstream) dimension than in the transverse (width) dimension, but not enough to alter the primary conclusion of length being much greater than width.

The repetition of aligned subcomponents many times across the landscape, with approximately regular spacing, suggests the possibility of representing the Ridge and Slough landscape using a standard "building block" or "unit cell" such as is depicted in plate 23. In such a representation the deterministic block or unit cell would be reproduced with slight random variations across the landscape, and with a deterministic adjustment to the block orientations, reproducing the slowly bending alignments visible in figure 11.12.

Yet another way of considering the pattern of ridges and sloughs found in this landscape is suggested by the parallel planes discussed earlier, one intersecting the surfaces of the ridges and another the bottoms of the sloughs. This conceptualization raises the question of whether this landscape is more appropriately considered as: (1) a sawgrass surface with sloughs cut into it or as (2) a shallow lake, dotted with sawgrass ridges (and tree islands) rising from it. In the visual arts, this would be equivalent to asking of sloughs and ridges, which is the "figure" and which the "ground"?

At this point there is insufficient information to answer the implied questions of geomorphological origin. However, it does appear possible, with the help of aerial photographs, to address an associated question of continuity, specifically of landscape and hydrologic continuity. In the previously mentioned area, considered here to resemble predrainage conditions most closely (west of the Miami Canal and south of Alligator Alley, I-75, in Water Conservation Area 3A), ridge outlines, when traced, mostly form distinct, individual shapes. Sloughs, in contrast, mostly form a continuous network, albeit with occasional "narrows" where two ridges may appear to "touch" and vegetatively close the gap and the slough. In other portions of the Ridge and Slough landscape, where postdrainage water management has resulted in drier conditions, the frequency of "ridge touching" and apparent slough closure increases. Experienced Everglades airboat drivers are familiar with this phenomenon. They learn to read the ridge shapes when traveling parallel to the ridge/slough orientation and will head for the former narrows, knowing that the sawgrass there will be least dense (Bill R. Kline, pers. comm., 2004–2006; Robert R. Johnson, pers. comm., 2007).

The existence of strong ridge-and-slough patterning formed from an inherently impermanent substrate such as peat naturally raises questions regarding its temporal stability. Has such patterning been present since peat began accumulating in the Everglades some five thousand years ago? Has the size and position of the landscape as a whole remained constant? Have individual ridges and sloughs always been in their present positions? While answering such questions is likely to be important for understanding the geomorphic needs of landscape restoration, they are beyond the temporal scope of this study. It can be pointed out that pollen and morphological analyses of soil cores from sloughs (e.g., fig. 59 in Spackman et al. 1976; Willard and Holmes 1997, p. 13; Cohen et al. 1999, pp. 26–28; Willard, Holmes, and Weimer 2001, p. 49; figs. 7–8, 11–13 in Bernhardt et al. 2004) tend to show several alternating bands of sawgrass and water lily peat, suggesting that both ridges and sloughs have been present for at least two thousand years; each persisted for several hundred years at a time; and a given point location may have alternated between being a ridge and being a slough. A spatial network of cores would be needed to determine if the alteration back and forth over time reflects lateral shifts in ridges or in situ changes in state.

Directionality

A striking aspect of the self-organization and structure in the Ridge and Slough landscape is the systematic orientation of the sawgrass ridges and strand tree islands. Thousands of sawgrass ridges and hundreds of tree islands, filling the landscape at fairly uniform spacing, were aligned in parallel (plates 26, 27 top, 24 left third). Perhaps even more striking, this parallel alignment of subcomponents persists, when mapped over the entire Ridge and Slough landscape, through long, slow bends similar to those made by iron filings aligned within a magnetic field (fig. 11.12). In geostatistical terms, this corresponds to a high degree of spatial autocorrelation. In the Ridge and Slough landscape, this gently bending alignment apparently followed the gradual changes in the downstream direction of the peat land surface.

Curiously, the directionality of the Ridge and Slough landscape was frequently remarked upon by predrainage explorers. Despite the absence of any aerial perspective, early explorers often described a definite orientation to the landscape, frequently also noting that the current flowed in the same direction:

> [This] part of the area [i.e., the Ridge and Slough landscape along the North New River Canal] is very gently undulating or hummocky, owing to a series of ridges and sloughs with local differences of 2 to 3 feet in elevation, having a general northwest-southeast direction.
>
> . . . The natural drainage conditions of the Everglades followed the general southeastward slope through the sloughs which trend in that direction . . . (Baldwin and Hawker 1915, p. 752)

The explicit mention of directionality by predrainage explorers is particularly striking because the landscape scale is such that neither directionality nor alignment of directionality is obvious to an observer standing in a given slough at present. King's photographs from 1917 do not suggest that directionality would have been any more obvious at that time either (plates 17–20). Two possibilities suggest themselves. First, that there was sufficient current under predrainage conditions that flow direction, and indirectly landscape direction, was more apparent than at present. Second, present-day experience indicates that the structure of this landscape does become more apparent when moving through it fairly quickly along a trajectory. Perhaps rapid movement—that is, passing sufficient lengths of ridges or crossing sufficient numbers of ridges within an easily recallable period—functions somewhat like the bird's eye perspective granted by aerial photography, where patterning and alignment become obvious. The reported travel rates of 8 to 15 miles (24 km) per day would likely have been sufficiently rapid.

Plates 26, 27 top, and 24 left one-third show aerial views of the directional pattern formed by sawgrass ridges and sloughs. These patterns of long, linear ridges and sloughs alternating with each other match the early descriptions of crossing the landscape at an angle approximately perpendicular to the ridges: "Chadwick [surveyor on the 1907 party across the Everglades] mentioned saw-grass ridges alternating with open leads of water running approximately in a southeastern direction" (Harshberger 1914, p. 164). The many sets of field notes from township surveys from 1910–1918, which describe ridges alternating with sloughs and tree islands, also appear to match well with aerial photographs of the better-preserved portions of the landscape.

In addition to the pattern formed by ridges and sloughs, the shape and alignment of the large strand tree islands also form a directional pattern across the landscape (plates 31, 32). Two early postdrainage maps show similar patterns of oriented tree islands and sawgrass ridges (King 1917a; Newman 1908-T50 R41). Even though they occur at different scales, the directional patterns formed by the tree islands and by the combination of ridges and sloughs, when compared, show essentially the same directionality.

The similarity and nature of those patterns suggest that some organizing force, not necessarily occurring at the same temporal scale for ridges as for tree islands, is responsible. The co-alignment of ridges, sloughs, and tree islands with the downslope direction and the direction of water flow has been noted since at least 1915 (Baldwin and Hawker 1915, pp. 752–753). It has been suggested that water flow is the organizing force that shaped this systematically aligned landscape (e.g., Parker et al. 1955, p. 148; Gleason and Stone 1994, pp. 171–172; Stone et al. 2002, pp. 87–88), yet authors have been equally quick to point out that the mechanisms are both unknown and puzzling, given the very shallow slopes (1×10^{-5} or 0.001 percent) and presumably very low predrainage velocities. Possible relationships between hydrology and Ridge and Slough geomorphology are presently an active area of research (e.g., Larsen et al. 2007). Flows are discussed further in chapter 11.

Uniformity

It was suggested earlier that the Ridge and Slough landscape was remarkably uniform across its full extent, an unusual situation for such a large area. The Herculean human effort that would be required to create an area 40 miles (65 km) wide, and yet level enough to maintain an even water depth of 3 feet (100 cm), plus or minus a few inches (3–10 cm) or even a few feet (30–60 cm), across the full width gives some idea of how remarkable the Everglades configuration originally was. Occasional occurrences of wet prairie species in sloughs suggest some degree of depth variation from place to place, but this variation does not appear to have been systematic. Similarly, the reported widespread similarity of ridge vegetation suggests that ridge heights above the water surface were very similar across the landscape. In the horizontal dimension, aerial photographs (e.g., plates 31 upper left, 24 left, 25 west of Miami Canal, 26) suggest that both the spacing in the cross-flow direction and the widths of sawgrass ridges were much more regular than random.

The regular, highly nonrandom aspect of those spatial observations and their repetition across the landscape argue for the presence of a physical organizing or structuring force that was remarkably uniform across such a large area. The leveling effect of gravity on the water surface and the strong influence exerted by water level on peat processes may explain much of the apparently self-organizing tendency toward uniform slough depths and ridge heights. The leveling effect of gravity does not, however, in itself suggest an obvious explanation for the regularity of ridge spacing and similarity of ridge widths. In the abstract, both of these seem more likely to have a flow-related, energetic origin, yet the large dimensions involved, the very low slopes, and the presumably very low velocities (and energies) make the potential causal processes hard to visualize.

The western portion of the Ridge and Slough landscape may provide some additional insight into the possible presence of a relation between flow and landscape pattern. As the western half narrowed by approximately a factor of three between Township 53 and Township 55, the vegetation remained the same, sawgrass ridges and open, water lily sloughs (fig. 11.3), suggesting that water depths did not change much under narrowing. However, the shapes of ridges and sloughs (in plan view) *did* elongate and become more streamlined (partially illustrated for Township 56, Ranges 36 and 35, in plate 36), much as would be expected if the narrowing cross section had resulted in increased flow velocities.

The drainage pattern found in the predrainage Everglades was also highly unusual. No central drainage channel was present, nor was any hierarchy of dendritic drainage-ways. Instead, both flows and depths within the Ridge and Slough landscape (within a particular subcomponent) were distributed evenly across the full width of the landscape. Within the cross section of the Ridge and Slough landscape, there is no evidence of portions with consistently higher or lower flows. While the flow arrows shown in figure 11.12 depict only flow direction and not measurements of magnitude, the parallel arrangement of the arrows (high spatial autocorrelation), except where they focus as the total cross section narrows into Shark River Slough, is strongly suggestive of spatially uniform flow magnitudes.

Hydrology

As the largest, topographically most varied, and deepest water landscape of the Everglades, the Ridge and Slough landscape significantly influenced Everglades hydrology. At the same time, as a patterned peatland, the regional hydrology strongly influenced the microtopography and spatial patterns of this landscape (plate 23). The interaction between landscape and hydrology, therefore, extended in both directions, with landscape directionality influencing direction of flow, and water flow perhaps helping to maintain the landscape. Similarly, water depths helped to maintain the peat microtopography, and the microtopography allowed the persistence of specific water conditions (fig. 8.3).

Focusing first on water flow, it is helpful to contrast the Ridge and Slough landscape with the Sawgrass Plains, where surface water covered the entire landscape in a sheet of more or less uniform thickness, and downstream movement of this sheetflow was uniformly retarded by the dense growth of sawgrass (chapter 6). In contrast to this even distribution, surface water in the predrainage Ridge and Slough landscape moved through a system of many parallel flow paths (sloughs), each of similar magnitude and seasonally separated by the peat bases of the long sawgrass ridges (fig. 8.3, plate 23). For this landscape and this flow regime, a term such as "sloughflow" might be more appropriate than "sheetflow." Even when surface water was present across the full landscape (i.e., covering ridge tops as well as filling sloughs), the sloughflow paths were still partially separated from each other by the shallower water and much higher plant stem density on the ridges. The topographic and vegetative differences between ridges and sloughs likely meant that flows were concentrated primarily within sloughs, with the degree of concentration varying between wet and dry season. Aerial photographs suggest that sloughs covered about 60 percent of the predrainage landscape.

Estimated predrainage water depths and hydroperiods for the Ridge and Slough landscape were based on a synthesis of information from many different sources. These included recorded predrainage water depths from varying locations, years, and seasons (appendix A); descriptions of predrainage slough vegetation (this section, appendix A, and accounts of cross-Everglades excursions); comparisons with postdrainage vegetation and accompanying water depths; descriptions of predrainage navigability of the landscape; estimates of relative predrainage slough, ridge, and tree island elevations; and soil core records of predrainage frequency of peat fires (e.g., Cohen 1984, p. 461). Additional discussion of the process of estimating predrainage Ridge and Slough hydrology is included in chapter 11.

We estimated long-term average end-of-season water depths in sloughs to be 1 foot (30 cm) at the end of the dry season and 3 feet (90 cm) at the end of the wet, based on synthesis of all of the preceding sources. Long-term average, end-of-season water depths on sawgrass ridges were assumed to have been the same as in the Sawgrass Plains (chapter 6), based on similarity in canopy height and stand density: 0.5 foot (15 cm) below ridge surface at the end of the dry season and 1.5 feet (45 cm) above at the end of the wet. Figure 8.3 illustrates these water depth estimates in relation to estimated heights of ridges and tree islands (table 8.1) and an assumption of level water surfaces over short distances ("level pool").

The estimated depths and elevations appear to be generally consistent with each other and with historical observations of the Everglades: that is, tree islands (heads) only slightly above or at water level part of the year but also free of surface water much of the year; sawgrass ridges drying slightly for a portion of the year (9- to 10-month hydroperiod; cf. fig. 11.2); and sloughs, on long-term average, containing water throughout the year (multiyear hydroperiods). The conclusion that predrainage sloughs were for the most part continuously aquatic is consistent with several aspects of predrainage slough flora and fauna as well as postdrainage changes in them. These aspects are discussed in detail in chapter 11 and appendix R.

Hindcasting

As a multimodal landscape with modes (subcomponents) differentiated by peat thickness and water regime, the Ridge and Slough landscape was particularly prone to anthropogenic changes once regional water depths were lowered by drainage. The subcomponent changes reported up to the 1940s (i.e., prior to the impoundment phase; chapter 2) are consistent with drier conditions (chapter 3). Tree islands became more subject to peat fires (Simpson 1920, pp. 126–127; Loveless 1959b, pp. 34–35; Craighead 1971, p. 126; Alexander and Crook 1974, p. 202) and peat oxidation, leading to declines in relative tree island elevation. Sawgrass ridges were invaded by woody species (Simpson 1920, p. 126; Johnson 1958, p. 29; Johnson 1974, p. 211; Robertson 1953, p. 80), blurring their original distinction from tree islands. Sloughs became shallower and open to invasion first by wet prairie species (Wallace 1955b, p. 5; Loveless 1959a, p. 7; Goodrick 1974, p. 185) and subsequently by sawgrass (Wallace 1955a, p. 2; Florida Game and Freshwater Fish Commission 1956, pp. 39–40; Loveless 1959a, p. 7; Olmsted and Armentano 1997, pp. 29–30). Invasion of sloughs by sawgrass then blurred the distinction between ridges and sloughs. The spatial distribution of these changes, especially the increases in sawgrass coverage, suggests a strong relation to changes in managed water depths. At the smaller scale of individual ridges and sloughs, the postdrainage changes in patterning suggest a possible influence of altered water flow as well (Science Coordination Team 2003).

The hindcasting task for the Ridge and Slough landscape consisted of identifying and reversing the altered portions of the Ridge and Slough and Ridge and Slough-like polygons mapped by J. H. Davis (1943b; plate 13) and by Jones et al. (1948a; plate 11). The altered areas were identified by comparing the predrainage Ridge and Slough landscape characteristics identified in this chapter and the postdrainage changes identified in chapters 2 and 3 with the postdrainage (1940s) descriptions of the polygons mapped by Davis and Jones et al. Comparison of the hindcast result (plate 5) with the 1940 maps (plates 11, 13; see also fig. 4.9) shows that the main areas of hindcasting were (1) the eastern and western edges of the landscape north of about Township 54, (2) the boundary between the Ridge and Slough and the Sawgrass Plains landscapes, and (3) the central portion of the landscape encompassing the area between and near the Miami and North New River canals ("1d" on fig. 4.9). Hindcasting of the eastern and western areas was addressed in chapter 4. The boundary with the Sawgrass Plains was mapped from index sheets of the earliest aerial photos (U.S. Department of Agriculture–Soil Conservation Service 1940), based on the location of the transition from unpatterned sawgrass to patterned ridges and sloughs.

The spatially and conceptually most extensive hindcasting was that of the central portion of the landscape encompassing the Miami and North New River canals. Analysis of expedition descriptions, soil information, and aerial photographs, among other sources, made clear that the Sawgrass Plains-like vegetation and surface-oxidized soil profiles seen there in the 1940s were postdrainage alterations of what had originally been Ridge and Slough landscape (see especially chapter 3). Remapping the area to its original condition resulted in a single, contiguous Ridge and Slough landscape (plate 5), extending across the full width of the Everglades. That configuration is consistent with a plethora of predrainage evidence, including (but not limited to) Harshberger's 1914 observations along the North New River Canal, Colonel Harney's 1841 expedition west of the New River, and Ingraham's 1892 expedition crossing the Everglades. Both Harney and Ingraham traveled through this central hindcasted area; compare plate 10 left with plates 11, 13, and 14 (1940s) and plate 5 (hindcast predrainage conditions). Those expeditions would have been possible with the Ridge and Slough landscape but not with the postdrainage, sawgrass-dominated conditions mapped in the 1940s. Harshberger's observations in 1913–1914 confirm the presence of distinct Sawgrass Plains and Ridge and Slough landscapes, with the transition recorded close to the landscape boundary mapped in this study.

9

Marl Marshes

The landscapes described in the preceding chapters were formed on substrates of thick peat soils (Histosols). The soil thickness masked the underlying bedrock and made the overall landscape topography, the microtopography, and the wetland water chemistry essentially independent of the bedrock. The Marl Marshes, three smaller landscapes in the southern Everglades, were different, with shallower soils and likely a greater bedrock influence (fig. 4.7). Those landscapes are often presently referred to as Marl Prairies, a name apparently given to them by John H. Davis during the highly drained 1940s, with "prairie" presumably reflecting drier conditions than a marsh, and "marl" reflecting the inorganic calcitic soil then visible at the surface. In light of their originally wetter hydrology we refer to these landscapes as marshes rather than as (wet) prairies, based on the distinctions drawn by Duever et al. (1986, p. 95). As will be discussed, it is possible that even the term "Marl" in the name Marl Marshes is misleading and does not accurately reflect the originally much wetter, and peat-forming, predrainage conditions in these areas. Geographically, the Ochopee and Rockland marl marshes are somewhat paired, with the Ochopee Marl Marsh flanking the west side of Shark River Slough and the Rockland Marl Marsh flanking the east side. The Perrine Marl Marsh was set somewhat apart, south of Long Pine Key and south of the Everglades Keys (plate 5). The Perrine Marl Marsh, which sloped downward from the Everglades Keys to the coastline, was hydrologically connected to the Rockland Marl Marsh by Taylor Slough and, seasonally, by the Marl Transverse Glades (fig. 9.1, plate 35; chapters 10, 11) and by the similar transverse glades crossing Long Pine Key (Craighead 1974).

Of the marl marshes, approximately half of the Rockland Marl Marsh and a quarter of the Perrine Marl Marsh have been developed (table 4.2), while almost all of the Ochopee Marl Marsh extent is still present. However, the remaining portions of all three of the Marl Marsh landscapes (plate 2) are arguably the most strongly altered areas within the remaining Everglades. As discussed in more detail later, all indications are that the predrainage Rockland and Ochopee marl marshes were wetlands with long enough hydroperiods to accumulate a shallow (and probably not complete) layer of organic peat soils, overlain on a base of marl (fig. 4.7). In these two wetlands flanking Shark River Slough, the present predominance of exposed marl appears to be the result of postdrainage oxidation of this shallow organic layer as illustrated in figure 11.5. In addition to leading to widespread peat oxidation, regional lowering of water tables appears to have shifted the vegetation communities drastically as well. Finally, the greater predrainage peat elevations within Shark River Slough, along with presence of peat in the Rockland and Ochopee marl marshes, meant that the predrainage differences between Shark River Slough and the two flanking landscapes were much less than at present. In short, the wetland changes in these two "marl marsh" landscapes appear to have been drastic and fundamental. The soils shifted from organic to primarily inorganic and the flow of water narrowed from the full width (Shark River Slough plus the Ochopee and Rockland marl marshes) to now little more than Shark River Slough.

The Marl Marsh landscapes were not frequently explored, limiting the number of available predrainage and early postdrainage descriptions. Nevertheless, the available information, supplemented with modern topographical information and emerging paleoecological studies, forms a consistent picture. This predrainage picture is quite different from the current condition of these landscapes.

Presently, the Rockland and Ochopee marl marshes flanking Shark River Slough differ from Shark River Slough because of higher elevation, inorganic marl soil rather than organic peat, shorter hydroperiods, and fewer aquatic plant communities. This postdrainage difference has a distinct and very clear signature in current satellite imagery (e.g., plate 4). The difference appears to be largely elevation driven, with the lower elevation Shark River Slough reflecting wetter hydrology. This postdrainage configuration of marl-forming areas flanking the peat-forming Shark River Slough is widely depicted, often with the implication that this was also the predrainage configuration (e.g., Davis 1946, figs. 1, 13; Jones et al. 1948a; 1948e, p. 14; Gleason and Stone 1994, p. 152, 171; Wanless et al. 1994, p. 203).

The actual predrainage configuration appears to have been one in which the differences between the marl marshes and Shark River Slough were much smaller. King (1917e, p. 414) in fact described the center of the Ochopee Marl Marsh—the very portion of the landscape most elevated relative to Shark River Slough—as "similar in character" to the Ridge and Slough landscape. King's assertion is borne out by his descriptions of the Ochopee Marl Marsh vegetation, fauna, and hydrology; it is corroborated by Mickler's township survey notes. The most frequently reported plants were dense stands of sawgrass and two plants characteristic of sloughs, white water lilies (*Nymphaea odorata*) and yellow water lilies (*Nuphar advena*). U.S. Deputy Surveyor William Mickler's numerous observations of deep water (15 to 30 inches; 40–75 cm) in the Ochopee Marl Marsh are significant, particularly as they were made well into the dry season (fig. 11.7).

Perhaps the strongest similarity to Shark River Slough (and the most striking difference from the current Ochopee Marl Marsh) was the reported presence of peat soil. King (1917e, p. 418) describes a significant surface layer of "muck" (peat) that was "0.5–1.5 feet" (15–45 cm) thick. Similar though less detailed indications of peat and of dense, peat-forming sawgrass stands were also reported in the Rockland Marl Marsh. A peat-based landscape with sufficiently long hydroperiods and deep water to support dense sawgrass, "open Everglades," and both white and yellow water lilies does indeed sound "similar in character" to the predrainage Ridge and Slough landscape. The primary differences seem to have been that the Rockland and Ochopee marl marshes were probably less distinctly patterned than the Ridge and Slough landscape, and that they sloped slightly upward away from Shark River Slough. The positions of areas of sawgrass (and sometimes other graminoid species), of patches of more open water, and of clumps (islands) of trees reflected the generally southwestward direction of water flow (southerly within Taylor Slough); traces of this overall directionality are still visible in satellite imagery. Compared to the Ridge and Slough landscape, patterning in the marl marshes was less clear and prob-

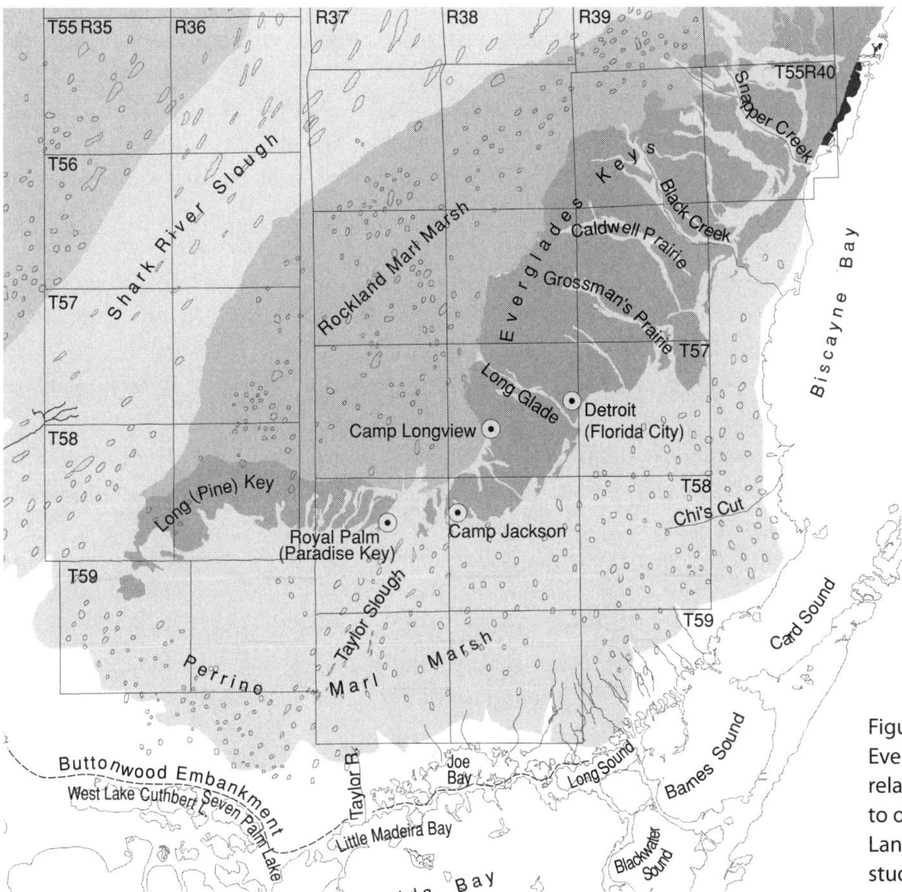

Figure 9.1. Map of the southeastern Everglades showing landmarks and relationship of the Perrine Marl Marsh to other landscapes and Florida Bay. Landscape extents derived from this study; Buttonwood Embankment adapted from Craighead (1964).

ably more random; sawgrass was more prevalent; and the hydroperiods may have been somewhat shorter.

Overall, the wetter predrainage Marl Marsh landscape conditions, along with the generally southwestward directionality seen in King (1917c) and (weakly) in current aerial photographs and satellite imagery, together indicate a predrainage flow configuration different from that assumed in previous studies. Rather than almost all flow having been focused into the narrows of Shark River Slough, as suggested by present-day satellite imagery (plates 2, 4), it appears more likely that during the wet season, and well into the dry season, a broad sheet of Everglades water covered (with varying depths) the full southern width of the southwestern Everglades. This wide sheet, all flowing approximately southwestward, would have covered Lostmans Slough, the Ochopee Marl Marsh, Shark River Slough, the Rockland Marl Marsh, and Taylor Slough. The highest velocities and flow volumes would still almost certainly have been centered within Shark River Slough, but flows and longer hydroperiods would have been present throughout the Marl Marshes and Lostmans Slough as well.

This predrainage scenario of fairly wet, flowing marl marshes, only weakly differentiated from Shark River Slough, begs the question of how the marl marshes became transformed after drainage into much drier, strongly differentiated landscapes. The twofold answer appears to lie in (1) overall postdrainage reductions in Everglades water flows and depths; and (2) changes in relative elevations due to differential peat subsidence. As Everglades hydroperiods shortened with the onset of upstream drainage in the 1880s, oxidation of any peats present also began, in the marl marshes as well as in Shark River Slough. However, because peats within Shark River Slough were thicker (fig. 4.7), and because the rate of peat subsidence is proportional to thickness (Stephens and Johnson 1951a, p. 218), the greatest loss of elevation would have been within Shark River Slough. The *percentage* of peat depth lost was likely greatest within the marl marsh areas, essentially 100 percent, but the actual *elevation* lost (subsidence) would have been greatest within Shark River Slough. The net result has been a postdrainage increase in the elevation differential (fig. 11.5; compare with fig. 4.7).

Present-day high accuracy topographic measurements made by the U.S. Geological Survey (e.g., Desmond 2003) appear to support the concept that much of the currently seen elevation difference between Shark River Slough and the Ochopee Marl Marsh is anthropogenic, having developed as a result of differential peat subsidence. Even at present, the elevation difference is small; increasing from a minimum of approximately 0.5 feet (15 cm) toward the south end of Shark River Slough to a maximum of approximately 1.5 feet (45 cm) northward, near Tamiami Trail. Such small differences are well within the range of subsidence losses recorded for Shark River Slough (e.g., fig. 3.13), confirming that subsidence can easily account for the differential. Additionally, the greater differential northward would be a natural result of the originally greater peat thickness northward—and hence greater potential for subsidence.

The degree of peat loss, and the timing of its occurrence, are among the larger uncertainties associated with the Marl Marsh landscapes, particularly the Perrine Marl Marsh. Observations made well before the onset of anthropogenic drainage (e.g., Army Dr. Jacob Rhett Motte's reports on the Lower East Coast in 1838, in Sunderman 1950, pp. 27–28]) established that exposed bedrock uncovered by peat was originally present in some locations, for example in the highest portions of the Rockland Marl Marsh. In the Perrine Marl Marsh, which differed by not bordering on a landscape of deeper peat, estimating the original degree of peat coverage is harder. The difficulty is increased by knowing that any peats present would have been thin and vulnerable to the anthropogenic alterations in Everglades hydrology that began in 1882. It is not clear, for example, whether photos or observations of exposed bedrock (absence of peat) made in the 1910s or 1920s can be assumed indicative of predrainage conditions or whether they are instead the result of drainage-induced oxidation of an original covering of peat (chapters 2 and 3).

In summary, for the Ochopee and Rockland marl marshes, the distinct postdrainage "valley" shown in plate 36 appears to have been largely absent prior to drainage. With a much flatter cross section, it is likely that the hydrologic and vegetational differences were also less. A flatter cross section likely explains surveyor Mickler's 1885–1886 notations of nearly identical water depths, and vegetation, within the Ochopee Marl Marsh and Shark River Slough (fig. 11.7). The flatter cross section, combined with greater predrainage quantities and depths of water, also suggests the concept of a vast, flowing sheet of water covering Lostmans Slough, the Ochopee Marl Marsh, Shark River Slough, and the Rockland Marl Marsh. Such a wide flowing sheet covering multiple landscapes, albeit with lateral variations in depth and velocity, in turn helps explain the widespread southwestward landscape directionality still visible to varying degrees across the full southwestern Everglades.

The available historical sources suggest that the current character of the so-called "marl prairies" found to the west and east of Shark River Slough on slightly elevated bedrock is largely a postdrainage phenomenon. A century of drainage and fire has removed not only the original water depths and hydroperiods but also the original thin layer of surficial peat. The loss of peat appears to have been almost complete within the Ochopee and Rockland marl marsh areas, where peat is now found only in solution holes where sawgrass persists. The use of "marl" in the name of these predrainage landscapes should eventually be changed to reflect predrainage conditions more accurately. The eastern and western "glades and hammock" landscapes, as they were referred to by Gallatin et al. (1958), would be more appropriate names. Until then, "Ochopee Marl Marsh" and "Rockland Marl Marsh" as used throughout this book are best thought of as place names rather than descriptions of their predrainage conditions.

Rockland Marl Marsh

The Rockland Marl Marsh was an intermediate wetland landscape positioned between the "Dade County" pines of the slightly higher Everglades Keys and upper Taylor Slough, and the ridges and sloughs of the slightly lower Shark River Slough (plate 5; fig. 4.7). It is underlain by bedrock that slopes gradually downward toward Shark River Slough (fig. I.2). As discussed in the preceding introduction, the lower portions of this slope were originally covered by peats; these in turn supported dense and tall sawgrass. In cross section, the peat-covered portion probably was of somewhat shallower slope than the highest, exposed bedrock portions of the landscape, although both slopes were slight. Water covered the entire landscape during the wet season, with water from the highest portions of the Rockland Marl Marsh flowing out through the Marl Transverse Glades (including those through the Atlantic Coastal Ridge and Long Pine Key), and from there into the Perrine Marl Marsh. Early observers reported a portion of the Rockland Marl Marsh landscape near and paralleling the pinelands (Everglades Keys) with adequate wet season water depths and sufficiently open vegetation to allow canoe passage southwestward from Miami River to Long Pine Key (fig. 9.1).

U.S. Deputy Surveyor George MacKay's description of the "southern Everglades" in an 1847 letter to Buckingham Smith encompasses aspects of both the Rockland Marl Marsh and the Everglades Keys:

> "I was forcibly impressed with the peculiarity of the southern Glades—a vast rock of interstices and partitions, something in the form of a honeycomb—the interstices or cells, filled with soil, saw grass or water, are barren [i.e., without trees], with islands of trees, traversable with canoes in high water, and upon horseback in low water, the whole presenting the most romantic view of grass, prairie, and water, with gems of islands of rare timber and shrubbery" (MacKay 1847, p. 65).

The honeycombed "vast rock of interstices and partitions" is now often called pinnacle rock. It was present within the Everglades Keys (fig. 10.5) and extended west into the highest portions of the Rockland Marl Marsh. (As discussed further in chapter 10, the Everglades Keys area, shown on plate 5, is now almost always referred to as the "Miami Rock Ridge" or as the southernmost portion of the "Atlantic Coastal Ridge" because drainage of the Marl Transverse Glades eventually blurred the original distinctions between keys and transverse glades [see Simmons and Ogden 1998, p. 6]. We use the names Miami Rock Ridge and Everglades Keys interchangeably. With increasing distance from the Everglades Keys, the elevation lowered and the exposed pinnacle rock became increasingly covered by soil. Harper (1910a, p. 284) mentions both peat and freshwater marl. In 1845 MacKay surveyed part of the north boundary of Township 56 Range 38, the eastern two miles (3 km) of which fell within the Everglades Keys (pinelands); the western four miles (6 km) were within the Rockland Marl Marsh. As was customary at this time, he abandoned the survey line two miles into the Rockland Marl Marsh (4th mile west) due to dense and tall sawgrass. Together, the progressive increase in soil thickness ("disappearance" of pinnacle rock) and the presence of dense sawgrass indicate that the soil was very likely peat:

> 2nd Mile: 60.00 chains: exit [pineland] to open level rock prairie.
>
> 3rd Mile: Open level rock becoming broken, with large holes—with saw grass
>
> 4th Mile: Rock broken up like honey comb, and gradually disappearing—saw grass more general, with islands of cocoaplum, [wax] myrtle and other timber not known—To the Eye the country descends West and ascends East—Saw grass to[o] high & thick to proceed. (MacKay 1845-T56 R38)

Another U.S. deputy surveyor, John Jackson, surveyed the same line two years later, with very similar results:

> 2nd Mile: 58.00 chains: Came to Everglades.
>
> 3rd Mile: Everglades very Rocky and about 2 inches deep of soil.

> 4th Mile: rough honey comb rocks with about 18 inches of water on Part of it.
>
> 5th Mile: Honey comb rocks Covered with about 2 ft of water and thick Saw grass. (Jackson 1847-T56 R38)

Jackson was able to survey an additional mile (total of 3 miles; 5 km) into the Rockland landscape but then, like MacKay, abandoned the line due to thick sawgrass. The dense sawgrass seen by both surveyors suggests peat deposits of some thickness, perhaps one or more feet (30+ cm). It is noteworthy that this dense sawgrass, and any associated peat, was much closer to the pines of the Everglades Keys (2–3 miles east) than to Shark River Slough (7–8 miles west). The downward slope of the bedrock toward Shark River Slough suggests that in the area extending west of the surveyor's point of relinquishment—that is, in the next 7–8 miles (11–13 km) of Rockland Marl Marsh—the peat deposits would have continued to increase in thickness.

Figure 9.2, an early postdrainage map of soil depths measured in 1918, illustrates this at a location farther north in the Rockland Marl Marsh, Township 54 Range 39. Peat thickness here increased from southeast, in Rockland Marl Marsh, to northwest, in Ridge and Slough landscape. Prior to drainage an additional 0.5 to 1 foot (15–30 cm) of peat would likely have extended southeast into the Rockland Marl Marsh (see fig. 3.16, next township to the east).

Water covered the full width of the Rockland Marl Marsh, even its highest portions adjacent to the pines, and with sufficient frequency to prevent encroachment of pines. The result was a distinct, widely recognized landscape edge (e.g., figs. 1.1 and 1.8):

> "At the border of the Everglades the rough surface of the oolite [i.e., the pinnacle rock present within the Everglades Keys, cf. fig. 10.5] becomes concealed for the most part beneath a mat of grass, and pines grow no further west. . . . In the zone of oscillation of water

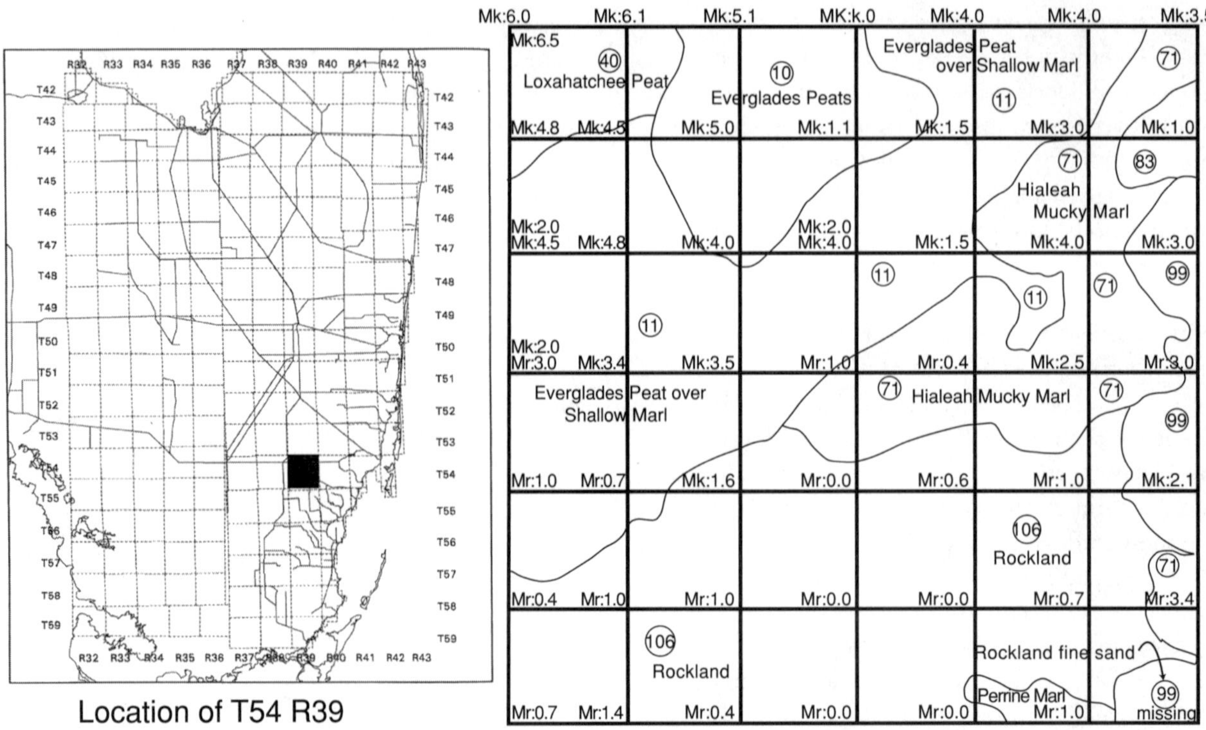

Figure 9.2. Soil thicknesses and soil types at the interface between the Rockland Marl Marsh and Ridge and Slough landscapes. Township 54 Range 39 includes an undeveloped remnant of the Everglades, the Bird Drive Recharge Area, and is bounded on the west by Krome Ave. (SW 17th Ave.), on the east by the Florida Turnpike, Homestead Extension, and on the south by N. Kendall Ave. The northern boundary is about one half mile north of Tamiami Trail (U.S. 41). Soil depths measured in 1918 are indicated in feet (Mk = muck/peat; Mr = marl); soil types (irregular polygons) mapped in 1940s by circled numbers. The southern third of the township is Rockland Marl Marsh (soil type 106; see plate 4), the northern two-thirds is Ridge and Slough (soil types 10, 11, 40, and 71). Some loss of organic soil had likely already occurred prior to 1918 (chapters 2, 3). Remaining peat in 1940 increases from zero in the southern third of the township to 5 to 6 feet (1.5–1.8 m) deep in the northwest corner. Sources: Southern Engineering and Construction Co. (1918-T54 R39); Jones et al. (1948a).

the grasses change in character from a small wiry variety near the pines to saw-grass six feet or more tall, and flags and cane growing in water" (Griswold 1896, p. 53).

As discussed at the beginning of this chapter, prior to drainage and subsidence the Ochopee Marl Marsh–Shark River Slough–Rockland Marl Marsh "valley" was very likely not as deep and narrow as implied by the 1980 measurements shown in plate 36 (see also Rose and Rosendahl 1983). As a result, prior to drainage the annual rise and fall of water levels that created the "zone of oscillation" defining the edge of the pineland probably resulted less from overflow of Shark River Slough waters and more from water flowing from upstream areas to the north, the present-day Northeast Shark River Slough and Water Conservation Area 3B. Stages in Shark River Slough would have been an important western boundary condition for this flowpath. The reports of "good open water" (Krome 1902g; see also M. A. Williams 1874-T57 R39) in the Rockland Marl Marsh, allowing travel from the Miami River southward along the western edges of the Everglades Keys, at least in some seasons, may be a reflection of a Northeast Shark River Slough to Taylor Slough flowpath.

The elevation of the western edge of the pineland (i.e., of the Everglades Keys) is about nine feet (2.7 m) above mean sea level. Figures 1.1 and 9.1 suggest that a number of the Marl Transverse Glades—low sawgrass-covered swales extending southeastward between the higher islands of pine (chapter 10)—constituted hydrologic connections between the Rockland and Perrine marl marshes. The Marl Transverse Glades, which were also at elevations of about 9 feet (2.7 m; plate 35), were water-covered for a sufficient fraction of the year to support sawgrass (figs. 1.1, 10.4). A portion of the water within the Marl Transverse Glades likely derived from the adjacent Everglades Keys. However, the very high bedrock porosity there ("and in the pine woods, between the Glades and the Bay of Biscayne, may often be heard the rippling sound of running water, and frequently, in the fissures of the rock, it may be seen at from 6 to 8 feet below the general surface of the ground" MacKay 1847, p. 65) suggests that the surface water flowing eastward from the Everglades (Rockland Marl Marsh) was also an important contributor to Marl Transverse Glades hydrology. This in turn reinforces the concept that under predrainage wet season conditions, the full cross section of the southwestern Everglades—that is, Lostmans Slough, the Ochopee Marl Marsh, Shark River Slough, and the Rockland Marl Marsh—all filled to the point of lateral overflow.

A concern expressed by early settlers near and south of Miami probably is a reflection of these overflows: "The Glades below Miami are rocky and are not of much value. The water should be held down to about the present low level to prevent overflow of east coast" (Richmond 1907, p. 45). Interestingly, despite regional and local drainage, the concern about flooding was still present as late as the 1960s. An unpublished but detailed map produced by the Central and Southern Florida Flood Control District (1960) labeled the great majority of the Rockland Marl Marsh—everything below approximately the 9-foot (2.7 m) elevation contour on the map—as being subject to "long-duration flooding." The remainder, a 1- to 2-mile (1.5–3 km) band adjacent to the Miami Rock Ridge, together with the Marl Transverse Glades, were labeled as "intermittent flooding without [Central and South Florida] Project influence." Only the individual highest areas making up the ridge—the rock "islands" originally known as the "Everglades Keys," and then only those above the 10-foot (3 m) contour—were labeled as free from flooding.

In 1874, due to the difficulty of walking on the pinnacle rock of the Everglades Keys, U.S. Deputy Surveyor Marcellus Williams sought a boat route to supply his survey work in Townships 57 and 58. Such a route, passing southwestward from the headwaters of the Miami River along the western edge of the Everglades Keys to reach Taylor Slough, seems to have been well-known to the Seminole Indians. Hugh Willoughby and his assistant Ed Brewer referred to this route in 1898, as did William Krome's rodman Wilson in 1902 (Krome 1902g). An observation of "very high stage of water in the 'Glades" in late November 1902 is probably a reference to this route as well (Krome 1902h). Unfortunately, conditions during the 1874 dry season (March) were exceptionally dry, preventing Williams from reaching Taylor Slough:

> I therefore proceeded to the Miami River and hired two Canoes and with them passed out of the Miami river into the Everglades. I was however told by the [Seminole] Indians that the water in the Everglades was unusually low and that I would find it impossible to pass through it—I however concluded to try it but soon found that this information was correct. I spent three days in this endeavour and returned—I will here say that there are portions of the Everglades that can be surveyed with water high enough to pass through the very many channels in there—Upon this trip I spent a night upon an Island about 4 miles South West of the outlet to the Miami,

which is planted by the Indians and is one of the richest pieces of land that I ever saw. (M. A. Williams 1874-T57 R39)

In January 1897, twenty-three years after Williams's attempt and fifteen years after the onset of Disston's drainage work, Hugh Willoughby and Ed Brewer apparently found the boat passage Williams had sought. Willoughby's description suggests an unexpectedly deeper water portion of the Rockland Marl Marsh adjacent to the Everglades Keys. "Round grass," if it does in fact refer to *Eleocharis*, would suggest deeper water conditions than those generally favored by sawgrass:

> At last we cleared the ridge, and the face of the rock began to get smoother, and more of the round grass [*Eleocharis* sp.?] and deeper water could be seen ahead. . . . Brewer thought that we would not have many more miles to travel before reaching a district that he had hunted over when going in from the east coast. He had followed down the pine-timber of the main land [i.e., the Everglades Keys] at a sufficient distance from it to find it insured deep enough water for the canoes. Finding better leads, we were now able to take the much-desired northeasterly direction. (Willoughby 1898a, pp. 158–159)

Early, and apparently substantial, drainage-induced changes make it difficult to determine definitively the vegetation that was present in the predrainage Rockland Marl Marsh landscape. Early reports refer to thick and tall sawgrass, along with more open areas with emergent aquatic species. Postdrainage descriptions tend to refer to sparser sawgrass, to graminoid species not usually found in dense sawgrass stands, and to scattered shrubs. Recognizing that mere omission is not the same as an explicitly stated absence, it seems significant that predrainage sources do not seem to mention shrubs as a landscape element (except when concentrated on tree islands), as in George MacKay's description of a "view of grass, prairie, and water, with gems of islands of rare timber and shrubbery" (MacKay 1847, p. 65). The omission seems particularly significant given the many examples in the Everglades of woody species increasing after drainage began lowering water levels.

Both pre- and postdrainage reports are consistent in reporting substantial tree islands with what seem likely to be tropical hardwoods. Smaller islands of more shrubby vegetation (e.g., cocoplum, myrtle, or bay trees) were also reported. Two early postdrainage township plat maps, King 1917-T54 R40 (fig. 1.8) and Newman 1908-T54 R40, show a directional, southwest orientation for tree islands within or directly adjacent to the Rockland Marl Marsh (plate 4), again reinforcing the impression of water across the full width of the southwestern Everglades having flowed as a sheet to the southwest (table 9.1). Several accounts suggest that the landscape was originally bimodal in appearance; either treeless, shrubless marsh or forested tree island. The documented postdrainage increase in shrubby species, especially willow (*Salix*), "changed the whole aspect" of the landscape. Robertson's (1953, p. 80) suggestion that the increase in woody vegetation was an effect of drainage, not fire, would be consistent with many other observations suggesting strong postdrainage lowering of water levels in this landscape.

At present, where the bedrock is not exposed, marl is the dominant soil of the Rockland Marl Marsh. However, both predrainage and postdrainage observations

Table 9.1. Township surveys within the Rockland Marl Marsh area

Township	Range	Comments	Citation
54	39	Strand tree islands oriented northeast to southwest	King (1917h)
54	39	Hammocks (tree islands) oriented northeast to southwest	Southern Engineering & Construction Co. 1918-T54 R39
54	40	Water current and tree islands oriented northeast to southwest	Newman 1908-T54 R40
54	40	Strand tree islands oriented northeast to southwest	King (1917h)
55	38	Never surveyed	Mayo (1959)
55	39	Border of Everglades; in transverse glades—"Sawgrass Prairie"	Jackson 1847-T55 R39
55	39	Border of Everglades; Rockland—"Rocky Ridgey Country," oriented northeast to southwest	King (1917h)
56	38	Border of Everglades; see appendix A	MacKay 1845-T56 R38
56	38	Border of Everglades; see appendix A	Jackson 1847-T56 R38
57	38	Border of Everglades; "Everglades very wet and rocky covered" with high sawgrass	Jackson 1847-T57 R38

suggest that prior to drainage and extensive fire, peat was more prevalent.

John Stewart (see chapter 1) reported that some parts of what likely was the Rockland Marl Marsh had enough peat to be harvestable: "From what I can learn the country in the vicinity of and south of Miami [i.e., the Everglades Keys] is rocky. Rock is full of pot holes in places in other places in sheets. The pot hole formation also occurs in the southern part of the Glades where the rock sticks out of the water, the holes being filled with muck. The muck is stringy and more of a peat and if thoroughly dry will burn. Several companies have been considering the advisability of pressing it into bricks for fuel" (Stewart 1907, p. 45).

By the 1940s, after decades of drainage, John H. Davis noted that this peat had been almost completely lost: "Moreover, large, shallow peat areas of the southern part of the Everglades that have neither been drained nor cultivated have lost their peat cover completely or nearly completely due to dry season fires, of which those during April and May 1945 were some of the most destructive. . . . In many places only bare rock now remains in some of these areas" (J. H. Davis 1946, p. 119). Davis's comment that these peat losses occurred in areas that had "neither been drained or cultivated" is correct in the sense of no local drainage canals, but he was apparently overlooking the significant effects of systemwide drainage (chapters 2 and 3).

Frank C. Craighead's observations suggest that a portion, perhaps a large portion, of the exposed pinnacle rock seen at present may be the result of postdrainage peat fires:

> Many great fires have been recorded (J. H. Davis 1943a, 1946; Robertson 1954). Early botanists frequently commented on fires in the 1920s (Small 1929a, p. 87). . . . Again, in 1945 to 1947, in 1951 and 1952, and in 1962 and 1965 fires destroyed tremendous expanses of saw grass. The top layers of peat containing the mat of rhizomes were burned to ashes. In some places the peat burned to bedrock or to marl soil. Much roughly eroded limestone formerly covered with peat was exposed in these great fires of the past fifty years. (Craighead 1971, pp. 141–142)

The predrainage presence of peat would be consistent with the reported vegetation (sawgrass as well as flags) and with water depths sufficient to provide navigability as well as to raise concerns about flooding in adjacent areas to the east.

Hydrology

Defining water depths for the Rockland Marl Marsh is complicated by the possible difference between its ground surface and water surface slopes and by the irregularity of the pinnacle rock, where exposed. The steeper slope of the ground surface than of the water surface in the westward direction means that water depths within the Rockland Marl Marsh were likely somewhat deeper toward Shark River Slough and shallower toward the Everglades Keys. A possible local exception to this is a narrow strip parallel to the west edge of the pineland (Everglades Keys), used as a boat passage. The estimates given here represent a mid-elevation point centered between the eastern and western borders of the landscape and are referenced to the soil surface or, where soil was absent, to the tops of the pinnacle rock.

Based on navigability, the presence of tall and thick sawgrass, and the presence of peat soil, we estimated a long-term average end-of-wet-season high water level of 2 feet (60 cm) above the soil surface. The end-of-dry-season low water is estimated at 0.5 foot below soil surface (-15 cm), implying a slightly wider vertical range than elsewhere in the Everglades. That range was influenced by the possibility of a degree of subsurface drainage, by the sloped landscape surface, and by possible drainage toward Taylor and Shark River sloughs. We estimated the hydroperiod—the duration of surface water above the average soil surface—as eight to nine months. This estimate is somewhat shorter than hydroperiods for other Everglades landscapes of similar water depths, based on the possibly stronger and slightly more rapid natural drainage in this landscape.

Hindcasting

The hindcasting changes we made to the extent of the Rockland Marl Marsh were small. The eastern border against the Everglades Keys was easily recognizable on township plat maps drawn well before drainage (1845–1847). This border matched almost exactly with the boundary line between the Rockland category (Rockland Marl Marsh) and the Rockdale complex (Everglades Keys) drawn a century later on the Jones et al. (1948e) soil map areas. We based the western edge of the predrainage Rockland Marl Marsh on the Jones et al. (1948e) boundary between Rockland category and Loxahatchee Peat (Shark River Slough). To compensate for postdrainage peat loss we relocated this boundary approximately 1 mile (1.6 km) eastward. It is possible that the actual predrainage boundary, though likely indistinct, was actually farther east—that is, Shark

River Slough wider and the Rockland Marl Marsh narrower—but we did not have historical sources upon which to base this.

We hindcasted an area enclosed and strongly drained by the Tamiami and Snapper Creek canals at the northern end of the landscape from Rockdale complex (Everglades Keys) back to Rockland Marl Marsh, slightly increasing the landscape extent. The hindcasting was based on photos of peat burns (Parker 1940), a township survey (Newman 1908-T54 R40), and a map of the Southern Drainage District (King 1917a; see fig. 1.8). Early drainage had caused this area to resemble and later be mapped in the 1940s as a part of the drier Everglades Keys.

A tongue of Perrine marl soil, shallow phase, present north of Taylor Slough (plate 11), was assumed to have been part of the surrounding Rockland Marl Marsh and hence was incorporated into it.

Ochopee Marl Marsh

We defined the Ochopee Marl Marsh (plate 5) as the unforested portion of the Ochopee marl soil extent mapped by Jones et al. (1948e; plate 15), specifically the open portion not covered by cypress (plate 14). As defined here, it also corresponds closely to the westernmost of the two areas mapped by J. H. Davis as "Marsh Prairies, Southern Everglades," but excludes the portion mapped by Davis north of the Tamiami Canal (plate 13; see also Davis 1943c, pp. 106–107, 259–260, 268–269; 1943b).

The Ochopee Marl Marsh likely resembled the Rockland Marl Marsh but with small distinctions due to a slightly different topographical context and a greater sand content in the soil (Gallatin et al. 1958, p. 12). The bedrock underlying the Rockland Marl Marsh sloped continuously upward from Shark River Slough all the way into the adjacent Everglades Keys. In contrast, the bedrock underlying the Ochopee Marl Marsh sloped upward only from Shark River Slough to about the middle of the landscape, then sloped downward slightly from there to Lostmans Slough (plate 2). Because of this differing configuration, current high accuracy elevation data (e.g., Desmond 2003), which in this area are measuring essentially bedrock surface, indicates that the Ochopee Marl Marsh bedrock does not rise as far above Shark River Slough as does that of the Rockland Marl Marsh. As discussed in the beginning of this chapter, prior to drainage and peat loss, the greater presence of peat meant that ground surface elevations within Shark River Slough were more similar to those of the Marl Marshes. Whether this also would have made the ground surface elevations of the two Marl Marshes more similar is not known.

The available early firsthand sources, discussed in more detail later, indicate that prior to the changes brought about by regional drainage, the Ochopee Marl Marsh was composed of three subcomponents, similar in appearance and arrangement to the subcomponents of the Ridge and Slough landscape. As in the Ridge and Slough landscape, these three subcomponents—a form of sloughs, a form of ridges, and tree islands or hammocks—were distinguished by increasing relative elevation. However, the sloughs and the ridges of the Ochopee Marl Marsh appear to have been somewhat drier, with shrubby species present on the ridges, especially wax myrtle (*Myrica cerifera*), and with sawgrass and "round grass" (likely *Eleocharis cellulosa*) present within at least some of the sloughs. The predrainage shapes and orientations of the tree islands are not known, but the Ochopee ridges and sloughs were originally elongated, parallel, and oriented southwestward, parallel to the direction of water flow, a directional configuration still visible in part today (William Loftus, pers. comm., 2009). There apparently was a degree of interaction with groundwater, with surface water reportedly disappearing into sinkholes. This characterization of the Ochopee Marl Marsh is based on early predrainage observations (township surveys by William Mickler made in 1885–1886; Lucas 1917?a, 1917?b; King 1917c, 1917e), supplemented by recent paleoecological investigations (Bernhardt and Willard 2006) and comparisons with recent vegetation mapping (McPherson 1973b). Although only a small number of observers appear to have left records concerning the Ochopee Marl Marsh landscape, in each case their information was based on extensive field experience in this landscape as well as the adjoining ones to the east and west. Similarities in the locations and nature of landscape features described by the various early observers, as well as consistencies with more recent vegetation mapping and paleoecological data, together strongly support the predrainage reconstruction presented here.

During November through January of the 1885–1886 dry season, U.S. Deputy Surveyor William Mickler surveyed the exterior boundaries of a large number of townships in southwestern Florida, in areas that now fall mostly within Big Cypress National Preserve and Everglades National Park (fig. 1.4). Mickler apparently never drew plat maps for the townships he surveyed, creating difficulties for formal land survey uses of his work. For ecological uses, however, his detailed field notes are an invaluable resource. Preserved in the same archives as those from other surveys shown in figure 1.4, his notes provide an essentially synoptic view of southwestern Florida. His descriptions of vegetation, terrain, and water depths clearly reflect the variety of landscapes he surveyed, including mangroves, Big Cypress, the Ochopee Marl Marsh, and the Ridge and Slough landscape. The geographic correspondence of Mickler's reported landscapes with current mapping confirms his field experience and the reliability of his descriptions.

Mickler's notes from approximately 100 linear miles (160 km) surveyed within the Ochopee Marl Marsh constitute one of the key sources for this landscapes. Surveyed just two years after the opening of Disston's canal connecting Lake Okeechobee to the Caloosahatchee River, Mickler's observations likely capture essentially predrainage landscape conditions. We examine Mickler's notes in detail here for the two townships that fell almost entirely within the Ochopee Marl Marsh (plate 4): Township 55 Range 35 and Township 56 Range 34.

Mickler's survey of T55 R35, conducted in late November and mid-December (i.e., one to two months into the dry season) of 1885, recognized all of it as part of the Everglades ("Land [=] Everglades" and "through Everglades"); that is, as an area clearly distinct from Big Cypress. The notes frequently state that the water was "too deep to erect Pits and Mounds." Quantitative water depths were given for all but one mile marker and included "12 to 28," "15 to 28," "15 to 30," and "18 to 30" inches, with an average of 15 to 30 inches (38–76 cm). Interestingly, this average water depth is only slightly less than water depths Mickler recorded within the adjacent Ridge and Slough landscape (18 to 30 inches, 46–76 cm in T57 R35; 18 to 31 inches, 46–80 cm in T53 R36; see also figs. 11.3, 11.7).

Regarding the vegetation of Township 55 Range 35, Mickler noted only very limited "timber" (i.e., marketable woody growth) and that the timber was restricted to the "few scattering islands"; likely the only locations sufficiently elevated and dry to support woody species. Mickler mentions "Hammocks," probably referring to dense growths of maritime hammock tree species, and individually mentions "Cabbage Palm" (*Sabal palmetto*), "Coco plum" (*Chrysobalanus icaco*), and "Bay wood" (*Persea borbonia* and/or *P. palustris*). The nonwooded majority of the township was described as "sawgrass" or "heavy sawgrass," alternating with "open Everglades." Mickler does not define "open Everglades," but it is clearly contrasted with the areas of sawgrass, and the open quality suggests some sort of slough. Interestingly, the railroad engineer William Krome used the same term, and apparently also to distinguish areas from less open, uniformly densely vegetated parts of the Everglades: "At that distance [edge of Shark River Slough, coming from East] we again strike open Everglades" (Krome 1903c). Mickler records water depths as deep as 15 to 30 inches (38–76 cm) in the "open Everglades" during the early dry season, so it seems likely that "open" refers to water deep enough to be unobstructed by dense emergent species; that is, a suggestion of something like open water sloughs. Mickler's summary that the township was "not fit for cultivation" presumably was based on the significant water depths present throughout.

The following representative excerpts from Mickler 1885-T55 R35 give a sense of the vegetation seen in this township:

> Timber is confined to Small Islands Scattering through the Everglades
> Timber none except a few Small Islands[,] considerable Saw Grass
> Timber none except Coco Plum and Bay wood on Small Islands
> Enter expanse of Saw Grass
> Through Saw Grass + water from 15 to 30 in[ches] deep
> Leave Saw Grass and enter Small Scrub Island
> Leave Island and reenter Saw Grass in Everglades
> Leave heavy Saw Grass and enter more open Everglades
> Leave Everglades and enter Small Scrub Island
> Leave Island and enter Everglades
> Timber none except on the Small Island there is a few small trees mostly Cabbage Palmetto. (Mickler 1885-T55 R35, pp. 59, 65, 172–175, 186)

Mickler's field notes also described the soil of Township 55 Range 35. However, his soil descriptions are puzzling and difficult to interpret. In addition to Mickler's surveys within the Ochopee Marl Marsh, his 1885–1886 work included townships within the Ridge and Slough landscape. Despite the unambiguous predrainage presence of peat soils within the Ridge and Slough landscape (in some cases even quite deep peat soils), Mickler's soil descriptions for these townships sound much more like shallow marls. For example, throughout T57 R35 within Shark River Slough and throughout T53 R36 within the present Water Conservation Area 3A (plate 4), Mickler recorded either "Soil mud 4th Rate" or "Soil mud from 6 to 10 inches then Sollid Rock 4th Rate." The term "mud" has in fact sometimes been used in early Everglades literature to refer to deep organic soil (i.e., peat): "upon this surface lies an immense accumulation of sand, alluvial deposits and decayed vegetable matter, forming a mass of sand and mud from two to ten feet or more in depth" (Florida Department of Agriculture 1911, p. 43; see also Ives 1856a, p. 20). Nevertheless, Mickler's usage does not seem to suggest peat, particularly as he substantially underestimates the soil thickness of the full profile: only 6–10 inches (15–25 cm). In view of Mickler's otherwise detailed and corroborated observations, his soil descriptions remain an enigma. Possibly the fact that he categorized all of these townships as "unfit for cultivation" on hydrological grounds influenced his soil descriptions, but this certainly does not fully solve the puzzle. In this context, Mickler's descriptions of the soil in the T55 R35 as "Soil none," "Soil mud from 8 to 10 in[ches thick] then Rock 4th Rate" (in a few cases "6 to 10" or "8 to 12" inches) very likely do not fully characterize the predrainage soil of this Ochopee Marl Marsh township. In particular, Mickler's descriptions cannot be interpreted as a reliable indication of an absence of predrainage peat within the Ochopee Marl Marsh landscape.

Mickler also surveyed Township 56 Range 34, the second township that fell entirely within the Ochopee Marl Marsh (plate 4). Although completed somewhat later (mid-January 1886), his observations are similar to those of T55 R35: "thick sawgrass" alternating with "more open Everglades" and "small islands" (Mickler 1886-T56 R34). The soil was recorded as "Soil mud 4th Rate" or "Soil mud from 6 to 10 in[ches]" (9–25 cm). Even at this somewhat later point during the dry season, water depths were reported as "18 to 30 inches" (46–76 cm) deep. The following excerpts give a sense of the vegetation:

> Through Everglades
> Enter thick Saw Grass
> Leave thick Saw Grass and enter more open Everglades
> No Timber
> 59.80 [chains] Enter Small Island
> 62.50 [chains] Leave Island and reenter Everglades." (Mickler 1886-T56 R34, pp. 416–418)

Mickler's "General Description" of township T56 R34 gives a clear sense that the Ochopee Marl Marsh was wet, both in the wet season and in the dry season. The conclusion of the land being "not fit for Cultivation," which he applied to T55 R35 as well, appears to be based on the absence of high, that is dry, land:

> "General Description:
>
> This Township is all inundated from 18 to 30 inches and in the Rainy Season the water is much deeper and the Soil is a Sort of lime mud from 6 to 10 in[ches] deep then there is a Sollid Lime Stone Rock; There is a few Small Hammock Island on it—I dont think that the largest of them will consist of an acre of High land it is not fit for Cultivation." (Mickler 1886-T56 R34, p. 423)

Figure 11.7 integrates water depth and vegetation observations from Mickler's township surveys, all made during the 1885–1886 dry season, and selected to form a rectilinear profile across the Ochopee Marl Marsh, Shark River Slough, and a small portion of the Rockland Marl Marsh. The profile suggests that water depths were slightly shallower within the Ochopee Marl Marsh than within Shark River Slough, but the distinction was weak. Overall, the impression from Mickler's surveys is that in 1885–1886, neither water depths nor vegetation differed greatly between Shark River Slough and the Ochopee Marl Marsh landscape.

Around 1917, approximately thirty years after Mickler's township surveys and after the onset of the first lowering of Lake Okeechobee water levels (chapters 2, 3, 5), the developer Captain Frank Jaudon of Miami purchased the "A. W. Hopkins Tract," also known as the "Chevelier Tract" (Ashworth 1919a; 1925?). This large area of 324 square miles (840 km^2), covering more than eight townships (T54–55, R31–34), spans portions of the Big Cypress as well as part of the Everglades. Jaudon engaged two civil engineers, S. J. E. Lucas and John W. King, to examine and report independently upon the A. W. Hopkins tract. S. J. E. Lucas apparently was instructed to focus on timber (board feet) and to report on drainage directions and soils. The date is either 1916 or 1917. He produced a map of considerable detail (Lucas 1917?a) and short narratives describing each of the townships (Lucas 1917?b). Civil engineer and naturalist John W. King was engaged to explore the eastern portion of the "A. W. Hopkins tract," an area that extended into the Ochopee Marl Marsh. After an initial reconnaissance visit, King returned to spend one month (February 11 to March 14, 1917) exploring the area. King reported his observations to J. F. Jaudon in a map dated March 20, 1917 (King 1917c; not shown) and in a report dated March 23, 1917 (King 1917e). Plate 37 reproduces photographs of the Ochopee Marl Marsh included in King (1917e).

Lucas's map of the A. W. Hopkins tract (Lucas 1917?a) makes a strong distinction between Big Cypress and the Everglades. Lucas draws a clear southwest-northeast line labeled as "Edge of Everglades Proper" diagonally through the approximate middle of T55 R33 and T54 R34. Interestingly, the position of Lucas's line, along the western edge of Lostmans Slough, approximately following the eastern edges of Dayhoff and Dixon sloughs (Duever et al. 1986, fig. 3.6), closely matches the boundary separating "Cypress Forests" from "Prairies" as drawn on a more recent vegetation map (McPherson 1973b). Additionally, Lucas's map visually reflects the distinction between Big Cypress and the Everglades: west of the edge, within the Big Cypress area, it differentiates numerous individual vegetation patches: four variations of cypress—"Scrub," "Dense," "Light," and "Small"—as well as "Cypress Strands"; "Prairie," "Pop Ash," "Pine," "Small Pine," "Small Pine and Cypress," "Saw Grass," "Saw Grass Prairie," "Scrub," "Rocky Scrub Hammock," and "Sloo" [slough] (Lucas 1917?a). In contrast, within the Everglades, Lucas's map shows very few individual vegetation patches.

Lucas also distinguishes two different areas of sawgrass vegetation within the A. W. Hopkins tract: (1) a coastal strip of sawgrass prairie, located inland from the 10,000 Islands "between the bays and the cypress," and (2) the area of Everglades proper, draining from northeast to southwest. Lucas described the coastal sawgrass prairie as extending from T54 R31 to the southwest corner of T55 R33, forming "a strip of prairie about three miles wide, extending from north west to south east. This prairie is covered with saw grass, inset with sloos [sloughs] and scrub hammocks and traversed by many shallow, sluggish, winding, tide flowing streams" (Lucas 1917?b, p. 1). These are likely the same sawgrass prairies referred to by Totch Brown, who usually approached them from the 10,000 Islands (Brown 1993, pp. 243–244). Lucas's reference to rich soil is almost certainly a reflection of sawgrass peat: "The soil in the [sawgrass] prairie is very rich, and could be made valuable by drainage" (Lucas 1917?b, p. 1), and "the southwest part [of T55 R33] is a boggy saw grass prairie the soil of which is much [muck]" (Lucas 1917?b, p. 8). Lucas does not clarify why he distinguished between the sawgrass of the coastal prairies and the sawgrass of the "Everglades Proper." It seems likely that the distinction was in part because coastal prairies were more uniformly sawgrass, while the Everglades (i.e., the Ochopee Marl Marsh) was more varied, with areas of sawgrass alternating with areas of more open water, as indicated by Mickler and by King (discussed later).

Lucas does not give much detail regarding the Everglades portion of the A. W. Hopkins tract, but he does provide several indications of regional water levels. Noting that T54 R33 was "perhaps the most elevated above tide of the entire tract" (p. 3), and also that it was the location, along with T55 R33, of the "so-called Lossman [Lostmans] River pine" (p. 3), Lucas describes soils that nevertheless reflected the presence of periodic surface water: "The soil in the pines [of T54 R33] is a rich sandy loam" (p. 3). Similarly, conditions in the prairies appear to have been sufficiently wet to accumulate peat rather than simply marl alone: "The soil in the cypress and adjacent prairies is very rich and the occasional streams form natural drainage" (Lucas 1917?b, p. 3).

Lucas's narrative description of the Everglades extent, in this case referring primarily to the Lostmans slough portion of the Ochopee Marl Marsh, resembles his map in being unambiguous about distinguishing the Everglades from Big Cypress: "The southern and eastern portion [of T54 R34] is everglades" (Lucas 1917?b, p. 4); "The southeastern portion [of T55 R33] being about 40% of the township is everglades" (Lucas 1917?b, p. 8); and "This entire township [T55 R34] is everglades except a little cypress and scrub hammocks in the northwest corner. This land like the township, north, slopes to the south and the southwest" (Lucas 1917?b, p. 9). Lucas correctly identifies the direction of water flow and makes the astute observation that the then planned efforts to increase drainage in areas eastward (near Miami) would be unlikely to provide the drainage needed to develop the A. W. Hopkins tract: "The natural drainage of the township is about southwest and I therefore believe that any improvement [i.e., drainage] to lands some distance east will not be effective on these lands [T54 R34]" (Lucas 1917?b, p. 4).

The observations of Captain Jaudon's second investigator of the A. W. Hopkins tract, John W. King, complement Lucas's observations and supplement them with additional ecological detail. There is no indication that King was familiar either with Mickler's field notes or with Lucas's work, so we assume that King's observations are independent of both Mickler and Lucas. It is likely however that the landscape seen by King (and by Lucas) had already been altered to some degree by drainage, as King himself remarked upon drainage-induced changes (King 1918, p. 150). Examination of natural features such as Chatham River, Alligator Bay, Lostmans River, and Sweetwater [Creek] drawn on King's map suggests that they are accurate to within 0.5 mile (0.8 km); certainly sufficient for evaluating the Ochopee Marl Marsh.

King classified the eastern portion of the A. W. Hopkins tract into five categories: "Cypress lands," "Pine lands," "Coastal Hammock," "Prairie," and "Everglade[s]" (King 1917e, p. 409). The Cypress lands, Pine lands, and most of the Prairie categories clearly form part of Big Cypress and are not relevant here. King's map indicates that the "Coastal Hammock" and "Everglade[s]" categories fall within the Ochopee Marl Marsh, primarily in

Township T55 R34 and in portions of Townships T54 R34 and T55 R33. These areas are essentially the same as the ones listed by Lucas (1917?a, b) as "everglades."

King's description of topography applies to the Ochopee Marl Marsh, at least in part. Of particular interest are his indications of water flow, of "natural water passages" (likely the "open Everglades" areas noted by Mickler), and the need for "permanent drying of the soil" if the area is to be developed:

> "TOPOGRAPHICALLY,—the Eastern portion of the A. W. Hopkins tract is apparently higher than accredited in my former report, from the dryness of the land [in Feb–Mar 1917], the sharply defined stream construction, rapidity of flow of water, etc. is indicative of sharper slopes of the surface soils, and better drainage than was supposed. With the running of a series of accurate and precise levels, elevation of 10' 0"–12' 0" or even 15' 0" may be found on the Eastern border. Therefore the cleaning and clearing out of the natural water passages will do much towards the ultimate and permanent drying of the soil, but could be facilitated by an adequate system of canals and laterals." (King 1917e, pp. 415–416)

King also described the bedrock of the Hopkins tract:

> "Underlying the entire area of this A. W. Hopkins tract is a bed of hard blue limestone known as the Lossmans [Lostmans] River Limestone, lying very near the surface, and in instances projecting through in the form of an outcrop. . . . This limestone in its natural formation, is a carbonate of calcium and is of a hard, blue, homogenious structure, much 'water worn' on the surface. It is unquestionably of considerable porousity and cavernous, as indicated by the number and extent of the sink holes found within this area. On the surface fragments of this rock show a high state of silicazation, wherein the more soluble Calcium Carbonates have been replaced by a precipitation of soluable silicas, creating an extremely hard, flinty structure.
>
> It is evident that a system of subterranean drainage exists throughout this entire area, from the dryness of the soil, the formation of streams, and the disappearance of same in sink holes of large or smaller extent." (King 1917e, pp. 414–415)

King's "Everglade" category, falling mostly within Township T55 R34, corresponds to the central and western portion of the Ochopee Marl Marsh, including a portion of Lostmans Slough. Within his Everglade category, King notes two distinct vegetative subcomponents: a lower, wetland subcomponent, which he refers to alternatively as "dry prairies" or as "sloughs," and a higher subcomponent referred to as "Myrtle Hammocks" or "Myrtle Ridges." From the wetland species that King associates with the slough subcomponent, slough seems an appropriate name. (William Loftus, pers. comm., 2009, notes that even at present, portions of Lostmans Slough include peat soil and vegetation similar to that found within Shark Slough.) King's reference to these sloughs as "dry prairies" seems to be more a reflection of a combination of natural and anthropogenic causes producing unusually dry conditions in February–March 1917: the dry season (late February), a long drought (per King), and the influence of drainage of Lake Okeechobee and of the upstream Everglades (chapters 2, 3, 5).

In the Ochopee Marl Marsh portion of his map (1917c), King draws a parallel, linear pattern of myrtle hammocks/ridges alternating with sloughs/dry prairies, both curving in the same downstream direction noted by Lucas: from almost north-south to northeast-southwest. The Ochopee Marl Marsh patterning is drawn parallel to the curved alignment shown for the Coastal Hammocks farther west and for the sawgrass ridges and slough of Shark River Slough, farther east. These directional patterns are still visible today (William Loftus, pers. comm., 2009). King's accompanying description suggests several resemblances to the Ridge and Slough landscape: alternating higher (drier) and lower (wetter) elements; similar spacing; and a linear, parallel arrangement. At the same time, his description suggests that the Ochopee Marl Marsh may have been somewhat drier than the Ridge and Slough landscape, whether due to a shorter hydroperiod, reduced water depths, or both. This is clearly reflected in the ridges, which instead of supporting sawgrass support the woody species myrtle (*Myrica cerifera*). The somewhat drier conditions in the Ochopee Marl Marsh were also reflected in the sloughs, which apparently included sawgrass ("covered with a growth of [sawgrass]") and very likely *Eleocharis cellulosa* ("a round water grass"):

> "The Southeasterly corner of the Hopkins Tract projecting Easterly of the Coastal Hammock enters into the *Everglades proper*. This area of Everglades [which he later refers to as 'sloughs'] is similar in character to other sections of the Everglades [i.e., the Ridge and Slough landscape with which King was already familiar] and covered with a growth of sawgrass and a round water grass, separated and divided by a number of Myrtle Hammock[s], very narrow and much elongated, following in parallel lines with the main

Coastal Hammock and ranging approximately 200 yards apart. From the ends of these long projecting Myrtle Ridges extends rocky reefs connecting with the preceding hammock. These rocky reefs project upwardly in sharply pointed finger-like masses, level with or above the surface of the ground [i.e., pinnacle rock]. The sloughs that in periods of high water [i.e., the wet season] spread over this ground, were at this time mostly dry, owing to the effect, among other causes, of a long period of drought, gather at irregular intervals into rivulets which usually lead to sink holes at the ends of some Myrtle Ridge and disappear beneath the surface of the ground." (King 1917e, p. 414)

In a table providing more detailed information for the five land categories he identified within the Hopkins tract, King included two columns for the "Everglade" (Ochopee Marl Marsh) category, reflecting the two subcomponents he found there: myrtle ridges, described in a column labeled "Myrtle," and the more open sloughs between them, described in a column labeled "Open." Although both the ridges and the sloughs of the Ochopee Marl Marsh were mostly dry at this time (February 23–25, 1917), King's descriptions suggest that he was well aware that at least the sloughs were typically wetter.

The woody species listed by King in the Myrtle column confirm that these ridges were drier than the sawgrass ridges of the Ridge and Slough landscape. King described the "character of the land" as "Myrtle [and] Rising" and "Dry" [apparently referring to that time point rather than in general]; he described the soil as "Mucky," 0.5–1.5 feet [15–45 cm] thick; with "Verdure—Trees" of "Myrtle [*Myrica cerifera*], Sweet Bay [*Persea borbonia*], Magnolia [*Magnolia virginiana*], [Jamaican] Dogwood [*Piscidia piscipula*], Willow [*Salix caroliniana*], Grape Vine [*Vitis* spp.], [a] Specie [of] Rattan [*Smilax*??], and Moon[flower, *Ipomoea alba*]" (King 1917e, p. 418). Game in the ridges was described as "But little animal life. Herons, Curlew, Birds of Prey" (King 1917e, p. 418).

King's "Open" column referring to the sloughs of the Ochopee Marl Marsh described them as: "Open [and] Level," "Dry" [i.e., not overflowed at that time], and the soil as "Muck," 0.5–1.5 feet [15–45 cm] thick, with no trees. The "undergrowth" was listed as "Saw Grass [*Cladium jamaicense*], Round Water Grass [*Eleocharis cellulosa*], Flag Blue [*Pontederia cordata*], Yellow Dock [spatterdock, *Nuphar advena*], and White Water lilies [*Nymphaea odorata*]" (King 1917e, p. 418). Sawgrass and *Eleocharis* can withstand seasonal drying. *Pontederia*, *Nymphaea* and *Nuphar*, however, are aquatic species. They, along with the "rich, loamy vegetable muck" found here by King (see later discussion), suggest that the sloughs of the Ochopee Marl Marsh were water-filled during most of most years. The clearly aquatic, long-hydroperiod fauna listed by King further support the hydrologic impression given by the vegetation and the wetland soils: "Alligators, Frogs, Snakes, Crawfish [*Procambarus* spp.], Shrimp [*Palaemonetes* spp.?], and Game Birds" (King 1917e, p. 418).

King's description of the sloughs as being "dry" in February–March of 1917, and his labeling of the sloughs as "Dry Prairies" on his map (King 1917c), appear to reflect anomalously dry conditions; that is, not typical of the flora, fauna, and soils described by King. The flora and especially the soils seem likely to be more reliable indicators of the long-term average hydrologic conditions.

In addition to his text, King's photographs show, and his captions mention, the sloughs (King 1917e, photos 1–4, pp. 403–405); two are reproduced in plate 37 of this study. The open appearance suggests water depths that might be similar to the "18 to 30 inches" (46–76 cm) and deeper water mentioned by Mickler. In the captions King mentions "breaking up" or "disintegration" of sloughs (photos 2–4). It is not clear exactly what King meant by this, although it is likely that he was contrasting the sloughs of the Ochopee Marl Marsh with his observations of sloughs in the Ridge and Slough landscape (Shark River Slough; King 1917f). "Breaking up" may refer to a possible blurring of slough to myrtle ridge transitions within the Ochopee Marl Marsh. Alternatively, and perhaps more likely, it may refer to decreased slough continuity in the downstream direction.

"The soil of this Everglades portion of the A. W. Hopkins tract consists of a rich loamy vegetable muck of irregular thickness, slightly more decomposed than in other places within the Everglades, and much drier." (King 1917e, p. 414). This sentence emphasizes one of the most striking aspects of King's report. Within postdrainage Everglades history, from J. H. Davis (1943a, b) forward, the slightly elevated wetlands east and west of Shark River Slough have been widely considered to be *marl* prairies (e.g., S. M. Davis et al. 2005). The accompanying assumption was that their hydroperiods were too short to support the accumulation of organic peat soil, allowing only the accumulation of inorganic marl.

King's 1917 soil description indicates that predrainage conditions in the Ochopee Marl Marsh landscape were sufficiently wet to be peat-forming, rather than marl-forming. Furthermore, these peat-forming hydrologic conditions must have been present for a number of centuries, as King states that the "muck" (a term commonly used for peat at that time) was "0.5–1.5" feet (15–45 cm) thick (King 1917e, p. 418).

King's description of the "rich loamy vegetable muck" of the Ochopee landscape as "slightly more decomposed than in other places within the Everglades, and much drier" merits comment. It is likely a comparison with the peat of the Ridge and Slough landscape that he had recently explored (King 1917b, 1917f). Although King does not indicate the basis for concluding that the peat was "slightly more decomposed," this might have been visible as a more amorphous, less fibrous structure and as a darker color closer to black than to brown. It seems likely that King's observation was accurate; it could reflect either a natural effect of slightly higher elevation and hence drier conditions than in Shark River Slough, or it might reflect a somewhat higher sensitivity in this landscape to the influence of early anthropogenic drainage. The indication that the muck was "much drier" seems less robust and perhaps a more temporary reflection of the particularly dry conditions present during King's fieldwork.

Several observations in King's (1917e) report support his observation of a noticeable layer of organic soil (0.5–1.5 feet; 15–45 cm) within sloughs. First, it is specific to the slough subcomponent of his Everglades (i.e., Ochopee Marl Marsh) category. King states "muck" for the sloughs, but "mucky" (i.e., mucklike) for the myrtle ridges, a realistic distinction given their higher elevation and correspondingly drier hydrology. At a broader scale, King drew soil distinctions among his five landscape categories, describing the soils of the "Cypress," "Coastal Hammock," and "Prairies" categories all as "Dark Veg. [etable] Loam"; the soil of the "Pine Islands" as "Rocky and sand"; and only the two "Everglades" subcomponents as "Muck" and "Mucky." King's indication of muck in the sloughs is also consistent with his descriptions of aquatic flora (e.g., water lilies and spatterdock) and of fauna generally associated with year-round or close to year-round presence of surface water: frogs, if pig frogs (*Rana grylio*), and (grass)shrimp. An organic soil is consistent with King's comment on the need for drainage: "ultimate and permanent drying of the soil . . . could be facilitated by an adequate system of canals and laterals" (King 1917e, p. 416). Finally, the apparently widespread presence of organic muck (King does not mention marl) within the central portion of the Ochopee Marl Marsh area is paralleled by the various reports of peat soil in the corresponding Rockland Marl Marsh (e.g., Cooley 1851, p. 43; Stewart 1907, p. 45) as well as by reports of tall and dense sawgrass there (e.g., MacKay 1845-T56 R38; Jackson 1847-T56 R38; Griswold 1896, p. 56). Overall, King's documentation of a noticeable layer of organic soil—within what in postdrainage times has been referred to as a marl landscape—fundamentally shifts our understanding of the landscapes flanking Shark River Slough: the Ochopee and Rockland "marl" marshes.

Field observations such as those of Mickler, Lucas, and King provide informative and spatially explicit "snapshots" of the Ochopee Marl Marsh landscape at particular points in time. These can be usefully complemented by paleoecological analyses of soil cores, creating an additional path for identifying predrainage conditions and postdrainage changes. Bernhardt and Willard (2006) analyzed soil cores from within the Ochopee Marl Marsh landscape, at locations that also fell within the former A. W. Hopkins tract, facilitating direct comparisons with the observations of King, Lucas, and Mickler. In 2003 most of the area was covered only by a thin layer of marl (<5 cm). Cores were therefore taken from solution holes as these were thought to have remained wet enough to preserve any organic sediments that might originally have been present. The cores were found to consist of peat with a thin surficial layer of marl (in one of the five cores only peat was present). Pollen evidence indicated that prior to about 1900, these sites had been sawgrass marshes that accumulated peat (Bernhardt and Willard 2006, p. 2). The pollen analysis thus corroborates Mickler's and King's reports of sawgrass, the presence of peat corroborates King's descriptions of "muck" (i.e., peat soil), and the accumulation of peat suggests dense sawgrass, consistent with Mickler's descriptions. Even more intriguingly, Bernhardt and Willard provide a mechanistic explanation of the difference between predrainage and current conditions in the Ochopee Marl Marsh landscape: "During the 20th century, vegetation changed to grass-dominated marl prairies, and calcitic sediments were deposited, indicating shortening of hydroperiods and occurrence of extended dry periods at the site. These data suggest that the presence of marl prairies at these sites is a 20th century phenomenon, resulting from hydrologic changes associated with water management practices" (Bernhardt and Willard 2006, p. 2). This explanation is similar to Craighead's description of drainage-induced losses of 4 to 12 inches (10–30 cm) of sawgrass peat, which exposed the underlying marl in the Marl Transverse Glades (Craighead 1974, p. 191).

These conclusions from Bernhardt and Willard's study provide a logical explanation for the strong contrast between botanist J. H. Davis's descriptions of the Ochopee and Rockland marl marshes in the 1940s (he considered them a pair; J. H. Davis 1943a, p. 259) and the preceding descriptions by George MacKay, John Jackson, William Mickler, S. J. E. Lucas, and John King. Davis apparently found little or no peat, only exposed marl or bedrock. Where the earlier observers had all reported thick or heavy sawgrass, dense enough for MacKay and Jackson to abandon their survey lines, Davis found primarily a remnant population of "generally low and sparse" sawgrass, an expected result of lowered water levels. The "round grass" (likely *Eleocharis cellulosa*) seen by King was still present, but it had been joined by numerous other graminoid species, probably the result of lowered water levels and possibly also of the change from peat to marl substrate. These vegetation changes are all consistent with the assumption of drainage-induced decreases in average water depths (pre-1940s; see chapters 2, 3) and with the vegetation-hydroperiod relationships found by Ross et al. (2006). Interestingly, Davis does not mention muhly grass (*Muhlenbergia capillaris* or *filipes*), a species now extensive in the marl prairies (Olmsted et al. 1980, p. 11) and thought to have increased under continued drainage since the 1940s (Werner 1975; Olmsted et al. 1980, p. 11; Armentano et al. 2006):

> "Southern Marshes and Wet-Prairies. These two marsh and wet-prairie areas both have sawgrass but it is generally low and sparse. In these areas some other sedges, grasses, rushes, and non-grass like herbs are usually abundant. Patches of the needle grass or spike rush, *Eleocharis cellulosa* Torr., cover large areas on parts of intermediate elevation, especially where there is marl. On the thinnest marl soils and on exposed rock the vegetation becomes a very sparse growth, figure 65. The beak rushes, *Rynchospora torreyana* H. Gray and *Rynchospora tracyi* Britton, and a number of grasses such as *Panicum tenerrum* Beyr., the switch-grasses, and a beard-grass, *Andropogon elliotti* Chapm., are usually present. Where the three-awn grasses, *Spartina* sp. and *Aristida virgata* Trin., known as the switch-grasses, and the black sedge, *Schoenus nigricans* L., are common there is usually more water than in the saw-grass or needle-grass marshes. Such switch-grass areas are considered wet-prairies rather than marshes." (J. H. Davis 1943a, p. 268)

In contrast to the dramatic hydrologic, vegetative, and soil changes that have occurred within the originally sloughlike subcomponent of the Ochopee Marl Marsh, the hammocks appear to have retained a greater degree of their predrainage character. This is not surprising, as the hammocks would originally have been the least aquatic landscape subcomponent and thus likely also the least sensitive to anthropogenically reduced water depths. Like MacKay (1847, p. 65), Mickler (1885-T55 R35), and King (1917e, pp. 412–413) before him, J. H. Davis in the 1940s was struck by these hammocks, the elevated areas that MacKay referred to as "gems of islands of rare timber and shrubbery": "The outstanding feature of these marsh and wet-prairie areas [i.e., the Ochopee and Rockland marl marshes] are the many cabbage-palm and tropical hardwoods hammocks that make these open plains beautiful. In the hammock forests are many trees and shrubs, ferns, vines, and epiphytes. The dense growth of woody plants in many of these hammocks south of the Tamiami Trail number over forty species" (J. H. Davis, 1943a, pp. 268–269). Davis notes a number of the same species that King (1917e) described in his Coastal Hammock category.

Hydrology

Based on the reported water depths for this landscape, the geographical similarity to the Rockland Marl Marsh, and the similar vegetation (thick sawgrass), soils (thin peat), and areas of exposed bedrock, we assumed that the predrainage hydrologic parameters estimated for the Rockland Marl Marsh apply here as well. As an average for the landscape, we estimated an end-of-wet-season high water level 2 feet (+60 cm) above the soil surface, an end-of-dry-season low water level that was 0.5 foot below the soil surface (-15 cm), and a long-term average hydroperiod of eight to nine months. The predrainage presence of aquatic species, such as white water lilies (*Nymphaea odorata*) and spatterdock (*Nuphar advena*), and other similarities to the predrainage Ridge and Slough landscape suggest that the hydroperiod and depth estimates presented here may both be underestimates.

Hindcasting

We drew the Ochopee Marl Marsh extent based on the extent of unforested marl soil mapped by Jones et al. (1948a). It is highly likely that the peat soil associated with Shark River Slough was originally both deeper and wider, suggesting that the Ochopee Marl Marsh would have been correspondingly narrower. However, given the absence of specific indications of the original boundaries, we simply adopted the soil extents, without any hindcasting.

Perrine Marl Marsh

This landscape is bounded on the north and west by the pinelands of the Everglades Keys and Long Pine Key, on the east and south by the mangrove fringe lining the shores of Biscayne and Florida bays, and on the southwest by Shark River Slough (fig. 9.1). Taylor Slough, a portion of the Ridge and Slough landscape running roughly north to south, bisects the Perrine Marl Marsh.

The Perrine Marl Marsh landscape has previously been referred to as coast(al) prairie (Harshberger 1914, pp. 178–180; Harper 1927, pp. 180–184), southern coast marsh prairie (J. H. Davis 1943c, pp. 110–111), southeast saline Everglades (Egler 1952, pp. 219–220), and southern marl-forming marsh (S. M. Davis et al. 1994, p. 435).

The Perrine Marl Marsh closely resembled the Rockland Marl Marsh in terms of plant species present but differed in being more influenced by hurricanes and salt spray and by tidal saltwater in the southernmost and easternmost portions. Pinnacle rock and solution holes are less common, probably because of deeper marl. The presence of "pygmy cypress forests," "sprinklings" of low mangrove bushes, and mangroves lining creeks further distinguish the Perrine from the Rockland Marl Marsh.

Landscape elevations decreased from about 9 feet (2.7 m) at the boundary with the Everglades Keys to close to sea level at the mangrove fringe (plate 35). Elevations dropped quickly near the Everglades Keys, with shallower slopes farther away. A slight and not fully continuous natural coastal levee, 1–1.5 feet (30–45 cm) high, often referred to as the "buttonwood embankment" (fig. 9.1; Craighead 1973, p. 56; see also Craighead 1964), somewhat delayed surface runoff and locally impounded water upstream from it (and still does; William Loftus, pers. comm., 2009). It may also have diminished somewhat the influence of salt water on the landscape.

Under predrainage (pre-1880) conditions, the Perrine Marl Marsh was connected hydrologically to the Rockland Marl Marsh by wet season outflows that passed through the Marl Transverse Glades (chapter 10) and by nearly year-round flows through Taylor Slough (chapter 8). Within Everglades National Park, a degree of connection to the Rockland Marl Marsh still persists, though greatly reduced.

Flows were augmented by direct rainfall and by any surface runoff from the Everglades Keys. Reported predrainage groundwater phenomena that disappeared after drainage—such as mounding underneath the Everglades Keys (fig. 11.9), coastal springs (e.g., U.S. Coast and Geodetic Survey 1892), and submarine freshwater springs in Biscayne Bay (fig. 10.7; Kohout 1966, p. 396; Kohout and Kolipinsky 1967, p. 488)—suggest that groundwater originating in the Everglades Keys or in the Rockland Marl Marsh may have contributed to Perrine Marl Marsh hydrology, but this cannot be confirmed from the historical record alone.

Two U.S. deputy surveyors surveyed much of the Perrine Marl Marsh (fig. 1.4), John Jackson in 1847 and Marcellus Williams in 1874, both well prior to the onset of anthropogenic drainage. Both surveyors found it necessary to wait until late in the dry season (March–May) to be able to survey. Table A.6 in appendix A records numerous descriptions from their field notes similar to the following: "very wet soft Prairie land," "Wet Marsh or Prairie land," "Wet Sawgrass Prairie," "1st rate Prairie land if drained," "rich wet Sawgrass prairie land," "very wet soft Sawgrass Prairie land," and "All Wet boggy Marsh." Even late in the dry season, saturated soils as well as a degree of surface water clearly complicated the surveyors' work. In fact, along a number of lines, conditions became sufficiently impracticable that the lines were abandoned: "Prairie is too soft and water too deep to proceed," "the Marsh became so boggy in places and in other places under water there were chasms of great depth—in fact it was impossible to extend the line farther," "Marsh too soft and water too deep to proceed," "Mud and Sawgrass very rich soft Prairie land too soft to proceed any further."

Although very few quantitative water depths were given in these field notes, the overall impression is one of a landscape with considerable water: wet soils that were "rich" in some places, "very soft" in many; and sawgrass that was common, if not ubiquitous. Unfortunately, the surveyors rarely stated whether the sawgrass was dense or sparse, nor did they clarify whether the descriptions of the soil as "very soft," "rich," and "boggy" referred to peat or to marl. The description "very soft" does not occur frequently in surveyors' accounts from the peat-based landscapes of the Everglades, suggesting that it may refer to wet, yielding marl. However, the term "boggy" was very common in peat areas, suggesting that significant amounts of peat may have been present, at least locally. "Rich" also tends to have been used primarily in conjunction with peat soils but not exclusively so. Surveyor Williams records a case of "having been baffled in every effort" to proceed with a survey, and deciding to wait a week: "The weather having continued without rain, and the sun quite hot. My hope was that the soft Mud had somewhat dried. In this I was not disappointed. . . . In many places the surface was parched and cracked and would bear the weight of a man, but in other places it was still boggy" (Williams 1874-T57 R39). Here again "soft Mud" suggests marl, but "parched and cracked" could be either marl or peat or a mixture.

The term "boggy" is more suggestive of peat or at least a marl-peat mixture.

In addition to Jackson's and Williams's Government Land Office surveys, a portion of the Perrine Marl Marsh was surveyed by William J. Krome in 1902 and 1903. Krome, working for Henry Flagler's Florida East Coast (FEC) railroad, investigated a potential route from Miami to Key West, first traversing the Everglades Keys and the Marl Transverse Glades (chapter 10) to the present-day location of Florida City and from there west across Taylor Slough and Long Pine Key, and then southwest through the Perrine Marl Marsh to Flamingo and Cape Sable. Krome's work is documented in a map (Krome 1903m; see also 1903g, h), a lengthy series of letters written in the field to FEC "Roadmaster" E. Ben Carter (Krome 1903l), and an engineering article summarizing his experiences (Krome 1904). Although Krome provides few measurements of water depths, his detailed documentation of his location (e.g., Krome 1903d, e, g) at a time when few or no roads were present south of Cutler (Krome 1903k) make his observations valuable. Krome was aware of Willoughby's 1898 trip and considered Willoughby's map (Willoughby 1898b) to be accurate (1903f). Krome was initially skeptical that the government township surveys (1903a) of the area had truly been executed, presumably because of the difficulty of the terrain, but came to respect them, even naming one of his camps in surveyor Jackson's honor (Krome 1903k).

Krome's observations suggest that the Marl Transverse Glades seasonally formed a strong hydrologic connection between the Rockland and Perrine marl marshes: he reported water in the "prairies" (Marl Transverse Glades) as being "waist deep" (October; Krome 1902d), "very high" (December; Krome 1902i), and "very deep" (September; Krome 1903i). The portion of Taylor Slough between the end of the "Biscayne pineland" (Everglades Keys/Miami Rock Ridge) and Long Pine Key also seemed to vary seasonally but was apparently wet enough to support deeper peat. One of his assistants at some time during the dry season found "bad bog and very little water" (Krome 1902h), while Krome himself found a "very fair crossing" in December (Krome 1902j) and in mid-February "about 3 or 4 miles of water deep enough to float the [supply] boats where they can be hauled along quite well." (Krome 1903c).

The numerous "prairies" or "water gaps" (Krome 1903e) that broke Long Pine Key into "a series of large pine islands and big hammocks" (Krome 1903a) apparently formed an approximately north-south hydrological connection between the Rockland and Perrine marl marshes. Krome did not report water depths but did map the numerous prairies in detail (Krome 1903m; see also 1902h, k; 1903d, e), contrasting the "rough and pot-holey pine woods" (pinnacle rock) with the "good depth of rich soil overlying the rock" in the hammocks and with the "fine vegetable land" of the prairies (Krome 1903d). Krome noted that the north-south prairies were similar to a one-mile-wide strip of "first class prairie" running east-west along the south side of Long Pine Key. Craighead (1974) maps the same transverse glades ("prairies") within Long Pine Key and notes postdrainage loss of the top soil of 4 to 12 inches (10–30 cm) of sawgrass peat; an additional indication of once substantial water and of a north-to-south hydrological connection.

Krome took soundings of soil thickness along the proposed railroad route, from the west end of Long Pine Key to Flamingo, finding that the bedrock sloped more steeply downward toward the coast than did the soil: "As a result of these soundings I expect to find the rock covered with about 4 feet of muck at a point 4 miles from the Long [Pine] Key reef. . . . From there towards Flamingo I am confidant of finding the muck grow deeper at the rate of about 8" to the mile until at Flamingo the rock is about 10 feet beneath the surface" (Krome 1903j). Craighead (1964, p. 4) notes a similar downward slope to the bedrock and corresponding southward increase in soil thickness, as does Gleason (1972, fig. 18). Whether "muck" refers to organic peat soil or to marl is not clear, but elsewhere Krome does distinguish between them and gives a general geographic distribution: "The whole interior of the peninsula is prairie. To the north it is black muck but along the Bay of Florida is a marl" (Krome 1903h). Krome also notes that at some times and places a 90-pound (40 kg) backpack of supplies could be carried, but in other places "pot-holes of deep muck [made] twenty-five pounds [11 kg] a wearying load" (Krome 1904, p. 11). Elsewhere Krome notes that "in many places the muck is almost bottomless and progress was very slow, 3 miles a day being all we could make at times" (Krome 1903b). Both peat and marl soil seem to have been present in the predrainage Perrine Marl Marsh, but the relative prevalence is not clear. Under current conditions, peat still forms around sawgrass growing within local depressions.

Pinnacle rock or some other form of bedrock was present underneath the soils, whether peat or marl. Curiously, pinnacle rock was mentioned only twice in all of the township survey field notes, and only in one township, T59 R38: "All wet Marsh with rocks, cracks and holes. A very bad [survey] line," "All wet boggy Marsh rocks under water" (Williams 1874-T59 R38). Since pinnacle rock is very obvious when exposed (fig. 9.3) and certainly would have been an impediment to surveying, its mention in only one township suggests that any pinnacle rock present in other surveyed townships of the Perrine Marl Marsh was covered by either marl or peat. Lack of reference to exposed pinnacle rock in Krome and in nearly all of the township surveys suggests that the extent of exposed pinnacle rock has increased since the onset of anthropogenic drainage. Although such an increase is not certain, if it has occurred, it would likely reflect postdrainage oxidation of a predrainage covering layer of peat.

Like the various soil terms used by the surveyors, the terms "Prairie" and "Marsh" frequently found in the surveyors notes similarly provide hints but incomplete information regarding the predrainage vegetation. Both were used in combination with sawgrass but do not provide a clear indication of either sawgrass density or height. It is worth noting that in the older literature, the terms "prairie" or "wet prairie" were applied not just to the flanks of Shark River Slough but also to areas of clear sawgrass and sawgrass peat (e.g., Anonymous 1841, p. 245; Florida Department of Agriculture 1911, pp. 44–45). The absence of references to tall sawgrass in the Perrine Marl Marsh landscape, something that was mentioned elsewhere, may indicate that sawgrass here was not tall. Harshberger describes the area as a "vast sawgrass marsh" (Harshberger 1914, pp. 178–180) but does not specify height.

Frank Egler, observing in the 1940s and 1950s, provides an indication that sawgrass density may have decreased, and frequency of other graminoid species increased, as water tables were lowered by drainage: "during the eight years of observation, the quantity of upland grasses among the sawgrass has markedly increased." (Egler 1952, p. 228). Egler ascribes this "change toward xericism" to "wholesale lowering of the water tables," which he in turn relates not only to local canal drainage but also to "the canal-building in the northern Everglades which removes the water that used to sweep southward" (Egler 1952, p. 228). (Totch Brown, who grew up in the 10,000 Islands in the 1930s, makes a strikingly similar comment: "The [upstream] damming [to drain land in the northern Everglades] also dried up the southern Everglades, and it had a terrible effect on the whole Everglades environment" [Brown 1993, p. 240].) Egler's observations suggest that both local and regional drainage (chapters 2, 3) significantly affected the Perrine Marl Marsh. His observation of xeric changes lend cre-

Figure 9.3. Exposed pinnacle rock with young stalks of sawgrass in the foreground in Perrine Marl Marsh, photographed in May 1918, after the onset of Everglades drainage. The location was 11 miles southwest of Royal Palm Hammock in Dade County (T59 R36?). Photograph from the J. K. Small Collection, Florida Photographic Collection, used by permission of the Florida State Archives.

dence to the idea that the terms "marsh" and "boggy," prevalent in the predrainage survey field notes, likely reflect a predrainage landscape of deeper water, more sawgrass, and less nonsawgrass graminoids.

In addition to sawgrass, trees were a distinguishing aspect of the Perrine Marl Marsh, occurring either in monospecific areas (cypress or red mangrove) or as hammocks or tree islands: "For the distance of twenty-odd miles southwest of Royal Palm Hammock and Long Key there is almost flat prairie. Its principal herbaceous growth is saw-grass (*Mariscus*) [*Cladium*]. There is often a scattered growth of pond-cypress (*Taxodium*) and of red-mangrove (*Rhizophora*) and scattered dense hammocks of a mixture of temperate and tropical trees" (Small 1923, p. 237). "The prairie landscape of the Everglades embellished with the unique cypress growth or the hammock islands or with both, especially with the rising or the setting sun, is both unique and enchanting" (Small 1934, p. 267).

Dwarf but very old cypress formed low stands in parts of the Perrine Marl Marsh (fig. 9.4). Traveling a few miles southwest from Royal Palm Hammock (fig. 9.1), Small states that "isolated, scrawny cypress trees appear in the landscape. . . . These pygmy cypress forests have the aspect of great age. . . . Where the marl is very thin there are forests of thousands of trees of this peculiar form of the pond-cypress varying from a foot to six or eight feet tall" (Small 1934, p. 261)

Mangroves were present as a continuous fringe along the southern and eastern edges of most if not all of the Perrine Marl Marsh. This fringe extended inland along the banks of creeks, as recorded by William Krome: "The mangrove is dense along the waterways but it seldom extends over a mile back and is often much narrower," presumably meaning that the mangroves usually extended much less than a mile upstream (Krome 1903h). Migration of these mangroves landward could occur in response to rising sea level, to reduced freshwater flows in the creeks as suggested by Glenn Simmons (Simmons and Ogden 1998, p. 7), or to both. Further analysis might yield a relationship between postdrainage increases in upstream mangrove extent and decreased freshwater flows.

Mangroves also occurred separately from the coastal and creek bank populations. These were small stature red mangroves (*Rhizophora mangle*) occurring as areas of low density, scattering individuals (fig. 9.5): "multitudes of red mangrove bushes, a few feet tall" (Harper 1927, pp. 182–183). Harper reported that these bush-sized mangroves were present in 1910 in the vicinity of Paradise Key (fig. 9.1), about 12 miles (20 km) from Florida Bay (Harper 1910b, p. 154). Red mangrove bushes are still present today, especially in Taylor Slough (William Loftus, pers. comm., 2009).

Hammocks or tree islands (figs. 9.5 and 9.7) were numerous and very obviously distinct from the surrounding sawgrass marsh: "[The Perrine Marl Marsh] is dotted with thousands of hammock islands. These islands

Figure 9.4. "Everglade prairie with diminutive pond cypress" in Perrine Marl Marsh, photographed near Royal Palm Hammock in December 1919, after the onset of Everglades drainage. Photograph from the J. K. Small Collection, Florida Photographic Collection, used by permission of the Florida State Archives.

Figure 9.5. "Diminutive mangroves in foreground and a hammock with *Paurotis* [*Acoelorraphe*] *wrightii* in the background" in Perrine Marl Marsh, photographed near Royal Palm Hammock in December 1919, after the onset of Everglades drainage. Photograph from the J. K. Small Collection, Florida Photographic Collection, used by permission of the Florida State Archives.

Figure 9.6. "Prairie and small hammocks," with cabbage palms and slash pines(?) protruding above hammocks in Perrine Marl Marsh, photographed southwest of Royal Palm Hammock in 1919(?), after the onset of Everglades drainage. Photograph from the J. K. Small Collection, Florida Photographic Collection, used by permission of the Florida State Archives.

represent the only spots in the Everglades that have not been seriously fireswept in prehistoric and historic times. Contrary to the popular opinion, they are not necessarily low spots. These are more frequently circular or irregular elevations of deeply eroded rock, often with a kind of moat which, together with the usually wet or moist accumulated humus in erosion holes and a consequent vigorous growth of broad-leaved shrubs and trees, served to stop the prairie fires that swept the Everglades through the ages" (Small 1934, p. 267). Harshberger (1914, p. 180) also noted that these "hammock[s] developed in a soil rich in organic matter," indicative of moist or wet conditions. The cabbage palms (*Sabal palmetto*), bays (likely *Persea borbonia, Ilex cassine*, and *Magnolia virginiana*), and myrtle (*Myrica cerifera*) reported in the surveyor notes—for example, "wet sawgrass prairie or Marsh interspersed with small Hammocks of Bay, Myrtle and Mangrove bushes" (table A.6, appendix A)—were probably found on these hammocks. Postdrainage imagery indicates that many of these islands are rather circular (fig. 9.7) with bays growing along their margins. Many have been burned out during fires that occurred since the 1980s (William Loftus, pers. comm., 2009), presumably an additional indication of drier, drained conditions having removed the originally protective moisture.

Although the landscape had changed by the 1940s, Egler noted the contrast between hammocks and the surrounding marsh: "Belt 3 [of the Perrine Marl Marsh] in general is a region of absolutely flat sawgrass swamp, dotted with impenetrable evergreen round hammocks that rise like cliffs from a sea" (Egler 1952, p. 233). The hammocks were found on elevated locations, whether due to rises in the bedrock or rises formed by organic material filling a depression: "Circular depressions and rises [in the level plain of Miami oolite] of approximately only 30 cms., and up to 150 m. or more in diameter, occur sporadically. On these depressions and rises are developed the hammocks" (Egler 1952, p. 234). J. K. Small's observation that the hammocks "flooded part of each year" (Small 1923, p. 237), combined with Egler's indication of 30 cm (1 foot) of elevation, provides an indication of water depths within the marsh. "Tails" extended downstream from the hammocks in the general direction of water flow (fig. 9.7): "Each hammock has its tail, sometimes fragmented and appearing like isolated rocks from out of the waters" (Egler 1952, p. 233).

Creeks were mentioned fairly frequently by the surveyors, and some were noted as being fresh water. William Krome, however, while searching for a maritime supply route as an alternative to the overland route through the Everglades Keys, found the creeks discharging into southern Biscayne Bay wanting, at least for transportation purposes: "Dusenbury, E. S. Frederick and myself . . . set out in a sail boat for Chi's Cut [a coastal creek in southern Township 57, northern T58; present-day Turkey Point cooling ponds]. We spent three days down there and thoroughly explored not only the cut but every other stream of any size at all that comes into the bay between Black Creek and the head of Card Sound. Chi's Cut is the only waterway that amounts to anything at all" (Krome 1902g). Elsewhere Krome notes that "the country south of Long [Pine] Key and east of Whitewater Bay consist[ed] of wet prairies, buttonwood [*Conocarpus erectus*] and mangrove swamps and countless lagoons and salt water creeks" (Krome 1903h). We note that these were saltwater creeks only to a railroad engineer and only at their lower end. For example, Taylor River runs fresh at the upper end for most of the year, as do small creeks east of it.

A letter from F. J. Pepper, agent of the Model Land Company, to Mr. F. S. Morse, gives a vivid impression of a trip into this landscape. The route can be followed on figure 9.1 and plate 4. Even in 1912, well after the onset of drainage and reductions in "the water that used to sweep southward," depths of 4–15 inches (10–40 cm) of water were still present at the *end* of the dry season:

> [On April 25, 1912, Mr. Goodson and I] left Camp Jackson at 5:30 with a pocket lunch and a machete: the distance [south] from the camp to the NE corner of the township [T59 R37] is about 4¾ miles along the range line. After going [south] over very rough pine land [i.e., the southwestern tip of the Everglades Keys] and arms of shallow rocky prairies for a distance of about one-half mile we struck the large Everglade prairie [i.e., the Perrine Marl Marsh] which continued to the corner of the township [T59 R37] unbroken except for an occasional cypress key; all of this prairie along the range line was covered with from four to fifteen inches of water with several deep sloughs running through [smaller side tributary sloughs draining westward into Taylor Slough; plate 4]; for a mile and half the walking was very difficult the water and mud being above our knees and in many places up to our waists [c.f., January 1902: "lots of mud and water now" Krome 1903a]. Most of this prairie over which we passed before coming to the township corner appeared to be of a good quality of marl running from six inches to two feet in depth; the lowest parts of this prairie is covered with a growth of saw-grass and the

Figure 9.7. Aerial photomosaic of eastern Perrine Marl Marsh in 1938, after the onset of Everglades drainage. Florida City, U.S. Highway 1, and Card Sound Road are visible. Heads and tails of tree islands (dark) illustrate landscape directionality. Reproduced from index for Dade County (U.S. Department of Agriculture 1938).

higher ridges with scattering mangrove bushes, with small cypress keys dotted here and there.

About a quarter of a mile north of the [T59 R37] township corner we found the soil beginning to get shallow, being about six inches deep at the corner. For a mile southeast of the corner we passed through scattering mangrove and a thin growth of a kind of wire grass; Mr. Goodson clim[b]ed a pine tree and from a height of about twenty feet as far as he could see with the naked eye west and south over the township it all looked the same, being covered with numerous small cypress keys and the stunted growth of scattering cypress. A great many of the keys near the northern line of the township contained scattering pine trees, and at a point about four or five miles west of the NE corner [of T59 R37] there appeared to be a pine island of considerable size penetrating down into the township. [Unclear what was seen here. The location indicated, 4–5 miles west extending southward into the north edge of T59 R37, would be exactly the location of Taylor Slough. Tree islands within Taylor Slough would collectively give a wooded appearance but seem unlikely to have been pines. William Loftus has suggested that it might have been Buzzard's Roost, a large island of large cypress that might have been mistaken, in the distance, for pine (William Loftus, pers. comm., 2009)]

As far as we went out into the township it was hard to find over four inches of soil except in an occasional hole which appeared to be filled mostly with partly decayed vegetation. On a great deal of it there was not over two inches of light marl underlaid with flat surfaced rock.

The soil was so shallow that we did not consider it necessary to take a separate sample of the top and one of the bottom. (Pepper 1912)

Collectively, historical sources give the impression that the predrainage Perrine Marl Marsh was somewhat drier than the Ochopee and Rockland marl marshes. This was perhaps because the latter two likely received flows directly from the Everglades farther upstream, whereas the Perrine Marl Marsh was more isolated by the intervening Everglades Keys. Marl was clearly present throughout the Perrine Marl Marsh; the degree of covering peat is less clear, with suggestions of peat in some locations but not in others. Sawgrass was likely present throughout, probably not dense and tall, but almost certainly not as sparse as the sawgrass stands seen after drainage lowered water tables. Other graminoid species were likely present but much less dominant than at present.

The species of trees present in this landscape and also their configurations (the "dwarf cypress forests," the areas of scattering, small stature mangroves, and the directional hammocks) all appear to have been present prior to drainage. It is certainly possible (but difficult to establish unambiguously) that their exact position may have shifted under the influence of extensive drainage, both of the Everglades and of the Everglades Keys.

The hydrology of the Perrine Marl Marsh has certainly been altered by regional and local drainage. Accordingly, postdrainage (post-1880s) observations and photographs cannot be assumed to be fully representative of predrainage conditions.

Hydrology

We estimated the long-term average annual hydroperiod in the Perrine Marl Marsh as eight to nine months and the average annual water levels as ranging from 1 foot below ground (-30 cm) at the end of the dry season to 1.5 feet (45 cm) above ground at the end of the wet season. The scarcity of predrainage water depth observations in this landscape means that these estimates may have been somewhat more influenced by postdrainage observations. Accordingly, this assessment may underestimate the original wetness of this landscape.

Hindcasting

The landscape extent, with the exception of the mangrove interface, was drawn as the extent of Perrine Marl delineated on Jones et al. (1948a). With the exception of the landscape's downstream border with the mangrove fringe, the extent of the Perrine Marl Marsh was assumed to have remained constant and therefore did not need hindcasting. We defined the southern extent of the Perrine Marl Marsh landscape as the transition from "Glade Land with Scattered Hammocks" to mangrove forest (U.S. Coast and Geodetic Survey 1937c). Coastal maps (U.S. Coast and Geodetic Survey 1928a, b, c; 1937a, b, c) were used to locate this boundary, which therefore reflects its position as of 1928, somewhat seaward of the present mangrove-marsh interface.

10

Landscapes Bordering the Everglades

The landscapes bordering the Everglades include the Eastern Flatwoods, the Everglades Keys, and the Marl Transverse Glades to the east and the Western Flatwoods and Big Cypress to the west. The predrainage state of these landscapes is hindcasted and described in this chapter. Mangroves also bordered much of the Everglades to the south but are not addressed in this work.

Eastern Flatwoods

Much of the Eastern Flatwoods landscape no longer exists as it has been drained and converted to urban/suburban developments or agriculture. Extensive drainage beginning in 1915 lowered water tables about 6 feet (2 m; Thomas 1970, pp. 68, 78, 85; Mock, Roos, and Associates 1990). In its present state, it is hard to imagine that this landscape was once so wet that it was navigable by boat from the Jupiter Inlet to the Hillsboro Inlet and that during the rainy season canoes could regularly pass from the Atlantic Coast west to the Everglades (Ives 1856a, pp. 16–17; Pierce 1970, p. 165).

The Eastern Flatwoods landscape bordered the eastern side of the Everglades, from Lake Okeechobee south to approximately the New River (plate 5). This landscape occupied a sandy shelf, from 5 to 25 miles (8–40 km) wide, between the Everglades and the narrow, slightly elevated Atlantic Coastal Ridge. Particularly in Broward County, the separation between the Eastern Flatwoods and the Atlantic Coastal Ridge was indistinct.

Although this landscape is now often referred to as "pine flatwoods," pines (primarily the slash pine, *Pinus elliottii*) were originally only one of many components. Before drainage the Eastern Flatwoods landscape was a mosaic of upland plants but also water-tolerant plants. The scattered slightly elevated areas and depressions combined with plentiful surface water to create diversity within this otherwise generally flat landscape (plates 39, 38, upper right). The upland plant communities, which included palmettos (*Sabal palmetto* and *Serenoa repens*), mixed grasses, and hardwoods, occupied the higher ground, while the various hollows and low points were home to water-loving and water-tolerant plant communities such as those found in wet prairies, marshes, and ponds. Additionally, open pinelands and cypress domes or strands were also a component of the Eastern Flatwoods landscape.

The wet prairies scattered throughout the Eastern Flatwoods (and analogous Western Flatwoods) were abundant. In contrast to the postdrainage wet prairies reported within the Everglades after about 1940, wet prairies were originally present only in the open areas of the predrainage flatwoods landscapes. They were sand-bottomed, seasonal wetlands with characteristic plant species, often including a shrub, sand cypress (*Hypericum fasciculatum*), along with some combination of maiden cane, spikerush, and beakrush. Daniel Austin (1976, p. 232) noted that sand cypress (also called marsh St. John's wort), not cypress trees (*Taxodium* spp.), is what gave wet prairies the historical common name of "cypress bottoms."

The confusing application of the same term, "wet prairie," to the peat-based areas of maidencane, spikerush, and beakrush found growing in former sloughs of the Everglades began only after drainage. Although some of the species overlap, the substrates are very different. The so-called wet prairies within the Everglades are an artifact of canal drainage, reflecting Everglades conditions that were sufficiently dry to resemble the shorter hydroperiods of the flatwoods wet prairies. (See also the hindcasting section of the Ridge and Slough landscape, chapter 8).

In addition to being a mosaic at the local scale, the Eastern Flatwoods varied regionally, differing somewhat north and south of the Hillsboro Canal, generally between Palm Beach and Broward counties. In Palm Beach County the landscape was wider and flatter and dominated by pines rather than cypress. Cypress was present mostly in numerous scattered cypress domes. A peripheral cypress swamp, running north to south, formed the eastern border of the Everglades in Palm Beach County (figs. 1.6, 1.7). The following passage from 1907 describes the southern portion of this strand on the border of

Townships 47 and 48 in Range 41: "Apr. 3. . . . Where this line was run out from the pine timber, the pine is bordered by open scrub cypress, and small cypress heads; then there are strands of cypress and sawgrass which run out to the Glades, these strands being very thick and hard to get through. The cypress at one time has been continuous, but for some reason, probably fire, a great deal of it has died. The cypress island where we set the bench mark is in the edge of open Glades. The open Glades are flags, lillies, water grass and water" (Stewart 1907, p. 58).

On the eastern side a freshwater marsh 35 miles (56 km) long and 1–2 miles (2–3 km) wide occurred between the Eastern Flatwoods and the Atlantic Coastal Ridge. No name was found in the historic literature for this marsh. The label "freshwater ponds" (plate 5) was adopted from an early nineteenth-century map of Florida (Vignoles 1823a).

Early accounts of the Eastern Flatwoods report plentiful game, including deer, bear, panther, and wading birds (Pierce 1970, p. 210; Henshall 1884, p. 60; Dupuis 1954b, p. 133). The Corbett Wildlife Management Area is a remnant of this landscape; however, this area is now almost certainly drier than it was under predrainage conditions. Lamar Johnson, who began observing the Everglades in the 1920s, suggests that the Eastern Flatwoods, rather than the Everglades, may have been the primary habitat for deer: "The recurring emotional problem of the deer herd in the Everglades is a case in point. Few remember that the slough and sawgrass Everglades is not the natural habitat of the deer. A half century ago [i.e., the 1920s] a deer track could not be found beyond the cypress and wet prairie margins of the Everglades. The deer only went into the deep Everglades as a result of the press of civilization and increased hunting. I know from personal observation that there were no deer deep in the Everglades forty-five or fifty years ago" (Johnson 1974, p. 183).

South of Cypress Creek in Broward County, the Eastern Flatwoods and the Atlantic Coastal Ridge intergraded as the ridge became wider and lower. Farther south, combined upland areas of the Eastern Flatwoods and the coastal ridge became sufficiently dissected by the Peat Transverse Glades to form an archipelago of isolated, high ground islands (plates 40–43).

Hydrology

The Eastern Flatwoods landscape was only slightly elevated above the Everglades (fig. 4.6). When water levels rose, there were times when the Everglades and the flatwoods were continuously connected by water. According to George W. Potter, president of the Dade County Bank, even as late as 1907 it was sometimes possible to canoe directly from the Everglades into and across the Eastern Flatwoods: "West of Palm Beach, there is a ridge near [the] edge of Glades, but at high water Glades water comes over into Clear Lake and the Indians bring their canoes into Clear Lake" (Potter 1907, p. 44).

Fifty years earlier Ives had also described a navigable trail across the Eastern Flatwoods, most likely south of the Clear Lake trail: "Westward, a trail runs from the village [Chachi's Village, which is thought to have been located near Dreher Park, between Southern Boulevard and Summit Boulevard in West Palm Beach] to the swamp bordering the Everglades, the eastern boundary of the former being about seven miles distant. . . . There were indications that it had been frequently traversed in boats during high water" (Ives 1856a, p. 17).

The proximity of predrainage Everglades water levels to the Eastern Flatwoods land surface was recognized by early settlers as a drainage issue. J. T. Stewart noted that some of those opposed to large-scale drainage of the Everglades did, however, favor lowering of the water sufficiently "to prevent the flooding of the coast" (Stewart 1907, p. 56).

Natural surface water drainage of the Palm Beach County portion of the Eastern Flatwoods was poor. The Loxahatchee Slough (fig. 10.1), a wetland connection between the Everglades and the North Fork of the Jupiter (now Loxahatchee) River, may have provided some local drainage. The Loxahatchee Slough appears to have been sufficiently level that the direction of flow varied depending upon local rainfall (fig. 11.11). The coastal freshwater ponds, draining into the north prong of the Hillsboro River (M. A. Williams 1870-T47 R43), functioned as a collector along the eastern edge of the Palm Beach County portion of the landscape. Other than the Jupiter (now Loxahatchee) River, no coastal rivers were present. Groundwater flow would have had a major contribution to the water budget in this area.

In contrast, natural surface water drainage in the southern portion of the Eastern Flatwoods (i.e., Broward County) was very evident, extending eastward from the Everglades through to the Atlantic Coast. Pines were less prevalent in Broward County than in Palm Beach County. The landscape in Broward County was instead dominated by directional cypress strands following the eastward drainage from the Everglades (plate 42). Patches of wet prairie and pineland were scattered throughout the cypress strands. These strands drained water from the Everglades into the numerous rivers piercing the coastal ridge: Hillsboro Creek, Hills-

Figure 10.1. "Cypress swamp and marsh vegetation with open lagoons covered with water lilies, August 11, 1911, west of West Palm Beach. Original" (Harshberger 1914). This photograph was probably taken within a portion of Loxahatchee Slough (possibly within the present "Grassy Waters Preserve," a part of the West Palm Beach surface water supply), prior to the formation of the Lake Worth Drainage District. Together with the West Palm Beach Canal, the Lake Worth Drainage District lowered regional water tables by approximately 6 feet (Austin et al. 1977). Reproduced from Harshberger (1914).

boro River, Cypress Creek, the north and south forks of the Middle River, and the north and south forks of the New River (figs. 1.6, 1.7). The eastward drainage through these bordering uplands is reflected in the eastward and southeastward flow (figs. 11.11, 11.12) in the adjacent areas of the Everglades (Townships 48–49 Ranges 40–41). Together with the Peat Transverse Glades farther south, the cypress strands and associated coastal rivers of the southern portion of the Eastern Flatwoods formed an important part of the natural eastern outflow of waters discharging into the Atlantic Ocean from the Everglades.

The topographic variability formerly present in the Eastern Flatwoods makes it difficult to specify average water depths and hydroperiods for the entirety of the landscape. We used two soil groups associated with the Eastern Flatwoods, the wet sands and upland sands (appendix O), partially to capture the variations that were formerly present. Long-term water depths for the wet sands typically ranged from an average annual high of 1 to 2 feet (30–60 cm) above ground to an average annual low of 1 to 2 feet (-30–60 cm) below ground, with a hydroperiod of five to six months. Analogous values for the upland sands found within the Eastern Flatwoods area were zero to 1 foot (0–15 cm) above ground and 2 to 3 feet (60–90 cm) below ground, with a hydroperiod of up to two months.

Hindcasting

The predrainage extent of the Eastern Flatwoods was determined primarily by the predrainage extent of the bordering Sawgrass Plains and Ridge and Slough landscapes. The eastern border with the Atlantic Coastal Ridge was not delineated precisely for this study but can be determined from the township survey plat maps. The southern border of the landscape, where it intergraded with the Atlantic Coastal Ridge landscape, was somewhat indeterminate. Based on Matson and Sanford (1913b, p. 50), we included the area southward to the New River.

Atlantic Coastal Ridge and Coastal Rivers

The highest ground east of the Everglades was (and is currently) known as the Atlantic Coastal Ridge. The topographic highs of the ridge along with some of the rivers remain at present, but almost all of the original vegetation has been lost, as have the low areas between islands of the ridge and the Peat Transverse Glades. As suggested by the name, the east edge of the ridge sloped down to either the coastal beaches or the shore of Lake Worth. The ridge extended from the Jupiter Inlet south to the Miami River where it merged with the Everglades Keys. Based on the presence of sand pine (formerly called spruce pine, *Pinus clausa*) vegetation, the Atlantic Coastal Ridge was sometimes referred to as the "Spruce Pine Barren." As the only area not subject to flooding, it was the first area to be settled around the turn of the nineteenth century. Austin et al. (1977, p. 348) provide an ecological study of the plants from this landscape.

The Atlantic Coastal Ridge has two parts: a northern and a southern portion. The northern portion in Palm Beach County extends south to the Hillsboro River (now Hillsboro Canal), and continues south to Cypress Creek to form a single, distinct ridge, varying in width between 0.5 and 1.5 miles (0.8 and 2.4 km). The west side of this northern portion of the ridge merged with the Eastern Flatwoods landscape. With the exception of the Hillsboro River, no coastal rivers cut through this northern portion of the ridge.

In Broward County the southern portion of the Atlantic Coastal Ridge was pierced by a number of coastal rivers and also by the Peat Transverse Glades, forming a series of islands of high ground (plates 40–43; see also White 1970; Finkl 1994; Knetsch 1999). The collection of islands formed a band about 5 or 6 miles (8 or 10 km) wide. In this portion, the strip of Eastern Flatwoods was absent; here the Everglades bordered directly on the islands of the Atlantic Coastal Ridge.

Hydrology

Historical information about this landscape is consistent and clear-cut. Three aspects of this landscape are particularly relevant to estimating the hydrology of the predrainage Everglades: (1) the numerous short coastal rivers all originated in the Everglades; (2) currents in those rivers were strong, at least four sets of rapids or falls were present, and several rivers were strong enough to drive water wheels; and (3) the coastal rivers flowed throughout the year.

Predrainage photos of two of the coastal rivers in Miami-Dade County, Little River (now the NW 79th Street Canal) and Arch Creek, are provided in figures 10.2 and 10.3. The three points noted previously, made by different observers over numerous years, provide strong evi-

Figure 10.2. Little River in the 1890s. The Little River was described in January 1841 during a canoe expedition: "After passing up the bay seven miles, they entered the mouth of Little River, a tortuous and extremely rapid outlet from the Everglades, where they struggled against the current until after midnight, when they reached their first resting-place—the site of an old plantation—where they landed" (Brooks 1880, p. 246). Non-native coconut trees shown indicate a place of settlement. Photograph by Ralph M. Munroe, Ralph M. Munroe Collection, used by permission of the Historical Museum of Southern Florida.

Figure 10.3. Arch Creek in the 1890s. "The most romantic scene of all was the view from the natural bridge looking up the creek as it continued on to the Everglades, two miles beyond" (Parks 1977). In 1858 George Lewis and Robert Fletcher spent a year attempting to construct a water-powered coontie mill using a ditch diverted from Arch Creek (Dietrich 1987). MacKay and Blake (1839) indicate "Rapids" at the head of Arch Creek. Photograph by Ralph M. Munroe, Ralph M. Munroe Collection, used by permission of the Historical Museum of Southern Florida.

dence that water depths in the predrainage Everglades were sufficiently high, even at the end of the dry season, to maintain substantial year-round flows in the coastal rivers. Water tables on the Atlantic Coastal Ridge uplands were always below ground.

Hindcasting

The predrainage vegetation of the Atlantic Coastal Ridge has been almost completely lost to urban development, so present vegetation is not helpful for determining the spatial extent of the landscape. However, township survey maps, mostly from the 1870s, clearly delineate this landscape.

Peat Transverse Glades

The Peat Transverse Glades were eastward extensions of the Ridge and Slough landscape through the Atlantic Coastal Ridge to the coastal rivers. Those "arms" continued out from the main body of the Everglades with close to the same elevation as that of the adjoining Ridge and Slough landscape. This is consistent with Buckingham Smith's description in 1847: "Inside the basin, near the heads of the rivers of the eastern coast, there are covers or indentations in the shore of the Glades about 2 miles in depth. The nearest point touches the margin of the rim where the waters of the Glades approach the heads of the rivers, and where these waters are about a foot in depth. There are within the coves, channels, converging to such a point, and in which the waters are from 1½ to 2½ feet in depth. The shallow places between them are covered with mud and rank saw grass" (B. Smith 1848a, p. 49).

The network of Peat Transverse Glades, in combination with any groundwater outflow, provided sufficient outflow to prevent significant ponding west of the coastal ridge. As the Peat Transverse Glades cut through the Atlantic Coastal Ridge, they maintained an elevation similar to that of the Ridge and Slough landscape. No historical evidence was found of water depths in the Peat Transverse Glades being deeper than those that would be typical for the Ridge and Slough landscape, though some evidence indicates that water depths were, at times, slightly shallower. The portion of the Peat Transverse

Glades landscape adjoining the coastal ridge differed from the Ridge and Slough farther west mainly because the presence of an underlying sand wedge (fig. 3.14) meant that the overlying peat thinned as it neared the coastal ridge.

Between Fort Lauderdale and Miami (Townships 51–53) a network of low, shallow valleys extended from the Everglades basin to the Atlantic shore (plate 43; Townships 51–52 Range 42). Those valleys were old drainage courses cut through the limestone of the Atlantic Coastal Ridge before the Everglades were formed. In the 1800s, prior to drainage, peat elevations and water elevations in the Everglades were sufficient to allow water to flow out through those valleys. Predrainage Everglades water levels were, in fact, sufficient to promote sawgrass growth and the accumulation of from 0.5 foot to several feet of peat within the valleys. As a result, a series of miniature Everglades were formed: long, flat marshes of dense sawgrass, one-quarter to one mile in width, bordered by higher-lying areas of pineland or hammock. The valleys, the Peat Transverse Glades, were narrow lateral arms, through which areas of sawgrass, peat soil, and water flow extended continuously from the main body of the Everglades to the East Coast. The presence of elongated tree islands in some of the Peat Transverse Glades gave them a resemblance to the Ridge and Slough landscape of the main Everglades. These features were recognized as islands of "myrtle and bay" on the land use and land cover map of Jones et al. (1948a). One such island is shown on plate 43 in Township 51, straddling the border between Ranges 41 and 42 (Sections 25 and 30, respectively). At least until 1990, despite intensive residential development, the eastern portion of this island (Range 42, Section 30) was still visible on aerial photographs as a stand of trees, likely bay trees.

Hydrology

Water flowed from the Everglades through the marshes of the Peat Transverse Glades in a combination of sheet flow and deeper water channel flow. In some of the transverse glades the channels coalesced to form the headwaters of short coastal rivers (fig. 3.18). The former presence of Seminole settlements on the edge of the uplands bordering some of the Peat Transverse Glades suggests that water depths were high enough during much of the year to permit navigation by dugout canoe between the settlements. Some examples and locations of such settlements include Big City Island in Township 51 Range 41 (plate 43); Aleck's Town and Tiger Tail's Town in Township 51 Range 41, Section 35 (see M. A. Williams 1870-T52 R41, Section 2); and Little Tiger's Town (Williams 1870-T52 R41, Section 23). See also Carr and Steele (1991); Carr and Sandler (1992); Carr et al. (1992); Carr (2002); Callsen (2008). Many of the coastal canals, in part, follow courses of the former Peat Transverse Glades.

Predrainage soils of the Peat Transverse Glades were mapped by Jones et al. (1948e) as Everglades peats over limestone or sand. The elevation of the upper surface of the peats (i.e., ground surface) in the transverse glades was likely similar to the peat elevations in the sawgrass ridges in the adjacent Ridge and Slough landscape. All of the peat elevations in the transverse glades would have been controlled by long-term average water levels in the main Everglades. With drainage, the original six inches to several feet of peat were lost to oxidation to such an extent that by the 1940s, many parts of the Peat Transverse Glades were mapped as sandy rather than as organic soils (fig. 3.18; see also chapter 3).

The importance of the Peat Transverse Glades to this study lies in their function as outlets for water flow from the Everglades and in their indication, through the presence of peat soils, of water elevations in the main Everglades. The peat depths of several feet present in the predrainage Peat Transverse Glades could only have been maintained if predrainage water elevations in the main Everglades were sufficient to flood those lateral glades during a substantial portion of most years.

Hindcasting

The Peat Transverse Glades were identified on the Jones et al. (1948a) soil map through synthesis of several sources of information such as township plat maps, bedrock topography, soil type, characteristic polygon shapes on the soil map, and adjacent upland soil types and vegetation. In some cases the original Everglades peats soils and sawgrass vegetation were still present in the 1940s and were mapped as such by Jones et al. (1948a). These Everglades peat polygons were left unaltered in the hindcast vegetation map (plate 5). In other cases the soils had oxidized under drainage, exposing the underlying sands, and the areas were mapped as Davie fine sands, Davie mucky fine sands, or in a few cases as sands of the wet sands group (table 1.2). We hindcasted these oxidized soils to the Everglades peats group.

Everglades Keys and Marl Transverse Glades

The Everglades Keys curve southwestward, separating from the Atlantic Coastal Ridge south of Township 55, and ending in Long Pine Key (fig. 9.1; now within Everglades National Park). Prior to drainage, when water

Figure 10.4. A marl transverse glade in the Everglades Keys. "Transverse marl prairie bordered on both sides by tall pine forest [Everglades Keys] near Princeton, Fla. August 18, 1911. Original" (Harshberger 1914). The vegetation in the glade still appears to be sawgrass in water, as it was sixty years earlier (Jackson 1847-T56 R39). The photograph was probably taken in "Caldwell Prairie," Township 56 Range 39, Section 10 (25.572, -80.426. Caldwell Prairie was replaced by the present C-102 canal. Reproduced from Harshberger (1914).

levels were considerably higher, the inundated Marl Transverse Glades (fig. 10.4) separated the high ground into a series of "islands," giving rise to the term "Everglades Keys" (plate 35).

The Everglades Keys and Marl Transverse Glades landscape correspond to the East Coast portion of the Slash-pine Formation, Prairie Formations (portions), Banana Hole Associations, and High Hammock Formation (Harshberger 1914); the Miami Limestone Region and transverse glades portion of Marl Prairie (Harper 1927); the Everglades Keys (Small 1929a); and the Miami Rock Ridge (J. H. Davis 1943a).

The archipelago of islands known as the Everglades Keys formed the southeastern portion of the rim of the Everglades basin. The northern portions of the rim were formed by the Atlantic Coastal Ridge and the Eastern Flatwoods. Without this rim, water and organic soils could not have accumulated within the Everglades basin. Elevations of the cuts, or low points, in this rim con-

Figure 10.5. Vegetation of the Everglades Keys. "Pine forest of *Pinus caribaea*, saw-palmettos and coontie, one mile west of Miami, December 27, 1910. Original" (Harshberger 1914). Note the exposed limestone rock. Within the pine forest were also a number of dense tropical hammocks. Similar vegetation can still be seen in the Larry and Penny Thompson Park in Miami-Dade County (T55 R39, Section 35) and on Long Pine Key within Everglades National Park. Reproduced from Harshberger (1914).

trolled the maximum possible elevations of water within the southern ("lower") Everglades. The Marl Transverse Glades were essentially eastward extensions of the Everglades, having similar soils, hydrology, and vegetation as the Rockland and Perrine marshes (fig. 9.1; plate 35).

Formed from a slightly elevated ridge of highly porous Miami oolitic limestone, the Everglades Keys differed from the Atlantic Coastal Ridge in surface soil, vegetation, and distance from the ocean. Except for some areas of sand, little soil was present, so pinnacle rock was visible at the surface. Subsequently, "rock plowing," a technique for pulverizing the surface rock, was invented and applied to form a thin layer of soil for agricultural use.

The Everglades Keys were vegetated by an open pine forest with an undergrowth of saw palmetto (*Serenoa repens*) and coontie (*Zamia integrifolia*; see fig. 10.5). A number of distinct tropical hammocks were present (e.g., Brickell, Castellow, Ross, and Nixon hammocks), containing tropical hardwoods, sinkholes, ferns, and a layer of rich organic soil. Bill Robertson noted that these large hammocks were always associated with [Marl] Transverse Glades (William Loftus, pers. comm., 2009).

Hydrology

The Everglades Keys, which eventually became known as the Miami Rock Ridge after about the 1940s, were punctuated by a series of shallow cuts, the Marl Transverse Glades. Figure 10.6 shows a 1938 aerial view of some of the Marl Transverse Glades, by this time sufficiently dry to allow farming. Leon Griswold's description, typical of predrainage times, is based on his explorations in 1896. At that time these transverse glades, here called "broad undulations," were still inundated during the wet season:

> The oolite [of the Everglades Keys] rises to an elevation of about twenty feet, and then maintains a generally level surface, with the exceptions of occasional broad undulations, having axes about east to west or north to south, and sinks. The hollows of the broad undulations are filled with water during the rainy season, so that pines do not grow in them; in the dry season grass grows luxuriantly, and some of the sinks are cultivated. Some soil has accumulated in these "prairies," whereas among the pines there is scarcely any soil to hide the jagged rock surface and loose fragments. (Griswold 1896, p. 53)

Griswold's description of "luxuriant grass" between the pines matches a 1911 photograph (fig. 10.4). Numerous observations from township survey field notes made in 1847 indicate that the "grass" was sawgrass (appendix A).

The cuts or outflows were dry enough in the dry season to walk on (Motte 1838b) and farm (Krome 1902f; Model Land Company 1918; Pepper 1918; Morse 1918; Ingraham 1918). In the wet season, the Marl Transverse Glades rewetted as water levels rose in the Everglades (Simmons and Ogden 1998). In at least some years they became quite deep, as in October 1902 during the Florida East Coast railroad survey of a potential route from Miami to Cape Sable and Key West: "The prairies [Marl Transverse Glades] are now about waist deep in water in most places and are steadily rising" (Krome 1902d). These transverse glades were likely in Township 56, west of Cutler (Krome 1902c, e). Water levels were still high in early December: "The water in the prairies is very high and as there is no road at all, it requires considerable work to get across them" (Krome 1902i). The same conditions were present the following year in the Marl Transverse Glades between Cutler and the present Florida City: "The prairies are very deep in water and it is next to impossible to haul any supplies from Cutler" (Krome 1903i).

Additional outflows took the form of offshore submarine springs (e.g., fig. 10.7), which are discussed in the following chapter.

Elevations

Plate 35 indicates that the elevation of most of the Marl Transverse Glades is between 8 and 9 feet (2.4 and 2.7 m) above sea level, with perhaps a slight ledge where their headwater ends met the Rockland Marl Marsh. Griswold, who explored south and west of Miami in 1896, described what appears to be the eastern edge of this same ledge: "The rivers [the Miami and adjacent rivers] have a well marked fall line on the limestone,—a few hundred yards of swift water just west of the pines, and representing the progress made by the rivers in cutting back from their mouths" (Griswold 1896, p. 53).

The elevations of the Marl Transverse Glades and of the ledge, combined with the knowledge that the Marl Transverse Glades flowed in the wet season, help establish likely wet season water elevations in Shark River Slough and in the bordering Marl Marshes (fig. 4.7). The typical annual high water level appears to have been approximately 9 feet (2.7 m) above sea level; the mean annual low was approximately 7 feet (2.1 m) above sea level. The high water elevation would bring Everglades water directly to the edge of the Everglades Keys, corresponding to the following description given by Griswold in 1896: "At the border of the Everglades the rough surface of the oolite becomes concealed for the most part beneath a mat of grass, and pines grow no further west. . . . In the zone of oscillation of water the grasses

Figure 10.6. Aerial photomosaic of Marl Transverse Glades (lighter; divided into agricultural fields) separating the individual Everglades Keys (higher, darker ground with pine forest), 1938. The elevation of most of the Marl Transverse Glades was about 8 feet above mean sea level, with a slightly higher ledge behind the headwaters (see plate 35). Prior to mechanized drainage lowering Everglade water levels, water in the Marl Transverse Glades would rise sufficiently each year to support vegetation described as "wet sawgrass prairies" or "sawgrass ponds" (see also fig. 1.1). Aerial photographs used for the mosaic were taken by the U.S. Department of Agriculture (1938).

Figure 10.7. An offshore freshwater spring in Biscayne Bay. The platform and hand pump were installed sometime after 1871. "The abundant spring water off the shore, while organically pure, and beautifully clear and tasteless, was somewhat hard" (Munroe and Gilpin 1930, pp. 118, 218). "All the [many freshwater springs in this part of South Florida] disappeared in the early twentieth century when the draining of the Everglades lowered the water table" (Parks 1977). Photograph by Ralph M. Munroe, Ralph M. Munroe Collection, used by permission of the Historical Museum of Southern Florida.

change in character from a small wiry variety near the pines to saw-grass six feet or more tall, and flags and cane growing in water" (p. 53). The high elevation of the water would also be consistent with observations by township surveyors of a distinct edge to the pinewoods (i.e., the Everglades Keys), beyond which the Everglades and standing water could be directly seen (as shown in Jackson 1847-T55 R39).

Western Flatwoods and Big Cypress

Pine flatwoods and the Big Cypress swamp formed the western border of the Everglades. Like the Eastern Flatwoods, the Western Flatwoods have a very slight regional slope. Local shallow depressions collected water during the wet season; elevated areas were drier. As a result of being more topographically diverse than the adjacent Everglades peatland, this landscape supported a mosaic of diverse vegetation communities including those found in pinelands, ponds, cypress heads, and long cypress strands. Generally, flatwoods were predominant to the north and cypress was predominant to the south, with some exceptions, such as Lostmans Pines. William Loftus notes that Lostmans Pines floods each wet season (pers. comm., 2009; see also C. R. Stone 1979, pp. 35–36). Duever et al. (1986) mapped the natural vegetation of this area using geographic information system (GIS) analysis combined with historical information.

We did not conduct an independent analysis of that area for this study. However we provide historical observations for this landscape (Historical Observations for chapter 10 on the DVD) that present a qualitative impression of water regimes in areas of the Western Flatwoods and Big Cypress that border the Everglades. The hydrologic separation between the Everglades and Big Cypress Swamp was not sharp, particularly during wet years. Historically, water from the northern part of Big Cypress flowed through Mullet Slough (Township 50) into the Everglades. A discussion of this and other Everglades inflows and outflows follows in chapter 11.

11

Predrainage Hydrology

The Everglades as a Whole

Like pieces in a puzzle, the hydro-ecological components described in the previous chapters fell into place, revealing an internally consistent picture of the predrainage Everglades. Although some pieces were less precise than others, the overall picture is apparent. This chapter recounts the piecing together of the evidence and what was learned from that process. It concludes with a detailed conceptual model of the predrainage landscapes and their hydrology.

Historical observations of the plants, animals, soils, fires, and water flow through the Everglades raise questions that must be addressed to reconstruct the natural hydrologic system, including the following:

- How deep were the waters of the Everglades within each landscape and overall?
- Apart from rainfall and evapotranspiration, where did water enter, where did it leave, and at what elevations did outflows occur?
- What were the regional water flow patterns and volumes?
- How did rainfall seasonality affect the flows, depths, and hydroperiods across the landscapes?
- How did water flows, depths, and seasonality influence landscape patterns?

These and related questions are addressed throughout this chapter through synthesis of the full suite of information contained in historical observations of the original Everglades. In this chapter the diverse collection of historical sources introduced in chapter 1, the postdrainage changes presented in chapters 2 and 3, and the individual predrainage landscapes discussed in chapters 5 to 10 are combined with a historically informed interpretation of the postdrainage, managed Everglades. Aerial photographs and satellite imagery, even though postdrainage, were particularly helpful in providing a spatial context for historical observations. Those sources taken together provided the foundation to build a conceptual model of predrainage Everglades hydrology, including flow directions, spatial patterns, water depths, and flows in and out of the Everglades. The time period covered by the conceptual model matches the rest of this study; the middle 1800s. It probably applies to the preceding one or two centuries as well. No attempt is made here to describe Everglades hydrology prior to that period.

During development of this conceptual model, it became apparent that relatively high levels of detail and precision were embedded within the historical and postdrainage data despite the absence of formal time series measurements of stages or flows. This detail exists within the hydrologic information implicit within the original vegetation communities, soils, and landforms and the assumed absence of large-scale hydrologic turbulent effects in these flat peatland landscapes.

One important assumption is that the vegetation and peat microtopography of the predrainage Everglades were in equilibrium with, or closely "tuned to" the original hydrologic driving forces. The concept of tuning, particularly of vegetation being adjusted to water depths, was suggested or recognized by a number of Everglades observers (e.g., Harshberger 1914, p. 61; J. H. Davis 1943a, pp. 116–117, 241; Egler 1952, p. 218; Loveless 1959a, p. 2; McPherson 1973a, p. 8; Alexander and Crook 1973, p. 18; Schortemeyer 1980, pp. 10–11; Ross, Ruiz, et al. 2006; and Ross, Sah, et al. 2006). The Sawgrass Plains is one such example of this "tuning" to hydrologic driving forces. The presence of an extensive plain dominated by sawgrass (*Cladium*) depended on a characteristic water regime that included an average annual water depth range from about 1.5 feet (45 cm) above ground surface to about 0.5 feet below (-15 cm; table 11.4); the period without standing water generally did not exceed two to three months. Firsthand reports from the time of first drainage (chapter 3) reinforce the impression of close tuning: within only a few years of lowering water levels, species requiring drier habitats such as willow, elder, and careless weed quickly spread into the sawgrass, and the sawgrass stands were reported to "weaken." The often repeated observation that "changes in elevation of only a few inches can have a great effect on the flora of Florida" (e.g., Craighead 1971, p. 46; McPherson 1973a, p. 8; Lodge 1994, p. 15) is another manifestation of this close tuning.

The tuning of this vegetation community to hydrologic conditions is particularly significant within the Everglades, considering that it occurred on a sloped land surface from which water was continually draining and evaporating. The predrainage water depths, and their annual rise and fall, developed from a balance between inflows and outflows; inflows from rainfall and Lake Okeechobee, and outflows as surface runoff and evapotranspiration. The presence of the particular water depths that favored sawgrass in the Sawgrass Plains thus depended, at a minimum, on the slope of the land, hydraulic resistance of the vegetation, atmospheric energy balance, and the duration and intensity of the rainy season. According to this tuning assumption, had any of those parameters been different, the resulting water depth regime would have favored a species other than sawgrass.

General characterizations of predrainage Everglades hydrology from previous studies (Parker et al. 1955; Parker 1974; S. M. Davis et al. 1994; Fennema et al. 1994) have highlighted the significance of outflows from Lake Okeechobee, southward flows of surface water, seasonal increases and decreases in water depth, and landscape-wide sheet flow. However, important details were elusive in those previous studies, including predrainage water depths, topography, flow directions, and the nature of the edges of the Everglades. Previous descriptions of predrainage hydrologic conditions ranged from an Everglades that dried out sufficiently for the peat soils to burn frequently (every five to fifty years) to an Everglades that was a vast flowing lake, retaining sufficient water even in dry years for large-scale peat fires to have occurred only once or twice in its five-thousand-year history.

This study reinforces most aspects of these earlier studies but reaches different conclusions because of the use of significantly improved descriptions of predrainage topography and landscape extents and explicit recognition of the directional, three-dimensional microtopography of the Ridge and Slough landscape. In many cases, rather than introducing completely new concepts, we have simply resurrected old concepts prevalent among early observers but that became less obvious as the Everglades were altered, often drastically, by anthropogenic drainage.

Water Depths

Studies of wetlands and peatlands around the world suggest that water depths, water flow, and water chemistry are almost certainly the principal hydrologic driving forces that maintain, and perhaps originally created, the Everglades. Estimation of predrainage water depths is therefore one of the key aspects for understanding the Everglades. Fortunately, despite the absence of predrainage stage gage time series measurements, a diverse wealth of other information was available to us from which to estimate water depths. Sources include numerous firsthand depth observations recorded by government land surveyors, depths recorded in the nightly logs of various cross-Everglades expeditions, and general descriptions from other firsthand predrainage observers. Figure 11.1 maps the locations of these observations, appendix A catalogs observed depths and associated information, and appendices E through N document Everglades expeditions. Additionally, we inferred depths from other sources, including the drafts and travel speeds of boats used in predrainage explorations and military navigation, predrainage soil types, hydrologic requirements and tolerances of wetland vegetation, and the frequency of predrainage peat fires as estimated from soil cores and narrative records.

The climate and geography of South Florida suggested ways that we could synthesize these historical observations of water depths to characterize the hydrology of the individual Everglades landscapes. The strongly seasonal rainfall pattern suggested, and historical observations confirmed, that Everglades water levels rose and fell annually. This suggested that the annual cycle could be characterized by an annual minimum, an annual maximum, and trajectories between those two extremes.

We focused on long-term averages to define further the hydrologic parameters suggested by the seasonal climate. Specifically, a focus on long-term averages defined fixed dates for the annual minimum and maximum water depth. Assuming that drainage did not shift the climate, those two dates can be estimated from postdrainage weather data. We assumed that the annual minimum occurred at the end of the dry season, in mid-May, and the annual maximum at the end of the wet season, in mid-October. A focus on long-term averages also implies a fixed value of the range, the difference between the long-term average minimum and maximum water depths. We address how that range was estimated later in this discussion.

Recalling the focus on long-term averages, rather than on individual hydrologic years, we assumed simple linear trajectories between the annual minima and maxima. During the wet season, the assumption of linearity is equivalent to assuming that the long-term rainfall rate was essentially constant throughout the wet season. Similarly, during the dry season, a linear decline assumes

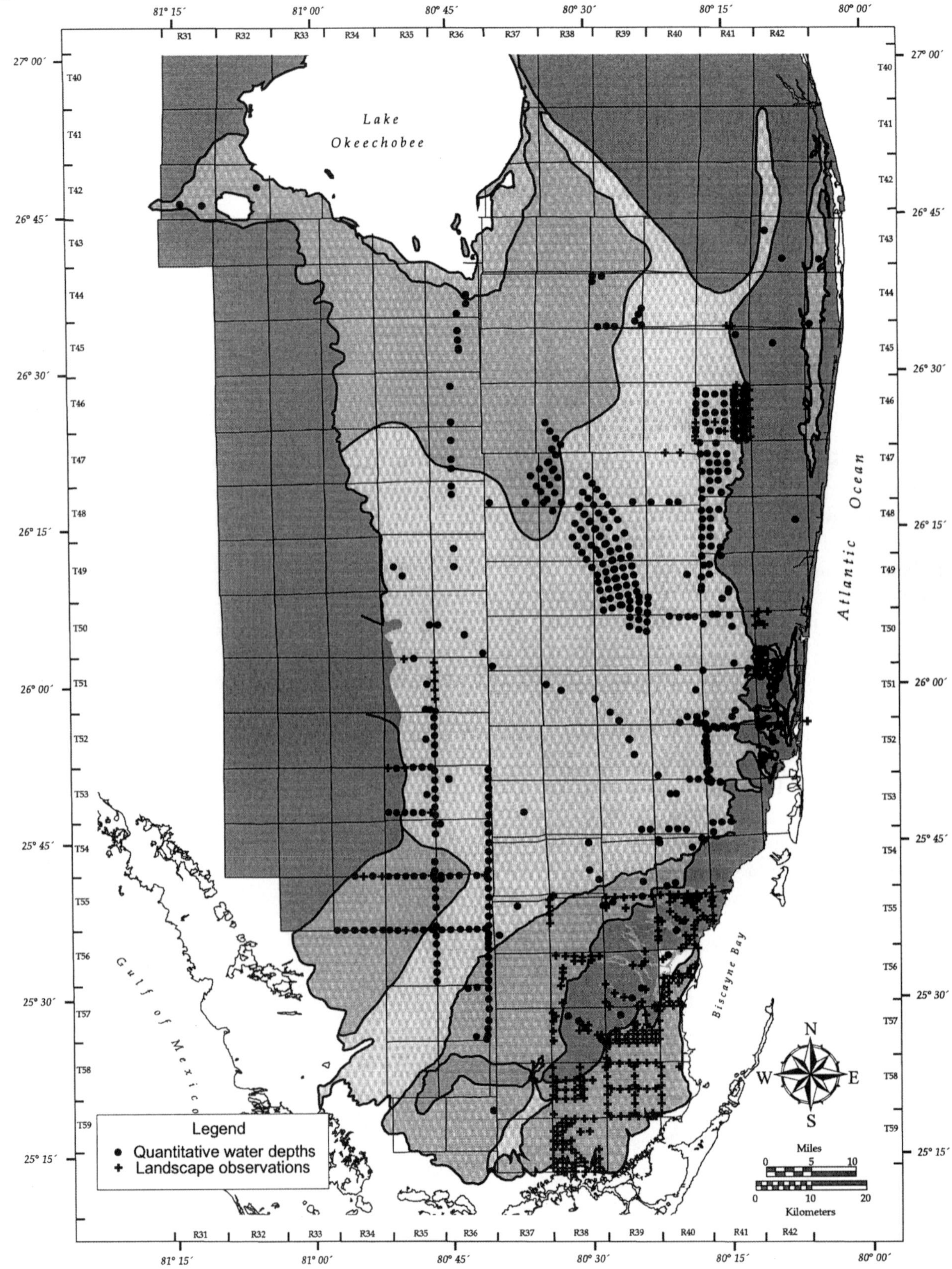

Figure 11.1. Locations of the predrainage point observations of water depths and vegetation recorded in appendix A. Circles = quantitative water depths; plus signs = qualitative and/or indirect depth information (e.g., from soils, vegetation). Sources include government township surveys (fig. 1.4); Everglades expeditions (plate 10 left; appendices E–M); and early scientific studies (appendix A).

that the rates of evapotranspiration and drainage (downstream flow) were approximately constant.

We were well aware of the importance of deviations from the long-term averages. As noted by Waldin (1912, pp. 31–32) and J. H. Davis (1943a, p. 278), "typical" or "average" years were the exception rather than the rule. Nevertheless, we focused our characterization of Everglades water depths on long-term averages for two reasons: first, because it seemed likely that such averages were the primary determinants of Everglades soils and vegetation; and second, because the more systematic predrainage data sets required to quantify higher order moments were not available.

Peat accumulates over decadal and greater time scales (Mitsch and Gosselink 1993), suggesting that peat processes are buffered, low-frequency processes influenced more by long-term average water depths than by interannual weather-driven variations. Similarly, the observed stability of sawgrass ridges and aquatic slough patterns over as much as fifty years in some postdrainage locations (plate 26; see also McPherson 1973a, pp. 40–50) suggests that the species forming these patterns are also somewhat buffered from interannual water depth variations. In other postdrainage locations, where hydrology has changed, the obvious pattern changes (plate 27; McPherson 1973a, pp. 33–39, 52–53) suggest sensitivities to shifts in long-term average depths. Observed postdrainage higher-frequency shifts between wet prairie species (e.g., *Eleocharis* spp., *Rhynchospora* spp., and *Panicum hemitomon*) and aquatic slough vegetation (primarily *Nymphaea odorata*) appear to be a primarily postdrainage phenomenon associated with reduced water depths in sloughs. A focus on long-term average hydrology seemed appropriate and useful to us. We note also that a combination of hydrologic modeling and postdrainage measures of weather variability may provide a way to estimate higher order moments associated with predrainage water depths.

The annual hydrograph that arises from the seasonal climate and the focus on long-term averages is quite simple; two line segments of slightly different slope, rising and falling each year between the average annual high and average annual low. As noted previously, the dates on which those occur (on average) are fixed, and the maximum is also fixed, being the sum of the minimum and the range. The only missing information pieces are the water depths corresponding to the long-term average minimum and to the long-term average range.

We assumed that the South Florida climate is regional; that is, that all of the Everglades experienced approximately the same climate. From this we could assume that the dates of the annual minima and maxima would be the same for all Everglades landscapes. In addition, we tentatively assumed that the water depth range (maximum minus minimum) was the same for all Everglades landscapes. These assumptions imply that different Everglades landscapes would differ only in the vertical offsets of their hydrographs; that is, differing values of the annual minimum. This is illustrated for two hypothetical Everglades landscapes in figure 11.2. Landscape 1 contains surface water throughout the year—the water surface never drops below the peat surface, and the hydroperiod is (theoretically) infinite; this landscape never dries out. In Landscape 2, the water level drops below ground (peat) surface for a portion of each year, and the hydroperiod is defined as the length of the period when surface water is present. For landscapes similar to Landscape 2, where the water level drops below ground, soil porosity causes an apparent amplification of the range and distortion of the hydrograph shape, but the annual range—in terms of equivalent free water—is the same.

Under the assumptions outlined here—fixed and known dates of occurrence of the long-term average annual minimum and maximum, and a single long-term average range applicable to all of the landscapes—characterization of the predrainage water depth regimes for

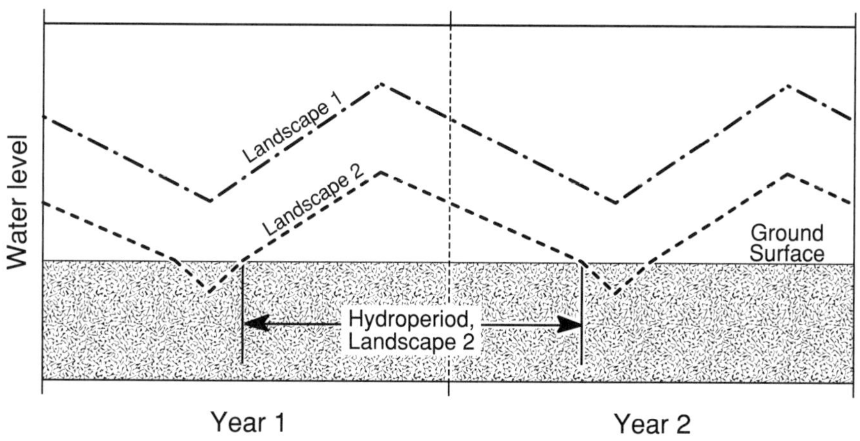

Figure 11.2. Long-term average hydrographs for two hypothetical Everglades landscapes. As drawn, Landscape 2 has a 10-month hydroperiod, while Landscape 1 is continuously water covered; that is, it has a continuous (multiyear) hydroperiod.

the different landscapes of the Everglades is simply a matter of determining the long-term average minimum for each; that is, their relative offsets.

In this discussion we have implicitly assumed that water depths at any given time were uniform throughout any given landscape. In a landscape with systematic microtopography this translates into the landscape subcomponents experiencing the same depths throughout the landscape: sloughs the same, ridges the same, etc. This seemingly strong assumption is the result of Everglades wetlands having been formed by peat accumulation. The assumption of uniform depths within a landscape is equivalent to stating that the planes of the peat ground surface and of the water surface above it were parallel to each other, an assumption also made by Hartwell (1970, p. 11). For level wetlands, this is intuitive. However, the Everglades were not level, instead sloping from an elevation of more than 20 feet (6.1 m) above sea level at Lake Okeechobee down to sea level at Florida Bay. Admittedly, the slope was very gradual, occurring over the course of more than 100 miles (160 km), but the elevation difference was sufficient to make it clear that the plane of the water surface could not have been level. It must have been sloped as well, at least to some degree.

We assume that the water slope in the downstream direction was in fact parallel to the ground slope. We assume this because the predrainage vegetation indicates that it was. Sawgrass was reported from top to bottom of the Sawgrass Plains; water lilies were reported in sloughs along the full length of the Ridge and Slough landscape. Had water depths not been constant, a spatial gradient of vegetation communities would have been observed; for example, deep-water species such as yellow or white water lilies grading into mid-depth communities of emergent wet prairie species and finally into sawgrass in the shallowest water. No evidence for such downstream vegetation gradients within the predrainage landscapes has been found. Figure 11.3, a downstream transect through two-thirds (55 miles; 90 km) of the Ridge and Slough landscape, illustrates observed water depths and vegetation under very nearly predrainage conditions. The consistency of both along the transect confirms the assumption of parallel water and ground surface planes.

The approach described in the preceding pages reduces estimation of predrainage water depths to estimation of a long-term average annual range in depth, applicable to all of the landscapes, followed by estimation of landscape-specific minima. We estimated the annual range of water depths from a synthesis of observed predrainage depths, vegetation tolerance, and several topographic considerations. Table 11.1 lists narrative descriptions of predrainage Everglades water depths, generally not assigned to individual landscapes but in almost all cases reported by known firsthand observers. Some observers specifically stated an annual range; others gave only an annual low and/or an annual high. These rough observations loosely suggested an annual range of 2 to 3 feet (60–90 cm) as well as a need to refine the estimate further.

The 2- to 3-foot (60–90 cm) estimate of the long-term average annual range was refined and reduced by comparison with other known aspects of the predrainage Everglades, in particular the Ridge and Slough landscape microtopography (fig. 8.3); recorded predrainage Ridge and Slough water depths (fig. 11.4); cross-sectional elevations of Shark River Slough, the Rockland Marl Marsh, and the Marl Transverse Glades (predrainage: fig. 4.7; postdrainage: fig. 11.5); and township "General Descriptions" submitted by federal surveyor William Mickler (table 11.2). Each of these aspects, discussed individually below, suggested that the long-term average annual range was closer to 2 feet (60 cm) rather than 3 feet (90 cm).

An annual range as high as 3 feet (90 cm) would be problematic because the resultant slough maximum of 4 feet (120 cm) would have overtopped a number of the tree islands in most years (fig. 8.3) and would have resulted in deeper conditions than were typically reported for predrainage tree islands. An estimated range of 3 feet (90 cm) also would have placed 2.5 feet (75 cm) of water on top of the sawgrass ridges at the end of the wet season. Field observations (J. H. Davis 1943a, p. 278; Wallace 1955b, p. 2; Andrews 1957, p. B-30; J. H. Davis 1957, p. C-4; McPherson 1973a, p. 32; Alexander and Crook 1975, p. 299; Toth 1987, p. 9), and controlled experiments (Newman et al. 1996, p. 779), suggest that 2.5 feet (75 cm) at the end of the wet season is deeper than optimal for sawgrass and would likely have favored a shift to slough species. Therefore, a value of 2 feet (60 cm) for the typical predrainage annual range seems most consistent with predrainage microtopography, vegetation, and tree island heights.

All available recorded predrainage water depths from the Ridge and Slough landscape found during the course of this study are plotted in figure 11.4. The points do not in any way represent a systematic study but rather are a necessarily random collection from many different observers, different years, and from an unidentified mix of ridge, slough, and possibly a few tree island locations. As a result, even the seasonal pattern of rise and fall is obscured. Nevertheless, most of the depths cluster at 1 to 3 feet (30–90 cm), reinforcing the impression that the annual range was most likely about 2 feet (60 cm).

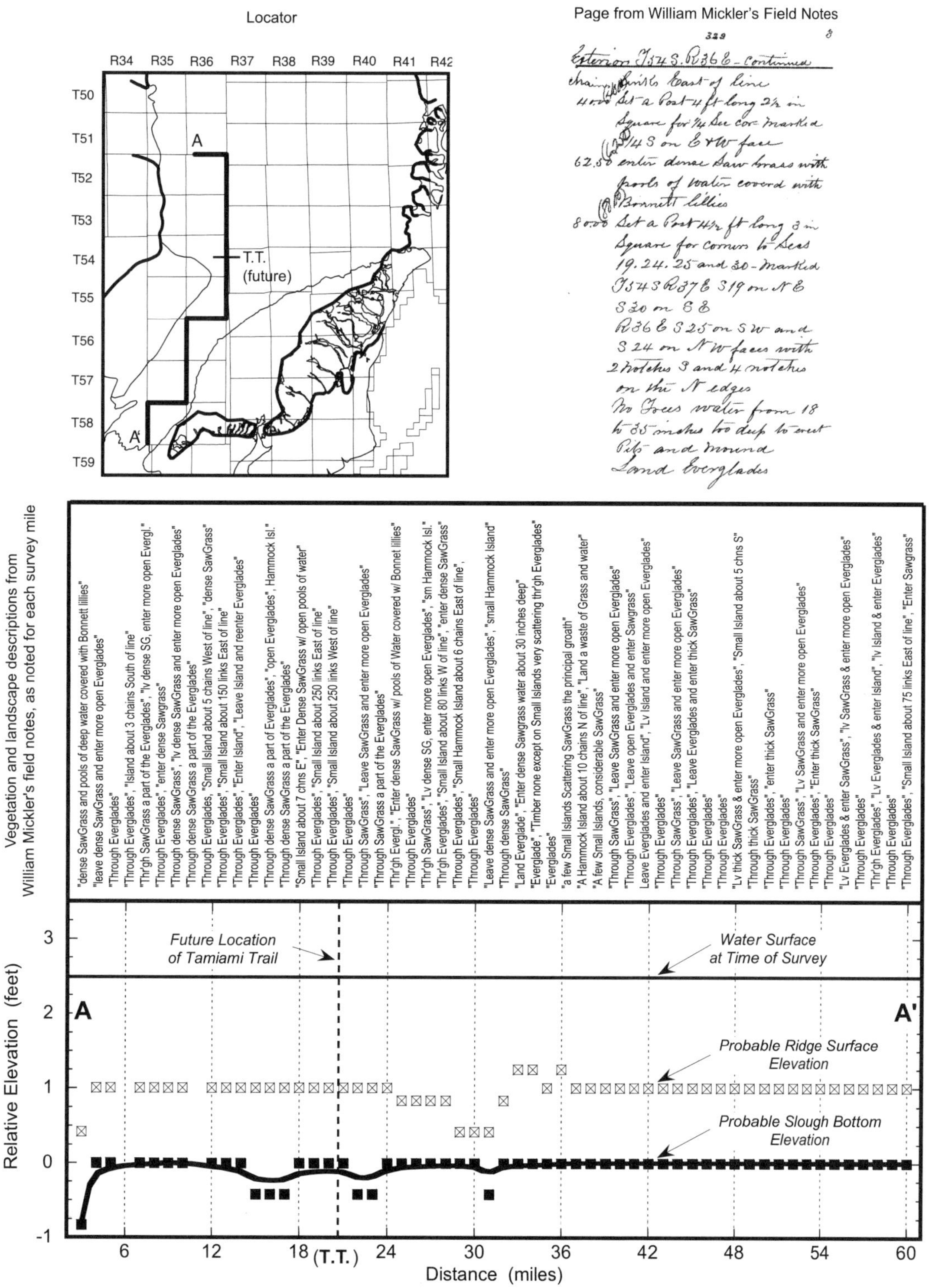

Figure 11.3. A "downstream" transect through the Ridge and Slough landscape under near predrainage conditions (see text), assembled from observations by U.S. Deputy Surveyor William Mickler from November 26, 1885, to January 28, 1886 (first half of the dry season). Top left: Locations of survey portions used to create the transect. Top right: Sample of Mickler's notes on vegetation/landscape, used for summaries shown in the graph. Bottom: Mickler's ranges of water depths (e.g., "water 18 to 30 inches deep") observed along each mile. The smaller numbers are assumed to reflect shallower water depths on sawgrass ridges; the larger numbers, deeper water in sloughs. Note the similarity of predrainage conditions north and south of the future Tamiami Trail. Note also that temporal extrapolation of Mickler's slough depths (cf. fig. 11.2) yields an annual high depth of 3 to 3.5 feet, very similar to the long-term estimate of 3 feet (table 11.4). Compiled from surveys by Mickler (1886-T52 R36; 1886-T53 R36; 1886-T54 R36; 1885-T55 R36; 1886-T56 R35; 1886-T57 R35; 1886-T58 R35).

Table 11.1. Early estimates of Everglades water depths: low (dry season), other (unspecified season), high (wet season), and annual range (high-low)

Observer's Name and Occupation	Year	Low (ft)	Other (ft)	High (ft)	Range (ft)	Citation
Charles Vignoles, civil & topographical engineer	1823	0.5–2				Vignoles (1823a)
John L. Williams, attorney & explorer	1837	2–4				J. L. Williams (1837)
Anonymous engineer in Col. Harney's command	1841	2–4				Brooks (1880)
Buckingham Smith, attorney and U.S. Treasury agent	1848	3–6				Smith (1848a)
Gen. Harney, Seminole Wars	1848	2.5–6				W. S. Harney (1848)
Lieut. Rogers, Seminole Wars	1848		0.5–5			Rodgers (1848)
S. R. Mallory, collector of customs	1847		4–6[a]	2[b]		Mallory (1847)
J. C. Ives, topographical engineer	1856	3			2–3	Ives(1856b)
J. A. Henshall, U.S. Fish. Comm.	1882			4–6		Henshall (1883)
W. Mickler, U.S. deputy surveyor	1885			2–5[c]		Mickler surveys in 1886, see table 11.2
N. S. Shaler, geologist/geographer	1890				5–8	Shaler (1890)
L. S. Griswold, geologist	1896				2	Griswold (1896)
William Dupuy, writer	1908				3[d]	Dupuy (1908)
Samuel Sanford, geologist, railroad engineer	1913				2; max 4	Matson and Sanford (1913b)
John W. Harshberger, plant ecologist	1914				2–3	Harshberger (1914)

Notes: a. In channels only (unspecified time of year).
b. On the "immense grassy plain . . . of dense saw grass" (average for the wet season).
c. Regarding Ochopee Marl area in western Everglades (T56 R36, T57 R36).
d. "A difference of nearly three feet in its levels in the two seasons."

The relative elevation of the Marl Transverse Glades, which received outflows from the Rockland Marl Marsh, also constrains estimates of the annual range. The marl soils and sparse sawgrass present in the predrainage Marl Transverse Glades (chapter 10) suggest that the maximum average annual depth was approximately 1 foot (30 cm). Figure 4.7 suggests that a 1-foot (30 cm) depth in the Marl Transverse Glades could be consistent with an average annual range of about 2 feet (60 cm) in the Rockland Marl Marsh and in Shark River Slough.

Information on water depths and ranges was provided by U.S. Deputy Surveyor William Mickler, who surveyed a number of townships within the Rockland Marl Marsh, Ridge and Slough, and Ochopee Marl Marsh landscapes during the 1885–1886 dry season as well as within Big Cypress and into the mangrove fringe (fig. 1.4). Letters from Mickler to the U.S. Surveyor General's Office in Tallahassee (Mickler 1859, 1861, 1885) suggest that by 1885 he was very familiar with the Florida wet and dry seasons and their effects on water levels in South Florida. His descriptions recorded at the end of each survey mile consistently match vegetation patterns known from other sources. As he completed each township, he wrote a "General Description" of its trees, soils, and water depths present at the time of the survey. In five of those descriptions he also included an estimate of wet season water depths, providing an additional way to estimate predrainage ranges (table 11.2). The estimated annual range extrapolated from Mickler's descriptions is 2.3 feet (70 cm).

Based on all of the above data and reasoning, we estimated the long-term average annual predrainage range in water depths as 2 feet (60 cm). This estimate may also be compared with informal estimates of postdrainage ranges (table 11.3; see also fig. 11.6). The postdrainage, post-impoundment ranges from the Water Conservation Areas tended to be somewhat greater than the predrainage estimate of 2 feet (60 cm), as would be expected from their use as impoundments. Overall, the estimate of 2 feet (60 cm) for the long-term average annual predrainage range in water depths appears to be consistent with many features of the predrainage and postdrainage information.

The final step in estimating predrainage water depth

regimes for the various landscapes of the Everglades was estimation of a long-term average minimum water depth for each landscape. Estimates were needed for sloughs of the Ridge and Slough landscape, ridges of the Ridge and Slough landscape, the Sawgrass Plains, the Custard Apple Swamp, and for the Rockland, Ochopee, and Perrine marl marshes. In some cases geographic and/or vegetative aspects allowed estimates for one landscape to be linked to estimates from another. The estimate for sloughs formed a root parameter for many of the other landscapes. Additionally, observed predrainage and early postdrainage water depths from each

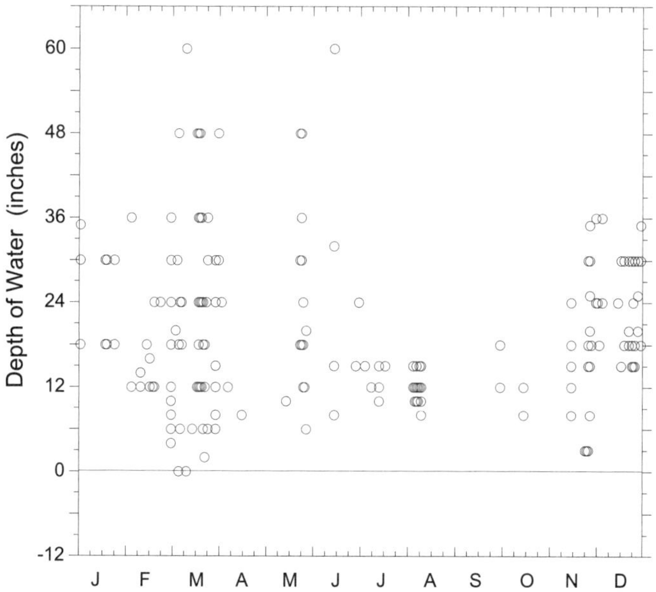

Figure 11.4. Depths of water within the Ridge and Slough landscape, measured between 1820 and 1920 and plotted against time of year. Measured depths may have come from either sloughs or sawgrass ridges; this was not reported. Point data from appendix A, table A.3; fig. 11.1.

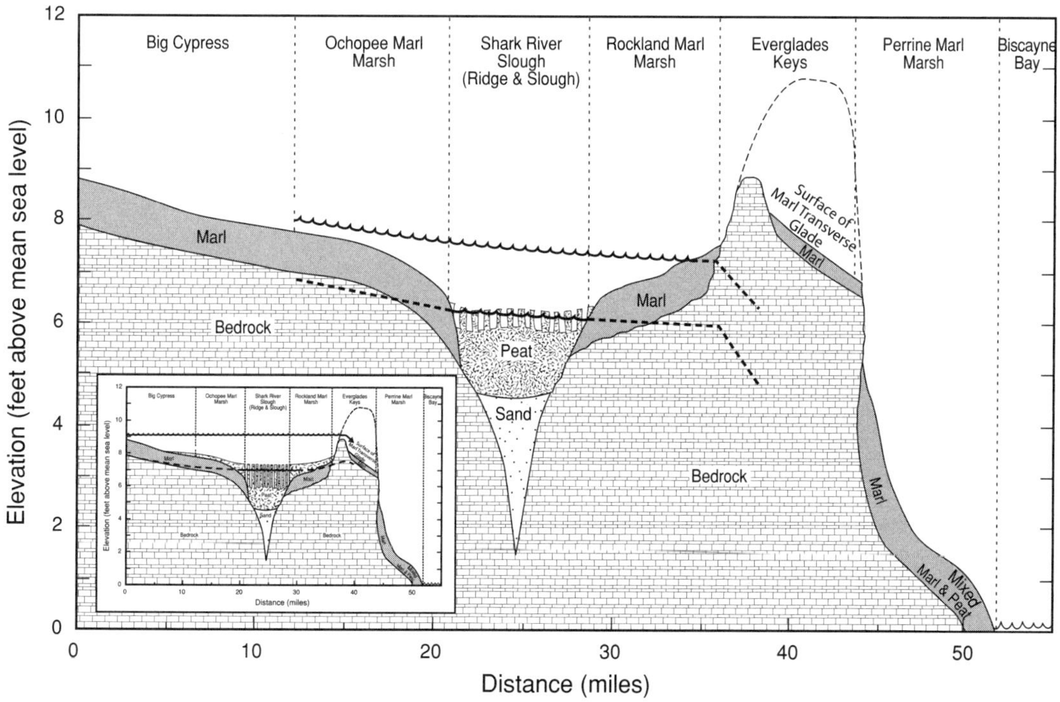

Figure 11.5. Approximate current (2010) cross section through the southern Everglades and the Everglades Keys. Location: see plate 35, dotted transverse line. Reduced flows south of Tamiami Trail along with strongly lowered water tables in the now urbanized Everglades Keys have reduced water depths (wavy lines) and shortened hydroperiods, causing almost complete oxidation of peats within the Marl Marshes (chapter 9), and a downward tipping of the water table to the east. Compare with plate 36 (1978–1979) and with contrasting predrainage conditions (fig. 4.7; reproduced here as inset). Sources: Marl ground surface from Everglades Depth Estimation Network (EDEN, Telis and Higer 2006); average annual high and low water surfaces (2000–2010) synthesized from regional simulations (South Florida Water Management District 2005, 2006; Lal et al. 2005; Senarath et al. 2008, 2010; EDEN) and vegetation studies (Ross et al. 2006).

Table 11.2. U.S. Surveyor William Mickler's observed mid–dry season and estimated wet season water depths in 1886, and two estimates of typical annual range

Landscape	Date	General Description	Range Est. 1 (ft)[a]	Range Est. 2 (ft)[b]
Ochopee Marl Marsh	Jan. 12–13	"This Township is all inundated from 18 to 30 inches and in the Rainy Season the water is much deeper. . . . There is a few Small Hammock Island on it" (Mickler 1886-T56 R34, p. 523).	—	—
Ridge & Slough	Jan 22–23	"This Township is all inundated from 18 to 30 inches and in the Rainy Season from 30 to 50 inches it has a few small Islands over it" (Mickler 1886-T56 R36, p. 495).	2.9	2.3
Rockland Marl Marsh	Jan. 19–20	"This Township is all inundated from 18 to 30 inches now and in the wet Season it is inundated from 3 to 5 feet. there is a few Small Island over it" (Mickler 1886-T57 R36, p. 485).	4.4	3.0
Ridge & Slough	Jan. 25–28	"This Township is all inundated from 18 to 30 inches and in the Rainy Season from 30 to 50 inches it has a few Small Islands over it" (Mickler 1886-T58 R35, p.522).	2.8	2.3
Rockland, Long Pine Key, Perrine Marl Marsh	Jan. 29–Feb. 1	"This Township is all under water from 18 to 30 inches and in the Rainy Season which Sets in in June it will be from 30 to 50 inches except a few Small Island w[h]ich does not consist of more than an acre of high land except the Pine Island in Section 13 which is a long narrow Island and in many places the lime Rock projects out of the ground on the Pine Island" (Mickler 1886-T58 R36, pp. 542–543).	2.7	2.3

Notes: a. Range Estimate 1 is estimated as the difference between Mickler's rainy season water depth (assigned to October 15) and an end of dry season depth (May 15), with the latter calculated by linear extrapolation of Mickler's observed water depth on the indicated date.
b. In Range Estimate 2, extrapolating Mickler's reported water depths and dates linearly forward yielded end of dry season estimated depths of -5 to +8 inches, which seemed low. The range estimates in this column were recalculated using an assumed end of dry season value of 12 inches. (Alternatively, if Mickler's reported depths, 18–30 inches, were for some reason underestimated by only 3 inches, the Estimate 1 ranges would be equal to the Estimate 2 ranges. For example, if the depths were actually 21–33 inches, Mickler might have rounded the observed depths to even, 6-inch multiples: 18–30 inches.)

landscape (e.g., appendix A; figs. 11.1, 11.4, 11.6), soils, vegetation, and microtopography (chapters 5 to 10), were all used in estimating the minimum water depths.

Sloughs

It is known that water depths in the sloughs of the Ridge and Slough landscape were typically deep enough to maintain open water (i.e., few emergent species), allowing easy, rapid canoe passage; that white water lilies (*Nymphaea*) were common throughout, with *Nuphar* less common but present; that the annual maximum slough water depth approached but generally did not continuously overtop the soil surface of adjacent tree islands; and that the sloughs typically did not dry out. The rarity of slough dryouts suggests that long-term average water depths were sufficient that sloughs retained a degree of water even during years of below average rainfall.

Sloughs did not typically dry out, but they did do so occasionally. The historical record is too sparse and too unsystematic to estimate a return frequency or even an exact definition of "dryouts," but we note the following five observations. These were the only reports of dry conditions that we encountered. William Cooley stated that in 1827, during "a season that was very dry," he rode on horseback 30 miles northwest from the New River "across the burnt tract," finding "large quantities of decayed vegetation," water "very scarce," but no "signs of . . . any place where the fire had taken hold of the soil" (Cooley 1851, p. 40). Buckingham Smith in 1947 recorded that "In *dry* seasons, as we were informed . . . it is said, persons have in seasons of drought walked on dry land 10 miles from the shore at some points" (Smith 1848b, p. 125). Three longtime observers interviewed by John Stewart in 1907 reported seeing at least a portion of the Everglades "dry." It is not known whether all three observers would have been reporting the same event: "Have known the Glades to be dry around the head of [Shark] river" (Roberts 1907, p. 39); "Glades were dry one season" (Stranahan 1907, p. 44); "Have seen Glades dry" (Graham 1907, p. 24). These dry periods apparently did not lower water tables sufficiently to dry the peat soil to the point of flammability, as predrainage peat fires were not reported (although postdrainage fires were widespread; chapter 3).

At the other end of the spectrum it is known that sloughs frequently filled to close to or equal to the

Table 11.3. Postdrainage estimates of average annual range in water depth (feet of water) for various locations in the Everglades

Range (ft)	Location	Time Periods	Comments	Citation(s)
1.5	Everglades	1940–1946	Water budget analysis	Langbein in Parker et al. (1955)
1.8	Tamiami Trail at 40 Mile Bend	1940–1946	From hydrographs	Parker et al. (1955)
1.6	Tamiami Trial at Krome Avenue	1940–1943	From hydrographs	Parker et al. (1955)
2	Lake Okeechobee Stage	1916–1936	From hydrographs	Schrontz (1942)
2	Southern WCA 3[a]	1951–1959	From hydrographs	Wallace et al. (1960)
2.5	WCA 2A	1980–1984		Worth (1988)
3.1[b]	ENP[c]; Ochopee Marls	1954–1985		Gunderson (1989)
2.3[b]	ENP; Perrine Marls	1954–1985		Gunderson (1989)
2.4[b]	ENP; Loxahatchee Peats	1954–1985		Gunderson (1989)
2–3	Gum Slough; Monroe County	1930–1950	Field observations	C. R. Stone (1979), pp. 35–36
1.7–2[d]	LNWR[e] (WCA1)	1953–1997	Figure 11.6 (1-7 gage)	This study
2.5–3[d]	WCA 2A	1953–1997	Figure 11.6 (2A–17 gage)	This study
2.5–3[d]	WCA 3A	1953–1997	Figure 11.6 (3A–28 gage = Site 65)	This study
1.5–2[d]	ENP	1953–1997	Figure 11.2 (P-33 gage)	This study

Notes: a. WCA = Water Conservation Area.
b. Not corrected for soil porosity.
c. ENP = Everglades National Park.
d. These estimates reflect only variation in above ground water depth; not water table.
e. Loxahatchee National Wildlife Refuge.

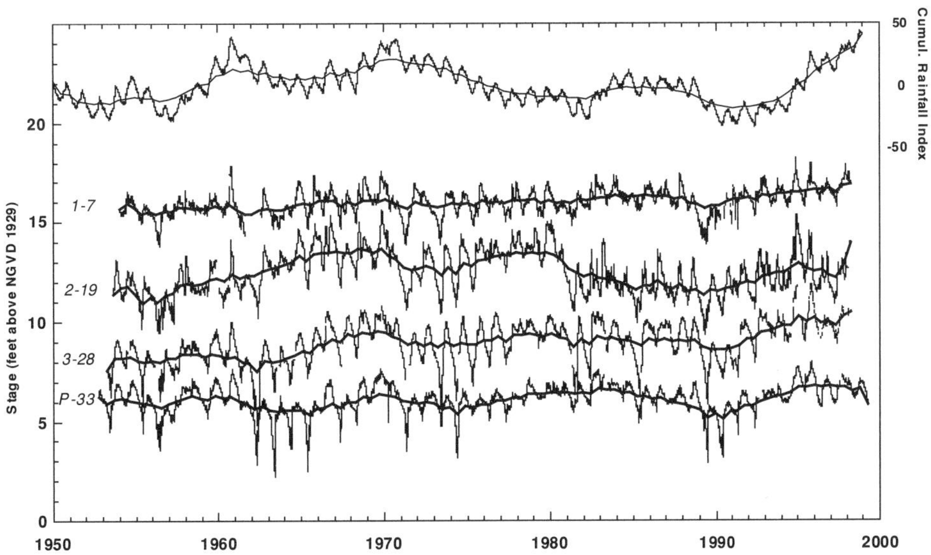

Figure 11.6. Daily water stages (1950–2000) for four compartments of the postdrainage Everglades: Loxahatchee National Wildlife Refuge (Water Conservation Area 1; stage gage "1-7"), Water Conservation Area 2A ("2-19"), southern Water Conservation Area 3A ("3-28"), and Everglades National Park (near bottom of L-67 Ext.; "P-33"). The daily rainfall index is shown (top) for comparison of trends. Note differences among the four compartments in variance (daily values) and in smoothed, decadal-scale lines. Compiled from South Florida Water Management District hydrologic records.

heights of tree islands (2 to 4 feet; 60–120 cm; appendix C). Predrainage military and explorer narratives frequently mention sleeping in their canoes or even making cook fires on them because there was no dry land on tree islands (Willoughby 1898a, p. 149, see appendix J; Sprague 1848, p. 354; Preble 1842, p. 31, see appendix G; Anonymous 1841, p. 251, see appendix F; Hopkins 1884b, see appendix H; A. P. Williams 1882, pp. 90, 92–93, see appendix H; Lamb 1975, pp. 1–2). For example:

> Scattered about in this sea of grass are islands of bushes and trees, called Keys. These Keys [tree islands] seem to owe their origin to an accumulation of vegetable matter which may appear some inches above water level, during the dry season becoming partially dry. During a night's sleep upon them one's bed is liable to settle to water level." (Griswold 1896, p. 54)

The observations of tree islands frequently being too wet for sleeping or cooking, along with a long-term typical annual range of 2 feet, suggests that the long-term typical minimum might fall in the range of 0–2 feet (0–60 cm).

From these specific features and from the records of observed water depths (figs. 11.4 and 11.6), we estimated a long-term average annual minimum water depth of approximately 1 foot (30 cm) for sloughs.

Sawgrass Ridges

We estimated the long-term average annual minimum water depth on the ridges of the Ridge and Slough landscape by assuming a level water surface across ridges and sloughs and an average sawgrass ridge height of 1.5 feet (appendix C; fig. 8.3). This yielded an estimated minimum of -0.5 foot. This miminum, indicating a drop of water level below the ridge surface for a short period during the dry season, is consistent with early field assessments of optimal conditions for sawgrass growing on peat (e.g., J. H. Davis 1943a, pp. 253, 278; Andrews 1957, B-30; J. H. Davis 1957, p. C-4; Loveless 1959a, p. 2).

Sawgrass Plains

We assumed that the long-term average annual minimum water depth estimated for sawgrass ridges applied as well to the Sawgrass Plains. This was based on the same species being dominant and the similarity in stand densities and height. The estimated minimum and corresponding hydrograph overlapped well with predrainage water depths observed in this landscape (appendix A).

Custard Apple Swamp

Custard apple (now called pond apple, *Annona glabra*) typically grows in wetter conditions than sawgrass (Michael Duever, pers. comm., 2004), so we assumed that the Custard Apple Swamp minimum was slightly (0.5 feet; 15 cm) wetter than the Sawgrass Plains minimum; that is, a minimum of zero feet (0 cm). Thus on average, water levels at the end of the dry season dropped to become even with the average Custard Apple Swamp ground surface.

Rockland and Ochopee Marl Marshes

We estimated the minima for the slightly sloped Rockland and Ochopee marl marshes (fig. 11.7) bordering Shark River Slough from the Marl Marsh–Shark River Slough cross-sectional topography (fig. 11.7), from the presence of shallow peat soils in at least parts of the landscapes, from the assumption that water levels dropped slightly below ground surface (and were accordingly distorted downward by soil porosity), and from the assumption that drainage in these Marl Marshes was perhaps slightly faster than within Shark River Slough. We estimated the minima as 0.5 foot below average ground surface (-0.5 foot; -15 cm).

Perrine Marl Marsh

Because the Perrine Marl Marsh was less directly connected hydrologically to the main portion of the Everglades (chapter 9), and historical accounts suggest that it was somewhat drier than the Rockland and Ochopee marl marshes (chapter 9), we estimated the long-term average annual minimum water depth in the Perrine Marl Marsh to be 0.5 foot (15 cm) lower than in the other marl marshes; that is, a minimum of -1 foot (-30 cm).

The minima for each landscape were combined with the overall estimated annual range of 2 feet (60 cm) to draw hydrographs for each predrainage landscape similar to those shown in figure 11.2. In turn, we used these hydrographs, along with all other landscape information (chapters 6 to 10) to estimate hydroperiods. The results of this synthesis of historical ecology are summarized in table 11.4 and mapped on plate 8. These water depth estimates also form the basis of the water surfaces shown on plate 7.

Overall, the water depths mapped in plate 8 and reflected in plate 7 match well the large body of predrainage observations and support the conclusion reached by Parker (1974, p. 29), namely that they "all add up to the judgment that, in predrainage days, the Everglades were generally either wet or flooded most of the time."

A Sample Evaluation: Ridge and Slough Landscape

We evaluated the water depth estimates shown in table 11.4 for the largest predrainage landscape by comparing them with a variety of indirect sources of present-day and predrainage information. Five evaluations were made, focusing on (1) white water lilies (*Nymphaea odorata*); (2) yellow water lilies (spatterdock or bonnets; *Nuphar advena*); and (3) comparison of two land surveys of the same transect, one from 1886 (predrainage), the other from 1955 (postdrainage); (4) landscape scale

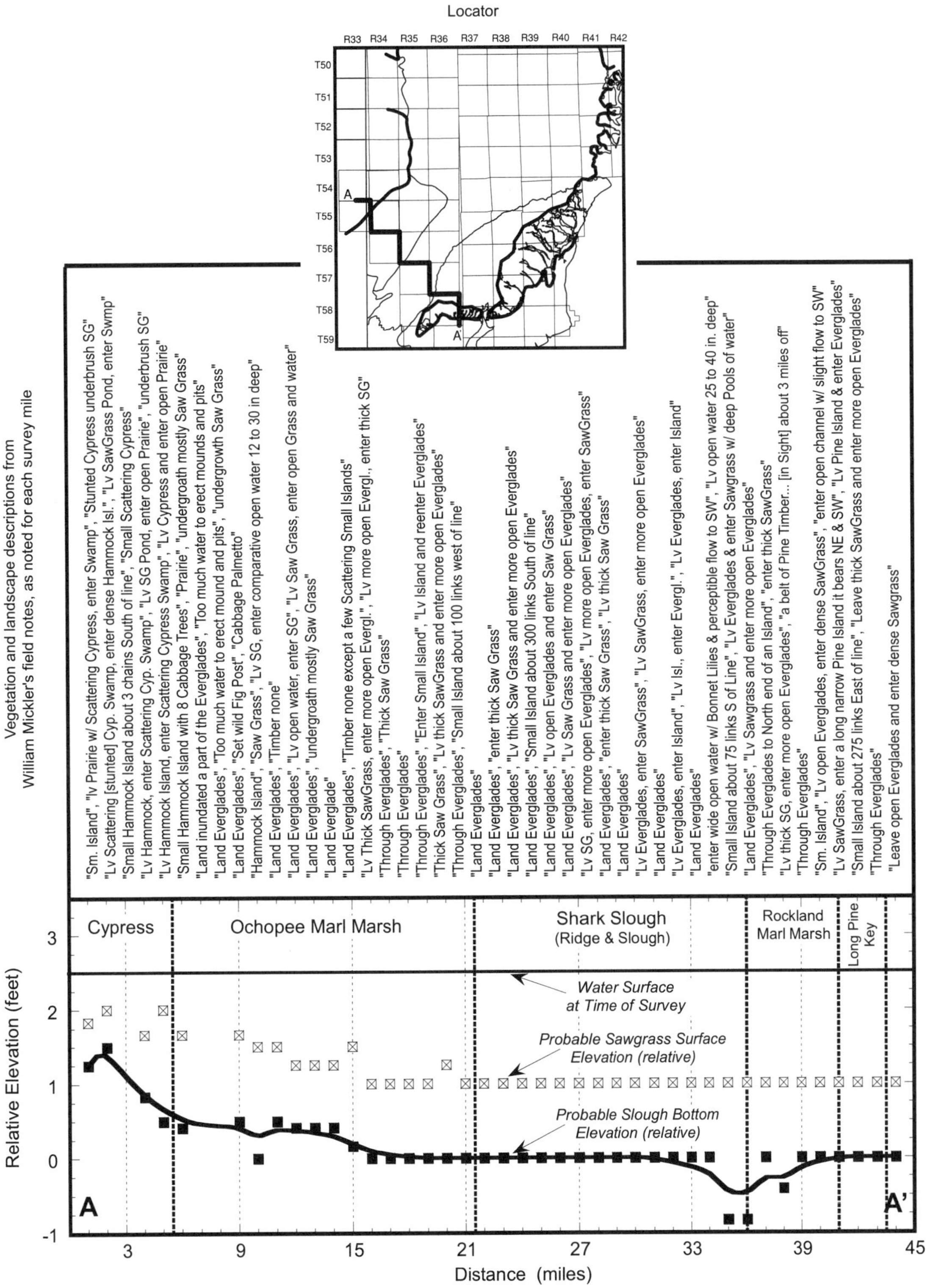

Figure 11.7. "Transverse" transect across portions of Big Cypress, Ochopee Marl Marsh, Shark River Slough, Rockland Marl Marsh, Long Pine Key, and Perrine Marl Marsh. Top: Locations along township edges. Bottom: Water depths and vegetation types summarized at end of each mile by U.S. Deputy Surveyor William Mickler. Townships were surveyed November 24, 1885, to Feb 1, 1886 (first half of dry season). Mickler's recorded ranges of water depths (e.g., "water 10 to 24 inches deep") likely reflect irregular limestone surface in Marl Marshes and ridge/slough differences in Shark River Slough. Note shallower depths toward the west edge of Ochopee Marl Marsh. Compiled from surveys by Mickler (1885-T55 R33; 1885-T55 R34; 1886-T56 R34; 1886-T57 R35; 1886-T57 R36; 1886-T58 R36).

changes in water depths and the relative elevations of sloughs, sawgrass ridges, and tree islands over three time periods (predrainage, ca. 1945, and ca. 2005); and (5) an individual slough-scale comparison of predrainage and current slough fauna.

White water lilies. In predrainage accounts, white water lilies (*Nymphaea odorata*) were frequently noted throughout the Ridge and Slough landscape (e.g., Dix and MacGonigle 1905, p. 522; Dimock 1907, p. 247; Dimock 1911, p. 116; see also appendices, this work). This ubiquitous distribution is consistent with the soil record: water lily remnants are characteristic of the Loxahatchee Peat soils (Evans and Allison 1942, p. 45; Jones et al. 1948e, p. 72; Gallatin et al., 1958, pp. 21–22), which underlie the entire length of the Ridge and Slough landscape, including Shark River Slough.

In contrast, at present, white water lilies are common only in the deeper, downstream portions of impoundments such as Water Conservation Area 3A (McPherson 1973a, p. 7; Schortemeyer 1980, pp. 56–61; Zaffke 1983, pp. B-1–B-9; David 1996, p. 19), suggesting that managed water depths in the shallower, mid to upper portions of Water Conservation Area 3A are less than predrainage depths were in those areas. South of Tamiami Trail, water lilies were originally widespread in Shark River Slough and Taylor Slough and would be expected there now. However, reduced inflows along with naturally unconstrained outflows have apparently lowered water depths too much to support water lilies other than in artificially deep areas such as airboat trails (Loftus and Kushlan 1987, p. 165). Water lilies have been uncommon, and reduced to a patchy distribution, within Shark River Slough at least since the 1970s (Alexander and Crook 1973, 1975; Wood and Tanner 1990, p. 136; Loftus and Eklund 1994, p. 465; Busch et al. 1998, p. 236; Ross, Ruiz, et al. 2006, pp. 11–12), and likely earlier. Alexander and Crook (1973, 1975) recorded *Nymphaea* in twelve quadrats within the three Water Conservation Areas but in no quadrats within Everglades National Park. In summary, comparison of the current distribution of *Nymphaea* with current water depths suggests that estimates of a typical predrainage annual low of 1 foot (30 cm) and high of 3 feet (90 cm) are accurate.

Yellow water lilies. Yellow water lilies (*Nuphar advena*), also called spatterdock and "bonnets," may be even more indicative of how much current water depths have decreased from predrainage ones. *Nuphar* tends to grow in deeper water than *Nymphaea* (Florida Game and Fresh Water Fish Commission 1956, p. 46; Loveless 1956, pp. 39–40; Cherry 2005, p. 105; Cherry and Gough 2006, p. 32). Under current conditions *Nuphar* was not observed in sloughs in any recent studies of Water Conservation Area 3A (McPherson 1973a, pp. 61–62; Schortemeyer 1980, p. 76; Wood and Tanner 1990, p. 136; Zaffke 1983, pp. B-1–B-9; David 1996, p. 22), nor in the Arthur R. Marshall Loxahatchee National Wildlife Refuge (Water Conservation Area 1; Jordan et al. 1997, p. 280), nor along two transects in Northeast Shark River Slough (Everglades National Park; Ross, Ruiz, et al. 2006, pp. 10–12). Olmsted and Armentano (1997, p. 40) found only a single occurrence of *Nuphar* in a willowhead and none in former sloughs. At present the only remaining *Nuphar* in the Everglades appears to be limited to locations that are deeper than sloughs, such as alligator

Table 11.4. Estimated predrainage (pre-1880s) long-term average annual water depths and hydroperiods

Everglades Landscape	Average Annual Low (feet)	Average Annual High (feet)	Average Hydroperiod (months)
Custard Apple Swamp	0	2	11–12
Sawgrass Plains	-0.5[a]	1.5	9–10
Ridge and Slough (sloughs)	1	3	12
Ridge and Slough (ridges)	-0.5	1.5	9–10
Ridge and Slough (landscape average)	(0.2)[b]	(2.2)[b]	n/a
Rockland Marl Marsh[c]	-0.5	2	8–9
Ochopee Marl Marsh[c]	-0.5	2	8–9
Perrine Marl Marsh[d]	-1	1.5	8–9

Notes: a. Negative values indicate distance below ground surface.

b. Calculated as area-weighted average, based on 50% sloughs, 45% ridges, and 5% tree islands.

c. Water depths across these landscapes were not uniform; values shown are for mid-elevation locations; i.e., about halfway from Shark Slough to upper edge of landscape.

d. Water depths along downstream axis of Perrine Marl Marsh may not have been uniform; values shown may apply only near mid-elevation locations.

holes ("deep edges of" alligator holes, Fogarty 1984, p. 215; "1–2 m" deep alligator ponds, Loftus and Kushlan 1987, p. 167) and the upper reaches of tidal creeks at the foot of Shark River Slough, such as Rookery Branch (William Loftus, pers. comm., 2009). More commonly, *Nuphar* is presently found in constructed features, primarily canals, ditches, and "borrow holes" next to artificial tree islands. The only recorded exception to these locations, found in only one out of thirty-five quadrats surveyed, was a single point within a deep, burned-over former tree island at the deep, downstream end of Water Conservation Area 1 (Arthur R. Marshall Loxahatchee National Wildlife Refuge; Alexander and Crook 1975, p. 329).

Predrainage and early postdrainage observations indicate that bonnets once occurred with some frequency in the Everglades, sometimes together with white water lilies (e.g., Mickler 1886-T52 R36, 1885-T54 R36, 1886-T57 R36; Dimock 1907, pp. 246–247; Dupuy 1908, p. 9895; S. R. Smith 1909, p. 8; Harper 1910b, p. 148; Ensey 1912-T51 R41; Harshberger 1914, pp. 168–169; Franklin 1914-T50 R40; Baldwin and Hawker 1915, p. 787; Small 1916, p. 168; 1917; 1921; Harper 1927, pp. 125, 127; J. H. Davis 1943a, p. 266). One observer noted: "The channels have a growth of small maiden cane, flags and numerous lilies, both white and yellow. The yellow lilies are locally known as 'bonnets.' The water is from 1 to 4 ft deep" (Stewart 1907, p. 47).

Surveyors also recorded bonnets. William Mickler made the following observations between December 30, 1885, and January 20, 1886, under essentially predrainage conditions. The water depths Mickler noted in the vicinity of bonnets tended to be somewhat deeper than his most common recorded depth of 18 to 30 inches (46–76 cm) for "Saw Grass" and "more open Everglades," perhaps reflecting a tendency of bonnets to occur in deeper water than white water lilies: "Through dense Saw Grass and pools of deep water covered with Bonnett lillies . . . water from 25 to 40 inches" (Mickler 1886-T52 R36); "enter dense Saw Grass with pools of water covered with Bonnett lillies . . . water from 18 to 35 inches" (Mickler 1885-T54 R36); "enter dense Saw Grass with open pools of water covered with Bonnett lillies . . . water from 18 to 35 in deep" (Mickler 1885-T54 R36); "enter wide open water with Bonnett lillies and a perceptible flow to the SW . . . [330 ft later] leave open water from 25 to 40 inches deep and enter Everglades" (Mickler 1886-T57 R36).

Twenty-five years later, surveyors working farther east and north also recorded bonnets in sloughs. Simultaneous mention of bonnets and white water lilies indicates that these surveyors distinguished between them. The shallower water depths compared to Mickler reflect a combination of the season (several were surveyed in May–June), the year (well after the first lowering of Lake Okeechobee), and the influence of the Hillsboro and North New River canals: "bonnet & lily slough . . . 10 [inches] water" (Franklin 1912-T47 R41); "bonnet slough . . . 15 [inches] water" (Franklin 1912-T47 R41); "E ½ of this mile passes in a succession of creek sloughs, lilly pad & bonnet ponds, alternating with small scrub myrtle islands . . . water 18 [inches] deep" (Ensey 1911-T47 R41); "1 [foot] water . . . bonnet slough" (Franklin 1914-T50 R40); "The country is open Everglade—Myrtle Bushes and sawgrass—Most of the mile lies in Bonnet and Lily pad ponds," "open prairie flat—lilies—Bonnets—Water 3 [feet] deep—boggy" (Ensey 1911-T50 R41); three references to "bonnet sloughs" in water 0.5 to 5 feet (15–150 cm) deep (Kackley 1912-T51 R40).

Similarly to the reduced postdrainage white water lily distribution, the reduction in frequency of yellow water lily (bonnets) suggests that predrainage slough water depths were deeper than at present and further supports the predrainage estimates of 1 to 3 feet (30–90 cm) deep.

A Six-Mile Transect, 1886 and 1955

A fortuitous overlap in federal surveys permits a side by side comparison of predrainage and postdrainage water depths. The transect, along the southern boundary of Township 58 Range 35, extends eastward from within Shark River Slough into the Rockland Marl Marsh (plate 4, appendix T, fig. T.1). U.S. Deputy Surveyor William Mickler's observations of this transect in 1886 likely represent close to predrainage conditions because Disston's canals, which initiated lowering of Lake Okeechobee levels, had been completed only four years earlier. When the south boundary of the township was resurveyed almost seventy years later by Norville Shearer, "Cartographer (Cadastral)" for the U.S. Department of Interior, the Everglades of 1955 were well into the period of the most intensive, uncontrolled Everglades drainage (chapters 2, 3). Field notes from the two surveys (appendix T) provide a valuable comparison for evaluating changes in water depths and vegetation. Remarkably, both surveys were conducted in the last two weeks of January, about halfway through the dry season.

The vegetation present and recorded at the two time points likely reflects a running average of antecedent water depths prior to each survey. As such, the vegetation adds robustness to the comparison by buffering it from possible weather anomalies in 1886 or 1955 that might have affected water depths. Shearer, using aerial photography from 1953, was able to recognize landscape features described by Mickler under close to predrainage conditions but noted that some of the vegetation present within each feature had changed:

> A transparency depicting the record topographic features of the original survey by William Mickler disclosed the close relationship between those record items of topography and similar features on the ground as depicted on the aerial photographs. . . . The Mickler field notes report a number of water courses and small island-like areas 1 to 2 feet above the general elevation of the open glades, covered with bay, cypress, and palms which he refers to as "islands." Evidence of the island areas and water courses as noted in the original record were conclusive as to their position. Areas of saw grass noted in the record were not so readily determined due to their destruction by fire and seasonal fluctuation of soil moisture. Soil characteristics inductive to [conducive to] more luxuriant growth were noted at the record positions, however, and shading of growth cover was very discernible upon the aerial photographs. (Shearer 1955-T58 R35, p. 3)

The "General Descriptions" given by the two surveyors suggest substantially drier conditions at the later, postdrainage time point. In the lower-lying western part, within Shark River Slough, Shearer (1955-T58 R35) found mangroves, as did Lamar Johnson (1958, p. 5) and Kolipinski and Higer (1969, p. 17), suggesting saltwater and tidal oscillation. As the ground level rose to the east by an estimated 2 feet (60 cm), Shearer found an estimated 6 to 12 inches (15–30 cm) of water during the wet season and scattered growths of pine and palms on "islands" of higher ground (Shearer 1955-T58 R35, p. 7).

Mickler, on the other hand, reported no mangroves (although he did report and map them in other contemporaneous township surveys), no tidal action, no trees, and considerably deeper water depths. He found the whole township, including the higher eastern portion, to be inundated with fresh water 18 to 30 inches (45–75 cm) deep in late January, halfway through the dry season, and 30 to 50 inches (75–125 cm) deep in the wet season (Mickler 1886-T58 R35, p. 522).

Mickler's observation of no trees, repeated in each of the summaries for the 2nd through the 6th mile, is consistent with Shearer's observation that "the major portion of the scattered timber growth along the line was determined to be of an age younger than the original surveys and consequently bore no survey marks" (Shearer 1955-T58 R35, p. 7).

Together, the two surveys suggest a substantial reduction in water levels, with dry season (January) water depths more than 12 to 18 inches (30–45+ cm) shallower in 1955 than in 1886. The wet season reductions in depths appear to have been even greater. The vegetation reported at the two time points suggests changes consistent with substantial reductions in water levels: "thick sawgrass" (1886) vs. "low saw and needle grass" (1955); "no Timber" (1886) vs. "pines, palms and [saw] palmetto" (1955). Additionally, the report of "mangrove and other saltwater-loving vegetation" (1955) vs. no report of mangroves in 1886 suggests not only higher predrainage water depths in 1886 but also higher freshwater flows keeping saltwater farther downstream.

Overall, the two surveys suggest a dramatic pre- to postdrainage decrease in water depths, even in an area far from any local or regional canals, a testimony to the regional drainage discussed in chapters 2 and 3. Like the changes in water lily distributions, the observed changes support the conclusion that the estimated predrainage water depths shown in table 11.4, deeper than present-day depths, are accurate.

Downstream Cross-section through the Ridge and Slough Landscape

Plate 30, drawn from measurements made in the current system and from predrainage information gathered in this study, illustrates changing ground and water levels along a transect through the middle of the remaining portion of the Ridge and Slough landscape. The 50-mile-long (90 km) transect follows a predrainage flow line (Flow Magnitudes and Directions section, this chapter; fig. 11.2) through the length of this landscape, passing through areas that became Water Conservation Areas 3A and 3B and Everglades National Park. The three time periods trace the shift from natural to managed water conditions, leading to significantly different hydrologic regimes: (1) predrainage, with no human management; (2) circa 1945, when anthropogenic drainage had significantly lowered water levels and affected peat elevations but impoundments had not yet been constructed; and (3) 2005 (current), when both water levels and peat surfaces had been affected by four to five decades of managed impoundments as well as by continued drainage.

The basic ground elevation in each cross section of plate 30 corresponds to the bottoms of the sloughs, to which the relative heights of sawgrass ridges and tree islands were added. (Ridge positions in upstream/downstream direction are approximate; and for graphic simplicity, only the heads of tree islands are depicted; near-tails and tails are lower and would extend much farther downstream.) Current (2005) ground elevation was synthesized from U.S. Geological Survey data, High Accuracy Elevation Data (HAED), measured throughout the Everglades on a 400 by 400 m grid (Desmond 2003). Ground elevation for 1945 was estimated from the subsidence history described in chapter 3 and by comparison with predrainage and current ground elevations. Ground elevation for predrainage conditions was based upon this study (Topography section, this chapter).

Relative heights of ridges and tree islands for predrainage conditions were based upon estimates from this study; 1.5 feet (45 cm) for ridges and 2 to 4 feet (60–120 cm) for tree islands (chapter 8). The relative heights of ridges for 2005 conditions were taken from field measurements (McVoy 2007; Ross et al. 2003, p. 291, for Shark River Slough). Similarly, the relative heights of tree islands for 2005 conditions were estimated from field measurements (McVoy 2007), from field observations of water table and tree island vegetation for the islands north of Alligator Alley (Highway I-75), and from Ross et al. (2003, p. 297) for islands in Shark River Slough. Relative heights of tree islands in Water Conservation Area 3A south of I-75 and for WCA 3B were estimated from current studies (Rutchey et al. 2009), which indicated that islands experience varying degrees of flooding, but at least some are free from flooding a significant portion of the year.

Ridge and tree island heights for 1945 were estimated with the assumption that both tree islands and sawgrass ridges were lower in absolute elevation (i.e., relative to sea level) in 1945 than they had been under predrainage conditions, but that they were *not* lower in absolute elevation than they were in 2005. In looking at the tree island heads, it is helpful to consider both relative and absolute elevations. For example, the two tree island heads upstream from the L-67 levees appear to have become taller by 1945 and especially by 2005 than they were under predrainage conditions. Their height above surrounding sloughs has in fact increased (actually sloughs lost elevation), but their absolute elevation, relative to sea level, has decreased. All elements of the landscape have lost elevation, but in this particular area, the tree islands have lost less, probably because of effects of the surrounding deeper waters created by impoundment.

Reviewing the overall plate indicates that the slough bottoms, the basic ground elevation, decreased by 1 to 1.5 feet by 1945, with greater decreases toward the northern end due to thicker peats there. By 2005 the slough bottoms had declined between 1 and 3.5 feet (between 30 and 105 cm), again with the greatest losses at the northern end due to thicker peats there and the lowest post-impoundment water depths. By 1945 the height of the ridges had declined by one-third, from 1.5 to about 1 foot (45–30 cm) high; by 2005 they ranged between 1 foot (30 cm) high and almost complete loss, resulting in little or no elevation difference from surrounding sloughs. Interestingly, the field measurements showed a decline in relative ridge height not only in the drier upstream portion of WCA 3A, where severe oxidation would be expected, but also in the very deep downstream end of WCA 3A. A mechanistic reason for this is not presently known.

As expected, tree islands lost more elevation than the sloughs, between 1.5 and 3 feet (between 45 and 90 cm) by 1945. By 2005 tree island elevation loss relative to predrainage conditions varied from 1.5 feet to almost 5 feet (0.45 to 1.5 m), with the most extreme losses in areas with the greatest declines in absolute water surface elevation.

The water depths shown in plate 30 for 2005 were derived from field measurements made along this transect in 2005–2006 (McVoy 2006). Predrainage water depths reflect the estimates given in table 11.4. Water depths shown for 1945, ranging from a maximum of a little under 1.5 feet (45cm) to a below ground minimum of about -0.5 foot (-15 cm), were estimated from the information presented in chapter 3, including the widespread occurrence of peat fires.

It is worth emphasizing that the water depth configuration shown for predrainage and 1945 conditions, a water surface forming a plane parallel to the ground (slough) surface, is very much a nonequilibrium condition. That is, it is a condition that could be maintained only as a result of continuous inflow from farther upstream in the landscape and continuous outflow downstream. In contrast, the water surfaces shown for 2005 conditions within Water Conservation Areas 3A and 3B closely approximate equilibrium conditions, with the water surface approaching level, reflecting the low to zero flow conditions present in these impoundments.

Considering plate 30 as a whole, several features emerge:

- The current relationship of landscape elements to water depths is very different from the predrainage relationship.
- Drainage and impoundment have both affected the landscape, but in different ways.
- The extent of tree island elevation loss has likely been influenced by surrounding water levels.
- Under predrainage conditions, tree island heads were above surrounding water levels most of the

year but not much above the long-term annual maximum water elevation.

- The maintenance of the ridge-to-slough elevation differential present under predrainage conditions probably depended on the predrainage water depth regime.
- Predrainage descriptions of the Everglades as half land, half water reflect the sawgrass stands extending above the water surface throughout the year as well as the associated peat ridges that were above water at the end of the dry season.

Considering the cross section for 2005 conditions, the origin of the opinion that current water depths in Water Conservation Area 3A are "too deep for the Everglades" is readily found. It can also be seen, however, that the impression is correct only for the southern portion of Water Conservation 3A. The northern portion, in fact, dries out too much, exacerbating peat subsidence there. The 1945 cross section provides additional insights into possible origins of a sometimes expressed view that Everglades water depths should be no more than 18 inches (45 cm). Observers familiar with the Everglades in the 1940s and 1950s—after substantial drainage but prior to creation of the impoundments—saw a landscape with low overall water depths (maximum of about 18 inches; 45 cm) compared to predrainage conditions, and with tree islands unnaturally high above the surrounding water levels. Both the low water depths and the rarity of tree island inundation during that period might have contributed to the impression of 18 inches (45 cm) being a "good depth," and by extension, also the likely predrainage maximum depth.

Plate 30 suggests that the predrainage water depth estimates shown in table 11.4, when considered in conjunction with the predrainage relative heights of sawgrass ridges and tree islands, are consistent with a number of other known characteristics of the predrainage Ridge and Slough landscape. It also provides some possible explanations of currently observed characteristics, perceptions, and depth targets.

Comparison of Pre- and Postdrainage Sloughs

Another approach to evaluating the depth estimates presented in table 11.4 is through qualitative comparison of pre- and postdrainage abundances of a variety of animal and plant species influenced by slough water depths. In the Everglades this approach is constrained by the absence of intentional, regular measurements of pre- and postdrainage abundance. However, for at least some species, sufficient information is available to estimate whether populations have increased or decreased since drainage. Consistent trends across multiple species, if present, would indicate whether predrainage sloughs were wetter or drier than at present, allowing predrainage depths to be estimated by comparison with the more systematically measured current water depths. Appendix R is an extensive compilation of such pre- and postdrainage comparisons. The comparisons, while generally focused on Shark River Slough, would also apply in many cases to the northern portion of Water Conservation Area 3A, where the current water regime and ridge-to-slough height differential are fairly similar to current Shark River Slough conditions (plate 30 bottom).

Plate 29 depicts the results found in appendix R: a typical predrainage slough and ridge (top panel) and the appearance of the same slough and ridge under present-day, postdrainage conditions. Prior to drainage, sloughs were self-maintaining habitats with generally open water and water lilies. The openness and absence of dense growths of emergent wet prairie plant species permitted easy and rapid canoe navigation in most years. Slough depths were sufficient to permit multiyear hydroperiods, supporting species such as bass and other larger fish that now depend on the deeper water of canals. Trading post records of shipments of tens of thousands of alligator hides and hundreds of otter pelts (Kersey 1975) suggest that these species were once abundant. Open sloughs would have benefited both alligators and otters, and the greater predrainage relative height of ridges would likely have provided good nesting sites for alligators (Frank Mazzotti, pers. comm., 2006). However, greater slough depths would not have benefited all species. Wading birds such as White Ibis, Snowy Egrets, and Wood Storks would typically have found water depths too deep for feeding throughout most years. At the other end of the spectrum, appendix R suggests that phenomena associated with shallower slough depths, such as dryouts, cessation of downstream flows, and peat fires, would have been rare under predrainage slough water depths.

John Lamb's observations of the Hillsboro Lakes area in the 1920s and '30s (Lamb 1975, pp. 1–5), when the water was still deep enough for this area to resemble the predrainage Ridge and Slough landscape, describe many aspects depicted in the predrainage panel of plate 29. Tree islands were present, "but there were very few islands with dry ground for campsites" (Lamb 1975, p. 1). "Very few of the [wading] birds used the interior of the area as a feeding ground except on all edges, due to the water depth" (p. 5). "Deer were plentiful along the eastern edge [likely refers to the Eastern Flatwoods] though none were found in the interior [of the Hillsboro Lakes area] at that time" (p. 2). Species that benefited from deeper, open water thrived: "While boating down the deep runs we could see great schools of bass and pan fish with the headlights" (p. 2). "Alligators were very abundant in the Loxahatchee at that time" (p. 4). "We gigged all the ducks we wanted by blinding them with the lights as they roosted on the sloughs by the thousands" (p. 2). While deer were not present in the Loxahatchee Ridge and Slough landscape, Lamb did report abundant deer in other parts of the Everglades that had been more strongly affected by drainage, for instance the area that later became northern Water Conservation Area 3A (pp. 2–3), where even by the 1940s, drainage had converted large areas of Ridge and Slough landscape into essentially pure sawgrass (chapter 3).

Overall, the observations in appendix R confirm that predrainage sloughs were substantially deeper than what is presently seen in Shark River Slough. Comparison of the predrainage slough characteristics shown in the top panel of plate 29 with field conditions presently seen in Water Conservation Area 3A suggests that water depths presently (2010) found in mid- to southern Water Conservation Area 3A (south of Highway I-75) likely resemble predrainage slough depths.

In summary, each of the above five ways of evaluating the estimated predrainage slough and ridge water depths shown in table 11.4 has imperfections and limitations. Nevertheless, taken together, these evaluations suggest that the estimates certainly do not overestimate predrainage water depths and are likely accurate.

Ecological Implications of Predrainage Water Depth Regimes

This section focuses on the links between the predrainage water depths shown in table 11.4 and plate 8 and the likely predrainage occurrence of macroinvertebrates, small fish, large fish, and wading birds. Detailed predrainage information on these faunal groups is lacking, so this section relies heavily on faunal studies of the current Everglades, extrapolated to predrainage conditions. It will be helpful to refer to plates 29 and 30.

Sawgrass Plains

The water depth regime of the Sawgrass Plains suggests fairly straightforward ecological implications. The ground surface was extremely uniform, so when the water table dropped each year below ground surface, few if any points of standing water remained, resulting in an annual dryout throughout the landscape. Fauna of the Sawgrass Plains would therefore have been restricted to animal species that either recolonized the reappearing surface water each year or could survive the late dry season by burrowing into the moist peat soil. Fauna and flora were additionally restricted by the high stem density and low light levels present beneath the tall, dense sawgrass canopy. The annual dryout of the Sawgrass Plains landscape may have had a significant influence on Everglades biogeochemistry, as the upper layer of peat and any compressed flocculant material would have been exposed to aerobic decay for a few months each year. An annual pulse of nutrients released into the water column as water levels rose again would be expected.

Ridge and Slough Landscape

The fact that the water levels in the predrainage Ridge and Slough landscape rose and fell seasonally, combined with the fact that the elevation of the ground surface varied spatially (i.e., ridges, sloughs, and tree islands), suggests that one way of considering this landscape is through comparison of the spatial mean and range of the ground surface elevations (microtopography) with the temporal mean and range of the water surface elevations (fig. 8.3). Significantly, for predrainage conditions, the vertical ranges of water and ground elevations were of very similar magnitude; about 2 feet (60 cm) for water, and about 3–4 feet (90–120cm) for ground surface. Even more significantly, the means of the ground and water surface elevations, measured relative to sea level, were also similar, with the difference being only 1 to 2 feet (30–60 cm).

Figure 8.3 illustrates the significance of the overlap in these numbers. Annual fluctuations in water levels occurred in the same range as that of the different landscape component elevations. The ecological importance of this can be appreciated by imagining alternative scenarios. Assuming the same microtopography and the same seasonal range of water depths but hypothetically increasing the mean water depth by as little as 2 feet (60 cm) would have changed the Ridge and Slough landscape entirely; from a mix of aquatic, semi-aquatic, and essentially terrestrial habitats into a purely aquatic environment, essentially a large lake. Similarly, assuming the same microtopography and water depth ranges but hypothetically lowering the mean water depth by only 2 feet (60 cm) would have yielded a permanently terrestrial landscape with water accumulating seasonally in sloughs. A little lower still and the landscape would have been purely terrestrial throughout the year. Alternatively, one can imagine a hypothetical scenario where the microtopography and mean water depths stay the same, but the range in water depths is increased or decreased. With only a 1- to 2-foot (30–60 cm) increase in annual range, the whole landscape would have become only seasonally aquatic, drying out each year, with a depauperate fauna restricted to species that either burrowed or annually recolonized, similar to the predrainage Sawgrass Plains described above.

These thought experiments suggest considerable significance to the actual water and ground elevation regimes that were present within the predrainage Ridge and Slough landscape. Water levels typically oscillated from somewhat above slough bottom to right around tree island surface, resulting in a mix of continuously aquatic (sloughs), semi-aquatic (ridges), and essentially terrestrial (tree islands) habitats as well as a seasonal oscillation in landscape status ranging between 50 percent and 95 percent inundated. These considerations and figure 8.3 suggest that considering this landscape simply as "Everglades marsh" significantly obscures key physical and ecological aspects. Similarly, although 95 percent of the area was water-covered for a portion of the year, describing the landscape as lakelike would also overlook important aspects.

The ecological importance of distinguishing ridge from slough habitats is underscored by the significant faunal differences that have been noted between ridges and sloughs under current conditions. The presence of faunal differences is significant even under the current, reduced elevation differential of less than 0.7 feet (20 cm; McVoy 2007; Jordan et al. 1997, p. 279; Ross et al. 2003, p. 291). Faunal differences between ridges and sloughs, when the predrainage elevation differential was more than twice as great (1.5 feet; 45 cm), would have been even greater. Quantitative comparisons of ridge and slough biology have been limited by the challenge of sampling within dense, tall sawgrass stands, but population or foraging densities have been quantified for current conditions for the two main macroinvertebrate prey for fishes and wading birds, grass shrimp and crayfish; for a fishing spider; for small fish assemblages; and for large fishes, alligators, and wading birds.

Jordan (1996, pp. 40–41) found that prawns (grass shrimp; *Palaemonetes paludosus*) were five times more abundant in sloughs (45 per m^2) than in sawgrass ridges (8 per m^2). In contrast, crayfish were five times more abundant in sawgrass ridges than in sloughs (5 vs. 1 per m^2; Jordan 1996, p. 41). Jordan (1996) referred to *Procambarus alleni*; however, these may in fact have been *P. fallax* (Hendrix and Loftus 2000, p. 197). Jordan et al. (1994, p. 239) found higher densities of the fishing spider *Dolomedes triton* in sawgrass than in either wet prairies or alligator holes, both of which are more similar to sloughs than to sawgrass. Densities of small fishes (<8 cm standard length) varied by species, but as an assemblage, they were more common in sloughs than in the sawgrass ridges (42 vs. 15 per m^2) of the Arthur R. Marshall Loxahatchee National Wildlife Refuge (Jordan 1996, p. 67). Trexler et al. (2002, p. 157) found very similar comparisons between wet prairie sites (shallow sloughs) and sawgrass ridges in Water Conservation Areas 3A and 3B, Shark River Slough, and Taylor Slough: 39 versus 21 fish per m^2.

A number of observers (Dineen 1984, p. 261; Loftus and Kushlan 1987; Jordan 1996, p. 68; Loftus and Eklund 1994, p. 477) indicated that large fish (>8 cm standard length) are found primarily in sloughs or wet prairies, and that they avoid sawgrass ridges, probably due to the high density of sawgrass stems (Loftus and Eklund 1994, p. 477; Chick et al. 2004, pp. 653–654). As a result, it is likely that sawgrass ridges act as predator refuges for small fishes and prawns, protecting them from large predatory fish (Loftus and Eklund 1994, pp. 461, 477; Jordan 1996, pp. 49, 73). Jordan (1996, pp. 45, 73), however, notes that prawns and small fishes may be more subject to predation by crayfish in sawgrass ridges due to the higher crayfish density there. Under current water management practices, all observers note that the largest fishes and largest concentrations of large fishes are found in canals open to the Everglades, indicating that canals currently provide refugia for large fish when sloughs dry out, as illustrated in the bottom panel of plate 29 (Dineen 1984, p. 266; Loftus and Kushlan 1987; Loftus and Eklund 1994, p. 477; Florida Fish and Wildlife Conservation Commission 2006, p. 22).

Wading birds also distinguish between sloughs and ridges. Bancroft et al. (1992, p. 395) reported that Wood Storks "feed in the most open areas, where their tactile foraging habits are least affected by submerged or emergent vegetation," and noted that they feed in "drying water-lily sloughs" at the south ends of Water Conservation Areas 1 and 2, and also in water-lily sloughs at the southern end of WCA 3A, but only when water levels are very low (c.f., plate 30 bottom panel). Hoffman et al. (1994, pp. 585–586) noted that Great Egrets, Great Blue Herons, White Ibises, and Wood Storks tended to avoid dense sawgrass. In the most quantitative study, Bancroft et al. (2002, p. 274) found that slough habitats in particular "appeared to be extremely important to wading birds." The tendency of wading birds to forage in sloughs and avoid dense sawgrass suggests that sawgrass ridges were doubly important as predator refuges for small fish and prawns; both from large, predatory fishes and from wading birds (Loftus and Eklund 1994, p. 461).

In addition to recording differing population densities or foraging intensities in ridges versus sloughs, observers have also recorded migration of prey species back and forth between the two habitats. Despite working in an area with diminished differentiation of ridges and sloughs (Shark River Slough), Kushlan and Kushlan (1980, p. 402) observed large numbers of prawns migrating during annual drying periods from dense vegetation into more open, deeper water areas. Loftus and Kushlan (1987) recorded a number of fish species that were widely dispersed during the wet season but concentrated in deeper water areas during the dry season. Perhaps most significant, Jordan, working in a peat-based area of sawgrass ridges, sloughs, and wet prairies (Loxhatchee National Wildlife Refuge), recorded not only increasing slough densities of fish and prawns during a drying event in May 1992 but also a subsequent "rebound," when both slough and ridge population densities returned to close to pre-drying event levels as water levels rose again (Jordan 1996, p. 75). Jordan interpreted this rebound as redispersal of the concentrated slough populations back into the surrounding sawgrass. Concentration of prawns was about tenfold during the May 1992 drying event, raising the density to 425 per m^2 (Jordan 1996, pp. 46–47) and about sixteenfold for small fishes, raising fish densities to about 400 per m^2 (Jordan 1996, pp. 75, 68). Cook (Sklar et al. 2006, pp. 6–14) found crayfish migration from ridges to sloughs during a drying period in created macrocosms, although results from a less detailed study (Jordan 1996, p. 44) suggested that in the Everglades, crayfish burrowed into the peat-based ridges rather than migrating. (The differing results may be habitat-specific as Cook studied crayfish in created macrocosms, whereas Jordan studied them in the Everglades; in addition, Jordan did not distinguish between *P. alleni* and *P. fallax*.) Given the function of sawgrass ridges as refuges from both predatory fish and wading birds, the annual migrations of fishes, prawns, and possibly crayfish from drying sawgrass ridges and into sloughs seems likely to have had important ecological consequences.

The critical dependence of wading birds, wading bird nesting success, and other predators on prey concentration during receding dry season water levels in the Everglades is widely accepted (e.g., Kushlan et al. 1975; Bancroft et al. 1992, p. 395; Frederick and Spalding 1994, p. 659; Ogden 1994, p. 566; Bancroft et al. 2002, p. 265; Gawlik 2002, p. 341), suggesting an obvious link to ridge-to-slough migrations. Although less well studied, there appears to be a similar relation between seasonality, large predatory fishes, and their prey. Large predatory fishes feed most heavily on small fishes and prawns when dry season water levels drop below ground surface of sawgrass ridges, forcing these prey species into and concentrating them in open sloughs and 'gator holes (Loftus and Kushlan 1987; Loftus and Eklund 1994, p. 477). Under present managed water conditions, in areas where sloughs dry out, the displacement of prey and predators into deeper waters is instead into canals.

With the preceding review of faunal studies of the current Everglades in mind, the implications of predrainage ground elevations and water depths can be assessed. To review, under predrainage conditions the tree islands—that is, the tree island heads, the highest points—are estimated to have been 2–4 feet (60–120 cm) above slough bottom, sawgrass ridges 1.5 feet (45 cm) above slough bottom, and sloughs to have experienced water depths characterized by a long-term average annual low water depth of 1 foot (30 cm), on average in mid-May, and an analogous long-term average annual high of 3 feet (90 cm), generally in mid-October (table 11.4; levels illustrated in plate 30 top panel). Those relations are assumed to have applied quite consistently throughout the Ridge and Slough landscape (chapter 8; plate 8). Water depths in any given year would have varied from these long-term average values.

Tree island heads would have experienced a range of conditions due to variation in height from one to the next; some would rarely have been covered with water, others would have been inundated as much as half the year, similar to bayhead swamp tree islands at present in Shark River Slough (Ross et al. 2003, p. 297). Because of the 2-foot (60 cm) annual range in water depths, most island heads would have been at least 1–2 feet (30–60 cm) above the water surface for a portion of each year. At the same time, the lower portion of the Gandy peat soil forming the heads would have been continuously wetted by lateral contact with the surrounding slough water, preventing peat fires, preventing tree drought stress, and likely reducing tree island fires through high soil and microclimate humidity (Egler 1952, pp. 227–228; Craighead 1971, p. 126). Total tree island area was on the order of 5 percent of the landscape.

Sawgrass ridges were inundated during most of the year, but on average lost surface water each year as the water table dropped to about 0.5 feet (15 cm) below soil surface. This drydown would have been accompanied by small fish, prawn, and possibly crayfish migration into sloughs. Accumulated flocculant organic material would have been compressed and likely incorporated into the peat during each drydown. High canopy and stem densities generally prevented foraging by wading birds and large, predatory fishes, providing smaller species with refuge from predation throughout much of the year. The distinct height of the ridges above sloughs likely made them attractive for alligator nests (Fogarty 1984, p. 215; Frank Mazzotti, pers. com., 2006). Ridges covered about half of the landscape surface area (chapter 8).

Sloughs were open water areas with white water lilies (*Nymphaea odorata*; plate 22; see also plates 17, 19), very likely bladderwort (*Utricularia* spp.) in the water column, and in some places yellow water lilies (*Nuphar advena*). Most sloughs would have had peat substrates. Hydroperiods would have been essentially continuous. Complete dryouts were rare, occurring only in the driest years. As a result, outflows from the landscape, whether to the Atlantic coastal rivers or to Florida Bay, continued throughout the year except in the driest years. Flocculant organic material (plant detritus) almost certainly was present as a loose layer at the bottom of the water column. Predrainage flow velocities may have transported this flocculent material downstream, either in response to average flow velocities or possibly in response to transient higher velocities associated with local rainfall or tropical storms.

Slough populations of large fishes were fostered by the multiyear hydroperiods. The deeper water of sloughs likely served the function now served by canals when sloughs dry out. Populations of small fish would likely also have been greater than at present. Jordan (1996, p. 44) noted that populations of small aquatic organisms in the Everglades appeared to increase under prolonged inundation such as would have been associated with the predrainage depths of 1 to 3 feet (30–90 cm). Jordan ascribed this increase in part to greater accumulation of detritus and a more productive detrital food chain (Loftus et al. 1990, p. 20) as well as to uninterrupted reproduction (Loftus and Eklund 1994, p. 463). However, on the basis of pull trap data, Kushlan (1976, p. 821) came to the opposite conclusion regarding prolonged inundation and small fish populations, indicating that predation by increased populations of large fish reduces small fish populations and that annual slough drydowns were therefore needed to limit large fish populations (Kushlan 1976, p. 824; Kushlan 1987). Reanalyzing Kushlan's pull trap data set with an overlapping throw trap data set, Loftus and Eklund (1994, p. 474) refuted Kushlan's conclusion, finding it an artifact of the trapping method. The throw trap data set showed high and increasing densities of small fish during a multiyear period without a drydown and did not show a shift from small to large species. Loftus and Eklund (1994, p. 480) concluded that both large- and small-fish assemblages respond positively to prolonged inundation, with dense vegetation (e.g., sawgrass ridges) likely providing a predator refuge for the smaller species. See Trexler et al. (2005) for additional information.

As a whole, the predrainage Ridge and Slough landscape may be thought of as a vast "lake country," an expanse of many, many long, narrow lakes, each only 100–200 m (300–700 feet) wide (plates 17, 19, 22). Each lake was bordered on the long east and west sides by sawgrass ridges functioning essentially as littoral zones (Loftus and Eklund 1994, p. 477). Those littoral zones dried in most years, causing annual spikes in the slough populations of species that migrated from the ridges, such as small fishes and prawns. The ratio of littoral area to "lake" area was large, approximately 1:1 (chapter 8), so the influence of the annual littoral zone dryouts would have been substantial. As noted earlier in regard to the Sawgrass Plains, the period of aerobic decay that would have occurred during the dryouts may have been followed by an annual nutrient pulse into the water column as water levels rose again.

The interpretation of the predrainage Ridge and Slough landscape presented here differs substantially from that of Turner et al. (1999, p. 907), who suggested that (1) "the longest-hydroperiod region of the drainage [i.e., the Ridge and Slough landscape] may have supported a fish community similar to that found in regions of WCA-3A that now experience near-lacustrine conditions"; and (2) "there is little reason, however, to anticipate that the biomass of fishes per unit area estimated from our study differs from that of historical Everglades marshes of similar hydroperiod in years lacking marked dry-down." Both statements imply a stronger than actual resemblance of current to predrainage conditions and tend to overlook the influence of a landscape with thoroughly interwoven but strongly differentiated ridge and slough habitats, with ridges regularly drying out and sloughs rarely doing so. Comparison of the top and bottom panels of plate 30 helps clarify the strong differences. The "near-lacustrine conditions" currently found in southern Water Conservation Area 3A (plate 30 bottom) reflect strongly altered conditions: diminished ridge-slough elevation differential, absence of ridge dryouts, overly deep sloughs, and greatly diminished flows—all making this area a poor analog for predrainage conditions. South of Tamiami Trail, the Shark River Slough portion of Everglades National Park is an even poorer analog, with water depths much shallower than predrainage, diminished ridge-slough elevation differential, a reduced or absent peat layer covering the bedrock, and much more frequent dryouts.

Turner et al. (1999, pp. 905–906) found the remaining Everglades to have high algal biomass but low standing stocks of invertebrates and fish when compared with other wetlands. The considerations discussed in this chapter so far suggest that this relationship would likely have been quite different for the predrainage Everglades, when standing stocks of fish were likely higher. Similarly, it is possible that Dineen's (1984, p. 258) observation that individual fish from the present-day Everglades do not grow as large as individuals of the species elsewhere in Florida or farther north might not have been true under predrainage conditions.

The preceding paragraphs have focused on the characteristics of small areas, essentially points within the landscape. It is helpful to shift focus to a more regional scale, to the landscape as a whole, considering especially the implications of the assumption of parallel planes of water and ground surface. One obvious implication, already referred to earlier and by Loftus and Eklund (1994, p. 478), is that under predrainage conditions, the Ridge and Slough landscape north and south of Tamiami Trail (Water Conservation Area 3A and Everglades National Park) would have been essentially indistinguishable (fig. 11.3; plate 30 top). Only flow rates might have been distinct, with somewhat higher rates in Shark River Slough (Everglades National Park) due to a narrower cross section and funneling by the Marl Marsh landscapes. A second highly important implication of the parallel planes concept is that all of the preceding considerations of ridges and sloughs would have applied throughout the extent of the landscape.

One ecological aspect of this uniformity is that across the full width of the landscape, the ground surface does not appear to have been "dished"; that is, the landscape was not deeper in the middle or shallower at the edges. Instead, Ridge and Slough landscape extended directly up to the bordering landscapes (see appendix D). Ecologically, this means that at the regional scale, water receding during the dry season did not concentrate fauna toward the middle of the landscape, nor did the landscape dry particularly earlier along the edges than in the middle.

Similarly, in the downstream direction, predrainage vegetation indicates that the concept of parallel planes also applied, such that even during the dry season lowering of the water level, the plane of water remained parallel to the ground surface plane (plate 30 top). Therefore, both ridges and sloughs down the entire length of the landscape dried essentially simultaneously, rather than sequentially from upstream to downstream. This has far-reaching ecological significance for the timing of seasonal phenomenona in the predrainage Everglades and implies a strong contrast with current Everglades conditions.

Comparison of the Water Conservation Area 3A portion of the top and bottom panels of plate 30 clarifies the difference. Under currently impounded conditions, the plane of the water surface remains approximately parallel during drying, but it is parallel to the slope of the high water plane and clearly *not* parallel to the ground surface. Thus under current, impounded conditions, as waters continue to recede, a drying zone forms upstream and moves progressively downstream through the landscape. It is not known whether this moving zone concentrates fishes and prawns by "sweeping" them downstream ahead of the drying front, but simply by producing shallow depths it does create a situation of "slow-moving [foraging] patches . . . [that] move across the landscape" (Gawlik 2002, p. 341). Depending on initial water depths and the rate of drying, the downstream movement across the landscape can continue until the patches reach the bottom of the impoundment at Tamiami Trail. A highly significant ecological aspect of this moving zone, an impoundment-derived phenomenon, is that it temporally lengthens the period of availability of optimal foraging conditions while also concentrating them spatially. Within the predrainage Ridge and Slough landscape, exactly the opposite would have been true: the period of optimal foraging would have been temporally shortened but spatially expanded or "diluted"; that is, occurring simultaneously across the whole landscape. This dramatically different situation between predrainage and current Ridge and Slough drying conditions may be a partial explanation for the comment that the "current reliance of reproductive success [of wading birds] on rapid drying rates is to a large extent unrepresentative of the historical Everglades" (Frederick and Spalding 1994, p. 660). It may also be a contributor to the postdrainage shift in wading bird colony location from the lower end of Shark River Slough upstream into the Water Conservation Areas, believed to have occurred sometime between the 1940s and 1970s (Ogden 1994, pp. 533–534).

As Gawlik (2002) demonstrated in an elegant field experiment, conveying fish to wading birds (prey availability) requires more than prey density alone; prey vulnerability to capture is equally important. Furthermore, as Gawlik (2002, p. 330) points out, in hydrologically pulsed wetlands such as the Everglades, water depths may have opposing effects on prey density (e.g., deeper depths reducing frequency of dryouts and yielding greater prey populations) and prey availability (deeper depths reducing wading bird access to prey).

Given that in the current Everglades, wading birds primarily use sloughs (as discussed earlier), it is appropriate to compare the available information on wading bird feeding depth thresholds with the predrainage estimates for sloughs (table 11.4); that is, a typical annual low water depth of 1 foot (30 cm) and high of 3 feet (90 cm). The comparison is thought-provoking. Table 11.5 suggests that White Ibis, Wood Storks, and Snowy Egrets, which Gawlik (2002, p. 339) refers to as "searchers," would have had difficulty feeding in the predrainage Ridge and Slough landscape during years with typical water depths. Water depths would have been deeper than the thresholds for these species throughout the year, even at the end of the dry season. However, during years with water depths approximately 1 foot (30 cm) lower than typical, there would have been a window of foraging opportunity for those species during the mid to late dry season. The prey densities and biomass during such years would very likely have been higher than seen at present as a result of the generally long (multiyear) hydroperiods.

The previous scenario refers to the Ridge and Slough landscape. A different form of prey concentration still occurs in the numerous creeks that drain water from the coastal prairies (e.g., Perrine Marl Marsh) and in southern sloughs from early to late dry season, offering wading birds numerous locations with concentrated prey (Jerome Lorenz, pers. comm., 2009). Similarly, Craighead noted a retention of water and prey species within both an inner, freshwater mangrove zone and an outer, brackish water mangrove zone: "The ponds and mangrove flats of the outer zone likewise retain water when the glades are dry. This water may vary from nearly fresh in the fall to more saline than sea water toward the end of winter. . . . The great reservoir of fresh to brackish water in these two zones is potentially rich in aquatic life, and provides a major source of food for much of the wildlife, especially ducks, wading birds, otters, and alligators" (Craighead 1964, p. 12).

Less detailed threshold information is available for the other five species in table 11.5, but size of the individuals and observations by Gawlik (2002, p. 339) suggest that Great Egrets and Great Blue Herons tolerate the deepest water, perhaps up to 1.5 or 2 feet (45 or 60 cm). It seems likely that the last five species in the table

Table 11.5. Approximate optimal or maximum feeding water depths for Everglades wading birds

Wading Bird Species	Depth, h (ft)	Depth, h (cm)	Source
White Ibis	$0.6 < h < 0.9$	$19 < h < 28$	Gawlik (2002, p. 338)[a]
White Ibis	0.3	10	Bancroft et al. (2002, p. 271)[b]
White Ibis	0.2–0.3	5–10	Kushlan (1986, p. 157)[c]
Wood Stork	$0.6 < h < 0.9$	$19 < h < 28$	Gawlik (2002, p. 338)[a]
Wood Stork	0.7	20	Bancroft et al. (2002, p. 271)[b]
Wood Stork	< 1.3	< 40	Kushlan (1986, p. 157)[d]
Snowy Egret	$0.6 < h < 0.9$	$19 < h < 28$	Gawlik (2002, p. 338)[a]
Glossy Ibis	< 0.9	< 28	Gawlik (2002, p. 337)[e]
Little Blue Heron	> 0.9	> 28	Gawlik (2002, p. 337)[f]
Tricolored Heron	> 0.9	> 28	Gawlik (2002, p. 337)[g]
Great Egret	> 0.9	> 28	Gawlik (2002, p. 337)[h]
Great Egret	1	30	Bancroft et al. (2002, p. 271)[b]
Great Blue Heron	> 0.9	> 28	Gawlik (2002, p. 338)[i]
Great Blue Heron	2.6	80	Bancroft et al. (2002, p. 271)[b]

Notes: a. "a depth threshold between 19 cm and 28 cm that seem to constrain the use of foraging sites"

b. "plateau or threshold of maximum use"

c. "usual" = 5–10 cm; "maximum" = 30 cm

d. "up to 40 cm"

e. "significantly more abundant in the shallow [10 cm] and medium [19 cm] than the deep [28 cm] treatments"

f. "did not appear to be constrained by deep water [28 cm]"

g. "used only the shallow treatments early in the experiment, but eventually moved into the medium- [19 cm] and deep-water [28 cm] treatments.

h. "they foraged in the deep water [28 cm] but their initial preference was for the shallow treatments [10 cm]"

i. "the density of Great Blue Herons was relative constant throughout the experiment . . . and showed no significant response [to water depth] at the levels provided [maximum of 28 cm]" would have been able to forage in predrainage sloughs during the latter half of the dry season and that they also would have experienced better foraging during drier years.

In this context, it is appropriate to recall the earlier discussion of implications of parallel planes of water and ground in the predrainage system, and that all areas of the Ridge and Slough landscape experienced similar water depths at the same time. Recalling the substantial predrainage elevation difference between ridges and sloughs, it is logical to wonder whether there might not have been ecotones of intermediate elevation and lower plant density along the edge between sloughs and sawgrass ridges. If present, such areas would have had shallower water and—given the large numbers of ridges—might collectively have formed a significant area. However, figure 8.5 and the accompanying discussion in chapter 8 suggest that the predrainage elevation transitions were abrupt, occurring over only a few feet (< 1 m). Even in the current postdrainage system, where previously sharp distinctions have tended to blur, several authors have remarked on the sharpness of the transitions between sloughs and sawgrass ridges (McPherson 1973a, pp. 32, 50; Wood and Tanner 1990, p. 131; Jordan et al. 1997, p. 277; see also plate 22). Older photographs of the landscape (e.g., plate 17), narrative descriptions, and the field notes from thousands of slough-ridge transitions measured and recorded in the township surveys (1911–1918) all also suggest sharp transitions.

In summary, it seems unlikely that Gawlik's (2002, p. 343) scenario of ideal wading bird foraging habitat would have been present within the predrainage Ridge and Slough landscape, at least under weather conditions corresponding to long-term averages. Predrainage water depths would have prevented sloughs from drying down to low enough to leave small pools with concentrated prey, and zones of such pools would not have moved across this landscape over time, as they do in the current system. The northward, postdrainage shift of wading bird colonies from the coastal areas to the Water Conservation Areas suggested by Ogden (1994, pp. 533–534) would have been unlikely to occur under predrainage conditions, when the Ridge and Slough landscape there generally provided poor foraging conditions (deep water with low prey vulnerability). Predrainage water depths would also mostly have been too deep for the characteristic "wet prairie" species of *Eleocharis cellulosa*, *E. inundata*, *Rhynchospora tracyi*, and *Panicum hemitomon*. Instead, numerous township surveys from 1911 to 1918 indicate that the distribution of water lily peat ("Loxahatchee") soils (Jones et al. 1948a, e; see also chapters 2, 3) and paleoecological studies (e.g., Willard et al. 2001) all indicate that most of the areas now described as wet prairies would originally have been open, white water lily sloughs.

Ochopee and Rockland Marl Marshes

As discussed in chapter 9, these two marl marshes probably sloped slightly upward to the west and east of Shark River Slough. Figure 11.7 suggests that the predrainage water surface extending across the two Marl Marsh landscapes and Shark River Slough was approximately level, producing uneven water depths, deepest in Shark River Slough and somewhat shallower at the outer edges of the marl marshes. In contrast to the predrainage Ridge and Slough landscape, where parallel water and ground surfaces produced uniform drying conditions across the landscape over time, the slight slope of the marl marshes would have created both spatial and temporal movement of drying conditions. Whether the receding dry season water levels would have concentrated fish regionally (Regional Concentration Model, Trexler et al. 2002, p. 177), from the marl marshes all the way into Shark River Slough, or whether they would instead have stranded concentrated populations of fish in local depressions (Local Concentration Model, Trexler et al. 2002, p. 177) is unknown, but either way a regionally systematic pattern of increasing prey availability and prey density might typically have occurred each year in the Shark River Slough–Marl Marsh area (now often referred to as the southern Everglades). In this multi-landscape area, exactly the ideal hydrologic conditions indicated by Gawlik (2002, p. 343) to "maximize the spatial extent and seasonal duration of high-quality [wading bird foraging] patches" would have been present, namely "all short-hydroperiod marshes [i.e., the full extent of the Marl Marsh landscape] inundated at the start of the dry season, followed by a large decrease in water levels over the course of the drydown." The predrainage 2-foot (60 cm) annual range would have accomplished this, drying out much of both marl marshes by the end of the dry season, but leaving water in the sloughs of Shark River Slough.

In addition, the originally deeper water and longer hydroperiods in the marl marshes (8–9 months, table 11.4) would have fostered higher production of more species of fishes and invertebrates and allowed more dry season survival in local refugia. It would also have been more favorable for amphibians and alligators and other reptiles.

The Everglades and Its Surroundings

The overall picture of predrainage water depths, wading bird prey concentration, prey density, and sustained availability of areas of high prey vulnerability throughout the dry season probably involved the full range of southern Florida habitats, including ones outside of the Everglades. One can imagine that at the beginning of the dry season, when water levels might still have been too deep for optimal foraging in the Marl Marsh landscapes, water levels in the numerous seasonal ponds within the Eastern and Western Flatwoods and in Big Cypress might have been very suitable. For a long period during the middle of the dry season, wading birds might have followed drying conditions down the slopes of the marl marshes, and longer-legged wading birds could have hunted in the sloughs of the Ridge and Slough landscape. Finally, in dry years, there might have been exceptionally good foraging during late dry season in the sloughs of the Ridge and Slough landscape. Large populations of prey, including large multiyear fishes accumulated over multiple years, would suddenly have become available as water depths dropped below normal depths.

Inflows to the Everglades

The principal inflows into the Everglades were rainfall and outflows from Lake Okeechobee. Early maps, the few available narrative accounts, and analysis of directional landscape patterns (U.S. Department of Agriculture–Soil Conservation Service [USDA-SCS] 1940 and satellite imagery) suggest that water also entered the predrainage Everglades from the Big Cypress, possibly in substantial volumes. Inflows from groundwater are mentioned by several observers (this chapter) and may have been significant, especially from the Everglades Keys, but are not quantifiable from historical sources. Additional water may have flowed into the Everglades from the Eastern and Western Flatwoods, but landscape patterns (USDA-SCS 1940) suggest the volumes were small. The Loxahatchee Slough may historically have provided water flow into, or received outflow from, the Everglades at different times (plate 11 in Parker et al. 1955).

Inflows from Lake Okeechobee

The predrainage vegetation bordering the southern half of Lake Okeechobee (plate 5) and its "tuning" to hydrology suggests that water from Lake Okeechobee flowed continuously into the Everglades during much of each year. The portion of the lakeshore that overflowed—70 miles along the southern shore, from Fisheating Creek on the west to Pelican Bay on the east—was formed by the accumulation of organic, wetland soils. Those soils, as much as 10 or 12 feet thick prior to drainage (Kreamer 1892; see also fig. 3.1), provided the "plug" or dam that allowed the waters of Lake Okeechobee to accumulate (Gleason et al. 1984, p. 291). Two different Everglades landscapes contributed to the formation of this broad lakeshore: the Sawgrass Plains bordering the southwestern shoreline and the Custard Apple Swamp bordering the southern and southeastern shoreline (plate 5).

In the Sawgrass Plains (chapter 6), we showed that along the western and southwestern portion of the shoreline, sawgrass extended directly to the lake edge. The presence of sawgrass right up to the shoreline implies that water depths along this shoreline must typically have varied within the range that sustains this community. This indicates the absence of an elevated rim between the lake and the Sawgrass Plains. As might then be expected, the water surface within the sawgrass marsh during much or most of the year was simply an extension of the water surface of the lake, as reported here for 1907: "The men waded the swamp [south of Lake Hicpochee] and continued the [survey] line in the direction of Lake Okeechobee until a point was reached where there was no slope in the surface of the water for several miles. This condition extended to the open water of Lake Okeechobee" (Wright 1909c, p. 153).

Sackett (1888, p. [5]) reported very similar conditions for 1878 and 1886. The configuration profiled in figure 5.3, parts A and B—a rimless border, water levels continuous with the lake, a sawgrass shoreline, and deep, continuous profiles of sawgrass peat (Dachnowski-Stokes 1930)—suggests that the water depth requirements of sawgrass can provide a quantitative measure of typical predrainage lake levels. Thus, along the western and southwestern shoreline, lake levels would typically have risen to a maximum of about 1.5 feet (45 cm) above the peat surface by the end of the wet season and dropped annually to about 0.5 feet (15 cm) below peat surface by the end of the dry season (table 11.4). During the nine to ten months of the year when lake levels typically exceeded the peat surface, water would have flowed from the lake into the Everglades: "The overflow from Lake Okeechobee floods the Everglades except in the dry season of the year" (Wright 1909c, p. 159). Even as late as 1912 F. C. Elliot, the chief drainage engineer, noted that in November of that year, "the lake continues to overflow all along its south shore" (Elliot 1912, p. 558).

Although this logic does not provide a measure of the velocity of outflow, it does suggest that the outflow cross section (height times width) for the southwestern, Sawgrass Plains portion of the shoreline (roughly 20 miles or 32 km long), ranged between zero toward the end of the dry season and 1.5 feet (45 cm) high during the wet season maximum, or 160,000 ft^2 (15,000 m^2).

Chapter 7 suggested that, like the southwestern Sawgrass Plains, it is unlikely that the southern Custard Apple Swamp formed an elevated rim under predrainage conditions. The descriptions of a custard apple rim began appearing in postdrainage accounts (chapter 7) and probably resulted from differential postdrainage soil subsidence (chapter 7). To estimate the timing and cross-sectional area of outflows through the southern shoreline, the strip of Custard Apple Swamp was considered to be a shallow extension of the lake, such that outflows would be controlled by the Sawgrass Plains just south of the swamp (C in fig. 5.3). Outflows would occur when lake (and Custard Apple Swamp) water elevations exceeded the ground elevation of the Sawgrass Plains. Estimates of cross-sectional outflows can be calculated by assuming that elevations in this portion of the Sawgrass Plains were equal or very nearly equal to the elevation of the more westerly portion of the Sawgrass Plains and by assuming that peat accumulation in both areas was controlled by lake water levels (chapter 5). This relationship is consistent with historical observations (e.g., Wright 1909c, p. 162; Dachnowski-Stokes 1930, p. 93) that the whole southern shoreline overflowed at the same time. Since the Custard Apple Swamp portion of the shoreline was approximately 50 miles long (chapter 7), the typical maximum outflow cross section of the southern shoreline was nearly 400,000 ft^2 (37,000 m^2). Adding this to the previous estimate for the southwestern (sawgrass) shoreline yields a total of 560,000 ft^2 (52,000 m^2) for the cross-sectional area flowing at the end of a typical wet season.

If this simplification is correct, and predrainage narratives, vegetation, and soils appear to suggest that it is (chapters 5 to 7), then this predrainage interconnection of lake and marsh, extending along a border that stretched over 70 miles (110 km), was a remarkable geomorphological phenomenon, compared to the vast majority of lakes. For most lakes, the shoreline is composed of mineral sediments with elevation formed and varying independently from the lake's water surface. Typically only a narrow area of low elevation is available, and as a result the widths of the distributaries are small relative to the diameter of the lakes that feed them. In the Everglades, it appears that the width of the outflow area was actually *wider* than the diameter of the lake, leading to radial outflow.

The explanation for this atypical geomorphology lies in the dynamics of the organic sediments that accumulated and decomposed along its broad shoreline, in equilibrium with the long-term average lake water surface. An annual period of aerobic decomposition that occurred when the lake water surface declined below the shoreline may have helped maintain this equilibrium.

The maximum cross-sectional area calculation does not fit as "cleanly" into this conceptual model as would direct, predrainage measurements of discharge volumes or velocities, if available. Estimating these velocities is challenging for a number of reasons. First, estimates would have to consider that at least near the shoreline, the hydraulic gradient might have been even less than the gradient representative of the Everglades as a whole (ca. 3 inches per mile; 0.00005). Close to the lake, the slope of the Everglades water surface would likely have been reduced by the lake influence, and the ground surface slope might also have been reduced (fig. 5.3). Second, the density of the predrainage sawgrass stand would have created substantial hydraulic resistance.

Third, in accordance with the tuning hypothesis that similar vegetation reflects similar water depths, the width of the southern shoreline (70 miles or 113 km) was greater than the width of the Sawgrass Plains (40 miles or 64 km), suggesting that the velocity near the lake was lower than farther downstream. This would also be consistent with radial flow directions found near the shoreline vs. parallel flow directions found downstream (see Flow Magnitudes and Directions section, this chapter).

Fourth, it is likely that variations in the ground surface elevations occurred at various scales along this broad shoreline of Lake Okeechobee. Tussocks or hollows in the peat might have raised or lowered the ground elevation locally as much as 0.5 feet (15 cm). Rather than attempt to capture this complexity, the cross-sectional approach assumes rectangular uniformity across the shoreline, a peat surface parallel to the water surface, and negligible outflow volumes from the short "dead" rivers that dissipated within the first few miles downstream into the Sawgrass Plains (Hopkins 1884b, p. 100; Wintringham 1964, p. 61; Hardin 1915-T44 R35, 1915-T44 R36, 1916-T43 R37). Reports of the whole southern shoreline overflowing at much the same time (Wright 1909c, p. 162) are consistent with this piece of the puzzle, but further research and modeling might yield a measure of the uncertainty associated with this approach. Until that time, it appears that predrainage velocities from Lake Okeechobee into the Everglades were lower than predrainage velocities in the sloughs of the Ridge and Slough landscape.

The outflows described above were essentially steady flows that varied little from day to day. They may have been supplemented by short-duration, exceptional flow events. Lake Okeechobee's very long fetch, as well as historical accounts, suggest that lake seiches may occasionally have contributed significant pulses of water into the Everglades. Predrainage and pre-dike accounts record near-shore high water marks of 5 feet (1.5 m; Wintringham 1964, p. 61, observed in 1883; Williams et al. 1911, p. 198) and land near the lake shore being "subject to overflow of the Lake 4 or 5 ft" (Hancock 1907-T41 R32, p. 32). The high levels presumably reflect exceptional lake level rises, most likely from seiches associated with hurricanes or other sustained high winds. Stage measurements taken during the 2005 hurricane-induced seiches (South Florida Water Management District data) suggest several points: (1) probably not the full 70 miles of southern shoreline would have been affected—40 miles might be more typical; (2) durations would likely be 6 to 18 hours; and (3) seiche elevations could be as much as 6 feet (1.8 m). These recent numbers reflect a lake encircled by the Hoover dike; under unconstrained predrainage conditions, seiche elevations might have been lower.

The tuning hypothesis suggests that lake outflow typically lasted as much as three quarters of the year. The uniformity of vegetation along the downstream axis of the Sawgrass Plains and Ridge and Slough landscapes indicates that water depths were also close to uniform along the downslope axis. This suggests the presence of an upstream inflow from Lake Okeechobee that, along with local rainfall, was sufficient to counterbalance the downstream outflow.

Predrainage descriptions made by early drainage engineers also support the idea that Lake Okeechobee outflows continued throughout the majority of the year during most years. George Hills, who "literally crawled on . . . hands and knees" for many miles of the pre-construction survey for the West Palm Beach Canal, states:

> Under those natural conditions the normal elevation of Lake Okeechobee was approximately that of the surface of the muck lands along its shores. Any increase in height of the lake resulting from floods in its tributaries resulted in the discharge of such flood waters upon the muck lands of the Everglades. . . . Under such conditions it is apparent that the muck lands of the Everglades were continuously saturated and constantly subject to inundation by waters of outside origin. (Hills 1931, p. 3).

Ben Herr, the chief drainage engineer for the Okeechobee Flood Control District, stated:

> The excess of water from rainfall and inflow over evaporation spilled over the low southern shore into the Everglades. This water together with the rainfall kept the Everglades flooded most of the year. (Herr 1943, pp. 12–13)

Both the long duration of annual inflows and the occasional pulses from seiches help explain the early drainage engineers' strong emphasis on lowering Lake Okeechobee levels as a first step in draining the Everglades:

> This [5-foot water mark] shows conclusively that the Glades can not be drained to prevent overflow until the lake is lowered and converted into a reservoir. . . . The intention of the drainage operations is to lower the level of this lake 6 feet. (Williams et al. 1911, pp. 197–198)

The importance to the Everglades of Lake Okeechobee outflows along the whole 70-mile shoreline cannot be overstated and helps explain the appearance of changes in the Everglades as early as the 1890s, when for a time Disston and Kreamer's drainage efforts succeeded in lowering Lake Okeechobee levels. Kersey attributes the decline of the Indian trade in alligator and otter hides to this early drainage: "The second reason for Brown's [sale of his store in February 1908] was that he had foreseen the beginning of the end of the truly profitable trade in pelts, plumes, and hides as a result of the drainage of the South Florida wetlands which was just getting under way" (Kersey 1975, p. 70). Kersey's *post facto* interpretation was supported by a contemporary (1910) quote from a medical missionary at Everglades Cross Mission, Dr. W. J. Godden: "[They make their living] now by hunting the alligator and otter and selling their hides, but soon the hunting season will be a thing of the past, as a means of livelihood" (quoted in Kersey 1975, p. 71). Another trader in Miami, Isidor Cohen, who in the 1890s noted "periodic visits of Indians . . . in canoes laden with racoon, otter, mink and alligator hides,") later remarked that "the partial drainage of the Everglades and its penetration by the whites have deprived the red man of his heritage, namely, the game . . ." (I. Cohen 1925, pp. 27, 205). Significantly, the declines in Everglades water depths implied by these observations were caused solely by drainage of Lake Okeechobee. This is evident as these declines precede any construction of drainage canals within the Everglades itself.

Inflows from the Eastern and Western Flatwoods and Big Cypress

Surface water flow from the flatwoods east and west of the Everglades does not appear to have been large. Directional vegetation patterns that might suggest significant flows from the adjacent flatwoods communities into the Everglades are not visible on aerial photographs (USDA-SCS 1940) or on historical maps. However, exceptionally high water levels might occasionally have allowed water to flow from the adjacent pine and cypress landscapes into the Everglades. Ives (1856a, p. 17) reported a well-traversed trail through the Eastern Flatwoods, passable by canoe during high water and leading to "the swamp bordering the Everglades."

Along much of the boundary between the Eastern Flatwoods and the Everglades in Palm Beach County (present-day boundaries), cypress strands were present. Usually these strands were oriented north-south, but occasionally east-to-west. East of the cypress border, the transition to flatwoods appears fairly sharp on aerial photographs (USDA-SCS 1940). The presence of numerous isolated and circular cypress domes within the flatwoods suggests the absence of a continuous surface water connection to the Everglades but that high water tables were present during the wet season.

Along the western boundary, tree island orientation, landscape directionality visible on 1940s aerial photographs, and directionality visible on recent satellite imaging (plate 25) suggest that surface water flowed from the Mullet Slough portion of Big Cypress (fig. I.2) into the Everglades. Flow directions and watershed boundaries estimated for current and predrainage conditions in Big Cypress also suggest that flows were directed through Mullet Slough toward the Everglades during wet conditions (Duever et al. 1986, p. 76; Michael Duever, pers. comm., 2003).

Inflows from Groundwater

A number of early and credible accounts discuss the presence of visible groundwater discharges into the Everglades. This is puzzling, as such discharges are not observed at present, nor does there seem to be evidence of downstream plumes that would be expected from the discharge of calcareous groundwater onto softwater peatlands. (One could speculate, however, that King's 1917 soil profiles of intermixed marl and peat along the future route of Tamiami Trail might be features related to groundwater discharge; see fig. 1.9 and King 1917d.)

The following observations should be considered when reconstructing the predrainage hydrology.

In the early part of the 1800s, J. L. Williams, a lifelong student of Florida geography, wrote that he gave credence to

> a statement made by a respectable gentleman who resides near the border of the Glades, and who has often visited them. He states that not far from the center of the Everglades there is an immense spring rising from the earth, covering an extent of several acres, and throwing up a large quantity of water with great force, and supplying the Everglades with all the water flowing through them. (Williams 1847, p. 64)

Such a statement might be dismissed as pure speculation except that (1) Williams himself was a detailed and careful observer; (2) the "respectable gentleman" is most likely William Cooley who lived near the edge of the Everglades and often made trips into it between 1827 and 1838 (Knetsch 1989, p. 40); and (3) other predrainage, firsthand explorers have made similar statements. MacGonigle, in a detailed *National Geographic* article, thought that water balance argued for an important contribution from "subterranean streams or springs":

> The evident elevation of the area above the east, west, and south coasts precludes the idea of drainage from surrounding areas, and we must look elsewhere for the sources of the water. These, I think, are found in part in precipitation, and, in part, in subterranean streams or springs. The rainfall over this vast area of three million acres must be very great. But when we remember that all the creeks and rivers lead out of and not into the glades, the rainfall, which possibly approximates an average of ten inches per month during the months of June, July, and August, is not sufficient. From what we know as to the subterranean relations between the lakes in the lake region, from the well-known conditions of Silver, Blue, and De Leon springs, as well as from data acquired by drilling for artesian water, it is reasonable to infer that the volume of water due to precipitation is materially increased by an underground supply. (MacGonigle 1896, p. 388)

Willoughby, writing after his trip across the Everglades, also thought that groundwater played a significant role in Everglades hydrology. His firsthand observations, while not quantitative, make it hard to reject the possibility that groundwater contributed to the predrainage water budget:

> Sometimes pools would be crossed eight or ten feet wide and five feet deep. . . . Occasionally, in the centre of these pools, a dark hole a few inches in diameter could be seen; down one of these I could push my pole to a long distance, and the water was coming out from it with quite a little head. They are to be found all over the Everglades, and are, I believe, one of its greatest water-supplies. All this moving water cannot be accounted for by the rain alone, and the water is too hard for rain-water, so that in all probability more comes from below than above. (Willoughby 1898a, pp. 119–120)

While exploring part of the Ochopee Marl Marsh during the dry season, engineer and naturalist John King described a surface-to-groundwater connection. His descriptions refer to surface water entering rather than leaving the ground, but Willoughby's observations suggest this would have been reversed in the lower elevation, non–marl marsh majority of the Everglades:

> This area of the Everglades [SE corner of the Hopkins tract; i.e., Ochopee Marl Marsh] is similar in character to other sections of the Everglades and covered with a growth of sawgrass and a round water grass, separated and divided by a number of Myrtle Hammock[s], very narrow and much elongated, following in parallel lines with the main Coastal Hammock and ranging approximately 200 yards apart. From the ends of these long projecting Myrtle Ridges extends rocky reefs connecting with the preceding hammock. These rocky reefs project upwardly in sharply pointed finger-like masses, level with or above the surface of the ground. The sloughs that in periods of high water spread over this ground, were at this time mostly dry, owing to the effect, among other causes, of a long period of drought, gather at irregular intervals into rivulets which usually lead to sink holes at the ends of some Myrtle Ridges and disappear beneath the surface of the ground. . . .
>
> It is evident that a system of subterranean drainage exists throughout this entire area, from the dryness of the soil, the formation of streams, and the disappearance of same in sink holes of large or smaller extent. (King 1917e, pp. 413–416)

In summary, some type and some degree of groundwater–surface water interaction certainly was present in the predrainage Everglades, but it does not appear possible to use historical observations to quantify the role of groundwater. It may have been a more important component of the Everglades water budget under predrainage conditions when water levels in the surrounding areas—Lake Okeechobee, the Eastern and Western Flatwoods, and even the Kissimmee Valley—were all higher than at present. Groundwater inflows from the Everglades Keys (Miami Rock Ridge) and/or higher water tables in that area also may have been an important aspect of Everglades hydrology to the east of Taylor and Shark River sloughs.

Outflows from the Everglades

In addition to evapotranspiration, surface water left the Everglades through five principal outflows. In clockwise direction (plate 5; see plate 15 for township locations), these were: (1) the East-West Cypress Strands (Townships 48, 49, 50); (2) the Peat Transverse Glades and associated coastal rivers (Townships 51, 52, 53); (3) the Marl Transverse Glades (Townships 54, 55, 56); (4) Taylor Slough (Townships 59, 60); and (5) Shark River Slough (Townships 59, 60). These outflows are known from historical accounts and are visible on older maps, soil maps, and aerial photographs. Additionally, the relation of these outflows to flow patterns within the Everglades can generally be seen in maps of surface flow directions, such as the three discussed in more detail in the Flow Magnitudes and Directions section (figs. 11.10, 11.11, 11.12).

While not a principal outflow, the Loxahatchee Slough (plates 2, 38) does appear to have provided a continuous aquatic connection between the northeastern Everglades and the Loxahatchee River. The long length of this slough, coupled with almost flat elevation, suggested to Parker et al. (1955, p. 143 and their plate 11) that net flows were small, with the flow direction alternating as a function of local and changing hydraulic heads.

These outflows may be classified into two types: (1) "spillways," in which discharge occurred only when a threshold elevation was exceeded; or (2) "flow-ways," which lacked a threshold and in which discharge varied as a function of water depth, resistance of the vegetation, and overall slope of the watercourse. Flow-ways flowed throughout the year, so long as there was surface water on the landscape; spillways flowed only when water levels exceeded a threshold elevation. Flow rates within flow-ways were controlled primarily by the internal resistance of the landscape itself.

As defined here, flow-way ground surfaces slope continuously downward, while the spillway ground surface rises near the threshold, then slopes downward toward the ocean. In the postdrainage Everglades, soil subsidence converted some outflows from flow-ways into spillways, as upstream subsidence led to the formation of an artificial ledge or threshold (chapter 3). This conversion occurred in the Peat Transverse Glades, to some extent in the East-West Cypress Strands, and in the headwaters of the more southerly coastal rivers (especially Little River and Miami River). Comparison of the peat losses described in chapter 3 with the predrainage topography estimated later in the present chapter indicates that in all three cases this flow-way to spillway conversion resulted from postdrainage peat subsidence in the Ridge and Slough landscape west of the coastal ridge.

Outflows through the Cypress Strands

The principal East-West Cypress Strands were present in the southern portion of the Eastern Flatwoods (see chapter 10) in present-day Broward County, north of Ft. Lauderdale (plate 5, Townships 49, 50), and to a much lesser extent, in Palm Beach County. These strands formed funnel-shaped areas with the wider side facing the Everglades and the narrower end oriented eastward and terminating in coastal rivers (plates 41, 42). Flow directions in the Ridge and Slough landscape adjacent to the cypress strands appeared to turn southeastward in Townships 49 and 50 (see Flow Magnitudes and Directions section, figs. 11.11 and 11.12). This southeast turn may reflect some influence of the eastward flows through the cypress strands. Such a relationship between Everglades, cypress swamp, and coastal rivers was described in 1823 by Charles Vignoles:

> New river and all the branches [i.e., the North and South Forks of the Middle River, and part of Cypress Creek] discharging through its bar originate in the Great Glade, running through pine lands and heading in cypress swamps; which have previously been inundated from the Glade. (Vignoles 1823b, p. 49)
>
> Cypress swamps are mostly near the head of rivers, and in a continued state of inundation; little or no underbrush, but only crowds of the cypress shoots or knees, which point up like small pyramids. (Vignoles 1823b, p. 90)

The ground surface of the cypress strands must have been slightly elevated above the Everglades peat surface immediately to the west, thus providing some resistance to outflow. However, most of the resistance was probably due to the long, shallow slope through the cypress area and to vegetation within the cypress strands, which included a mixture of deeper areas of open water and shallow areas with dense vegetation (chapter 10). The combination of these resistances apparently prevented the cypress strands from becoming a major outflow from the Everglades. At the same time, Vignoles (1823b, p. 91) and others suggested that very little peat accumulated in the cypress strands: "The land in the cypress swamps here appears to be neither rich nor deep, being apparently but pure silex [silica] with an admixture of [organic] sediment" (Mallory 1847, p. 63).

Possible explanations for the limited predrainage accumulation of peat within the cypress strands relative to the Everglades include lower primary productivity, faster decomposition of organic materials, flushing during occasional high-flow events, and scarcity of graminoid root mats that might otherwise have stabilized peat in the shady understory.

Outflows from the cypress strands terminated in Cypress Creek, the North and South Forks of the Middle River, the North Fork of the New River, and possibly one of the southern forks of the Hillsboro River.

Outflows through the Peat Transverse Glades and Coastal Rivers

The Peat Transverse Glades (plates 2, 40) were an important part of the predrainage hydrology of the eastern axis of the Everglades, transferring water to the coastal rivers, Biscayne Bay, and ultimately into the Atlantic. Predrainage land surveys and the postdrainage soil map (Jones et al. 1948a) accurately define the spatial aspects of these rivers: widths, lengths, and location between the Everglades and the headwaters of the coastal rivers (plates 40, 41, 43, 46). Knowledge of predrainage soil types and vegetation helps identify the possible water depths and clarify the original topographic relation to the Everglades. Narrative descriptions of rapids, of year-round water flow, and of canoe travel provide further information.

Prior to drainage and subsequent subsidence, the Ridge and Slough landscape west of the Peat Transverse Glades sloped southeastward toward the coast at a rate of about three inches to the mile (Newman 1907, p. 46). As travelers approached the coast from this landscape, pines, or farther north, cypress, formed a welcome line along the horizon (Buker 1975, p. 53; Preble 1842, p. 48; Moses 1892, p. 25; Willoughby 1898a, p. 164; Dimock 1907, p. 248). Breaks in this line served as landmarks (Willoughby 1898a, p. 164; Pierce 1970, p. 160), indicating places where the Everglades extended farther eastward between the "islands" of higher, pine-covered ground (plates 43 and 40). These extensions were the Peat Transverse Glades; the "coves and indentations" noted by Buckingham Smith (1848a, p. 49). These Glades—open areas between the taller canopies of the pines—were marshes of sawgrass growing on 1–3 feet

(30–90 cm) of peat soil (Jones et al. 1948a; predrainage land surveys; chapter 3). In addition to sawgrass, some of these glades included tree islands (plates 44, 45) and possibly a ridge and slough–like mosaic (plate 46).

Historical information, vegetation reported on township surveys (Williams 1870-T51 R42, 1870-T52 R42, 1870-T52 R41), and soils mapped in the 1940s (Jones et al. 1948a) all suggest that the transition from Everglades to Peat Transverse Glades was smooth; that is, the southeastwardly sloping plane of organic soils in the Everglades simply continued eastward into the Transverse Glades (plate 7, fig. 4.6, bottom). As a result, both water depths and vegetation in the Transverse Glades resembled those of the Everglades. Only the progressive narrowing of the vista by pinelands to the north or south would have indicated a departure from the open Everglades (plates 7, 40, 41). This configuration suggests two parallel planes of peat surface and water surface sloping *downward* within the Peat Transverse Glades at the same time that the bordering upland ground surface rose *upward* on the pine islands.

Two to four miles east of the open Everglades, the Peat Transverse Glades typically ended at or enclosed the headwaters of a coastal river. Those headwaters included braided rivulets or rapids, or both, indicating a breakpoint in the slope, with the river channel falling more steeply to the ocean than did the transverse glade. As settlers explored the connection between Everglades and the coastal rivers (e.g., Pierce 1970), usually by canoe, the individual Peat Transverse Glades became known by the names of their respective rivers: "Undefined waters of Arch Creek," "Glades," "Valley of Snake Creek," "Waters of Snake Creek" (plate 41). The label on one branch of Snake Creek, "Principal Passage to the Everglades," indicates the predrainage boat travel.

Two early observers, a drainage engineer and a biologist, note free and rapid flow through rivers draining the Everglades. J. O. Wright, formerly supervising drainage engineer at the U.S. Department of Agriculture and later chief drainage engineer for the State of Florida, noted that

> there are numerous channels [on the East and West coasts] through which the surface water flows quite freely from the Everglades, both into the Atlantic Ocean and the Gulf of Mexico. In many places these channels are worn down several feet, but do not extend far beyond the rim into the interior. The water is brought from the margin of the 'Glades in small rivulets to the heads of these streams, which increase in size as they approach their outlets. The difference in elevation between sea level and the source of these streams gives many of them sufficient fall to cut out large and deep channels. The streams on the east coast, beginning at Rockledge and going south, are as follows:
>
> Sebastion River, St. Lucie River, Loxahatchee River, Hillsboro River, Cypress Creek, Middle River, New River, Snake Creek, Arch Creek, Little River, Miami River and Snapper Creek. These streams are shorter, and have more fall per mile than those on the west coast. (Wright 1912, p. 24)

Harshberger (1914), the biologist, noted the more rapid flow in the eastern rivers, likely because of the shorter, steeper slopes mentioned by Wright:

> The [natural] drainage of the Everglades is by the short rivers previously mentioned that empty into the Atlantic Ocean. Part of the surplus water of the 'Glades finds its way into the Gulf of Mexico by several short rivers, such as the Harney, Rodgers, Lostmans, Shark and Chatham Rivers, but the drainage in this direction must be much more sluggish than through the eastern streams that empty into the Atlantic Ocean. . . . The short rivers that flow from the Everglades into the Atlantic Ocean are characterized by rapids where they flow from the 'Glades. (Harshberger 1914, pp. 54–55)

Taken together, these descriptions of the Peat Transverse Glades and the immediately adjoining Ridge and Slough landscape suggest that water did not typically impound west of the portion of the Atlantic Coastal Ridge between the New River and the Miami River. This lack of impoundment in turn suggests that the Peat Transverse Glades continued the downward slope of the Everglades and that the coastal rivers apparently included sufficient conveyance capacity to export the water flowing eastward from the Everglades (see also plate 7). This conclusion is consistent with the absence of predrainage vegetation reflective of deeper, impounded conditions (appendix D) and also with the southeastward and eastward-oriented flow directions discussed in more detail in the Flow Magnitudes and Directions section and illustrated in figures 11.10, 11.11, and 11.12.

As the eastern part of the Everglades was drained by deep canals, the situation changed substantially. A subsidence trough developed in the originally thick peats just west of the coastal ridge (plate 16). Instead of a peat surface sloping continuously *downward* from Everglades through the Peat Transverse Glades, the subsided peat surface now begin to slope *upward* toward the ridge. As a result, when water levels rose in 1947 to reflood close to the original extent of the Everglades (fig. 11.8), the area west of the coastal ridge and east of Water Conservation Area 3B ("Area B"; Kohout and Hartwell 1967, p. 57) flooded to a depth of "about 3 to 5 feet above the land surface" (Kohout and Hartwell 1967, p. 27; see also Schroeder et al. 1958, pp. 32–36). Such depths—in excess of predrainage ones—are reflective of the postdrainage subsidence of the peat-based land surface (see also Kohout and Hartwell 1967, p. 45).

Outflows through the Marl Transverse Glades

In contrast to the unobstructed, rimless, and continuously flowing Peat Transverse Glades, the Marl Transverse Glades were raised "spillways" as defined earlier in this chapter, receiving water from the Everglades only during the wet season, and then only when water levels reached the upper edge of the Rockland Marl Marsh. The significance of the Marl Transverse Glades for understanding predrainage Everglades hydrology lies not in their volumes of outflow but instead in their indication that Everglades waters from the Rockland Marl Marsh typically rose high enough each year to flow out through the Marl Transverse Glades. Once again assuming a level water surface, this provides additional information for estimating the typical annual maximum elevation of waters covering Shark River Slough and the Rockland Marl Marsh (figs. 4.7, 11.7; plate 35).

Previous studies have not generally distinguished between the Marl and Peat Transverse Glades—perhaps because of their similar shapes and similar sawgrass vegetation. However, historical observations, topography, and the soils discussed in chapter 10 suggest that they functioned quite differently. While the Peat Transverse Glades maintained several feet of peat soil and typically remained inundated, the Marl Transverse Glades were only seasonally inundated and supported only marl soils. These differences in hydrology and soils are consistent with the differing slopes of the adjacent Everglades landscapes: the Ridge and Slough landscape sloped *downward* toward the Peat Transverse Glades (plate 7), whereas the Rockland Marl Marsh sloped *upward* to the Marl Transverse Glades. The Marl Transverse Glades could fill only after water had risen to the upper (eastern) edge of the Rockland Marl Marsh.

Drainage has so drastically lowered water levels both within the Marl Transverse Glades and within the Everglades that it is now nearly impossible to imagine waters high enough to overflow through these former Transverse Glades. Name changes in this area help picture the original hydrology. By the 1940s, after years of drainage, the eastern rim of the Everglades basin south of Miami had become known as the "Miami Rock Ridge," suggesting a continuous, dry ridge forming the edge of the Everglades. Prior to drainage, the area formed a series of slightly elevated rock islands separated by the individual Marl Transverse Glades (figs. I.2, 9.1) and therefore was "generally known as the 'Everglades Keys' because they are islands of pine or hammock forests often surrounded by water during wet seasons" (Davis 1943b, p. 53; see also Harper 1910b, p. 46; Small 1929b, p. 46; Beard 1938, p. 9). They were surrounded not only by water but by Everglades: the Rockland Marl Marsh to the west, Perrine Marl Marsh to east, and in between by strips of the "Everglades in miniature" (Simmons and Ogden 1998, p. 5)—that is, the Marl Transverse Glades (fig. 10.4).

Under postdrainage conditions, only the occasional high-water event or "flood" comes close to reproducing predrainage water levels. For example, from a human perspective, the high waters of 1947 constituted a devastating flood, contributing to congressional approval of the Central and Southern Florida Project for Flood Control and Other Purposes. Yet if the boundaries of the flooded area are compared with the predrainage boundary of the Everglades (fig. 11.8), they prove to be very similar, suggesting that the perceived "flood" simply returned water levels to what they typically had been under predrainage conditions. As Terence Thomas (1970, p. 5) noted, the 1947 flood was "generally considered to best approximate the predrainage conditions of the vast Kissimmee-Okeechobee-Everglades watershed." Figure 11.8 shows water fully covering the Rockland Marl Marsh, and once again extending up to the pines and pinnacle rock of the Everglades Keys/Miami Rock Ridge, just as described in 1896:

> At the border of the Everglades the rough surface of the oölite becomes concealed for the most part beneath a mat of grass, and pines grow no further west. . . . In the zone of oscillation of water the grasses change in character from a small wiry variety near the pines to saw-grass six feet or more tall, and flags and cane growing in water. (Griswold 1896, p. 53)

Even as late as the early 1960s, and despite expanding drainage, the Marl Transverse Glades still flooded seasonally, at least in wet years (Bill Robertson, pers. comm., 1997).

Like the map of the 1947 flood extent, a map of water table contours measured during a postdrainage "high stage" (fig. 11.9 left) also helps identify predrainage con-

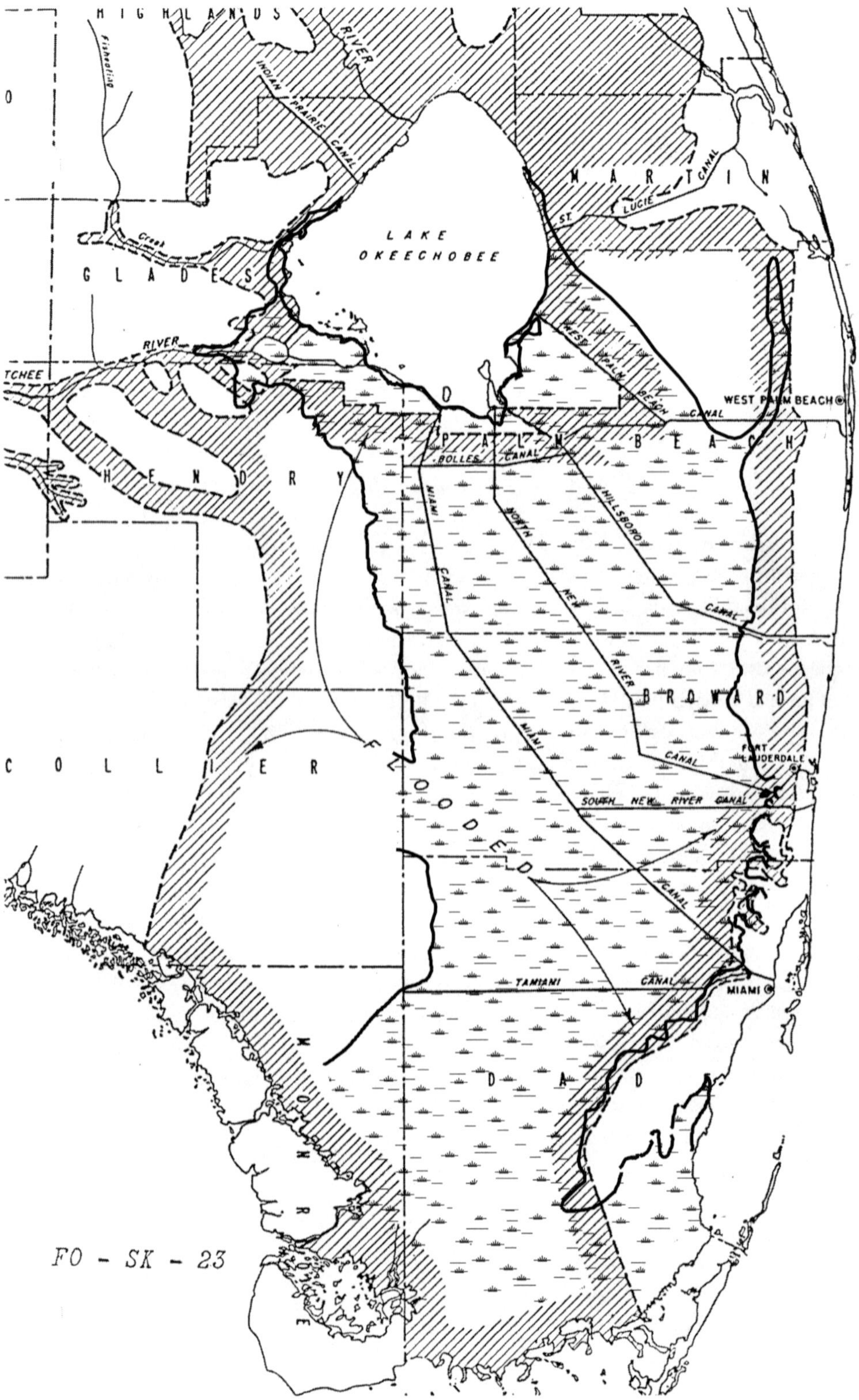

Figure 11.8. Comparison of extent of the flood of 1947 (border delineated with diagonal hatching) with extent of the predrainage Everglades. Note coincidence of southeastern edge of the flood with western edges of the Atlantic Coastal Ridge (South New River Canal to Miami Canal) and of the Everglades Keys. Similarly, the flood extent northwest of Lake Hicpochee also corresponds closely to the predrainage border of the Everglades. After seventy years of drainage, this event refilled the Everglades basin (and beyond), apparently to approximately predrainage water elevations. Modified from U.S. Army Corps of Engineers (1947; see also 1948).

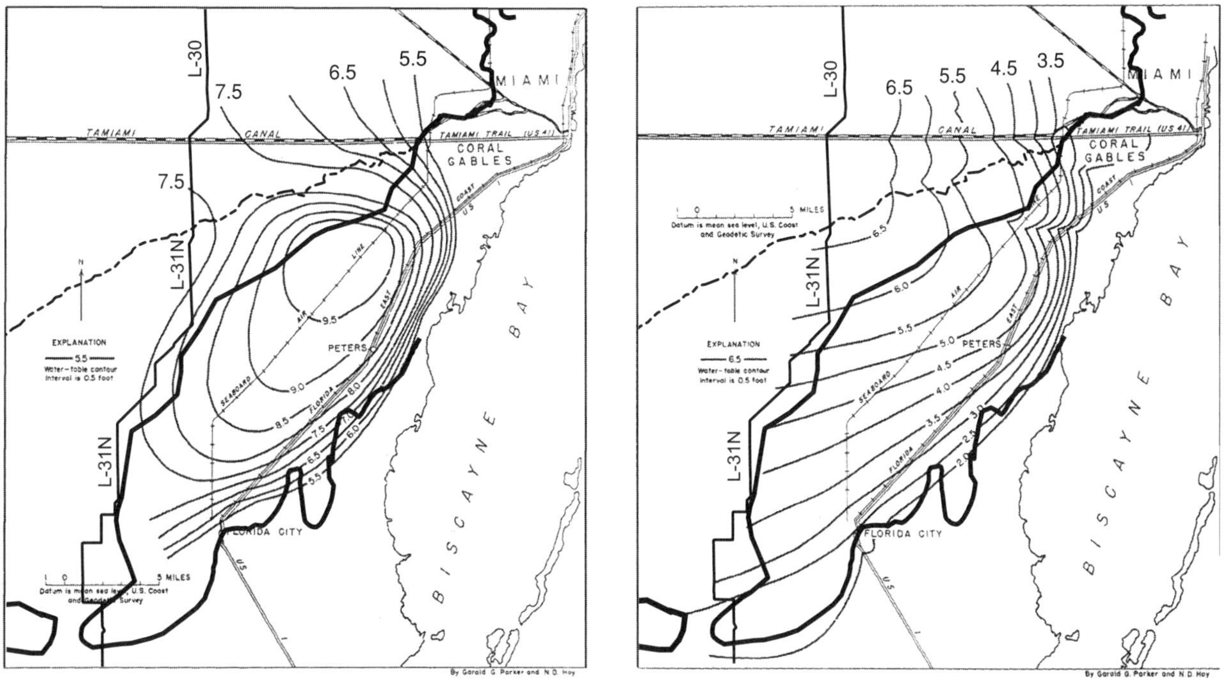

Figure 11.9. Postdrainage water tables in the Everglades Keys area. Contours are in feet above mean sea level. The heavy line delineates the Everglades Keys (Miami Rock Ridge; plates 5 and 35); the dashed line separates Rockland Marl Marsh (south) from Ridge and Slough landscape (north). The L-31N canal is shown only for location; it was not present in the 1940s. Right: The stage measured on March 17, 1941, and described as "medium," shows very strong influence of drainage. Left: The "high" stage on September 23, 1940 also reflects drainage, especially of the Everglades, but concentric contours centered on the Everglades Keys suggest partial resemblance to predrainage conditions. Water stages of more than 9 feet (2.7 m) found in the Rockland Marl Marsh suggest that predrainage stages there would have been even higher. Modified from Parker et al. (1955, pp. 210–213).

ditions in the Everglades Keys/Marl Transverse Glades area. There is no inherent reason that postdrainage high stage should be equivalent to predrainage typical stages, but several features of the contours for September 23, 1940 suggest that this is the case: (1) the similarity in shape of the water table contours (fig. 11.9 left) to the ground surface contours (plate 35); (2) the strong dissimilarity to lower "medium stage" contours shown in figure 11.9 right, which clearly reflect effective anthropogenic drainage; and (3) the 9.5-foot (2.85 m) elevation of the water table in the Everglades Keys; high enough to be consistent with predrainage descriptions of a water table close to the surface (MacKay 1847, p. 65; Harper 1910b, p. 45) and filling sink holes in the area (Harshberger 1914, pp. 101–102; Harper 1910b, pp. 45–46). Two other aspects of the September 1941 contours suggest that they may in fact be a little lower than predrainage: the decline in stage westward toward Shark River Slough and the below- rather than above-ground water table in the vicinity of the former rapids of the Miami River (ground elevation approx. 7 feet; 2.1 m).

The September 23, 1940 "high stage" contours (fig. 11.9 left) extend the information provided by the 1947 flood map and suggest that water depths along the border between the Everglades Keys and the Rockland Marl Marsh would have been about 1 foot (30 cm) or perhaps even more above the ground surface. It would have been these depths that helped provide water to the Marl Transverse Glades. Depths of 1 foot (30 cm) or more are consistent with the presence there of sawgrass (fig. 10.4), with the basin topography, and with Krome's observations (Krome 1902d; 1902i; 1903i). In addition to water from the Everglades, local rainfall and groundwater mounding may also have contributed water to the Marl Transverse Glades.

Plate 36, figure 4.7, and figure 11.5 together help illustrate predrainage conditions in this part of the Everglades by clarifying the relation of present to predrainage water depths. Plate 36 shows two postdrainage cross sections of water depths measured toward the ends of the 1979 dry season and the 1978 wet season. The upper portion of the plate illustrates the spatial extent of surface water at both times relative to the extent of Shark River Slough and the flanking Marl Marsh landscapes. At the end of the 1978 dry season, surface water was restricted to Shark River Slough, with depths between zero and 1 foot (30 cm), noticeably drier than the predrainage slough depth of 1 foot (30 cm; table 11.4). An even greater reduction from predrainage conditions was apparent at the end of the wet season, both in depth and ponded extent. The Shark River Slough water depth of about 1.5 feet (45 cm) was much reduced from the typical predrainage value of 3 feet (90 cm), and surface water extended only slightly more than halfway up the Rockland Marl Marsh. No water was present at the top of the Rockland Marl Marsh where it bordered the Everglades Keys and accordingly, in 1978, no water would have been flowing from the Everglades out into the (former) Marl Transverse Glades (incidentally, a result much desired by the drainage engineers). Note that the depth of the V-shaped cross section shown in the lower portion of plate 36, approximately 1.5 feet (45 cm), is very likely greater than under predrainage conditions, due to uneven postdrainage peat oxidation (chapter 9).

In contrast, figure 4.7 shows an estimated reconstruction of predrainage land and water surfaces along a transect near the measurements shown in plate 36. The elevations of the marl subsoil shown in fig. 4.7 are taken from current (2000) measurements of ground surface elevation (Telis and Higer 2006; Desmond 2003). Under predrainage conditions these marls were covered by a thin (0.5–1 foot; 15–30 cm) layer of peat soil, as discussed in chapter 9. Predrainage observations of water depths and vegetation (e.g., fig. 11.3) indicate that typical Ridge and Slough water depths were present within Shark River Slough; that is, a typical annual high of about 3 feet (90 cm) and a typical annual low of about 1 foot (30 cm) in the sloughs. Sawgrass ridges within Shark River Slough are assumed to have been about 1.5 feet (45 cm) above adjacent sloughs, as elsewhere within the Ridge and Slough landscape. We have drawn the water surface as flat across Shark River Slough and out across the Ochopee and Rockland marl marshes. In actuality, the water surface may have been somewhat concave, rising to a degree within the marl marshes. Water depths on the marl marshes were deep enough to support the sawgrass stands and sloughlike areas reported by John King and William Mickler (chapter 9), but water levels also dropped below ground surface almost every year. Regionally, the annual rise of the water levels was enough to overtop the Everglades Keys, seasonally discharging Everglades water into the Marl Transverse Glades (chapter 10).

Figure 11.5 shows current (2000–2010) soil and hydrologic conditions along the same transect illustrated in figure 4.7. The covering layer of peat present under predrainage conditions within the "marl" marshes has been lost to oxidation, except within local depressions (not shown), a loss that likely happened as early as in the 1930s or 1940s. Similarly, peat oxidation has lowered ridge elevations by a foot or more (30+ cm) within this portion of Shark River Slough. Model simulations, as well as the plant communities seen in this area, suggest that water depths now vary less between wet and dry season than under predrainage conditions and on average are considerably lower. The hydrologic regime no longer supports peat accumulation in the Ochopee and Rockland areas. As seen in the measurements from 1978–1979 (plate 36), water levels are well below the western edge of the Everglades Keys. These lower levels (along with the L-31N levee and canal) prevent eastward flow of Everglades water into the former Marl Transverse Glades.

Outflows through Taylor Slough

Taylor Slough is easily recognizable on satellite imagery (e.g., plates 2, 4; see also plate 7), topographic maps, vegetation maps (e.g., plate 13), and aerial photography (plate 34). It is a topographic low within the surrounding Perrine Marl Marsh. Field studies (Gleason 1972; Olmsted et al. 1980, p. 3; Willard et al. 2001, p. 50) indicate that peat or mixed peat and marl soils occur within Taylor Slough versus predominantly marl outside of it. Aerial inspection and photographs confirm the presence of Ridge and Slough landscape patterning, as suggested by plate 34. A map by coastal surveyors, Sketch F (Bache 1850), of the northern coast of Florida Bay indicates a sizable river connecting the interior to Florida Bay in the vicinity of the current Taylor Slough system. The river bisects the Perrine Marl Marsh into a western portion labeled "Prairies and Hammocks" and an eastern portion labeled "Rocky and Sandy Prairies mixed with Hammocks." From the position and size, there is little doubt that this is the Taylor River, also shown on a U.S. Bureau of Topographical Engineers (1856) map, and forming the termination of Taylor Slough.

At present the Rockland Marl Marsh acts as somewhat of a drainage divide between Shark River and Taylor sloughs. In contrast, historical accounts and landscape patterning visible on aerial photographs suggest that originally much of the Rockland Marl Marsh was a collecting area for flows through Taylor Slough (fig. 9.1, plate 7). The Everglades Keys to the east and Long Pine Key to the south may have helped direct water from the Rockland Marl Marsh (and ultimately from farther upstream in the Everglades) toward the low spot between them, west of Camp Jackson and east of Royal Palm Hammock (Paradise Key). These latter two points were well-known, slightly elevated landmarks separated by the waters of Taylor Slough (fig. 9.1). According to Harper (1910b), the levels and width of inundation have varied considerably here from as little as 10 feet (3 m) wide and only a foot (30 cm) deep to as much as a mile (1600 m) or more wide and 4 or 5 feet (120–150 cm) deep:

> About a mile east of Long [Pine] Key there is a shallow slough running north and south, which is one of the principal outlets for that end of the Everglades in the rainy season, and is a route frequented by hunters at times when there is not quite enough water for their boats elsewhere in the vicinity. This slough has given previous explorers by the overland route considerable trouble. One botanist whom I know was once wading across it a little ahead of his two companions, in water up to his armpits, when he stepped on a large alligator—or possibly a crocodile (for these ferocious beasts do inhabit the south coast of Florida)—which of course immediately raised a great commotion and almost scared him to death. Another (Dr. Small on his first trip to Long Key) found about six feet of water and mud, not to mention alligators and moccasins, in the same slough, and in order to cross it he had to go back nearly to Camp Jackson and get an old boat which had been abandoned there by the surveyors a year or two before. But where Mr. A. and I crossed it the first time the water was only about a foot (30 cm) deep (and the channel perhaps ten feet wide), and on the way back we found a place a little farther south where it was practically dry. (Harper 1910b, p. 149).
> Harper indicates that water levels at the time of his trip (March 25–30, 1910) were very low, perhaps even lower than what would typically be expected late in the dry season.

Under present-day, postdrainage conditions, upper Taylor Slough infrequently receives flows and then only when gates on the L-30 and L-31W (fig. I.1) are open. As a result, upper Taylor Slough flows are flashy and the area is heavily drained.

Under predrainage conditions, the configuration of the Ochopee Marl Marsh–Shark River Slough–Rockland Marl Marsh cross section discussed in chapter 9 suggests that Taylor Slough would have been a flow-way. With higher peat ground surface elevations in Shark River Slough and in the western portion of the Rockland Marl Marsh, the predrainage water surface was spread across the full cross section (chapter 9). Water flowed southwest through Shark River Slough and south through the eastern portion of this cross section; that is, into upper Taylor Slough. The contours shown for the Rockland Marl Marsh area on plate 6 and the artist's rendering shown in plate 7 suggest that the water reaching the headwaters of Taylor Slough came from farther upstream in the Everglades, entering Taylor Slough in a continuous descent, and without passing over a spillway threshold. Water tables and/or groundwater flows related to the Everglades Keys may have provided an important eastern boundary condition (pressure head) to this cross section. Some of the upper Taylor Slough water may have passed through the more open and slightly deeper navigation route originally running along the west edge of the Everglades Keys (Williams 1874-T57 R39; Willoughby 1898a, pp. 158–159; Harper 1910b, p. 149; see also chapter 9).

Taylor Slough has sometimes been considered the southernmost of the Marl Transverse Glades. The location of the upper portion of Taylor Slough between the last of the Everglades Keys and Long Pine Key (fig. 9.1) as well as the postdrainage, spillway-like configuration now present would reinforce this classification. However, important differences visible in the soils and landscape patterning indicate that Taylor Slough formed a Ridge and Slough landscape on a mix of marl and peat soils, while the Marl Transverse Glades grew mostly sawgrass on a base of marl only. The correspondingly deeper water depths and longer hydroperiod of Taylor Slough probably reflect its stronger flow-way connection to the Everglades upstream from it (plate 7).

Outflows through Shark River Slough

Shark River Slough was the largest single outflow from the Everglades. The land surface above and within Shark River Slough sloped continuously downward to within a few miles of the coast. Near the coast, the Buttonwood Embankment (Craighead 1971, e.g., pp. 41, 45, 49) may have formed a slight rise across some portions of Shark River Slough, but it was not continuous, being interrupted by the many headwater channels of Harney and Shark rivers. Despite the discontinuities, aerial observations made in the 1960s and 1970s suggested impounded water within the lower portions of Shark River Slough, extending some miles upstream from the embankment (John Ogden, pers. comm., 2005). It seems logical to assume that this phenomenon would also have occurred under predrainage conditions.

With a continuous downward slope and no upstream rim or threshold, Shark River Slough was primarily a flow-way. Water levels and flow rates were therefore controlled by the resistance to flow offered by vegetation, the slope of the land, the supply of upstream water, and sea level. The lack of a continuous dam or ledge at the bottom portion meant that Shark River Slough could dry out if upstream flow stopped. Under present water management, where inflows at Tamiami Trail are strongly restricted, Shark River Slough does, in fact, frequently dry out. Historical quotes indicate that Shark River Slough did occasionally become dry: "Have known the Glades to be dry around the head of the river [Shark River]" (Roberts 1907, p. 39). However, such observations are rare in the historical record compared to descriptions of ample surface water. For example, "We have plenty of water at present, and go along with a great deal of ease" and "wad[ing] through mud and water three or four hundred yards, up to our waists" (*St. Augustine News* 1841, p. 59; see also numerous township surveys by William Mickler in 1885–1886; fig. 11.3; table 11.2; appendix A, Ridge and Slough landscape; Willoughby 1898a, pp. 112–115; Dimock 1907, p. 247).

The historical record as a whole, and the predrainage soils and vegetation, suggest that prior to drainage, dryouts of Shark River Slough probably did not occur more frequently than once every twenty or thirty years. The frequency of such dryouts would have been regulated primarily by the storage capacity of the upstream portions of the Everglades, including perhaps Lake Okeechobee and the Kissimmee Valley. Within the Everglades itself, this capacity would have been principally a function of the upstream water depths and the annual range in water depths.

The landscape pattern visible on satellite images and aerial photographs of Shark River Slough appears elongated relative to other parts of the Ridge and Slough landscape. Longer, thinner ridges, sloughs, and tree islands have a more streamlined appearance (plate 36), suggesting that flow rates within Shark River Slough may have been greater than farther north in this landscape. Faster flow would be consistent with partial funneling of flow as the Ridge and Slough landscape narrowed south of Township 54, between the Ochopee and Rockland marl marshes. Funneling would have been only partial because the marl marshes contributed to the flow cross section as well (cf. fig. 11.7; chapter 9). Mass balance requires either an increase in water depths or an increase in flow rate, or some combination of both, as the cross-sectional area decreases south of Township 54. Both the absence of a depth increase south of Township 54 (fig. 11.3) and the similarity in vegetation suggest that depths were similar but flow rates faster within Shark River Slough than elsewhere in the Ridge and Slough landscape.

Paired, predrainage measurements of flow rates within and outside of Shark River Slough do not exist. The few available qualitative descriptions of flow in Shark River Slough suggest it was fast flowing at times, but whether flows were actually faster than elsewhere in the Ridge and Slough landscape is not clear from historical records:

> Has seen a good current 6 [typed characters unclear: "5"?, "26"?, "25"?] miles north of the head of Shark River. At high water period saw same current throughout Glades. (Graham 1907, p. 24)
>
> Shark River has a rocky channel and a swift current at its head in the high water period. At low water the tide runs nearly to the head where the channel suddenly disappears into the Glades. At the high water period have traveled by boat 30 miles northeast of the head. (W. A. Roberts 1907, p. 39)

Outflows through Lostmans Slough

Very little documentation was found for this area, but satellite imagery suggests that this may have been a minor, secondary outflow from the Ridge and Slough landscape, parallel to the main western outflow, Shark River Slough (e.g., plates 2, 4). Some degree of landscape directionality aligned with the northeast-southwest axis of Lostmans Slough furthers the impression of secondary outflow (plate 36 upper left corner). Overall reduction of postdrainage Everglades water levels and flows and construction of Tamiami Trail have likely reduced flows through Lostmans Slough; in contrast, the significant westward compression of flow by the L-67 levees may have increased flows. Too little information is available to make a definitive predrainage to postdrainage comparison. Comparisons are further complicated by the postdrainage losses of peat in this area (chapter 9).

In summary, under predrainage conditions, the Marl Transverse Glades were natural spillways and the East-West Cypress Strands, Shark River Slough, Taylor Slough, Lostmans Slough, and the Peat Transverse Glades were natural flow-ways.

Outflows as Groundwater

The existence of predrainage groundwater outflows from the Everglades seems well established from numerous observations similar to the following:

> The rock [of the Everglades Keys] is usually wet, even in the driest times. In fact, under the limestone ridge there are channels of water running from the Everglades and bubbling out in the form of springs along the shore of Biscayne Bay. (Gifford 1911, p. 17)

Many historical accounts refer to submarine springs along or off the shore of Biscayne Bay (e.g., De Brahm 1773; Romans 1775, p. 192; Motte 1838a, p. 27; Griswold 1896, p. 53; MacGonigle 1896, p. 388; U.S. Coast and Geodetic Survey 1892; Willoughby 1898a, p. 65; cf. fig. 10.7), as do postdrainage studies (e.g., Parker et al. 1955, p. 582; Kohout 1966, p. 396; Kohout and Kolipinski 1967, p. 488; Kohout et al. 1973; Gaby 1993, p. 151):

> Small springs, recognized by the birefringent mixing action of the waters, are reported by local residents to have existed as far as three-quarters of a mile seaward from the shoreline in the Cutler area. Near shore, freshwater springs welled up through the bottom of Biscayne Bay as large boils; one such potable spring just north of the study site is marked by the words 'fresh water' on Coast and Geodetic Survey Navigation Chart No. 166, published in 1892. The early mariners and spongers customarily lowered kegs to the spring orifice to obtain fresh drinking water. (Kohout and Kolipinski 1967, p. 488)

Several early observers (e.g., MacKay 1847, p. 65; Shaler 1890, pp. 145–146; Griswold 1896, p. 53; Dix and MacGonigle 1905, p. 521) relate these springs to waters originating in the Everglades and passing through the porous rock of the Everglades Keys:

> And in the dry season I found that the course of these currents was, owing to numerous rock basins, in many instances perforated with holes in the bottom like a colander, into which these currents poured and disappeared; and in the pine woods, between the Glades and the Bay of Biscayne [i.e., the ridge formed by the Everglades Keys, most likely in Township 54 Range 41 or 40], may often be heard the rippling sound of running water, and frequently, in the fissures of the rock, it may be seen at from 6 to 8 feet below the general surface of the ground. (MacKay 1847, p. 65)

Other historical observations of the Everglades Keys (chapter 10) also suggest an active groundwater connection between the predrainage Everglades and the Atlantic Coast.

As water levels within the Everglades were lowered by drainage, the springs largely disappeared:

> Springs in Dade County, such as Mangrove Spring at Coconut Grove, Miami Springs, and various bayside springs which were reported to flow in earlier years, no longer flow owing to lowered water tables in the area. Mangrove Spring which supplied water for the United States Fleet at Havana in 1898, was reported to flow at a rate of 100 gallons per minute in 1903. (Ferguson et al. 1947, p. 65)

Springs of Florida (Rosenau et al. 1977) confirms the results of Ferguson et al. (1947), listing no springs for Dade County, nor any submarine springs off the Dade County coast. Rosenau et al. (1977) do note three constructed artesian wells in [Miami-]Dade and Monroe counties that "have been identified by the public as springs," including "Mineral Springs," a well 1,244 feet deep then flowing at around 1,000 gal/min in Grossman's Hammock within the Everglades (Township 55 Range 37E, Section 25).

As was the case for inflows from groundwater into the Everglades, the historical record does not provide enough information to quantify predrainage groundwater outflows. The record does however clearly document their existence, locations, and their reduction or elimination after drainage began.

Flows within the Everglades

At this point in the reconstruction process, the reader is reminded that some pieces are more speculative than others and that current hydrologic terms may not fully capture the broad, long-term, holistic overview that is the specific goal of these final sections. The term "sheet flow," which is widely used in popular and technical discussions of Everglades hydrology (e.g. Marshall 1980; Restrepo et al. 1998, p. 764), may not be appropriate for all landscapes of the Everglades. Differences in flow regimes between landscapes were strong and perhaps important enough to merit separate terms. The extreme breadth and thinness of the flowing "sheet" of water (40 miles wide by 2 feet deep; a ratio of 1:100,000) gives credence to the term "sheet flow" in the Everglades, especially in the Sawgrass Plains landscape. However, within the Ridge and Slough landscape, where ridges presented a resistance to flows quite different from sloughs, an alternate term such as "slough flow" may be more appropriate.

To define differences between these two flow regimes it is helpful to distinguish three different aspects of Everglades land surface conditions: slope, levelness, and microtopography (table 11.6). Slope is defined as relative change in elevation along the upstream-downstream axis. Levelness is defined here as the presence or absence of topographic curvature in the transverse direction, perpendicular to the downslope axis. Levelness was determined from the estimated predrainage contours shown in plate 6 and discussed in detail in the Topography section of this chapter. It reflects the absence

Table 11.6. Topographic aspects of Everglades landscape surfaces

Aspect	Possible Conditions			Axis	Scale
Slope	Horizontal	vs.	Sloped	Downstream	Regional
Levelness	Level	vs.	Concave or convex	Transverse	Regional
Micro-topography	Present ("uneven")	vs.	Absent ("flat")	n/a	Local

Table 11.7. Classification of Everglades landscapes according to table 11.6

Landscape	Slope	Levelness	Micro-topography	Substrate	Patterned Peatland
Custard Apple Swamp	Horiz.(?)	Level	Flat	Peat/silt	No
Sawgrass Plains	Sloped	Convex to level	Flat	Peat	No
Ridge & Slough	Sloped	Level to convex	Uneven	Peat	Yes
Marl Marshes	Sloped	Level	Uneven	Marl, rock, peat	No

("level") or presence of transverse convexity or concavity (e.g., a "ridge" or "valley") at the scale of the full width of the landscape, and parallel to the downslope axis. Microtopography refers to variations in elevation across horizontal scales smaller than the full length or width of a landscape; that is, here on the order of hundreds of feet (tens of meters). In principle these three surface conditions could occur in any combination. An Everglades landscape could therefore be simultaneously "sloped," "level," and "uneven" or "horizontal," "convex," and "flat."

Table 11.7 applies these three land surface definitions to the four main Everglades landscapes, also noting which landscapes could be considered patterned peatlands (H. E. Wright et al. 1992).

All landscapes were sloped (plate 6, Topography section of this chapter), with the possible exception of the narrow Custard Apple Swamp. In the transverse direction, the convexity of the upper Sawgrass Plains was likely the result of the transition from the semi-circular southern shoreline of Lake Okeechobee (all at essentially the same elevation) to the downward slope of the Everglades south of the lake (plate 6, Topography section). With increasing distance from the lake, the influence of the shoreline dissipated and the topography tended toward level in the transverse direction. This transition from convex to approximately level with distance southward appears to be reflected in the flow directions simulated by the Natural System Regional Simulation Model, version 3.3 (plate 3a). The opposite downstream transition within the Ridge and Slough landscape, from level to slightly convex, is discussed further in the Drainage Patterns section below.

The key landscape feature highlighted in table 11.7 is the distinction between the Sawgrass Plains and the Ridge and Slough landscapes. Although both formed on peat substrates, the upstream landscape was unpatterned, while the downstream Ridge and Slough landscape was distinctly patterned, with the directional patterning aligned with the downslope directions.

Sheet Flow in the Sawgrass Plains

The Sawgrass Plains formed a remarkably uniform stand of dense, impenetrable sawgrass (chapter 6). To the degree that other species were present, they appear to have been intermixed within the sawgrass stand rather than forming distinct patches. According to the tuning hypothesis, the predominance of sawgrass suggests that despite the regional slope, water depths must have been relatively uniform across the landscape. If microtopographic features had been present—that is, areas of consistently different water depths—these would have been reflected in the vegetation as broad-leafed aquatics or emergent species in the deeper areas; and shrubs in the shallower ones. These were not reported, nor do aerial photographs (e.g., USDA-SCS 1940) suggest they were present. A thin, unusually uniform layer of surface water flowed across the Sawgrass Plains, making "sheet flow" a highly appropriate description here.

Slough Flow in the Ridge and Slough Landscape

Sheet flow may not be the most effective term for describing water movement through the predrainage Ridge and Slough landscape. At the local scale, flow characteristics associated with ridges, sloughs, and tree islands differed.

Tree islands, especially the tails but also the heads, were at times covered with water (e.g., explorers reporting no dry place to camp); however, the combination of shallow water and dense vegetation would have kept flow rates lower than those present in predrainage sloughs. Similarly, during part of the year, surface water fell below the peat surface of the sawgrass ridges. Even when water levels rose again during the wet season to cover the ridges, the flow rates (on average) were likely less than those in sloughs, due to shallower water depths and friction caused by the high density of sawgrass stems.

The term "slough flow" is proposed to emphasize that within the Ridge and Slough landscape, the directional microtopography and corresponding pattern of plant growth (plates 24, 25, 26) meant that neither water depth nor resistance to flow was uniform over the landscape. Use of this term captures the conjecture that water flowed preferentially through the generally continuous network of sloughs. Given the differences in depth and resistance to flow, and that sloughs covered somewhat more than half of the Ridge and Slough landscape, one can assume as a first approximation that most of the water passed through sloughs; that is, through the larger portion of the landscape cross section (plate 23; see also Restrepo et al. 1998, p. 770).

Physically, chemically, and biologically this created a very different hydrologic environment from the much more uniform sheet flow of the Sawgrass Plains. Water moved more readily through the sloughs, which presented little flow resistance, whereas water moved more slowly and at times not at all over ridges. Water in the sloughs would transport dissolved constituents while water in the ridges would move more slowly, facilitating greater exchange of dissolved constituents.

In general, the ridges created a degree of physical and chemical differentiation between ridges and sloughs and between neighboring sloughs. Additionally, studies of the current system—despite reduced postdrainage elevation differences between ridges and sloughs—suggest significant biological differences between the two habitats, as reviewed earlier in this chapter (see Ecological Implications of Predrainage Water Depths).

Finally, a key distinction between sheet flow and slough flow is that resistance to flow in the Sawgrass Plains was assumed to be isotropic (resistance the same in all directions), but resistance was assumed to be anisotropic in the Ridge and Slough landscape. Anisotropy indicates that the resistance to flow in the direction parallel to the ridges and sloughs was much less than the resistance in the transverse direction. Under predrainage conditions, it appears that the direction of least resistance was aligned with the downslope direction (compare fig. 11.12; plate 3a; and plate 6). The effects of anisotropic resistance to flow therefore may have been primarily local, for example, elongating the dissipation of a local storm-induced bulge in the hydraulic head. As drainage and soil subsidence altered the land slope, newly emerged differences between the downslope direction and the microtopographic direction of least resistance may have given rise to altered postdrainage flow patterns and, eventually, to degradation of the original landscape patterning.

Flow Magnitudes and Directions

There exist very few quantitative estimates of Everglades flow velocities by firsthand, predrainage observers, and these few observations are clearly estimates rather than actual measurements. Nevertheless, considerable qualitative information is available, from a wide range of observers, times, and locations. These, in combination with modern measurements in the postdrainage, managed system, provide helpful insight into predrainage flow conditions. It is also useful to distinguish between magnitude and direction. While quantification of predrainage flow velocities would have required careful measurements, quantification of direction is straightforward. We can have very high confidence in our knowledge of predrainage flow directions.

Appendix S documents a small number of observations of very slow flow velocities within the Everglades, e.g., "current very sluggish," and "current so slight as to make it difficult to say if any existed." However, such observations are in the minority and they give the impression of having been recognized as exceptions. More commonly, most predrainage observers instead remarked on the absence of stagnation, on "perceptible flows," on "decided" or "marked" currents, and in several cases even noted currents "sufficiently strong to bend the sawgrass." Variation with season (i.e., water depth) was also noted: "at high water period saw same [good] current throughout Glades." Parker et al. (1955, p. 154) even recorded organic sediment transport, albeit under hydrological conditions of unknown resemblance to predrainage ones: "some of the [better-established drainage courses] have become small streams in which the underlying organic deposit has been stripped away in their channels and the bedrock exposed." Loftus also noticed visible organic floc movement in the water column when throw traps for fish stirred up floc (William Loftus, pers. comm., 2009). Some flow observations were clearly the result of local magnification of flows ("In going from one glade [slough] to another through a narrow channel we found quite a strong current"), but the overall impression is one of widespread sheet flow at visually perceptible rates.

Recent interest in postdrainage loss of the original Ridge and Slough landscape geomorphology (e.g., Science Coordination Team 2003), and in the potential role of flows and sediment transport in maintaining and/or restoring landscape geomorphology, has spurred a number of research efforts. Flow rates have been measured over multiyear periods, in areas both impounded (Water Conservation Area 3A) and unimpounded (Everglades National Park). Within this impoundment, Ridge and Slough patterning is still partially present, but water surface slopes are much diminished from predrainage conditions, and slough vegetation densities are likely somewhat higher. Flow rates in the impoundment averaged 0.3 cm s^{-1}, and ranged from 0.02 to 0.8, with a several hour peak of 5 cm s^{-1} after Hurricane Wilma (Harvey et al. 2009). In Shark River Slough (Everglades National Park), free from impoundments but with much reduced water depths and much increased emergent plant stem densities, average flows were four times higher than in WCA 3A, averaging 1.2 cm s^{-1}, with 90 percent of the flows between 0.5 and 2.3 cm s^{-1} (Schaffranek 2004). The potentially transportable, low density organic floc has been characterized (Noe et al. 2007; Larsen et al. 2009a), and the bed shear stresses required for floc entrainment have been determined (1×10^{-2} Pascals [Pa]; Larsen et al. 2009a) and found to be much lower than required for mineral floc entrainment. Larsen et al. (2009c) calculated that mean flows in the present managed system do not achieve the bed stresses needed to entrain floc, a partial explanation for topographic flattening (reduction of the difference in relative elevation of ridges and sloughs; Larsen et al. 2007). The absence, in the current system, of flow velocities sufficient for entrainment is not altogether surprising, given the changes caused by water management and the now measured effects of these changes on flow. For example, Leonard et al. (2006) found that flow rates measured in Shark River Slough decreased with decreased water depth, while Larsen et al. (2009b) and Harvey et al. (2009) found that both bed shear stress and velocity were highly sensitive to water surface slope and to stem densities. Leonard et al. (2006) also found that higher densities of floating periphyton and of *Utricularia* within the water column reduced flow velocities.

Collectively, these measured dependencies are important steps toward understanding predrainage sheet flow and floc transport. It is clear that *all* currently measured flow velocities are less than predrainage ones (plate 3). In the areas of the current system where landscape morphology and water depths resemble predrainage conditions (WCA 3A), the water-surface slope is greatly reduced by impoundment (Telis and Higer 2006). In areas where the surface slope approaches predrainage levels (e.g., Everglades National Park), the water depths are much lower and vegetation densities much higher. We conclude that (1) there are *no* predrainage flow velocity analogs in the currently remaining system, and (2) predrainage mean velocities certainly exceeded, perhaps substantially, the 1.2 cm s^{-1} rate currently measured. It remains to be determined whether actual predrainage velocities, once estimated, will prove large enough to have achieved bed shear stresses sufficent to entrain Everglades organic floc.

It is of interest to compare the few quantitative predrainage velocity estimates with the careful measurements from the current managed system. Early observers described flow rates of 0.5 to 1.5 miles per hour, corresponding to 22 to 67 cm s^{-1}. Even taking into account that some of these flows were observed in local high velocity channels, and that predrainage slopes, depths, and vegetation densities all point in the direction of higher predrainage velocities, these values seem high. They should probably be taken as a confirmation of the qualitative observations: widespread sheet flow, occurring at visually perceptible rates.

A number of sources of information are available for estimating the predrainage directions of water flow in the Everglades. The sources used here are: (1) historic predrainage and early postdrainage point observations of flow or landscape directionality that can be georeferenced to within a small radius relative to the size of the landscape; (2) narrative descriptions that refer to areas rather than specific points; (3) studies of flow patterns by informed observers; (4) landscape directionality derived from 1940s aerial photographs; and (5) extant directionality of the patterned Ridge and Slough peatland and of lenticular tree islands.

The Ridge and Slough peatland appears to have been so strongly directional that even without the benefit of an aerial perspective, many early observers remarked upon it (chapter 8). Several also noted that the direction of flow (e.g., Rose 1898, p. 456; Dix and MacGonigle 1905, p. 521), the orientation of the landscape elements (e.g., Harshberger 1914, p. 164), and the downslope direction were all one and the same (Baldwin and Hawker 1915, p. 752; Parker et al. 1955, p. 152):

> The 'grain' of the Everglades . . . is believed to be developed entirely on fresh-water peat and muck and apparently does not reflect an underlying pattern of marine bars. It merely represents a drainage pattern produced on a very gentle sloping surface of organic deposits. The 'grain' is composed of tree islands and

> swales [sloughs] that trend parallel to the regional slope, just as one would expect in an area of consequent drainage. (Parker et al. 1955, p. 152)

Figure 11.10 maps individual observations of directions of water flow and landscape orientation made between 1827 and 1913 and tabulated in appendix S. The following representative note, made during the dry season (March) of 1884, refers to what is now western Water Conservation Area 3A:

> The ["12–18 inches" of] water running in the glades [sloughs] is pure, and tends toward S.E. There is every appearance that the water runs S.E. in the wet season. (Sollie 1884-T51 R35)

Within the eastern half of the Everglades, figure 11.10 shows southeastward flows directed toward the numerous natural outlets piercing the Atlantic Coastal Ridge, specifically Cypress Creek, Middle River, New River, Snake River, Arch Creek, Little River, and Miami River (see also plates 40–42 and 7). The field observations match the many narrative descriptions of southeastward flow "in the direction of the outlets" (e.g., Rose 1898, p. 456; Roberts 1907; Frederick 1907; Stewart 1907, p. 55; Dimock 1907, p. 249; Wright 1912, pp. 18, 24; Harshberger 1914, pp. 54–55; Baldwin and Hawker 1915, p. 752; Beard 1938, p. 46; J. H. Davis 1943a, p. 119; U.S. Army Corps of Engineers 1960b, p. B-3; Leach et al. 1972, p. 8; Parker 1974, p. 18). The southeastward flow directions recorded on figure 11.10 are also consistent with several surveys into the Everglades that reported downward slopes toward the coast (Rose 1898, p. 456; Newman 1907, p. 46; Ashworth 1919a). Farther west, the field-observed directionality mapped in figure 11.10 appears, from north to south, first south-southeast, then southward, and finally southwestward south of Township 53, also consistent with narrative accounts (e.g., Willoughby 1898a, p. 138; Dimock 1907, p. 249).

Figure 11.11, drawn by Garald Parker as part of an extensive study of the Everglades, is attributable to an informed but not predrainage observer (Parker's studies began around 1940). There is some question as to whether his "directions of surficial drainage" refer to observed postdrainage or estimated predrainage conditions. However, Parker's figure includes arrows in areas already drained by the 1940s and was also republished in a later publication focused on predrainage hydrology of the Everglades (Parker 1974). The directionality shown on figure 11.11 resembles that noted by historical observers, as shown in figure 11.10, as well as the shape of the longitudinal axes discussed in chapter 4.

As part of this study, directionality of ridges and sloughs was estimated for the entire Everglades landscape (fig. 11.12). A 2-mile by 2-mile grid was overlaid on index sheets of the earliest set of aerial photos covering all of the Everglades (USDA-SCS 1940). In each grid an arrow was aligned parallel to the ridge and slough grain. The arrows in figure 11.12 show a spatially consistent pattern with a strong spatial correlation between cells. Overlay of the 1940 directions shown in figure 11.12 on current satellite imagery (not shown) indicates very close correspondence with current Ridge and Slough landscape directionality still visible within the remaining Everglades.

Two assumptions are implicit in the use of figure 11.12 to map predrainage flow directionality: first, that the landscape directionality visible in the 1940s photographs is the same as the predrainage directionality; and second, that under predrainage conditions, landscape directions were the same as flow directions. This is particularly important for the urbanized eastern portion of the Everglades, roughly Townships 50–53 and Ranges 39–41 (plate 4), where the photographic record consists primarily of a single, postdrainage time point, 1940. The assumption made here, of constancy in the urbanized portion, has both a mechanistic and an empirical basis. Mechanistically, studies of postdrainage landscape pattern throughout the Ridge and Slough landscape (e.g., Sklar et al. 2004, pp. 6-34 to 6-38; Wu et al. 2006) suggest that the clarity of the predrainage pattern can initially persist even though the elevation differential between ridges and sloughs is declining. At some point, ridges begin breaking into smaller segments; at this stage the landscape directionality can still be discernible. At a final stage, possibly when the ridge to slough elevation differential has reduced to zero, all pattern disappears and directionality can no longer be detected.

It is theoretically possible that anthropogenic flows, oriented differently from predrainage ones, could cause this final "erasure stage" to be followed by a "rewriting stage," one in which anthropogenic flows create new ridge and slough patterns in a new direction. However, to date this has never been observed in the Everglades. We assumed that the 1940 directionality reflects predrainage directionality because: (1) the patterns are similar to those from other areas of partially degraded ridge and slough patterning; (2) no areas of suspected rewriting have been found elsewhere in the Everglades, even adjacent to canals where forces for rewriting would have been greatest; (3) the postdrainage water depths and flow energies in this area would have been less than

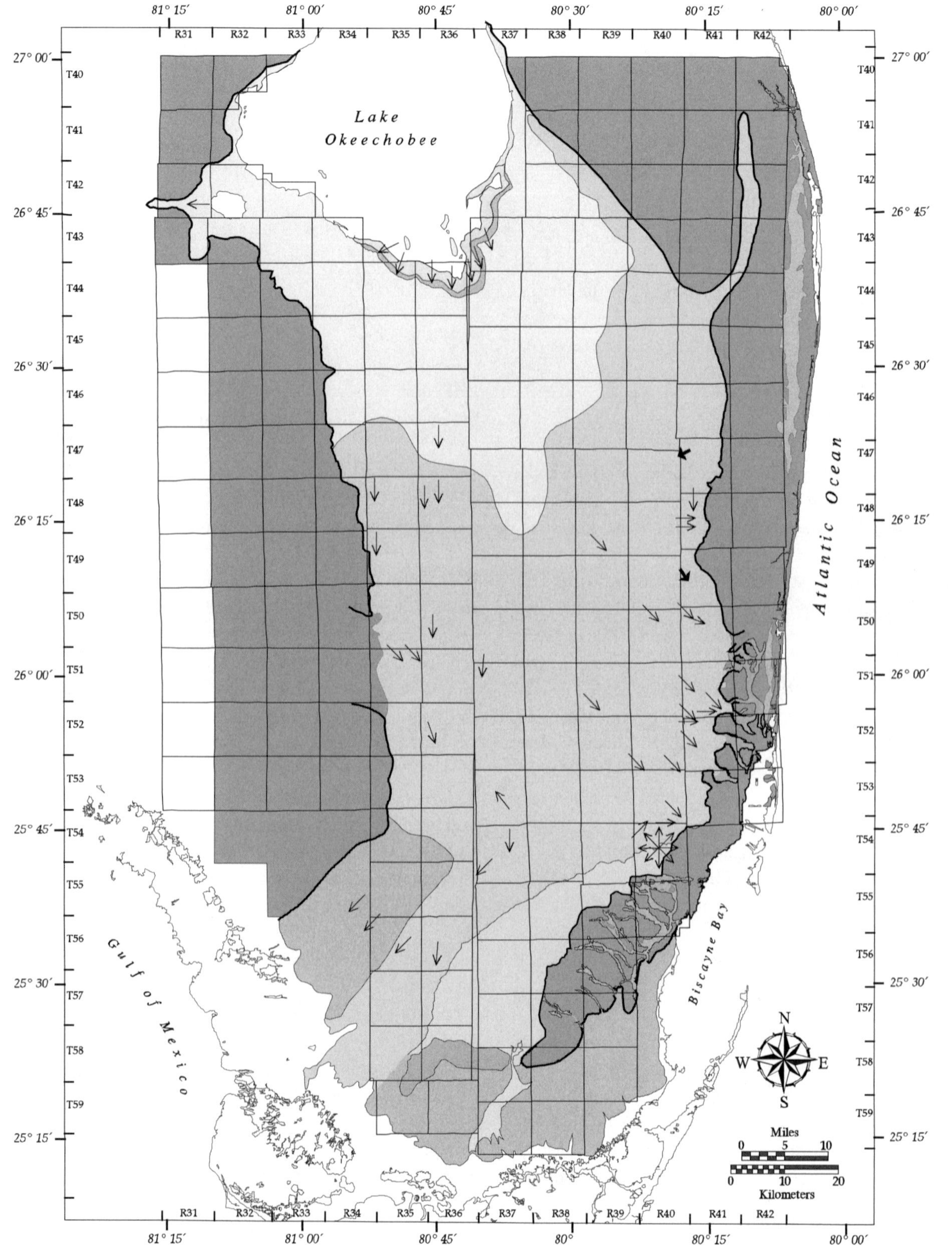

Figure 11.10. Directions of observed surface water flow recorded prior to 1913. Thicker arrows in Townships 47 and 49 reflect an average of seven to eight adjacent observations. Sources include township surveys, soils surveys, and cross-Everglades expeditions (appendix S).

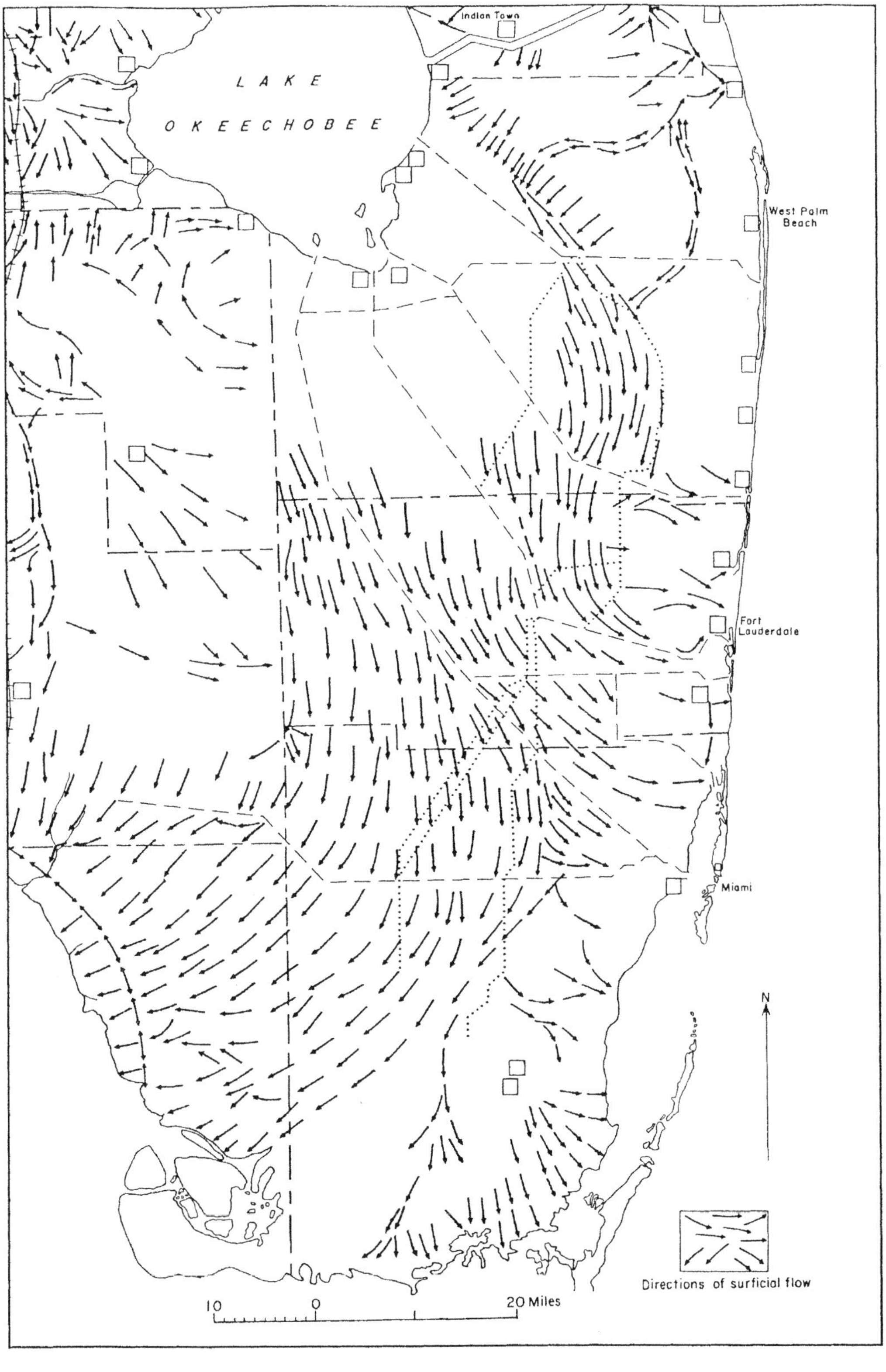

Figure 11.11. Directions of surface water flow estimated by G. G. Parker, likely representing predrainage conditions. Modified from plate 11 in Parker et al. (1955) and Parker (1974) by addition of major pre-1950s canals (dashed lines) and 1990s canals or levees (dotted lines).

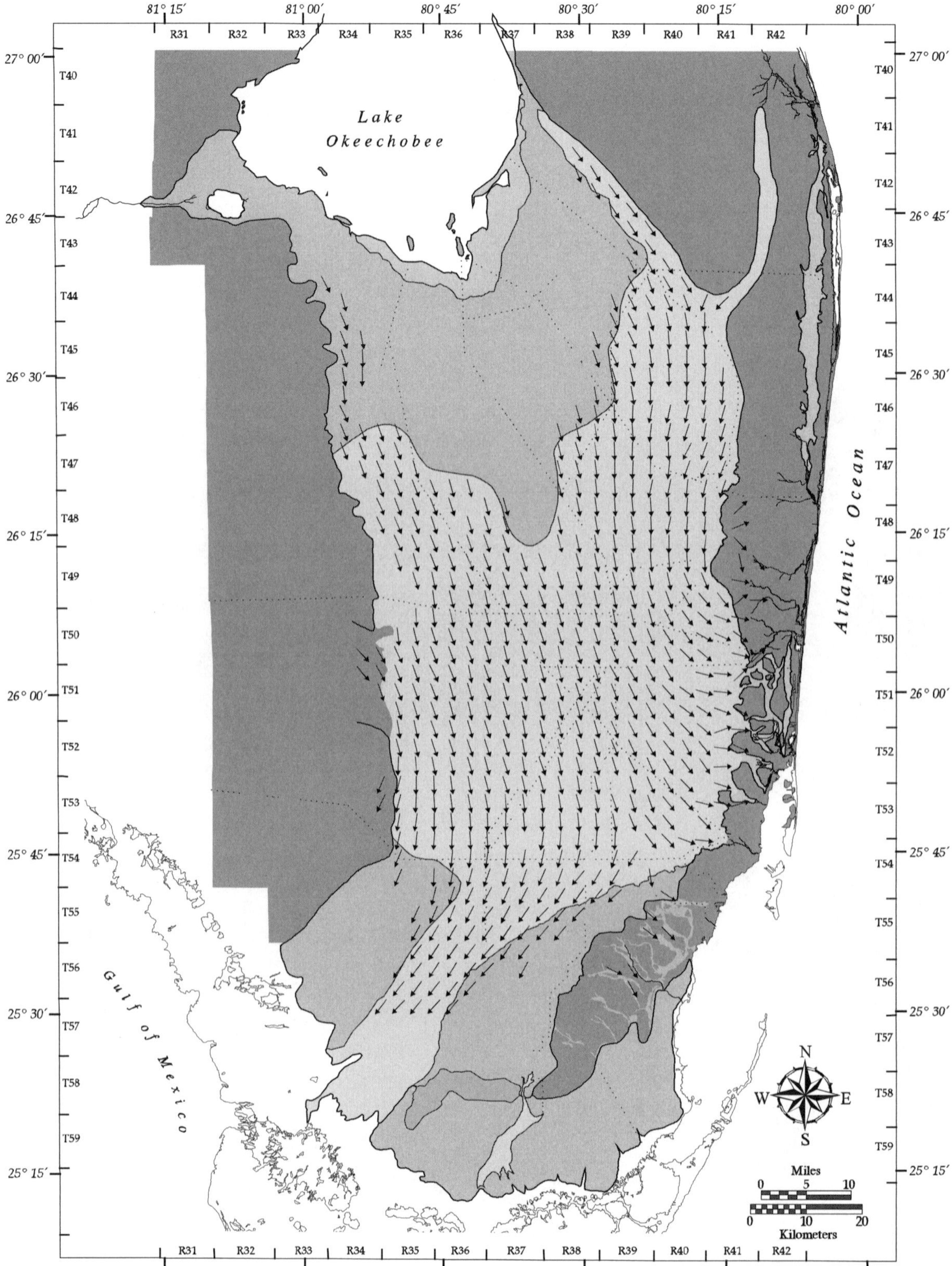

Figure 11.12. Directions of predrainage surface water flow within Ridge and Slough landscape. Directions were estimated by measurement of the angle of landscape "grain" formed by parallel sawgrass ridges and sloughs visible on aerial photographs (USDA-SCS 1940). Angles were measured on a grid of 2 × 2 miles (3.3 × 3.3 km).

predrainage ones, hence less likely to write new patterns; (4) the directions in this area seamlessly extend regional patterns; and (5) the directions drawn from the 1940 photographs match those noted by predrainage observers (fig. 11.10).

Figures 11.10, 11.11, and 11.12 as well as the tree island directionality visible in plates 31 and 32 show very similar patterns, with no apparent abrupt changes in directionality. Recent simulations of predrainage surface water flows also show similar patterns (plate 3a). One has the impression of a self-organized field of gradually bending, parallel paths, reminiscent of iron filings organized by a magnetic field. Together the consistency among narrative observations of predrainage explorers, the directions shown on figures 11.10, 11.11, and 11.12, and the directionality visible on current aerial and satellite imagery suggest a general but strong relationship between landscape directionality and flow directionality in the predrainage Everglades. As Parker et al. (1955, p. 152) wrote, "a drainage pattern produced on a very gentle sloping surface of organic deposits."

Drainage Patterns

As a first approximation, and as viewed in the context of more typical drainage systems, the Everglades watershed behaved as a single basin: a continuous wetland lying between higher lands to the east and west (plate 7). However, estimates of predrainage flow fields within this wetland may also lead to a more complex picture, including the existence of sub-basins, as presented here for discussion and future analyses.

The generalized Everglades predrainage pattern, although continuous, discharged water through two separate and distinct groups of outlets, Shark River Slough to the southwest, and the eastern coastal rivers to the east and southeast. Flow directions over wide areas angled toward these two outlets. The pattern of flow directions, the relation of the pattern to the outlets, and the very wide, shallow geometry of the Everglades suggest two parallel, longitudinal flow paths. Although it certainly does not constitute independent evidence for these flow patterns, it is relevant to note that similar ones appear in a regional simulation of predrainage surface water flows (plate 3a).

We assume that the western flow path originated as radial flow out of the southwestern shore of Lake Okeechobee between township Ranges 32 and 36; then ran southeastward for as much as 40 miles (63 km); turning due south around Township 54; and finally southwest through Shark River Slough to discharge into Whitewater Bay. Similarly, it is assumed that the eastern flow path originated as southeastward flow out of the southeastern lakeshore (Townships 44 to 42); ran more or less south through what are now Water Conservation Areas 1 and 2; and started curving southeastward about Township 49, discharging to the Atlantic Ocean and Biscayne Bay through Cypress Creek and the Middle River (cf. plate 42, Township 49) and the various breaks in the Atlantic Coastal Ridge between the New River and Miami River (plate 41).

There was no dividing wall or barrier between the eastern and western flow paths (plate 7). Rather, the pattern of flow directions shown in figures 11.11 and 11.12 suggests that from approximately Township 44 to Township 49, the two flow paths ran roughly in parallel and that the landscape was about "level" in the transverse direction (table 11.7). Contours would have been approximately straight. Only the great width of the landscape, the absence of consistent lateral driving forces (e.g., similar water depths and similar hydraulic heads), and the numerous sawgrass ridges would have kept these long-term regional flows separated.

Farther south (Townships 50 to 53), the divergence of the flow directions, southwest versus southeast, suggests a slight and very broad convexity in the original peat topography, with the eastern flow path sloping southeastward down toward the Peat Transverse Glades and the western flow path sloping southwestward down toward Shark River Slough (plate 7). Curved contours corresponding to such a convexity were in fact drawn on a 1935 map (Board of Commissioners of the Everglades Drainage District 1935). A number of early Everglades observers mention a central ridge (Willoughby 1898a, p. 138; Moses 1907, p. 43; Frederick 1907; p. 44; Roberts 1907, p. 52), which has sometimes been imagined as a localized, narrow central strip, partitioning the eastern and western flow paths. However, while a gradual divergence in flow directions does appear to have been present, there is no evidence of a narrow, local ridge, and such a feature would be an unexpected anomaly within an area formed by peat accumulation. Instead it seems more likely that the early observers were perceiving a broad convexity similar to the shape shown on Board of Commissioners of the Everglades Drainage District (1935), with its highest point somewhere near the center of the Everglades (see also plate 6).

The concept of eastward and westward flow paths was noted by a number of early observers well before the availability of aerial photographs or maps of flow directions. The first concluding point in Daniel Beard's report on his extensive exploration and reconnaissance for the projected Everglades National Park was clear: "1. The water in the Everglades drainage basin used to flow in two main channels, one to the east and one to the west" (Beard 1938, p. 50). In the main text he elaborated:

> There were two main flows of this water through the Everglades—one to the southeast and the other to the southwest. The eastern flow broke through the slight ridge along the Atlantic in places such as the New River at Fort Lauderdale and Miami River at Miami. It also went through between the Everglades Keys and emptied into Biscayne Bay or Florida Bay.
>
> We see, then, that all the water in the Everglades drainage basin did not come through the proposed park, but the park got the bulk of the western flow and some of the eastern flow that went through the Everglades Keys (Taylor River from Paradise Key to Madeira Bay Section). Probably the flow through the park was not quite half of the total water in the basin, although this is just a guess. (Beard 1938, p. 46)

His description matches the conclusions of this study, except for the fact that the "eastern flow . . . through the Everglades Keys" mentioned by Beard would actually have been lateral outflows through the Marl Transverse Glades and would have originated from the western flow path. Recalling that Beard in 1938 had at his disposal neither a soil map to identify the location of the Rockland Marl Marsh nor topographic contours to discern the marl marsh's elevation relation to Shark River Slough, his description is understandable.

Beard's description is similar to an earlier one by MacGonigle: "The Everglades consist of two great basins lying between lake Okeechobee and the extreme southern point of the peninsula" (MacGonigle 1896, p. 387).

W. R. Moses, secretary of the Ingraham Expedition that crossed through both the western and eastern flow paths in 1892 (plate 10 left), was even more explicit in describing two flow paths with southeastern and southwestern outflows, including the concept of a separating ridge: "[Moses] thinks the water in the Glades divides running to the southeast and southwest. . . . The dividing ridge where they [the Ingraham expedition] crossed was about ten miles wide" (Moses 1907, p. 43).

Willoughby, who during his trip passed through both the eastern and western flow paths (plate 10 left), also thought there was a division between eastern and western waters: "This showed me clearly that we were very near the dividing water-shed which runs down the center of the State and has its terminus in the heart of the Everglades" (Willoughby 1898a, p. 138).

J. S. Frederick was a surveyor from Miami, likely the surveyor employed by Rose in 1898 to run survey lines westward into the Everglades and definitely the surveyor who participated in the FEC railroad survey of the portions of the Rockland Marl Marsh area (Krome 1902a, b). He told drainage investigator J. T. Stewart he thought there was a dividing ridge from which water should be drained to the southwest and southeast: "J. S. Frederick, Miami, Florida, surveyor. Thinks there is a ridge in Glades which divides the water. Better to open all little streams in the Glades than one large one. Should drain southeast and southwest" (J. S. Frederick 1907, p. 44).

Thomas Roberts, another early Miami resident interviewed by Stewart and considered experienced in the Everglades, thought there was a central ridge separating flows to the southwest from those to the east: "Ridge all the way down the center which divides the water flowing east and southwest. At high water period water covers this ridge and from a point 10 miles south of Brown's [Store] there are strands of water [sloughs] crossing this ridge in many places. Water leads [sloughs] in Glades are in direction of outlets" (Roberts 1907, p. 52).

Comparison of a map of existing and proposed canals (fig. 11.13; see also Willey 1911, p. 430) with figures 11.11 or 11.12 suggests that the drainage engineers of the early 1900s may well have sought to route the main canals down the original downslope directions and to make use of the original eastern and western outflow locations.

To summarize, the earliest aerial photography and historical data together make a case that water in the 40-mile-wide predrainage Everglades usually flowed in two parallel, regional flow paths, and that these ultimately diverged to discharge through separate outlets. Although no distinct dividing barrier separated the paths, we assume that the extreme width to depth ratio, the absence of persistent lateral driving forces, and the resistance to lateral flow posed by the many sawgrass ridges would all have limited the opportunities for lateral mixing between the two paths. Transient periods of lateral mixing would have occurred when lateral head gradients became non-zero, for example if local, heavy rainfall fell in only one of the two paths. Altogether, this suggests that the historical Everglades consisted of two de facto sub-basins: eastern and western.

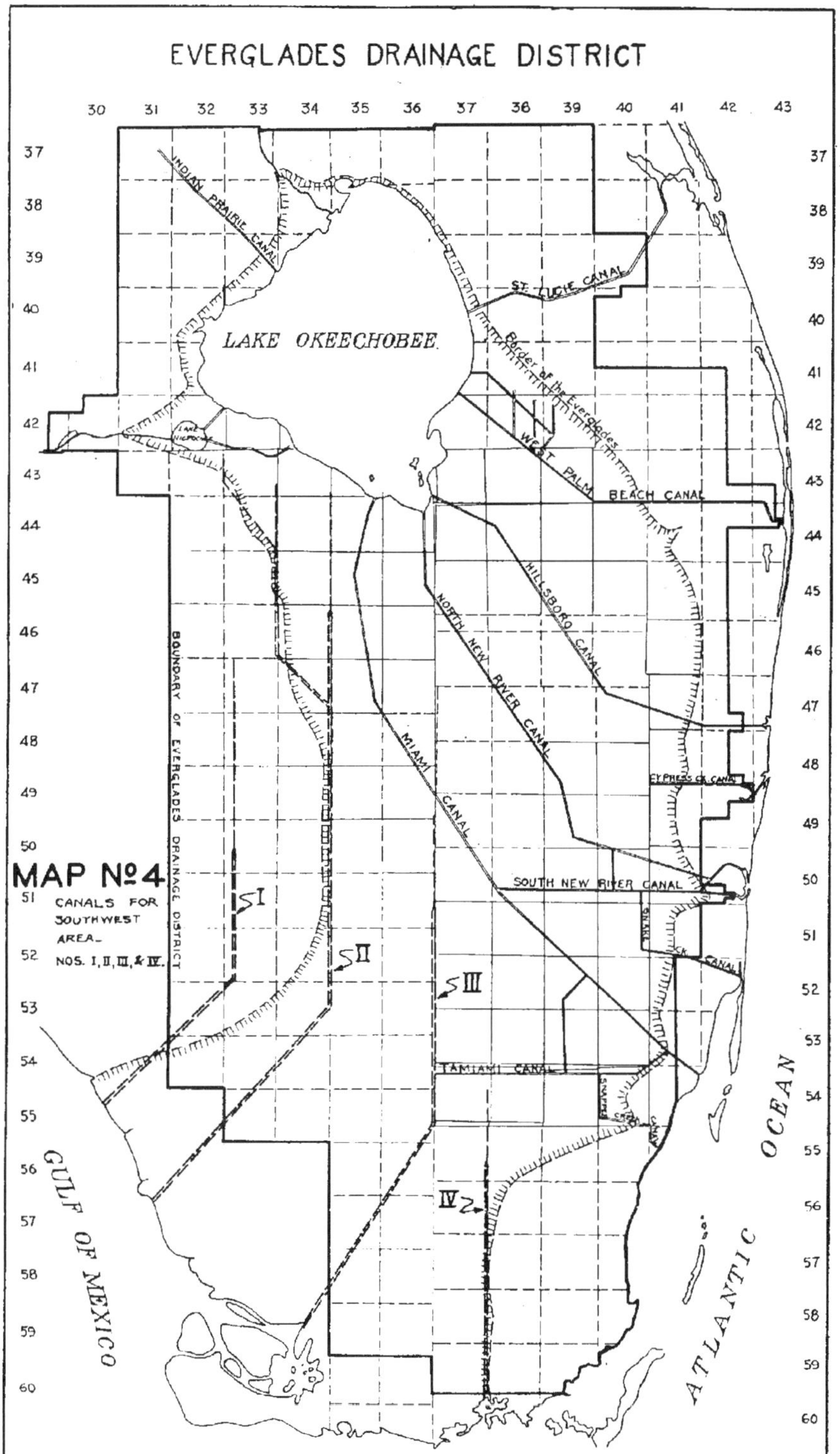

Figure 11.13. Map from 1927 showing existing canals (solid lines), draining to the southeast, and proposed canals (dashed lines numbered I, II, and III), which would have drained to the southwest. The locations and orientations of the two sets generally coincide with the southeastward and southwestward sets of flow directions seen in figures 11.12 and 11.13. Similar sets of proposed canals draining southwestward and southeastward are seen in Willey (1911, p. 430) and in Florida Everglades Engineering Commission (1914, pp. 12–13). Reproduced from Elliot (1927, p. 64).

The sub-basins were not separated by a distinct divide but instead were defined by overall flow patterns noted by historical observers and arising from a broad, very slight topographic rise that directed water generally southwestward toward Shark River Slough or generally eastward toward the Atlantic Ocean and Biscayne Bay. This configuration is reflected in the artist's rendering of the Everglades shown in plate 7 and, separately, in the model simulations shown in plate 3a.

Uncertainties in Predrainage Flows

Two questions arise from the hypothesis of regionally divergent flow paths: (1) How were the outflows partitioned—that is, what fraction flowed southeast vs. southwest? (2) Was the partitioning constant throughout the year, or did it change as a function of water level or rainfall pattern?

The relative proportions of water flowing southeast vs. southwest under predrainage conditions cannot be determined from historical narratives. A first-order estimate of the partitioning can be made using the following simplifying assumptions: that the point of separation between the two flow paths was constant over time; that velocities and water depths were the same within each path; and that average rainfall was evenly distributed across the watershed. Flow in each path would then be simply proportional to path width. Drawing a cross section in figure 11.12 from Forty-Mile Bend in Tamiami Trail (near T53 R35) to the South Fork of the New River (near T50 R41) suggests that, very roughly, 40 percent of the outflow might have flowed eastward and 60 percent westward.

To address the second question, temporal variation, one must again return to the tuning hypothesis and assume that the similarity of vegetation within the Ridge and Slough landscape was indicative of spatially similar water depths and hydroperiods. This, in turn, implies that with the exception of local, transient deviations, the planes of water surface and ground surface in the predrainage Everglades were generally parallel to each other, that the direction of regional water flow was the same as the downslope direction of the ground surface, and that flow directions would tend not to vary over time.

In previous sections, we argued that landscape directionality was aligned with the predrainage downslope direction, suggesting that landscape directionality (temporally constant) would tend to diminish seasonal variations in regional flow direction. To assess this in more detail, one can compare seasonal extremes. When water levels dropped below the 1.5-foot-high (45 cm) ground surface of the ridges during the dry season, water would be restricted to the sloughs (plate 23), forcing the regional flow directions to align with the ridge and slough directionality. When water levels rose again above ridge ground surfaces during the wet season, only stem densities on the sawgrass ridges would have constrained transverse flow, so theoretically the regional flow direction could have deviated under these conditions from the ridge and slough directionality. However, as discussed earlier, given the absence of persistent transverse driving forces, transverse flows would not have been persistent.

It would appear that the constancy of three influential aspects—the ground surface downslope direction, the water surface downslope direction, and the landscape resistance to flow—all suggest a system in which regional flow directions would have tended to remain generally constant. Accordingly it also seems likely that outflow partitioning remained constant over the course of the year.

An important exception to the concept of constant flow direction could have arisen if there were areas where the predrainage water surface slope became non-parallel to the ground surface slope. Such a situation would be expected in the presence of impounding conditions. The lower portion of the eastern flow path, where it approached the coastal ridge and coastal rivers (Townships 50–54, Ranges 40–41; cf. fig. 11.12, plate 40), had the potential to create such conditions. If, for example, the combined outflow capacity of the coastal rivers was less than the eastward slough flow, then water could have been impounded behind the coastal ridge and could have risen sufficiently to shift flow directions within the impounded area from southeast to south, or even southwest, directing "eastern water" into the western flow path.

The combined capacity of the breaks in the coastal ridge (the New River, several branches of Snake Creek, the two branches of Arch Creek, the Little River, and the Miami River) is not obvious from simple inspection and might have been either greater or less than the combined flows coming southeastward from the Everglades. To address this uncertainty, a quantitative analysis to estimate predrainage river capacity by calculating open channel discharges from historical predrainage estimates of slope, length, width, and roughness is an important next step. However, as noted in chapter 10, historical records suggest that: (1) these rivers and creeks flowed year-round; (2) the current in most was strong enough to run water wheels; (3) within each Peat Transverse

Glade no evidence was found for a threshold elevation limiting outflow from the Everglades; and (4) neither vegetation nor narrative observations suggested the presence of deeper (impounded) water in the area upstream from the coastal ridge. Thus qualitatively, the historical record provides little evidence that the outflow capacity of the breaks in the coastal ridge formed a constriction that caused water to impound behind the coastal ridge.

Topography

The original predrainage topography of the Everglades is now obscured by the diverse and, in some areas, large changes in elevation caused by postdrainage peat oxidation and subsidence (chapter 3). Unfortunately, because the earliest systematic surveying of the Everglades (Jones et al. 1948e) began about 1940 when most subsidence had already occurred, predrainage topography must be estimated based on a synthesis of indirect methods.

The smoothness (high spatial autocorrelation) of the downslope flow patterns seen in figure 11.12 suggests that the shape of the predrainage Everglades land surface can be estimated by drawing contours perpendicular to the arrows. This approach, along with elevation data from early canal surveys and data from the best available surveys of the unsubsided, nonpeat soil areas along the eastern and western borders, was used to estimate predrainage topography of the Everglades.

The earliest available elevation data for the interior of the Everglades were estimated by digitizing Florida Everglades Engineering Commission (FEEC) surveys of land surface made in 1913 along the routes of the recently constructed canals: the Hillsboro, North New River, South New River, and Miami canals (FEEC 1914, oversize plate of canal profiles). The results are shown in figure 11.14. An additional survey was also available for the proposed route of the West Palm Beach Canal (FEEC 1914, profiles 5–13). The Hillsboro, North New River, South New River, and Miami Canal surveys were referenced to an unknown "Ft. Lauderdale datum." The 1913 FEEC surveys were re-referenced to sea level (National Geodetic Vertical Datum of 1929, NGVD 29) by comparing the portions of the West Palm Beach and Hillsboro canals excavated through mineral soils—that is, through the unsubsided coastal ridge—with NGVD 29–based 1940s survey data (Jones et al. 1948e; also reproduced in U.S. Army Corps of Engineers 1960b). From this approach it was concluded that Ft. Lauderdale datum = NGVD 29 - 0.5 foot.

For the upper portion of the Everglades, down to about elevation 11–12 feet above NGVD 29, contours plotted from the FEEC (1914) survey data produced regular, slowly bending contours as expected from the landscape directionality. However, in the southeastern portion of the Everglades, where the canals approached the coastal ridge, the survey-derived contours made very sharp bends that did not match either landscape directionality or narrative descriptions of predrainage topography. Inspection of the surveys themselves (fig. 11.14), particularly of the North and South New River canals, showed steep drops in elevation as the coastal ridge was approached, suggesting that peat elevations had changed substantially by 1913 and suggesting an explanation for the sharp contour bends. Historical accounts from the time of canal construction documented dramatic lowering of water tables in this area and very rapid drying and exposure of formerly flooded lands after the canals were opened (McCutcheon et al. 1907, pp. 110–111; *Miami Metropolis* 1907; Jennings 1907, pp. 122–125; Newman 1908-T50 R41; Trustees of the Internal Improvement Fund 1908a, p. 280; West et al. 1909, p. 122; Wright 1910, p. 332). Early and extensive soil subsidence and soil decomposition were also reported in slightly later accounts (Wright 1910, p. 332; Gifford 1911, pp. 8–9; Baldwin and Hawker 1915, p. 753; Mitchell 1932; Stephens and Johnson 1951). Figure 3.1 illustrates the extent of water table drawdown already present in 1915 at both the north and south ends of the North New River Canal. Subsidence appears to have been most severe along the eastern half of the South New River Canal, perhaps because this area was drained not only by this canal but also by the adjacent North New River Canal upstream and Miami Canal downstream (plate 16).

Subsidence corrections for the canal surveys were estimated from the slope of the unsubsided portions of the surveys, from estimated amounts of subsidence reported in the early accounts mentioned, and from available information regarding predrainage elevations at the eastern edge of the Everglades (e.g., Bache 1850, p. 47; MacGonigle 1896, p. 392; Rose 1898, p. 456; Newman 1907, p. 46; U.S. Congress 1911; U.S. Army Corps Of Engineers 1960b; Gaby 1993, p. 2). The upstream ends of the canals also seem to have experienced some subsidence, approximately 0.5 foot, presumably due to the lowering of Lake Okeechobee stages beginning in the 1880s. The elevation of the Lake Okeechobee shoreline from a few miles south of Fisheating Creek to approximately Bacom (Cypress) Point was about 20.5 feet above NGVD 29 (about 22 feet above the Punta Rassa datum typically used for Lake Okeechobee prior to

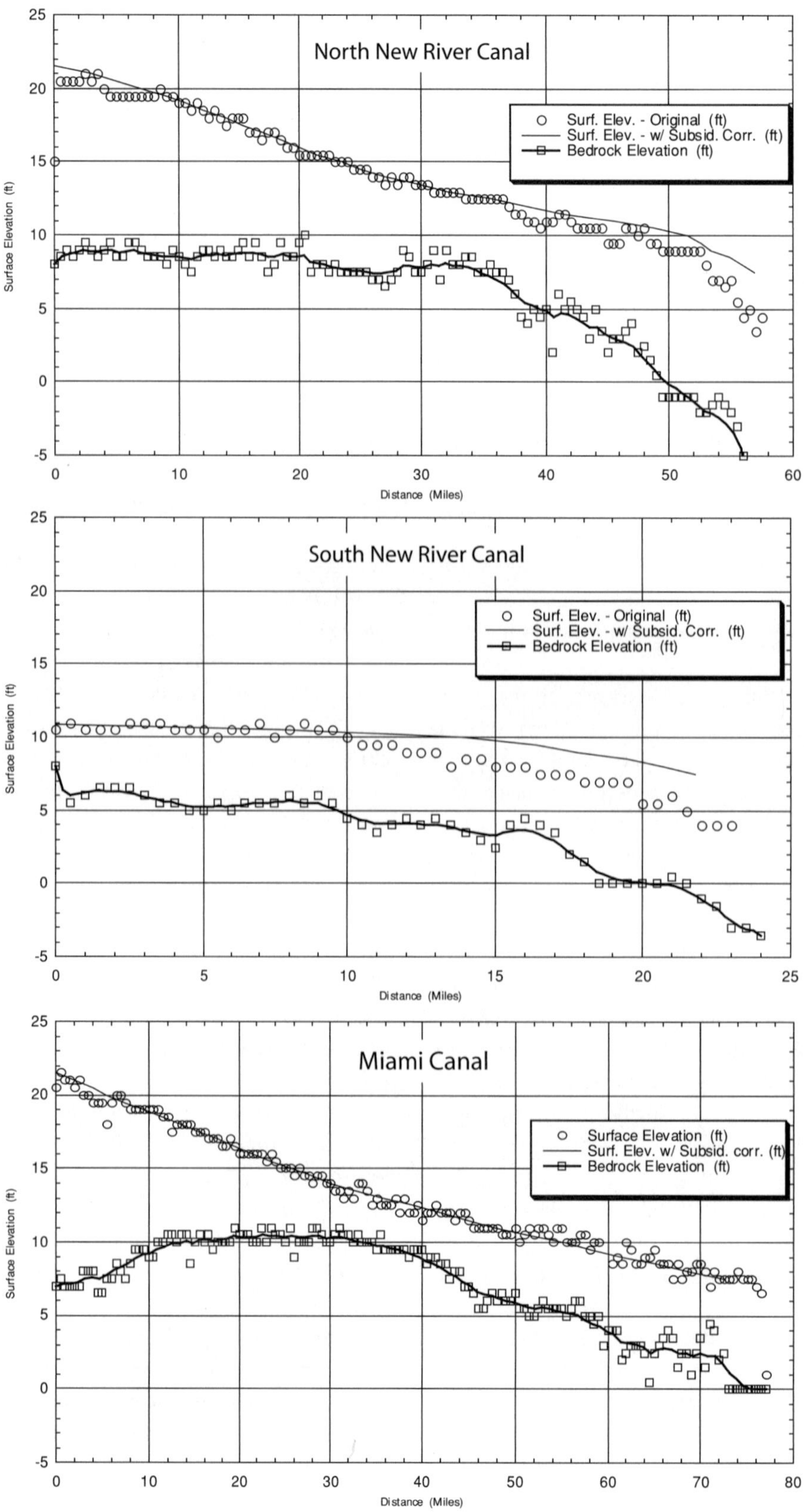

Figure 11.14. Surface and bedrock elevations along early canals, 1913 measurements. Surface elevation lines reflect estimated corrections for peat subsidence. Sources: Florida Everglades Engineering Commission (1914, unnumbered plates) and this study for subsidence-corrected predrainage estimates; see text.

the 1920s), based on observations from Meigs (1879, p. 870), Sackett (1888, p. 5), Kreamer (1892), and Slattery (1913).

Elevations of the unsubsided upland areas, available from the U.S. Army Corps of Engineers (1960b) and for the unsubsided Marl Marshes from recent U.S. Geological Survey data, along with the subsidence-corrected 1913 canal surveys and the predrainage directions of flow (fig. 11.12), were used to draw the estimated predrainage contours shown in plate 6. The three-dimensional diagrams shown in plate 1 compare the estimated predrainage topography with current topography, highlighting the contrast between the originally smooth and continuous Everglades slopes and the stair-stepped elevations of the present-day, compartmentalized system of Water Conservation Areas.

Pattern and Process: Microtopography and Flow

A glance at a larger satellite image of South Florida (e.g., plates 31, 32, 36) shows a pattern of tree islands: elongated ovals or teardrop shapes, all aligned in parallel, and with the teardrop shapes suggesting an upstream and downstream direction. Closer inspection of more detailed images shows a second level of pattern: the even more elongated, thin strips of the sawgrass ridges (plate 25, west of Miami Canal and especially the first 10 miles south of I-75). These do not seem to have obvious upstream or downstream ends, but they are definitely aligned parallel to each other and to the tree islands. While the heads of some of the tree islands have a bedrock base, recent measurements suggest that the peat microtopography forming sawgrass ridges has no relation to variations in the underlying bedrock (Sklar et al. 2003, pp. 33–34). Thirteen miles of early elevation survey along the North New River Canal (Mitchell et al. 1928) similarly show much greater variability in bedrock elevation than in peat surface, reinforcing the impression that bedrock variability became masked by peat accumulation. All indications are that the directional patterning seen in the Ridge and Slough landscape of the Everglades is a surface feature of the peat substrate, formed "from the top down," rather than a reflection of patterns in the underlying bedrock (plate 23).

The shape of the tree island and sawgrass ridge patterning, the origin of this patterning in peat microtopography, the regular spacing and dimensions of the pattern, and the orientation of the pattern parallel to the downslope direction all strongly suggest that water flow has somehow shaped the peat surface of the Ridge and Slough landscape. This impression is not new (Parker et al. 1955, p. 152; Gleason and Stone 1994, p. 172; Science Coordination Team 2003), but it is enigmatic because the very shallow slope of the Everglades suggests very low energies and thus flow velocities that were not obviously erosional or depositional. On the other hand, the presence within the water column of a bottom layer of particulate organic flocculent material may prove to be a link between low energies and flow-based patterning. Larsen et al. (2007) have suggested that further advances in Everglades geomorphology may come from comparative studies of the sand-dominated, ridge-forming anabranching rivers of Australia (Wende and Nanson 1998; Tooth and Nanson 2000; and Nanson 2004). Those river systems tend to have similar patterns of vegetated ridges, separated by channels or swales, spread across the floodplain and aligned in the direction of river flow. Although the widths of such river systems are much narrower, from 0.3 to 1.2 miles (0.5–2 km), the energies higher, and the sediments are mineral rather than organic, the pattern of ridges and intervening swales is remarkably reminiscent of the Everglades Ridge and Slough landscape.

The mechanisms responsible for the directional pattern of microtopography originally present throughout the Ridge and Slough landscape are still uncertain. Several aspects can however be distinguished. Parallels can be drawn to the patterned peatlands ("string bogs") of boreal climates (Glaser 1992, p. 5), which also form on peat substrates of shallow slope and include similar components: open water, aquatic "flarks"; herbaceous "strings," and tree islands. At the same time, string bogs differ from the Ridge and Slough landscape in two important ways: the slopes are somewhat steeper, and the orientation of the long, linear flarks and strings is crosswise rather than parallel to the direction of flow. The pattern-forming mechanisms in string bogs are also not well understood, but the radically different orientation present in the Everglades suggests differences from the bog processes.

In considering the mechanisms responsible for Ridge and Slough patterning in the Everglades, distinctions can be made among initial *formative* processes, subsequent *maintaining* processes, and postdrainage degradation processes. These three are not necessarily identical and may involve quite different energies and time scales. A distinction can also be made between different possible causal mechanisms as a function of their energies and return frequencies. These range from the very low-energy but high-frequency (continuous) processes associated with typical predrainage Everglades flow velocities to the high-energy but very low-frequency processes associated with hurricanes. Transport mechanisms of intermediate energy may be associated with convective storms that cause transient, localized water depth increases and flows. Formative processes may be associated with high-energy, infrequent events. Maintenance of the Ridge and Slough pattern may be associated with low-energy but more continuous "events." However, such associations are only speculative at this point.

Distinction can also be made between hypotheses of a specific flow-related process, such as the transport of flocculent organic carbon out of sloughs, versus a more general hypothesis that water flow, through as yet unknown mechanisms, is responsible for the directional Ridge and Slough patterning. Examples might include the transport of seeds or the differentiating effects of faster-flowing (slough) and slower-flowing (ridge) areas on sediment deposition, nutritional environments, plant growth rates, or decomposition rates.

Although the specific mechanisms are not understood, it is possible to list characteristics of the patterned Ridge and Slough peatland and to compare this list systematically with possible driving forces—wind, peat fires, bedrock patterns, cycling of water depths, and water flows. Inclusion of two time periods, pre- and postdrainage, helps elucidate links, as both landscape and driving forces changed between the two periods. Such a comparison suggests that changes in water depths and flows can explain most of the differences between original patterning and postdrainage changes.

The importance of gaining a better understanding of roles of water depths and water flows is underscored by two observations. The first is that the predrainage Everglades included two quite different types of peatlands—patterned and unpatterned. The unpatterned Sawgrass Plains, with a closed, dense canopy, less diversity of habitats, and annual drydown was less hospitable to fish, mammals, and wading birds. The patterned Ridge and Slough landscape, which maintained a more diverse range of habitats, from wooded terrestrial to aquatic, appears to have supported a greater quantity and diversity of species. The second and critical observation is that after drainage, some portions of the Ridge and Slough landscape have been losing their original patterning, becoming more like the Sawgrass Plains (fig. 8.6; plates 26, 27; Sklar et al. 2000, p. 26; Science Coordination Team 2003; Sklar et al. 2003; p. 33).

These two observations suggest that the presence of patterning was, and is, very sensitive to hydrology. Whether this means that the Everglades is "fragile" is arguable, but it does suggest a strong degree of tuning to the original hydrology. Development of an understanding of the hydrological driving forces and conditions that originally created and maintained the patterned Ridge and Slough landscape—preventing it from becoming unpatterned sawgrass—is an important part of ensuring successful restoration of the remaining Everglades.

A Conceptual Model of Predrainage Everglades Hydrology

This conceptual model uses the data, assumptions, and conclusions discussed to synthesize a vision of the hydrology of the Everglades as it existed during the 1800s, prior to anthropogenic drainage. The description may apply as well to the preceding century or two but makes no attempt to describe the full five-thousand-year period of Everglades development.

The predrainage Everglades was an enormous, seasonally pulsed wetland. The annual rise and fall in water levels over an area of more than 4,600 square miles (1.2 million hectares) was driven by the seasonal variation in direct rainfall, inflows from Lake Okeechobee, and inflows from Big Cypress. Inflows from Lake Okeechobee into the Everglades were sustained eight to ten months of most years, while the duration of inflows from Big Cypress is less certain.

Within the Ridge and Slough landscape, the presence of extensive microtopography also meant that anyone traversing the landscape from west to east or east to west encountered numerous rises and falls in the ground surface elevation—a corrugated landscape. This corrugation, when combined with the seasonal rise and fall in water levels, caused significant annual transformations of the landscape. These transformations typically eliminated surface water from approximately half the surface area of the landscape, the littoral zone–like sawgrass ridges, each year, effectively driving the more mobile species into the other half of the landscape, the more lake-like sloughs. While water depths in the sloughs were typically too deep to support foraging by White Ibis, Snowy Egrets, and Wood Storks, the multiyear hydroperiods and the presence of predator refuges within the sawgrass ridges likely sustained both small fish and large fish assemblages at higher population densities than seen at present. During dry years, when slough levels dropped sufficiently low, wading bird foraging would probably have been excellent.

The character of the Everglades as a seasonally pulsed, partly terrestrial, partly aquatic wetland was maintained by the particular combination of factors that together kept water stages fluctuating just above the average ground surface elevation. The uniqueness is further accentuated by the fact that the Everglades was sloped, so that all these relations were dynamic and in a constant state of flux. The basin was continually draining to the south and periodically refilling from the north and from above (direct rainfall). Without annual pulses of new water during each successive wet season, the Everglades would have gone dry, a condition that historical sources (this work), the pollen record (e.g., Willard et al. 2001, p. 47), and the charcoal record (A. D. Cohen 1984, p. 461) all suggest was rare. The average depth of water was therefore determined by a critical balance of rates of inflows, outflows, and of overland flow down the landscape. Parameters determining inflows and the rise of water levels included the duration of the wet season, the annual rainfall rate, and the annual volume of outflows from Lake Okeechobee and Big Cypress. Parameters determining outflows from the Everglades and the fall of water levels included the slope of the land surface, the hydraulic resistance to flow due to vegetation and surface microtopography, and the rate of evapotranspiration. According to this conceptual model, the dynamic positive and negative feedback mechanisms among flow rates, depths of water, plant growth, and plant-based hydraulic resistance to flow were all part of what sustained the Everglades.

Within the Ridge and Slough portion of the Everglades, these feedbacks also were instrumental—in combination with related processes of peat accumulation, peat loss, and possibly sediment transport—in maintaining this landscape's characteristically patterned peat microtopography. The resulting parallel, highly directional configuration of alternating ridges and sloughs, repeated many times across the full width of the landscape, was likely the result of the feedbacks interacting with both continuous (daily), lower energy flows and infrequent, higher energy flows.

Inflows to the Everglades from Lake Okeechobee formed a critical part of the timing and levels of water in the system. The 70-mile stretch of the southern shore of Lake Okeechobee and the bordering Sawgrass Plains and Custard Apple Swamp landscapes were at similar ground levels (fig. 5.3). Both vegetation types were underlain by deep organic soils forming a natural plug that held water within Lake Okeechobee (Gleason et al. 1984, p. 301). Under predrainage conditions, prior to regional soil subsidence, there does not appear to have been an elevated soil rim (chapters 5, 7). Instead, the water surface of the lake extended smoothly into the adjacent Everglades (fig. 5.3), and water flowed southward so long as the lake level exceeded the ground surface of these bordering landscapes. Lake outflows appear to have persisted throughout much of the year (this chapter), probably even after the end of the wet season, though at decreasing rates as the lake stage approached ground surface.

This conceptual model allows one to consider predrainage Everglades topography by following two imaginary flow paths, one beginning along the southeastern border of Lake Okeechobee, the other beginning along the southwestern border. Both flow paths extended through the Sawgrass Plains, southeastward and in parallel. About 60 miles (100 km) farther south, within the Ridge and Slough landscape, the two flow lines eventually diverged (compare with plate 3a). The western line continued south and then southwest, ending at Whitewater Bay in the Gulf of Mexico. The eastern flow path continued south and then southeast, dividing into the multiple Peat Transverse Glades and short rivers that breached the coastal ridge (plate 7). Both flow paths passed through two large peat landscapes, first traversing unbroken sawgrass and then, farther south, a mixture of sloughs, sawgrass ridges, and tree islands (appendices H and J). Water depths within a given landscape were approximately the same along the entire flow path (fig. 11.3, plate 30 top panel) as a result of the gradual slope, the uniformity of the landscapes, and the flow impedance provided by the vegetation.

The configuration of water surface parallel to ground surface within a landscape appears to have been largely preserved throughout the annual rise and fall in water levels (plate 30). The maintenance of a parallel configuration would imply that flow directions would be the same as the downslope land surface directions, and also that flow directions would remain constant during the seasonal rise and fall of water levels. Transient deviations in flow direction might well have occurred in response to transient mounding of the water surface caused by local rainstorms. In the downstream direction the concept of parallel water and ground surface planes applied individually to both flow paths. Laterally, in the latitude of Townships 51 to 53, where the two flow paths diverged, the water surface plane would actually have been slightly convex, mirroring the slight convexity of the ground surface.

The configuration of parallel planes within the Ridge and Slough landscape would have strong implications for wading bird feeding. Local ridge-to-slough concentration of prey likely occurred every year, but it occurred almost simultaneously throughout the landscape and over a fairly short time period. In the downstream direction the Ridge and Slough landscape is therefore quite unlikely to have provided a regional pattern of sustained, high quality wading bird foraging patches moving systematically down the landscape, as the impoundments do now. Similarly, in the transverse direction, the parallel planes meant that neither transverse concentration nor creation of high quality edge patches, moving progressively from the edges toward the center of the landscape, seem likely. Unless intermediate-depth, ridge-to-slough ecotones were widespread, it appears that while the predrainage Ridge and Slough landscape may have supported fish populations effectively, and even concentrated them annually, slough water depths would typically have been too deep for all but the tallest wading bird species. Optimal feeding in typical years likely depended on a wider range of landscapes, each successively offering advantageous foraging depths during the course of the dry season. In particularly dry years, however, foraging conditions in sloughs might have been remarkably good.

A local exception to the parallel planes of water and ground surface may have been present in the lower reaches of Shark River Slough. Here aerial observations made in the 1960s and 1970s suggest that then, and presumably also under predrainage conditions, there may have been some degree of natural impoundment that held water back slightly, creating an upstream wedge of water.

The configuration of parallel planes of surface water and ground surface does not appear to have applied to the Marl Marsh landscape flanking Shark River Slough. Here the water surface appears to have approached gravitational equilibrium in the transverse direction; that is, the water surface tended toward a level elevation across Shark River Slough and the Marl Marshes. As the ground surfaces sloped slightly upward away from Shark River Slough, the level water surface meant that water depths decreased in the Marl Marshes with increasing distance from Shark River Slough (figs. 4.7, 11.7). It is likely that the hydrology of the Marl Marsh landscape contributed significantly to wading bird feeding, both because of the shallower water depths present there and because of the regional pattern of annual drying, a regional pattern that appears to have been largely absent within the Ridge and Slough landscape.

A curious and unexplained geomorphological aspect of the Everglades is the simultaneous presence of two different peat-based landscapes—the unbroken Sawgrass Plains and the microtopographically varied Ridge and Slough landscape. There does not appear to be an obvious discontinuity to explain this landscape transition. The canal surveys from 1913 (FEEC 1914, oversize plate of canal profiles; fig. 11.14) do not suggest a distinct break in land slope. Changes in peat thickness seen in those surveys and in figure 4.8 (top) seem more reflective of bedrock slope and do not match the location of the Sawgrass Plains to Ridge and Slough transition. Figure 3.1 (bottom) shows an increase in both the mean and variance of water depth that does coincide with the transition into the Ridge and Slough landscape, but this seems more descriptive than causal. Figure 4.8 (bottom) shows a distinct increase in the frequency of points with higher percentages of inorganic material that coincides with the landscape transition, but again it is not clear that this would be causal.

One possible explanation for the differentiation of the two landscapes is that the water surface and ground surface planes were perhaps not perfectly parallel but were instead very slightly wedged, creating deeper water downstream. If present, a wedge configuration could differentiate the landscapes by the spatial pattern of soil drying. A wedge-shaped relationship between ground and water planes would suggest that during the dry season, after outflows from Lake Okeechobee ceased, a "drying front" would progress southward through the Sawgrass Plains. Under such a scenario, the interface between Sawgrass Plains and the Ridge and Slough landscape might correspond to the average southerly reach of such a drying front. However, this hypothesis would need to be reconciled with the hypothesis that the sawgrass ridges in the Ridge and Slough landscape became dry to the same extent as the Sawgrass Plains, and that within sawgrass areas in either landscape, water depths were similar.

Considering the predrainage Everglades conceptually suggests that the slow annual pulsing can be thought of as the rising and falling of a water surface parallel to the slightly sloped plane of the Everglades land surface. The rise and fall of the water surface plane raises several additional questions. One of these is: how often did sloughs dry out? As noted earlier in this chapter in the section on water depths, the average annual minimum water depth in sloughs was 1 foot (30 cm). The sawgrass ridges were approximately 1.5 feet (45 cm) higher than the bottom of the sloughs (chapter 8), meaning that the water surface on average descended to an annual minimum of about 0.5 foot (15 cm) below the ridge surfaces. This situation is consistent with early descriptions of the Everglades as alternating land and water—"half land, half sea" (Dupuy 1908, p. 9895)—in the dry season (Stuart 1891, p. 110; Dix and MacGonigle 1905, p. 513; Wright 1911, pp. 156–159).

Deviations from long-term average weather conditions during any given year would naturally cause corresponding deviations from the long-term average water depths. The predrainage occurrence of periods when the sloughs dried out completely would be related to the return frequency of drought conditions sufficiently extreme to remove an additional 1 foot (30 cm) of water from the normal end-of-dry-season water depth. This return frequency would also relate to the frequency of cessation of flow in the Atlantic coastal rivers supplied by the Everglades. The absence, so far, of evidence of either widespread ash layers from peat fires or of multiple ash layers within soil cores (A. D. Cohen 1984, p. 461), along with the predrainage indications of continuous outflow from the coastal rivers (e.g., Jones 1847, p. 61; Rodgers 1848, p. 61; Powell 1848, p. 60; *Miami Metropolis* 1905, p. 3), together suggest that complete drying out of sloughs was a very rare event (see also Tropical Bioindustries 1990, p. 15). This also suggests that the observations of postdrainage periods of zero flow through the Tamiami Trail outlets noted by Leach et al. (1972, p. 33), namely 114 days in 1956 and 189 days in 1961–1962, are most likely a postdrainage reflection of lowered water levels and reduced inflows to the Everglades. Similarly, these authors' comment that in the Everglades area there is "very little residual effect from one year to the next" (p. 33) likely applies only to postdrainage conditions. Under deeper predrainage conditions, a residual effect does appear to have been present because of surface water carrying over from one year to the next, reducing the frequency of slough dryouts and allowing the coastal rivers to continue flowing through most, if not all, years. In contrast, after drainage projects lowered water levels, dryouts and muck (peat) fires definitely occurred, destroying as much as several feet of soil over large areas (e.g., Parker 1940; Brown 1943; Bender 1943; Schortemeyer 1980, p. 22; Glenn and Kersey 1982, pp. 26–30; figs. 3.15, 3.16).

Ecologically, an important difference existed between the Sawgrass Plains and the Ridge and Slough landscapes. While the sloughs in the Ridge and Slough landscape would typically have functioned as multiyear aquatic refugia, such refugia were likely absent from the Sawgrass Plains, where the water level typically descended annually below ground surface. Surface outflows from the Sawgrass Plains obviously also ceased at this time. (However, some southward seepage of pore water from the peats of the Sawgrass Plains and the sawgrass ridges may have continued after surface flows had ceased, helping sustain water levels in the sloughs.)

The important distinction between the Peat Transverse Glades and the Marl Transverse Glades relates to the differing geometry of the contributing landscapes. The Marl Transverse Glades were connected to the upslope edge of the Rockland Marl Marsh. Outflows through the Marl Transverse Glades therefore occurred only when water levels rose sufficiently during the wet season to cover the Rockland Marl Marsh (fig. 11.7). Although water from the Everglades flowed through the Marl Transverse Glades regularly, these glades were often dry during a portion of each year (Griswold 1896, p. 53; Harper 1910b, p. 45; Waldin 1912, pp. 15–16; Morse 1918).

In contrast to the Marl Transverse Glades, the Peat Transverse Glades were simply downslope extensions of the Ridge and Slough landscape (compare plates 2a and 3a). The soils, vegetation, observations of predrainage water levels, and accounts of military road building (chapter 10) all suggest that the Peat Transverse Glades typically contained surface water throughout most years. This is consistent with the conceptual model that the contributing Ridge and Slough landscape also contained surface water throughout most years as well as with the notion that the Atlantic coastal rivers fed by the Peat Transverse Glades flowed throughout the year during most years.

To conclude this conceptualization, Taylor Slough behaved somewhat similarly to the Marl Transverse Glades but differed in being a flow-way with no threshold separating it from the waters of the Rockland Marl Marsh and from farther upstream in the Everglades. Flows into Taylor Slough accordingly persisted for a much longer part of each year. Historical accounts (this chapter), the presence of peat soils, and the ridge and slough patterning there (chapter 8) indicate that Taylor Slough was much wetter than the Marl Transverse Glades. The topography of the Taylor Slough area likely contributed to its development as a distinct southward extension of the Ridge and Slough landscape.

Synopsis

The predrainage Everglades was hydrologically unique because of its particular combination of geometry and climate. The basin topography allowed the accumulation of an enormous body of peat soil, which in turn allowed the formation of a vast and exceedingly flat ground surface. The slight slope of the basin kept the accumulating peat surface slightly tipped from horizontal; 3 inches to the mile (5 cm/km). This slope and the balance between the energy of the flows and the structural coherence of the peat prevented formation of either a central drainage channel or a dendritic drainage pattern. These flows were sufficient to create and/or maintain systematic microtopographic relief, making a large portion of the Everglades a patterned peatland and creating a multicomponent landscape with sustained elevational differences. The stable elevational differences between sloughs, sawgrass ridges, and tree islands created thousands of semiterrestrial areas surrounded by persistent wetlands. The absence of a central or dendritic drainage pattern meant that water flow was distributed evenly among hundreds of similarly sized sloughs, spreading the flow field across the full 40-mile (60 km) width of the landscape.

The slope and flow meant that the Everglades was likely never in hydrologic equilibrium but instead was continually draining. The strongly seasonal rainfall distribution made the Everglades a seasonally pulsed system, with rainfall exceeding drainage during the wet season, each year reversing the declining water depths of the dry season. The balance between rates of inflow and rates of outflow was such that each year's rainy season typically arrived just before the system had completely dried out. Thus water depths within sloughs throughout the Ridge and Slough landscape typically rose and fell each year from a low of about 1 foot (30 cm) to a high of about 3 feet (90 cm), an environment that would support long-lived aquatic organisms. Sawgrass ridges were just high enough to be semi-aquatic and without surface water during part of each typical year. The annual cycle of wetting and drying on ridges may have been chemically and biologically important, releasing a nutrient pulse as the soil reflooded during the wet season, and concentrating populations of small aquatic fauna into sloughs during the dry season. Tree island peat surfaces were high enough to provide habitat for woody vegetation, and low enough to derive dry season water from surrounding sloughs.

In contrast to the Ridge and Slough landscape, a portion of the flanking Marl Marsh landscape typically dried below ground surface each year, perhaps leading to a more diverse flora. Survival of longer-lived fauna must have been supported by the aquatic refugia within the extremely irregular and porous limestone bedrock combined with recolonization from Shark River Slough and estuarine creeks.

Altogether, the unique combination of climate, geometry, peat, and vegetation of the Everglades created a "region of mystery" (Dix and MacGonigle 1905, p. 512) that in the words of Buckingham Smith (1848a, p. 51) was "in some respects the most remarkable on this continent," where the water was "pure and limpid and almost imperceptibly moves, not in partial currents, but, as it seems, in a mass, silently and slowly to the southward," (p. 51) and where it annually rose and fell to create a "region that [was] not exactly land, and . . . not exactly water" (Dix and MacGonigle 1905, p. 513)—as Marjorie Stoneman Douglas (1947) famously phrased it: a "River of Grass."

12

Summary

This research effort represents a robust, ecologically useful description of the predisturbance Everglades based upon a wide range of observations from the predisturbance (pre-1880) period in Everglades history. As scientists trained in soils, botany, and hydrology, we developed an internally consistent, spatially explicit, and scientifically detailed description of the predrainage Everglades from firsthand, though not specifically scientific, historical observations.

The predisturbance Everglades of the 1800s was one of the last American frontiers, a place sufficiently difficult to traverse that very few systematic, scientific measurements were made within its bounds. In contrast, by 2000, just over a century later, this same region had been so greatly altered by major anthropogenic disturbances that a massive restoration effort was initiated (U.S. Congress 2000). Restoration implies reconstruction of the original form. In order to restore the Everglades, an understanding, in as much detail as possible, was needed of the original predisturbance state, a condition for which scientific studies were lacking.

Previous work suggested that a scientific approach using historical sources could be fruitful. The historians Julian Dovell and Charlton Tebeau, in particular, had recognized that the Everglades had been strongly altered. In several publications (e.g., Dovell 1942, 1947a, b; Tebeau 1968, 1973), these historians identified a number of sources "of ecological importance" to understanding predisturbance conditions. However, their primary focus was on social aspects, so that they attempted "to show in a general way what was here," in most cases without enough technical detail or cross-checking to assist ecologists. Steve M. Davis et al. (1994) took that approach an important step further by using sources indicated by Dovell, Tebeau, and others to map the predrainage Everglades and its component landscapes; the work by S. M. Davis et al. (1994) is a direct predecessor to this study.

The primary focus of the present study remained on the truly predisturbance historical observations (i.e., from before 1880). However, during the course of this effort we concluded that with care, early postdrainage and even current information could in some cases contribute significantly to understanding predrainage conditions. While we recognize that the inclusion of postdrainage (postdisturbance) information can be problematic, two assumptions apply: (1) at least some parts of the current Everglades system bear at least a partial resemblance to the original system; and (2) historical information provides support for identifying which aspects of which areas most resemble the original.

The research we conducted for this study synthesized a wide array of available data from divergent but reliable sources. Those sources were validated by other sources from the same time or from subsequent periods and/or by recent sources such as peat cores. For example, the statement that the long-term average end-of-wet-season water depth in the Sawgrass Plains was 1.5 feet (45 cm) represents the distillation of hundreds of predrainage field observations of water depths within that landscape made by multiple observers across many years. As such, the statement summarizes measured data points, extending them to end-of-wet-season conditions. Similarly, the statement that "white water lilies (*Nymphaea odorata*) were ubiquitous in predrainage sloughs" is based on often repeated and consistent firsthand observations made by land surveyors, explorers, early settlers, and early agricultural scientists throughout the Ridge and Slough landscape. These observations are validated by the presence of traces of white water lilies in soil cores from throughout the same landscape.

The water depth and water lily examples are fairly straightforward. Other aspects of the reconstruction of the predrainage Everglades were similarly based on firsthand observations. Our approach throughout has been to reference and make available the original data, to point out consistencies (or inconsistencies) across different data sources, and to describe any synthesis or inference involved. As best as can be determined, our examination of the historical evidence covered the great majority of sources.

A surprising amount of firsthand information exists regarding the 2,800,000 acres (1,100,000 ha) of the predrainage Everglades (consider the large number of annotated Historical Observations assembled for chapters 5–10 on the accompanying DVD). One measure of the scientific usefulness of this detail is its applicability to numerical simulations of the regional predrainage hydrology of the Everglades. Several predrainage simulation models do, in fact, derive their main input data from the results of this study. The Natural System Model, version 4.5 (SFWMD Hydrologic Systems Modeling Division 1998), and the Natural System Regional Simulation Model, version 2.0 (Said et al. 2006), both derive the following spatially explicit predrainage model input data from this study: vegetation types (plate 5), vegetation density (chapters 6–9), topography (chapter 11, plate 6), landscape boundaries (plate 5), locations of inflows and outflows (chapters 10–11), and information regarding microtopography (chapter 8, plate 23). In addition to providing input data for these models, other results from this study provide benchmarks against which model output can be evaluated; that is, a form of model validation. Those benchmarks, discussed primarily in chapter 11, include the spatial distribution of predrainage flow directions (figs. 11.10, 11.12), the spatial distribution of predrainage water depths (plate 8), and the seasonal variation in water depths (table 11.4). As an example, the simulated flow directions shown in plate 3a compare favorably with figures 11.10 and 11.12. Additional, less direct benchmarks include the temporal presence and spatial distribution of obligate wetland species, and the locations and elevations of outflows (chapters 10–11, plate 7). The predrainage relationships between water depths and relative elevations of sawgrass ridges and tree islands (chapter 11, plate 30) provide additional validation benchmarks.

Prior to embarking on this study our goal was to develop a picture of the predrainage Everglades that was spatially explicit and scientifically detailed. Our initial premise anticipated that the results would be internally consistent and that the observers would present noncontradictory pictures of the predrainage landscape. To a remarkable degree this proved to be the case. We repeatedly observed: (a) consistency across multiple observers, (b) locational consistency between adjacent landscapes, (c) temporal consistency (same aspect reported at different times), (d) spatial consistency (same feature seen in the same place), and (e) combinations of the above. While there were a few inconsistencies between observations, overall they were rare. Exactly opposite to the ancient parable of the blind men and the elephant (who interpreted the appearance of the elephant by what portion they were able to touch), the many different observers of the Everglades, spanning more than a hundred years and a wide range of locations, in fact come together to weave a remarkably consistent description of the undeveloped Everglades. Unlike in the parable, we, the proverbial "blind men" who cannot see the original predrainage landscapes with our own eyes, are fortunate to have access to a large number and variety of firsthand descriptions that, when pieced together, provide a detailed picture of the whole.

An additional form of internal checking was suggested by the postdrainage changes described in chapter 3. Broadly, the present-day Everglades conformation is the result of superimposing the postdrainage changes described in chapter 3 upon the predrainage Everglades. When supplemented with additional post-1950 changes, not generally included in chapter 3, this simple perspective provides an important conceptual check on the picture of the reconstructed predrainage Everglades. Given the known aspects of the current system and the known types and degrees of postdrainage changes, we repeatedly asked ourselves whether the reconstructed predrainage picture was plausible. Overall, the answer appeared to be "yes," suggesting strong temporal and causal consistency.

The results of this study can be divided into three categories—two of emergent or overarching points and a third of specific points: (1) emergent points or concepts related to the predrainage Everglades and bordering areas, (2) emergent points or concepts related primarily to changes that occurred as a result of anthropogenic drainage and water management, and (3) specific details that relate primarily to a particular predrainage landscape.

Emergent Points Regarding the Predrainage Everglades

(A) *Water Depths.* Prior to initiation of drainage in the 1880s southern Florida was extremely wet. Standing water was rarely far away, and large areas, for example the sloughs of the Ridge and Slough landscape, were continuously water-filled during all but the driest years. Waist- or even shoulder-height water depths were observed in the Everglades, and depths of 2 to 3 feet (60–90 cm) were common. The nearly permanent presence of water in sloughs sustained the continuous outflows of the coastal rivers. In the surrounding "uplands"—that is, in the Eastern and Western Flatwoods—ponds were plentiful and some had sufficiently long hydroperiods to support peat accumulation and canoe navigation during the wet season.

(B) *Wetland Soils*. Organic soils ("peats" and "mucks") rather than mineral soils were widespread throughout the Everglades but also present in the bordering landscapes of Big Cypress and the Eastern and Western Flatwoods. These organic soils affected and were affected by the physical and biological conditions surrounding them. For example, predrainage hydrologic conditions made peat accumulation possible. These peat accumulations in turn developed very large, very flat expanses with nearly uniform hydrology.

(C) *Water Flows*. Almost all predrainage observers noted that water in the Everglades was never stagnant but instead "almost imperceptibly moves . . . in a mass, silently and slowly to the southward" (B. Smith 1848b, p. 28). At some points within the Ridge and Slough landscape visible currents were reported, probably where sloughs narrowed, focusing flows. Water flow constituted an organizing force, creating a downstream sequence of Everglades landscapes. This directional aspect of water flow had causal and temporal implications.

(D) *Temporal Buffering*. The predrainage Everglades appears to have included considerable hydrologic buffering, behaving much like a low-pass filter. This buffering was most visible in the greater similarity between wet and dry season hydrology than might be expected, given the strong difference between wet and dry season rainfall. Outflows from the Everglades (e.g., Shark River Slough and the eastern coastal rivers) continued throughout the year in most years. Water remained within the Everglades through most dry seasons, and there is some indication of carryover effects extending more than one year. Much of this predrainage buffering appears to have originated in slow releases of water from wetlands in the Kissimmee Valley, from Lake Okeechobee, from peat soils of the Sawgrass Plains, from peat soils of sawgrass ridges in the Ridge and Slough landscape, and from the sheer size of the Sawgrass Plains landscape.

(E) *Ecological Feedbacks*. Close, bidirectional feedbacks between vegetation and hydrology appear to have been present. Water depths, and possibly velocities, affected the type of vegetation community; simultaneously, the type of vegetation (species, geometry, stand density, etc.) affected hydraulic resistance and hence water depths and velocities. (Faunal feedbacks were likely present as well but were not readily documentable from historical sources.)

(F) *Spatial Complexity*. The physiographic complexity of the 2,800,000 acres (1,100,000 ha) of predrainage Everglades was very low, being divisible into only four major subunits, the "landscapes" of this volume; six if the Marl Marshes are differentiated. This high degree of uniformity, especially within each landscape ("vast expanses of sawgrass, stretching to the horizon"), was due to hydrologic uniformity (compare with plate 3a), itself the result of remarkably level peat substrates.

(G) *Biodiversity*. Plant biodiversity of the predrainage Everglades appears to have been very low. The vast Sawgrass Plains were strongly dominated by a single plant species, sawgrass, as were the sawgrass ridges of the Ridge and Slough landscape. Sloughs probably included more plant species but were still dominated by only a few. Tree islands, although comprising only about 5 percent of the total area, were likely the floristically most diverse component.

Emergent Points Regarding Anthropogenic Drainage

(A) *Time Point of Initial Disturbance*. In the Everglades, where conditions moved quickly from minimal to intensive human disturbance, the concept of distinct predisturbance and postdisturbance time periods emerged as a key organizing principle. The time of initial disturbance, corresponding to the first lowering of Lake Okeechobee water levels and the associated reductions in inflows to the Everglades, was identified as occurring in the early 1880s. As a corollary, only observations made prior to the early 1880s can fully and unambiguously reflect the natural, predisturbance Everglades.

(B) *Ubiquity of Drainage*. The pervasive extent, substantial degree, and widespread influence of man-made drainage in South Florida are remarkable. Essentially all of South Florida has been influenced by canal drainage with water tables inside and out of the Everglades boundaries drawn down by as much as 6 feet (2 m) or more. The ubiquity of man-made drainage is essentially a restatement of the abundance of water in predrainage South Florida.

(C) *Widespread Loss of Wetland Soils*. Drainage throughout South Florida exposed previously saturated peats to oxidation, leading to widespread soil loss. In bowl-shaped peat deposits, vertical thinning also led to decreased horizontal extents. In some places the peat was completely oxidized, exposing underlying sands and rendering the former wetland areas essentially indistinguishable from sandy uplands. Soil maps are all postdrainage and therefore cannot be used to map predrainage wetlands without additional information and interpretation.

(D) *Anthropogenic Spatial Complexity*. Construction of canals and particularly levees subdivided the Everglades into compartments, which over time developed internal depth and vegetation gradients. Differentiation within compartments as well as differentiation between separate compartments both artificially increased spatial complexity.

(E) *Differing Ecosystem Response Rates*. Historical accounts indicated that anthropogenic drainage did not uniformly and simultaneously affect all components of the Everglades ecosystems. Herbaceous species generally responded more quickly, while trees and soils responded more slowly. In some cases these differences allowed postdrainage observations to be used to estimate predrainage conditions.

Detailed Results

Methods (Chapter 1)

a. Numerous firsthand predrainage sources of diverse types and by numerous observers from diverse professions were located.
b. Sources reflect extensive spatial coverage: expeditions crossed in various directions and across all landscapes; township surveys conducted between 1845 and 1916 included all landscapes and approximately 70 percent of the Everglades.
c. Sources had wide temporal coverage, beginning in the 1830s, continuing into the early 1900s, and reaching up to the present.
d. Sufficient sources exist to allow cross-checks across time, space, and different observers.

Phases of Everglades Change (Chapter 2)

a. Drainage began early with the lowering of Lake Okeechobee water levels starting in the early 1880s.
b. Lowered lake water levels reduced inflows into the Everglades; this effect likely extended downstream through most or all of Everglades and occurred earlier than previously recognized.
c. Everglades drainage history includes three major phases: the Lake Okeechobee phase (1880s–1890s), the muck canal phase (1910s–1950s), and the impoundment phase (1950s–present).

Postdrainage Changes to the Everglades through 1940 (chapter 3)

a. Drainage succeeded, dramatically altering the Everglades water balance. The removal of water achieved by the 1920s and 1930s significantly altered virtually all aspects of the Everglades: hydrology, soils, vegetation, and microtopography.
b. As a result of uncontrolled drainage, the period 1920–1950 was the (unnaturally) driest in Everglades history, resulting in extensive muck fires.
c. Maps of the Everglades in the 1940s definitely do not represent the predrainage conditions.

Reconstruction of the Predrainage Everglades (Chapter 4)

a. The predrainage descriptions we developed may seem at times extreme. This is both understandable and accurate; the postdrainage changes *were* extreme.
b. The goals for the detailed landscape chapters were both geographic and archival. Specifically, these goals were to map the boundaries of the Everglades as it existed in the nineteenth century, to describe the soils and vegetation of the Everglades at that time, and to publish the key observations used to establish those boundaries and descriptions.
c. Excellent synoptic studies of the Everglades were completed in the 1940s, but this was well past the onset of drainage, and their authors, by their own accounts, acknowledge that the studies do not depict predrainage conditions.
d. More than a thousand different pre- and postdrainage sources were obtained for this study. These were then gleaned for observations of Everglades vegetation, soils, hydrology, topography, and fauna, with relevant quotations recorded in an annotated database.
e. As noted previously, each source was screened for ecological credibility. The large majority of observations passed these credibility tests. Consistency among numerous sources was the rule rather than the exception.
f. The differences between the results of this study and earlier historical ecological studies often can be traced to insufficient hindcasting of postdrainage changes in the Everglades.

Lake Okeechobee as Water Source for the Predrainage Everglades (Chapter 5)

a. As noted by others, Lake Okeechobee was probably a result of the natural dam that was formed by centuries of accumulated Sawgrass Plains peat.
b. Sawgrass and sawgrass peat extended directly up

to the lake on its southwestern and western shores (more than 30 miles, 50 km), forming a substantial fraction of the overflowing shoreline.

c. The water surface present in Lake Okeechobee extended smoothly into the surrounding sawgrass and custard apple swamp.
d. No elevated rim or "lip" was present along the outflowing shoreline; the "custard apple rim" reported after about 1910 was a postdrainage phenomenon due to differential peat subsidence (though a custard apple forest existed predrainage).
e. Lake Okeechobee's outflows into the Everglades continued throughout much of the year and occurred in most years.
f. Two parallel outflow paths existed from the lake: one directly into the Sawgrass Plains and another through the Custard Apple Swamp into the Sawgrass Plains.
g. Lake elevation originally averaged around 21 feet (6.4 m) above mean sea level (NGVD 29).
h. Lake levels likely varied little; deviations greater than 2–3 feet (0.6–0.9 m) above or below the average were probably rare. High stages were limited by the very wide outflow area. The absence of woody vegetation along the sawgrass shoreline suggests a rarity of low lake stages.
i. Sawgrass land bordering the lake sloped downward away from the lake as an outflow rather than upward as a littoral area or expansion zone.
j. The predrainage lake extent of Lake Okeechobee was similar to its present extent.
k. No predrainage rivers connected Lake Okeechobee to the Atlantic Coast or to the Gulf of Mexico; however, a marsh connected it to the headwaters of the Caloosahatchee River, and several short rivers ("dead" rivers) through the Custard Apple Swamp connected it to the Sawgrass Plains.

Sawgrass Plains Landscape (Chapter 6)

a. This landscape was dominated by tall (8–12 feet; 2.5–3.5 m) sawgrass, invariably described as dense and "impenetrable."
b. Seminole Indians rarely (if ever) crossed this landscape. The first crossing appears to have been made in 1884 and was accomplished by burning a pathway through the dense stands of sawgrass in order to proceed into the interior of the Sawgrass Plains.
c. The Sawgrass Plains were confined to the upper portion of the Everglades. Later mapping of a central band of a Sawgrass Plains–like area extending southward to the Miami River, splitting the Ridge and Slough landscape into two parts, was an artifact of severe drainage.
d. The Sawgrass Plains were an unpatterned peatland having an extremely flat peat surface plane and no systematic microtopography, and which sloped south-southeastward throughout.
e. The hydraulic resistance of the dense sawgrass to surface water flow was isotropic (nondirectional) within the Sawgrass Plains landscape.
f. The thickest Everglades peats were found in the Sawgrass Plains; 13–14 feet thick (4–4.3 m) near the lakeshore to about 7–8 feet (2–2.4 m) at southern end of this landscape.
g. Prior to drainage, the soil of the Sawgrass Plains was a brown, fibrous peat, apparently with greater vertical than horizontal conductivity, at least in the upper profile.
h. The typical annual water depths ranged between 1.5 feet (45 cm) and -0.5 feet (-15 cm). Annual (or near annual) absence of surface water had implications for aquatic fauna.
i. Slow releases of water from the peat during the annual dry season may have extended the duration of dry season flows into the Ridge and Slough landscape immediately downstream.

Custard Apple Swamp Landscape (Chapter 7)

a. The predrainage presence of this landscape is confirmed (i.e., it was not a postdrainage phenomenon).
b. It is the only Everglades landscape with continuous forest cover throughout.
c. The Custard Apple Swamp landscape extent may be related to the extent of the high mineral content (ca. 50 percent) of the underlying organic soil.
d. The higher mineral content of the Custard Apple Swamp soil prevented it from subsiding to the same degree as the adjacent sawgrass peat. No lake rim existed under predrainage conditions. Custard apple occupies wetter locations than sawgrass, so the Custard Apple Swamp landscape was slightly *lower* than the Sawgrass Plains landscape. Subsequent postdrainage dewatering, compaction, and oxidation of the more highly organic Sawgrass Plains soil resulted in an artificial Custard Apple rim.
e. The Custard Apple landscape extent is well documented by township surveys (ca. 1911–17).
f. Its extent was narrower than that mapped by J. H. Davis (1943b) and shorter than that mapped by Harshberger (1914).

g. The downstream border of willow and elderberry ("elder") probably was partially present prior to drainage but likely widened during the drier conditions resulting from drainage.
h. The landscape was pierced by approximately ten short rivers flowing southward from Lake Okeechobee into the Everglades. Within a few miles of the lakeshore, these rivers branched and disappeared into the Custard Apple Swamp or northernmost Sawgrass Plains.

Ridge and Slough Landscape (Chapter 8)

a. This was the largest predrainage landscape (55 percent of the total). Most of the present-day Everglades was originally part of this landscape. In the downstream sequence, it followed the Sawgrass Plains.
b. This landscape was a "patterned peatland" with a surface pattern (microtopography) that was highly nonrandom, forming a strongly linear, strongly directional pattern.
c. The directional pattern of this landscape paralleled the directions of regional water flow, suggesting a mechanistic relation between flow and pattern. The mechanism is not known.
d. The directional landscape pattern was formed by three principal elements: sloughs, the lowest; sawgrass ridges, approximately 1.5 feet (0.5 m) higher than sloughs; and tree islands, approximately 2–4 feet (0.6–1.2 m) higher than sloughs.
e. The sloughs were long, linear, interconnected aquatic habitats throughout most years, with characteristic white water lily (*Nymphaea odorata*) and occasionally yellow water lily (*Nuphar advena*) observed.
f. Sawgrass ridges were long, linear, generally separate, distinct units. They were fairly regularly spaced apart in the transverse direction (i.e., the direction perpendicular to flow). Dense sawgrass (*Cladium jamaicense*) dominated. Peat ridges were inundated much of the year but emerged during part of the dry season.
g. Tree islands occurred in several shapes: circular, teardrop, and possibly linear. Frequent predrainage reports of their being wet in the wet season suggest that tree island elevations, formed by peat accumulation, developed in equilibrium with long-term average water levels.
h. The presence of great numbers of sloughs, ridges, and tree islands provided a diversity of habitats in close proximity to one another.
i. While only limited information is available, it appears that elevation transitions between sloughs and ridges and between sloughs and tree islands were abrupt. Sloughs in cross section appear to have been flatter rather than concave. As a result, the *widths* of these transitions were narrow (1 m or less), while the great number of ridges meant that the *length* of these transitions was great.
j. The predrainage Ridge and Slough landscape extended continuously as a single unit across the full width of Everglades. The 1940s mapping of this landscape as two separate areas was an artifact of severe drainage.

Ochopee, Rockland, and Perrine Marl Marshes (Chapter 9)

a. These landscapes are presently covered with marl (mineral) soil, with little organic matter, and elevated (Ochopee and Rockland marl marshes) above Shark River Slough.
b. Predrainage descriptions, while limited, suggested the presence of thin peat soils and of predrainage water depths only slightly shallower than those originally found in Shark River Slough.
c. Field notes from township surveys conducted during 1885 and 1886 indicate that vegetation in the marl marshes was similar to that in Shark River Slough: "thick" or "heavy" sawgrass with contrasting areas of "open Everglades."
d. The plant species and hydrology of the predrainage Ochopee and Rockland marl marshes appear to have resembled the Ridge and Slough landscape (Shark River Slough) between them, but the marl marshes were much less patterned, and "round grass" (probably *Eleocharis cellulosa*?) was more common.
e. Postdrainage peat losses appear to have occurred in the Ochopee and Rockland marl marshes as well as in Shark River Slough; however, elevation losses were likely greater within Shark River Slough because the original peat thicknesses were greater there. Overall, peat subsidence appears to have increased the elevation difference between Shark River Slough and the Marl Marsh landscape.

Landscapes Bordering the Everglades (Chapter 10)

a. The originally extensive Eastern Flatwoods included thousands of ponds, some with hydroperiods long enough to support peat accumulation.

b. Predrainage township survey notes and maps indicated that the transition from flatwoods to Everglades (Ridge and Slough landscape) was abrupt, making the boundary of the Everglades well defined and easily recognizable. The broad transitional bands of bordering wet prairies that had appeared by the 1940s were artifacts, the result of anthropogenic drainage and peat loss.
c. Hydrologically, lands bordering the eastern side of the Everglades subdivided into three areas: Palm Beach County (present extent), which included no significant outflows from the Everglades; Broward County south to south Miami, where the Atlantic Coastal Ridge was thinnest and was pierced by the Peat Transverse Glades and multiple coastal rivers draining from the Everglades; and south of Miami, where seasonal outflows passed through the Marl Transverse Glades.
d. Outflows to the southeast through the coastal rivers were a significant part of the Everglades water budget, perhaps as much as 40 percent of predrainage flows.
e. On the western border of the Everglades, inflows appear to have come from Mullet Slough, but little predrainage information is available for quantifying flow.

Hydrology of the Everglades (Chapter 11)

a. Long-term average water depths in sloughs ranged from an annual low of 1 foot (30 cm) to an annual high of 3 feet (90 cm). These average depths exceeded feeding requirements for most wading birds. Corresponding water depths on sawgrass ridges and in the Sawgrass Plains were 0.5 feet below peat surface (-15 cm) and 1.5 feet (45 cm) above. Individual years deviated from these averages.
b. Predrainage vegetation and observations indicate that the plane of the Everglades water surface was essentially parallel to the plane of the peat surface. In other words, water depths along the downstream axes of Everglades landscapes were essentially the same, despite both planes being sloped. Accordingly, predrainage water depths in the present Water Conservation Area 3A and in Shark River Slough (Everglades National Park) were originally indistinguishable. The very gradually sloped water surface produced continuous and visible water flow.
c. The parallel configuration of water and ground surface planes appears to have persisted during annual dry season declines in water depth. Ecologically, this means that at any given time, foraging conditions for wading birds in sloughs would have been approximately the same in upstream and downstream sloughs of the Ridge and Slough landscape.
d. Historical accounts as well as descriptions of vegetation and soils along the southern shoreline of Lake Okeechobee indicate that predrainage Lake Okeechobee levels tended to oscillate slightly above and below the elevation of the shoreline peat.
e. Early accounts indicated that outflows from Lake Okeechobee into the Everglades occurred along the whole 70 or so miles (110 km) of the southern shoreline. These outflows appear to have occurred in most years and continued during a significant portion of the year.
f. The Kissimmee watershed, comparable in size to the Everglades, discharged into Lake Okeechobee, which essentially passed this water through to the Everglades because evaporation from Lake Okeechobee approximately equals direct precipitation, and because non-Everglades discharges from Lake Okeechobee were relatively small. Lake Okeechobee outflows into the Everglades formed a significant part of the predrainage Everglades water budget.
g. "Sheet flow" accurately describes the movement of a thin and extremely wide sheet of water across the Sawgrass Plains. In contrast, within the Ridge and Slough landscape, the deeper water column (+1.5 foot, +45 cm) and the lower plant stem densities found in sloughs made them preferential flow paths, suggesting the term "slough flow." Similarly, overall resistance to flow in this landscape was directional (anisotropic), with the lowest resistance in the direction parallel to the ridge and slough alignment.
g. Narrative descriptions indicate that at any given location within the Ridge and Slough landscape, landscape direction, topographic downslope direction, and flow direction were the same.
h. Flow directions were strongly autocorrelated. Large areas flowed in the same direction; changes in direction were smooth and gradual. This strong autocorrelation reflected the regional smoothness of the predrainage peat surface.
i. Between the latitudes (approximately) of Fort Lauderdale and Miami, flow directions gradually diverged; one flow turned southeastward to flow into the multiple eastern coastal rivers, and the other turned southwestward to flow into Shark River Slough. An estimated 40 percent of Everglades flows discharged to the Atlantic Ocean through the coastal rivers.

Epilogue

The predrainage Everglades pictured in this volume provides a window onto one of many human alterations of the Earth. The alterations of the Everglades were particularly dramatic because of their degree and their rapidity: from essentially pristine to highly altered within a matter of decades.

At the time of writing, society is preparing to reverse some of these alterations; that is, on the brink of "restoring the Everglades." The predrainage picture developed here provides a reference for that effort, a picture of the original condition. In addition this picture, when compared with the current condition, can help predict whether restoration is likely to succeed. Overall, the outlook is positive.

Causes for pessimism are readily found. The right side of plate 2b shows that half the original wetland extent has been drained and converted to cities or farms. The connection between Lake Okeechobee and the Everglades, originally 50 miles (80 km) wide, has been eliminated. Levees and roads fragment the remaining Everglades, creating impoundments that distort water depths and interrupt sheet flow.

The causes for optimism are equally strong and arguably more important. It is clear that the full *extent* of the original Everglades cannot be restored. The more relevant question, however, is not extent, but *degree;* how closely can a chosen remaining footprint be restored to its original condition? To address that question, the insights that emerge from this book cause us to be highly optimistic.

Comparison of original and current conditions in the remaining western portion of the Everglades—Water Conservation Areas 3A and 3B and Everglades National Park—suggests that excellent potential exists for a very high degree of restoration there. Key elements of this potential are location, size, and resilience.

Figures 11.10, 11.11, and 11.12 together suggest that predrainage flow through the western footprint was largely independent of flows through the now urbanized eastern portion, implying that the potential for western flow restoration remains undiminished. The potential is in fact increased because of requiring proportionately less water for this narrower footprint. The potential for western flow restoration is further enhanced by the observation that, despite peat subsidence, the peat surface retains a regional slope (plate 30).

The sheer size of the western footprint—more than 20 miles (35 km) wide and 70 miles (110 km) long—greatly increases the likelihood of restoration success. It is large enough to maintain the landscape identity despite the inevitable edge effects and long enough to preserve 98 percent of the original downstream run through the Ridge and Slough landscape.

The presence of "seed areas," portions of the western footprint that have been able to preserve many of the original landscape features, suggests both resilience within these areas and a likelihood of expanding recovery outside them once hydrologic driving forces are restored to predrainage levels.

The primary challenge to restoration of the western footprint will be re-creation of water inflows and water quality equivalent to those that originally came to this area from Lake Okeechobee.

Overall, the picture of the predrainage Everglades developed in this volume suggests a high potential for restoring the western portion of the remaining Everglades to something closely approximating its original, predrainage condition.

References

References to township surveys can be found in Township Survey References beginning on page 321.

Agassiz, A. 1890. Note appended to "The topography of Florida." Quoted in: Shaler (1890), p. 158.

———. 1896. The Florida elevated reef. *Bull Mus Comp Zool* 28: 27–62.

Alexander, T. R. 1967. A tropical hammock on the Miami (Florida) limestone: A twenty-five year study. *Ecology* 48: 863–867.

Alexander, T. R., and A. G. Crook. 1973. *Recent and long-term vegetation changes and patterns in South Florida (EVER-N-51). Part I: Preliminary report.* South Florida Ecological Study, Appendix G, NTIS PB 231939. University of Miami, Coral Gables, Fla. 215 pp.

———. 1974. Recent vegetational changes in southern Florida. In: *Environments of South Florida: Present and past*, edited by P. Gleason, pp. 61–72. Miami Geological Society, Coral Gables, Fla.

———. 1975. *Recent and long-term vegetation changes and patterns in South Florida. Part II: Final report.* South Florida Ecological Study, Appendix G, NTIS PB 264462. University of Miami, Coral Gables, Fla. 827 pp.

Allison, R. V. 1943. The need of the Everglades for a specific plan of development based on the physical and chemical characteristics of its soils and a rational handling of its natural water supply. *Soil Sci Soc Fl* 5-A: 126–131.

Allison, R. V., J. R. Baker, G. Bender, H. A. Bestor, C. K. Davis, N. S. Emery, F. C. Elliott, G. E. Ferguson, W. A. Glass, H. Gunter, B. Herr, J. R. Neller, G. C. Parker, J. Silver, C. R. Vinten, and W. T. Wallis. 1944. *Report by Advisory Committee on the present drainage system in relation to water control requirements of Everglades Drainage District.* May 1, 1944. Everglades Drainage District, Miami. 41 pp.

Allison, R. V., O. C. Bryan, and J. H. Hunter. 1927. *The stimulation of plant responses on the raw peat soils of the Florida Everglades through the use of copper sulfate and other chemicals.* Tech. Bull. #190. University of Florida, Agricultural Experiment Station, Gainesville, Fla. 33–80 pp.

Allison, R. V., and A. P. Dachnowski-Stokes. 1932. Physical and chemical studies upon important profiles of organic soils of the Florida Everglades. In: *Proceedings and papers of the second International Congress of Soil Science,* Leningrad-Moscow, July 20–31, 1930, pp. 223–245.

Anderson, S. R. 1907. [Interview with S. R. Anderson, February 17, 1907.] Quoted in: Stewart (1907), p. 29.

Andrews, R. 1957. Vegetative cover-types of Loxahatchee [National Wildlife Refuge] and their principal components. In: *Central and Southern Florida Project for Flood Control and Other Purposes. Part I: Agricultural and conservation areas, supplement 25, general design memorandum, plan of regulation for Conservation Area No. 1,* pp. B–25 to B–33. U.S. Army Corps of Engineers, Office of the District Engineer, Jacksonville, Fla.

Anonymous. 1840. [*Expedition through the Everglades, December 4–15, 1840.*] "Transcribed from a writer who accompanied Col. Harney on his expedition among the Everglades." Microfilm collection of the University of Florida; from unidentified source in Library of Congress, Gainesville, Fla. pp. 12–17.

———. 1841. [Extracts of account by an anonymous engineer accompanying Col. W. S. Harney on an expedition into the Everglades during the Seminole War, January 1–10, 1841.] Quoted in: Brooks (1880), pp. 245–252.

———. 1946. A canoe expedition into the Everglades of Florida in 1842. *Tequesta* 5: 30–51.

———. 1960. "Notes on the passage across the Everglades" from The News—St. Augustine: January 8, 1841. *Tequesta* 20: 57–65.

Antonini, G. A., D. A. Fann, and P. Roat. 2002. *A historical geography of southwest Florida waterways.* Vol. 2: *Placida Harbor to Marco Island.* Edited by P. Roat. Florida Sea Grant Program, Gainesville, Fla. 168 pp.

Armentano, T. V., D. T. Jones, M. S. Ross, and B. W. Gamble. 2002. Vegetation pattern and process in tree islands of the southern Everglades and adjacent areas. In: *Tree islands of the Everglades*, edited by F. H. Sklar and A. van der Valk, pp. 225–281. Kluwer Academic Publishers, Dordrecht, Netherlands.

Armentano, T. V., J. P. Sah, M. S. Ross, D. T. Jones, H. C. Cooley, and C. S. Smith. 2006. Rapid response of vegetation to hydrological changes in Taylor Slough, Everglades National Park, Florida, USA. *Hydrobiologia* 569: 293–309.

Ashworth, F. K. 1919a. [Map showing location of the 207,000 acre Chevelier Tract and the 12,000 acre Shell Mound Tract, with proposed railroad and highway connection on southwest coast of Florida.] Unpublished blueprint; "Compiled from maps and data for Capt J. F. Jaudon by F. K. Ashworth C. E. Dec, 1919." 20 × 30 inches. 1:280,000 (approx.). Jaudon

Collection, box 18, Historical Museum of Southern Florida, Miami.

———. 1919b. [Untitled map of southern Florida showing location of proposed Tamiami Railway.] Unpublished blueprint; "Compiled for Capt J. F. Jaudon by F. K. Ashworth C. E., Miami, Fla. June, 1919." 18 × 22 inches. 1:560,000 (approx.). Jaudon Collection, box 18, Historical Museum of Southern Florida, Miami.

———. 1919c. [Untitled map of southern Florida showing proposed Tamiami Railway.] "Compiled for Capt. J. F. Jaudon by F. K. Ashworth C.E., Miami, Fla. June, 1919." 9 × 22 inches. 1:630,000 (approx.). Jaudon Collection, box 17, Historical Museum of Southern Florida, Miami.

———. 1925? [Map of Chevelier.] "Compiled for Capt. J. F. Jaudon by F. K. Ashworth, C.E. ." 16 × 22 inches. 1:600,000 (approx.). Jaudon Collection, box 17, Historical Museum of Southern Florida, Miami.

Aumen, N. G. 1995. The history of human impacts, lake management, and limnological research on Lake Okeechobee, Florida (USA). *Arch Hydrobiol Spec Issues Advanc Limnol* 45: 1–16.

Aumen, N. G., and S. Gray. 1995. Research synthesis and management recommendations from a five-year, ecosystem-level study of Lake Okeechobee, Florida (USA). *Arch Hydrobiol Spec Issues Advanc Limnol* 45: 343–356.

Austin, D. F. 1976. Vegetation of southeastern Florida I. Pine Jog. *Fla Sci* 39: 230–235.

———. 1985. *Commelina gigas*: Rediscovered and lost. *Palmetto* 5: 11.

Austin, D. F., K. Coleman-Marois, and D. R. Richardson. 1977. Vegetation of southeastern Florida II–V. *Fla Sci* 40: 331–361.

Bache, A. D. 1850. *Letter from the Secy. of the Treasury, communicating the report of the Superintendent of the Coast Survey, showing progress of the work for the year ending November, 1849.* Senate Exec. Doc. no. 5, Dec. 27, 1849. 31st Congress, 1st Session, Washington, D.C. 68 pp.

Backus, E. 1838. [Untitled map (sketch) of Lake Okeechobee.] Accompanies letter of Capt. E. Backus of Sept. 12, 1838. Records of the Seminole War, National Archives, Washington, D.C.

Bailey, H. H. 1925. *The birds of Florida.* William and Wilkins, Baltimore. 146 pp.

Baldwin, M., and H. W. Hawker. 1915. Soil survey of the Fort Lauderdale area, Florida. In: *Field operations of the Bureau of Soils, 1915*, pp. 751–798. U.S. Department of Agriculture, Washington, D.C.

Bancroft, G. T., D. E. Gawlik, and K. Rutchey. 2002. Distribution of wading birds relative to vegetation and water depths in the northern Everglades of Florida, USA. *J Waterbird Soc* 25: 265–391.

Bancroft, G. T., W. Hoffman, R. J. Sawicki, and J. C. Ogden. 1992. The importance of the Water Conservation Areas in the Everglades to the endangered Wood Stork (*Mycteria americana*). *Conserv Biol* 6: 392–398.

Barbour, T. 1946. *That vanishing Eden: A naturalist's Florida.* Little, Brown and Company, Boston. 250 pp.

Beard, D. B. 1938. *Wildlife reconnaissance: Everglades National Park project.* U.S. Department of Interior, National Park Service, Washington, D.C. 106 pp.

Bender, G. J. 1937. Everglades fire control. Quoted in: Mayo (1940), pp. 88–91.

———. 1943. The Everglades Fire Control District. *Soil Sci Soc Fl* 5-A: 149–152.

Bentley, G. R. 1950. Colonel Thompson's "Tour of tropical Florida." *Tequesta* 10: 3–12.

Bernhardt, C. E., and D. A. Willard. 2006. *Marl prairie vegetation response to 20th century hydrologic change.* Open File Report 2006–1355. U.S. Geological Survey, Reston, Va. 12 pp.

———. 2009. Response of the Everglades ridge and slough landscape to climate variability and 20th-century water management. *Ecol Appl* 19: 1723–1738.

Bernhardt, C. E., D. A. Willard, M. Marot, and C. W. Holmes. 2004. *Anthropogenic and natural variation in ridge and slough pollen assemblages.* Open File Report 2004–1448. U.S. Geological Survey, Reston, Va. 47 pp.

Bestor, H. A. 1942. The principal elements of a long time soil and water conservation plan for the Everglades. *Soil Sci Soc Fl* 4-A: 90–99.

———. 1943. Reclamation problems of sub-drainage districts adjacent to Lake Okeechobee. *Soil Sci Soc Fl* 5-A: 157–165.

Blatchley, W. S. 1932. *In days agone: Notes on the fauna and flora of subtropical Florida when most of it was a primeval wilderness.* Nature Publishing Company, Indianapolis, Ind. 338 pp.

Board of Commissioners of the Everglades Drainage District. 1935. *Map of Everglades Drainage District of Florida.* 17 × 30 inches. 1:380,000. Board of Commissioners of the Everglades Drainage District, Miami.

Brandt, L. A. 1997. *Spatial and temporal patterns of tree islands in the Arthur R. Marshall Loxahatchee National Wildlife Refuge.* PhD diss., University of Florida, Gainesville, Fla. 142 pp.

Brooks, A. M. 1880. *Petals plucked from sunny climes.* Southern Methodist Publishing House, Nashville, Tenn. 495 pp.

Brooks, H. K. 1984. Lake Okeechobee. In: *Environments of South Florida: Present and past II*, edited by P. Gleason, pp. 38–68. Miami Geological Society, Coral Gables, Fla.

Broward, N. B. 1905. Message of Gov. N. B. Broward to the Legislature of Florida relative to reclamation of Everglades; dated May 3. Quoted in: U.S. Congress (1911), pp. 99–109.

Browder, J. A., P. J. Gleason, and D. Swift. 1994. Periphyton in the Everglades: Spatial variation, environmental correlates, and ecological implications. In: *Everglades: The ecosystem and its restoration*, edited by S. M. Davis and J. C. Ogden, pp. 379–418. St. Lucie Press, Delray Beach, Fla.

Brown, L. G. 1993. *Totch: A life in the Everglades.* University Press of Florida, Gainesville, Fla. 269 pp.

Brown, L. 1929. Memorandum to the Secretary of War. Subject: Report on agricultural possibilities of the Florida Everglades. Quoted in: Mayo (1940), pp. 48–51.

Brown, R. H. 1943. "Muck burn, Broward Co.," no. A-429 [photograph]. Private collection of Garald G. Parker. Files of C. McVoy, South Florida Water Management District, West Palm Beach, Fla.

Buker, G. E. 1975. *Swamp sailors in the Second Seminole War*. University Presses of Florida, Gainesville, Fla. 152 pp.

Busch, D. E., W. F. Loftus, and O. L. Bass. 1998. Long term hydrologic effects on marsh plant community structure in the southern Everglades. *Wetlands* 18: 230–241.

Callsen, P. (2008). *Glades period settlement patterns in the Everglades culture area*. Master's thesis, Florida Atlantic University, Boca Raton, Fla. 215 pp.

Canova, A. P. 1885. *Life and adventures in south Florida*. Southern Sun Publishing House, Palatka, Fla.; repr. 1906, Tribune Printing, Tampa, Fla. 158 pp.

Carr, R. S. 2002. The archaeology of Everglades tree islands. In: *Tree islands of the Everglades*, edited by F. H. Sklar and A. van der Valk, pp. 187–206. Kluwer Academic Publishers, Dordrecht, Netherlands.

Carr, R. S., and W. S. Steele. 1991. *Archival and other evidence on the location of Snake Warrior's Island, Broward County, Florida*. AHC Technical Report #45. Archaeological and Historical Conservancy, Davie, Fla. 23 pp.

Carr, R. S. and D. Sandler. 1992. *An archaeological survey of the Westridge Property [Pine Island], Broward County, Florida*. AHC Technical Report #48. Archaeological and Historical Conservancy, Davie, Fla. 70 pp.

Carr, R. S., R. Haiduven, and W. S. Steele. 1992. *An archaeological and historical assessment of the Miramar Oaks Sites, Broward County, Florida*. AHC Technical Report #53. Archaeological and Historical Conservancy, Davie, Fla. 85 pp.

Central and Southern Florida Flood Control District. 1957. *Central and Southern Florida Flood Control Project: Eight years of progress 1949–1957*. Central and Southern Florida Flood Control District, West Palm Beach, Fla. 45 pp.

Central and Southern Florida Flood Control District, Engineering Division. 1960. [Untitled map of Central and Southern Florida Flood Control District, with hand-drawn shading to indicate areas of long-duration and intermittent flooding, with and without Project influence.] Drawn on base map of District, Nov. 1958, Dwg. no. FO-23. 33 × 46 inches. 1:200,000. Central and Southern Florida Flood Control District, West Palm Beach, Fla.

Chadwick, E. W. 1911. [Letter to J. W. Harshberger, dated March 22, 1911.] Quoted in: Harshberger (1914), p. 164.

Cherry, J. A. 2005. *The effects of herbivory, competition, and disturbance on a wetland community with emphasis on the dominant aquatic macrophyte,* Nymphaea odorata *Aiton*. PhD diss., University of Alabama, Tuscaloosa, Ala. 176 pp.

Cherry, J. A., and L. Gough. 2006. Temporary floating island formation maintains wetland plant species richness: The role of the seed bank. *Aquatic Botany* 85: 29–36.

Cheves, M. 2007. Did surveyors lie (to avoid alligators and snakes)? *American Surveyor*, December, 28–35.

Chick, J. H., C. R. Ruetz, and J. C. Trexler. 2004. Spatial scale and abundance patterns of large fish communities in freshwater marshes of the Florida Everglades. *Wetlands* 24: 652–664.

Church, A. 1892. A dash through the Everglades [March 14 to April 16, 1892]. Quoted in: Marchman (1949), pp. 13–41.

Clawson, M., and C. L. Stewart. 1965. *Land use information: A critical survey of U.S. statistics including possibilities for greater uniformity*. Resources for the Future, Baltimore. 404 pp.

Clayton, B. S. 1936. *Subsidence of peat soils in Florida*. Report #1070. U.S. Department of Agriculture, Bureau of Agricultural Engineering (mimeog.), Belle Glade, Fla. 15 pp.

———. 1943. Nature and extent of the surface subsidence of the organic soils of the Everglades. Part I: Shrinkage and subsidence due to drainage and cultivation. *Soil Sci Soc Fl* 5-A: 118–120.

Clayton, B. S., J. R. Neller, and R. V. Allison. 1942. *Water control in the peat and muck soils of the Florida Everglades*. Tech. Bull. #378. University of Florida, Agricultural Experiment Station, Gainesville, Fla. 74 pp.

Cohen, A. D. 1984. Evidence of fires in the ancient Everglades and coastal swamps of southern Florida. In: *Environments of South Florida: Present and past II*, edited by P. Gleason, pp. 459–464. Miami Geological Society, Coral Gables, Fla.

Cohen, A. D., C. P. Gage, and W. S. Moore. 1999. Combining organic petrography and palynology to assess anthropogenic impacts on peatlands. Part 1: An example from the northern Everglades of Florida. *Int J Coal Geol* 39: 3–45.

Cohen, A. D., and W. Spackman Jr. 1984. The petrology of peats from the Everglades and coastal swamps of southern Florida. In: *Environments of South Florida: Present and past II*, edited by P. Gleason, pp. 352–374. Miami Geological Society, Coral Gables, Fla.

Cohen, I. 1925. *Historical sketches and sidelights of Miami, Florida*. Revised ed. Privately published, Miami. 213 pp.

Cooley, W. 1851. [Letter to Governor Thomas Brown narrating reconnaisances of the Everglades, dated August 11, 1851.] Quoted in: Knetsch (1989), pp. 41–44.

Cornwell, G., and E. C. Hutchinson. 1974. *An ecological analysis of an Everglades township in southwestern Palm Beach County, Florida*. Ecoimpact, Gainesville, Fla. 30 pp.

Cory, C. B. 1896. *Hunting and fishing in Florida*. 2nd. ed. Estes and Lauriat, Boston. 304 pp.

Costanza, R. 1975. *The spatial distribution of land use systems, incoming energy and energy use in South Florida from 1900 to 1973*. Master's thesis, University of Florida, Gainesville, Fla. 254 pp.

Crabtree, H. 1921. Profile of Tamiami Trail from Station Number 1176 +42 to Sta Number 1810 at west county line. Unpublished blueprint; "Surveyed by Howard Crabtree, County Engr. May 1921"; "Datum Mean low water Biscayne Bay." 21 × 52 inches. 1:4,800 (Horiz.); 1:48 (Vert.). Jaudon Collection, box 18, Historical Museum of Southern Florida, Miami.

Craighead, F. C. 1964. Land, mangroves, and hurricanes. *Fairchild Tropical Garden Bulletin* 19: 1–28.

———. 1971. *The trees of south Florida. Vol. 1: The natural environments and their succession*. University of Miami Press, Coral Gables, Fla. 212 pp.

———. 1973. The effects of natural forces on the development and maintenance of the Everglades, Florida. In: *National Geographic Society Research Reports 1966*, pp. 49–67. National Geographic Society, Washington, D.C.

———. 1974. Hammocks of south Florida. In: *Environments of South Florida: Past and present*, edited by P. Gleason, pp. 53–60. Miami Geological Society, Coral Gables, Fla.

Dachnowski, A. P. 1924. The chemical examination of various materials by means of food-stuff analysis. *J Agr Res* 29: 69–83.

Dachnowski-Stokes, A. P. 1930. Peat profiles of the Everglades in Florida: The stratigraphic features of the "Upper" Everglades and correlation with environmental changes. *J Wash Acad Sci* 20: 89–107.

Dachnowski-Stokes, A. P., and R. V. Allison. 1928. A preliminary note on blue-green algal marl in southern Florida in relation to the problem of coastal subsidence. *J Wash Acad Sci* 18: 476–480.

Dade County, Department of Environmental Resources Management. 1983. *Groundwater elevation record—Well G-614 for 1950–1979*. Dade County, Department of Environmental Resources Management, Miami.

Dancy, J. M. 1883. [Report of James M. Dancy, Civil Engineer, on lands reclaimed by the Atlantic and Gulf Coast Canal and Okeechobee Land Company. Orange Mills, Fla., July 12, 1883.] In: *Minutes of the Trustees, Vol. III (1904)*, pp. 242–246. Trustees of the Internal Improvement Fund, Tallahassee, Fla.

———. 1884. [Report of James M. Dancy, Civil Engineer, appointed to examine the southern or Caloosahatchee Division of the Drainage District. November 1884.] In: *Minutes of the Trustees, Vol. III (1904)*, pp. 332–338. Trustees of the Internal Improvement Fund, Tallahassee, Fla.

Daniel, J. J., W. H. Davison, and J. Bradford. 1886. Abstract of report by J. J. Daniel, W. H. Davison, and John Bradford, committee appointed by the governor of Florida on November 17, 1885. Quoted in: U.S. Congress (1911), pp. 21–24.

David, P. G. 1996. Changes in plant communities relative to hydrologic conditions in the Florida Everglades. *Wetlands* 16: 15–23.

Davis, C. K. 1943. Summary of three years of conservation work in the Everglades, and plans for the future. *Soil Sci Soc Fl* 5-A: 116–117.

Davis, J. H. 1943a. *The natural features of southern Florida: Especially the vegetation, and the Everglades*. Geological Bulletin #25. Florida Geological Survey, Tallahassee, Fla. 311 pp.

———. 1943b. *Vegetation map of southern Florida*. Accompanies J. H. Davis (1943a). 23 × 29 inches. 1:400,000. Florida Geological Survey, Tallahassee, Fla.

———. 1943c. Vegetation of the Everglades and conservation from the point of view of the plant ecologist. *Soil Sci Soc Fl* 5-A: 105–113.

———. 1946. *The peat deposits of Florida: Their occurrence, development, and uses*. Geological Bulletin #30. Florida Geological Survey, Tallahassee, Fla. 247 pp.

———. 1957. Probable effects of planned water level controls on the types of vegetation in the Conservation Area No. 1. In: *Central and Southern Florida Project for Flood Control and Other Purposes. Part I: Agricultural and conservation areas, supplement 25, general design memorandum, plan of regulation for Conservation Area No. 1*, pp. C-1 to C-6. U.S. Army Corps of Engineers, Office of the District Engineer, Jacksonville, Fla.

———. 1967. *General map of natural vegetation of Florida*. Circular S-178. 25 × 30 inches. 1:1,330,000 (approx.). University of Florida, Institute of Food and Agriculture Science, Gainesville, Fla.

Davis, S. M., E. E. Gaiser, W. F. Loftus, and A. E. Huffman. 2005. Southern marl prairies conceptual ecological model. *Wetlands* 25: 821–831.

Davis, S. M., L. H. Gunderson, W. A. Park, J. R. Richardson, and J. E. Mattson. 1994. Landscape dimension, composition, and function in a changing Everglades ecosystem. In: *Everglades: The ecosystem and its restoration*, edited by S. M. Davis and J. C. Ogden, pp. 419–444. St. Lucie Press, Delray Beach, Fla.

De Brahm, W. G. 1773. *Report of the general survey in the southern district of North America*. Edited by L. DeVorsey, Jr. Repr. 1971, University of South Carolina Press, Columbia, S.C. 325 pp.

d'Escalante Fontaneda, H. 1575. *Memoir of Do. d'Escalante Fontaneda respecting Florida*. Translated by Buckingham Smith, 1854. Edited by David True. Repr. 1944. University of Miami and Historical Association of Southern Florida, Miami. 77 pp.

Desmond, G. B. 2003. *Measuring and mapping the topography of the Florida Everglades for ecosystem restoration*. USGS Fact Sheet 021-03. U.S. Geological Survey, Reston, Va. 4 pp.

Dewhurst, W. W. 1927. *Paradise prairie* [Advertising brochure for 30,000 Acres in Township 59, Range 35]. St. Augustine Historical Society, St. Augustine, Fla. 3 pp. + 1 map, 1 photograph.

Dibthorpe, H. K. 1947. *Map of Collier County, Florida*. "Computed and mapped by H. K. Dibthorpe under direction of D. Graham Copeland. Approved April 7, 1947. Checked by D. Graham Copeland, Chairman, Board of County Commissioners, Collier County. Copyright 1947 D. Graham Copeland." 27 × 40 inches. 1:127,000.

Dickerson, R. E. 1942. Trend of Pleistocene ocean currents across the Florida Everglades. *Geog Rev* 32: 136–139.

Dickey, F. C. 1910. The survey of Lake Okeechobee in the year 1910. By Major F. C. Dickey, U.S. Army Corps of Engineers. Unpublished typescript. Records of J. Knetsch, Florida Department of Environmental Protection, Division of State Lands, Bureau of Survey and Mapping, Tallahassee, Fla. 10 pp. + photos.

Dieterich, E. P. 1987. Arch Creek: Prehistory to public park. *Tequesta* 47: 49–65.

Dimock, A. W. 1907. Crossing the Everglades in a power-boat. *Harper's Magazine*, January, 213–220.

———. 1911. In the depths of the Everglades. *Travel*, January, 116–119, 141–143.

Dineen, J. W. 1972. Life in the tenacious Everglades. *In Depth Report (Central and Southern Florida Flood Control District)* 1: 1–12.

———. 1974. Examination of water management alternatives in Conservation Area 2A. *In Depth Report (Central and Southern Florida Flood Control District)* 2: 1–10.

———. 1984. The fishes of the Everglades. In: *Environments of South Florida: Present and past II*, edited by P. Gleason, pp. 258–268. Miami Geological Society, Coral Gables, Fla.

Disston, H. 1895. [Communication to the Board of the Internal Improvement Fund regarding request by F. A. Hendry to have canals west of Lake Okeechobee and Hicpochee filled in with embankments. May 1, 1895.] In: *Minutes of the Trustees, Vol. IV (1904)*, pp. 331–337. Trustees of the Internal Improvement Fund, Tallahassee, Fla.

Division of Water Survey and Research. 1954. *Summary of observed rainfalls on Florida to 31 December 1952.* Water Survey and Research Paper #11. Florida State Board of Conservation, Division of Water Survey and Research, Tallahassee, Fla. 334 pp.

Dix, E. A., and J. N. MacGonigle. 1905. The Everglades of Florida: A region of mystery. *Century Magazine*, February, 512–527.

Dodge, C. R. 1894. Subtropical Florida. *Scribner's Magazine*, March, 345–362.

Dodson, P. 1973. *Journey through the old Everglades.* Trend House, Tampa, Fla. 75 pp.

Douglas, M. S. 1947. *The Everglades: River of grass.* Rinehart, New York. 406 pp.

Dovell, J. E. 1942. A brief history of the Florida Everglades. *Soil Sci Soc Fl* 4-A: 132–161.

———. 1947a. *A history of the Everglades of Florida.* PhD diss., University of North Carolina, Chapel Hill, N.C. 598 pp.

———. 1947b. The Everglades before reclamation. *Fla Hist Q.* 26: 1–43.

Duever, M. J., J. E. Carlson, J. F. Meeder, L. C. Duever, L. H. Gunderson, L. A. Riopelle, T. R. Alexander, R. L. Myers, and D. P. Spangler. 1986. *The Big Cypress National Preserve.* 2nd. ed. National Audubon Society, New York. 444 pp.

DuPuis, J. G. 1954a. *History of early agricultural relations in Dade County, Florida.* Privately published, Miami. 85 pp.

———. 1954b. *History of early medicine in Dade County, Florida.* Privately published, Miami. 140 pp.

———. 1954c. *History of early public schools in Dade County, Florida.* Privately published, Miami. 82 pp.

Dupuy, W. A. 1908. An air-line across the Everglades. *World's Work*, February, 9893–9897.

Duval, H. S. 1884. [Report of H. S. Duval, State Engineer, with appendix showing the townships reclaimed by the drainage operations of the Atlantic and Gulf Coast Canal and Okeechobee Land Company. Aug. 18, 1884.] In: *Minutes of the Trustees, Vol. III (1904)*, pp. 314–322. Trustees of the Internal Improvement Fund, Tallahassee, Fla.

Egler, F. E. 1952. Southeast saline Everglades vegetation, Florida, and its management. *Veg Acta Geobot* 3: 213–265.

Elliot, F. C. 1912. [Report of F. C. Elliot, Acting Chief Drainage Engineer, on work of dredges under Furst-Clark construction company's contract. Nov. 1, 1912.] In: *Minutes of the Trustees, Vol. IX (1913)*, pp. 554–558. Trustees of the Internal Improvement Fund, Tallahassee, Fla.

———. 1923. [Letter to Ernest Graham, Pennyslvannia Sugar Co., regarding construction of Tamiami Trail, dated June 23, 1923.] Quoted in: Graham (1951), pp. 45.

———. 1927. *Biennial report, Everglades Drainage District, Engineering Department of 1925–1926.* Everglades Drainage District, Tallahassee, Fla. 76 pp.

Elliott, C. G. 1904. Drainage investigation, 1904, by C. G. Elliot, Engineer, U. S. Department of Agriculture, in charge. Quoted in: U.S. Congress (1911), pp. 94–97.

Elliott, D. H. 1890. *A new sectional map of Florida.* D. H. Elliott, General Land Agent. 35 × 46 inches. 1:632,000. Associated Railway Land Department of Florida, Sanford, Fla.

Ensey, R. F., and F. C. Elliot. 1911. *Map of Everglades Drainage District.* "Issued by the Trustees of the Internal Improvement Fund. November 1911." 20 × 34 inches. 1:380,000. Trustees of the Internal Improvement Fund, Tallahassee, Fla.

Evans, C. B., and R. V. Allison. 1942. The soils of the Everglades in relation to reclamation and conservation operations. *Soil Sci Soc Fl* 4-A: 34–46.

Everglades Drainage District. 1947. *Tentative report showing estimated losses to agriculture, highways and drainage works in eleven counties comprising Everglades Drainage District Florida.* Everglades Drainage District, West Palm Beach, Fla. 10 pp.

Everglades Drainage District, Operations Department. 1928. *Ground water studies at Shawano (Brown Company Plantation) on the Hillsboro Canal.* Jan. 13 to July 20, 1928. "Figure 18." Everglades Drainage District, West Palm Beach, Fla.

Fennema, R. J., and T. N. Bhatt. 2003. Natural System Model review and calibration: Report prepared for Everglades National Park. May 31, 2003. Unpublished report. Dhasaan, West Palm Beach, Fla. 96 pp.

Fennema, R. J., C. J. Neidrauer, R. A. Johnson, T. K. MacVicar, and W. A. Perkins. 1994. A computer model to simulate natural Everglades hydrology. In: *Everglades: The ecosystem and its restoration*, edited by S. M. Davis and J. C. Ogden, pp. 249–289. St. Lucie Press, Delray Beach, Fla.

Ferguson, G. E. 1943. Summary of three years of surface water studies in the Everglades. *Soil Sci Soc Fl* 5-A: 18–23.

Ferguson, G. E., C. W. Lingham, S. K. Love, and R. O. Vernon. 1947. *Springs of Florida.* Geological Bulletin #31. Florida Geological Survey, Tallahassee, Fla. 196 pp.

Finkl, C. W. 1994. Disaster mitigation in the South Atlantic Coastal Zone (SACZ): A prodrome for mapping hazards and coastal land systems using the example of urban subtropical southeastern Florida. *J Coastal Res* Spec. Iss. no. 12: Coastal Hazards: 339–366.

Fish, J. E., and M. Stewart. 1991. *Hydrogeology of the surficial aquifer system, Dade County, Florida*. Water Resources Investigations Report 90-4108. U.S. Geological Survey, Tallahassee, Fla. 50 pp.

Florida Agricultural Experiment Stations and Soil Conservation Service. 1962. *General soil map of Florida*. 28 × 34 inches. 1:1,040,000 (approx.). Florida Agricultural Experiment Stations, Gainesville, Fla.

Florida Department of Agriculture. 1911. General classification of Florida soils. In: *Eleventh biennial report of the Commissioner of Agriculture, Jan. 1, 1909 to Dec. 31, 1910*, pp. 29–49. Florida Department of Agriculture (T. J. Appleyard, State Printer), Tallahassee, Fla.

Florida Everglades Engineering Commission. 1914. *Florida Everglades: Report of the Florida Everglades Engineering Commission to the Board of Commissioners of the Everglades Drainage District and the Trustees of the Internal Improvement Fund. State of Florida. 1913*. Senate Doc. no. 379. 63rd Congress, 2nd Session, Washington, D.C. 148 pp.

Florida Fish and Wildlife Conservation Commission. 2006. *South region freshwater fishing guide*. Florida Fish and Wildlife Conservation Commission, Tallahassee, Fla. 30 pp.

Florida Game and Fresh Water Fish Commission. 1956. *Recommended program for Conservation Area 2 by Florida Game and Fresh Water Fish Commission*. "Letter of transmittal signed A. D. Aldrich, Oct. 31, 1956." Florida Game and Fresh Water Fish Commission, Tallahassee, Fla. 120 pp. + 2 maps.

Florida Surveyor General's Office. 1849. *A plat exhibiting the state of the surveys in the State of Florida with references*. Approved R. A. Putnam, Surv. Genl. 24 × 27 inches. 1:1,140,000. Surveyor General's Office, St. Augustine, Fla.

———. 1853. *A plat exhibiting the state of the surveys in the State of Florida with references*. Approved John Wescott, Surv. Genl. 22 × 25 inches. 1:1,140,000. Surveyor General's Office, St. Augustine, Fla.

Fogarty, M. J. 1984. The ecology of the Everglades alligator. In: *Environments of South Florida: Present and past II*, edited by P. Gleason, pp. 211–218. Miami Geological Society, Coral Gables, Fla.

Fordham, E. P. 1906. *Personal narrative of travels in Virginia, Maryland, Pennsylvania, Ohio, Indiana, Kentucky; and of a residence in the Illinois territory: 1817–1818*. Edited by F. Ogg. Arthur H. Clark, Cleveland, Ohio. 248 pp.

Forman, C. 1979. The Forman family: Everglades pioneers. *Broward Legacy* 3: 2–9.

Forsaith, C. C. 1916. A report on some allochthonous peat deposits of Florida, Part I: Topographical. *Bot Gaz* 62: 32–52.

———. 1917. A report on some allochthonous peat of Florida, Part II: Morphological. *Bot Gaz* 63: 190–208.

Forthman, C. A. 1973. *The effects of prescribed burning on sawgrass,* Cladium jamaicense *Crantz, in South Florida*. Master's thesis, University of Miami, Coral Gables, Fla. 83 pp.

Frazier, W. B. 1884a. [Statement of Mr. W. B. Frazier to Col. H. S. Duval, State Engineer. Fort Thompson, Fla., July 28, 1884.] In: *Minutes of the Trustees, Vol. III (1904)*, pp. 322–323. Trustees of the Internal Improvement Fund, Tallahassee, Fla.

———. 1884b. [Statement made to James M. Dancy, State Agent, by Mr. W. B. Frazier. Fort Thompson, Fla., Nov. 5, 1884.] In: *Minutes of the Trustees, Vol. III (1904)*, pp. 338–339. Trustees of the Internal Improvement Fund, Tallahassee, Fla.

Frederick, J. S. 1907. [Interview with J. S. Frederick, Surveyor, Miami.] Quoted in: Stewart (1907), p. 44.

Frederick, P. C., and M. G. Spalding. 1994. Factors affecting reproductive success of wading birds (Ciconiiformes) in the Everglades ecosystem. In: *Everglades: The ecosystem and its restoration*, edited by S. M. Davis and J. C. Ogden, pp. 659–691. St. Lucie Press, Delray Beach, Fla.

Fuller, M. L. 1904. *Contributions to the hydrology of the eastern United States, 1903*. Water-Supply and Irrigation Paper #102. U.S. Geological Survey, Washington, D.C. 522 pp.

Gaby, D. C. 1993. *The Miami River and its tributaries*. Historical Association of Southern Florida, Miami. 193 pp.

Gallatin, M. H., J. K. Ballard, C. B. Evans, H. S. Galberry, J. J. Hinton, D. P. Dowell, E. Truett, W. L. Watts, G. C. Willson, and R. G. Leighty. 1958. *Soil survey (detailed-reconnaissance) of Dade County, Florida*. Series 1947, #4. U.S. Department of Agriculture, Soil Conservation Service, Washington, D.C. 56 pp.

Gallatin, M. H., and J. R. Henderson. 1943. Progress report on the soil survey of the Everglades. *Soil Sci Soc Fl* 5-A: 95–104.

Gawlik, D. E. 2002. The effects of prey availability on the numerical response of wading birds. *Ecol Monogr* 72: 329–346.

German, E. R. 2000. *Regional evaluation of evapotranspiration in the Everglades*. Water Resources Investigations Report 00-4217. U.S. Geological Survey, Tallahassee, Fla. 48 pp.

Gifford, J. C. 1911. *The Everglades and other essays relating to southern Florida*. 2nd. ed. Everglades Land Sales, Kansas City, Mo. 226 pp.

Givens, L. S. 1957. Water level management: Its effect on the ecology, wildlife and fisheries resources of Loxahatchee Refuge. In: *Central and southern Florida Project for Flood Control and Other Purposes. Part I: Agricultural and conservation areas, supplement 25, general design memorandum, plan of regulation for Conservation Area No. 1*, pp. B–5 to B–24. U.S. Army Corps of Engineers, Office of the District Engineer, Jacksonville, Fla.

Glaser, P. H. 1992. Peat landforms. In: *The patterned peatlands of Minnesota*, edited by H. Wright, pp. 3–14. University of Minnesota Press, Minneapolis, Minn.

Gleason, P. J. 1972. *Holocene Sedimentation in the Everglades and saline tidal lands*. Field-trip guidebook to Everglades National Park, American Quaternary Congress, Dec. 2–3, 1972. Central and Southern Florida Flood Control District, West Palm Beach, Fla. 47 pp.

Gleason, P. J., A. D. Cohen, H. K. Brooks, P. Stone, R. Goodrick, W. G. Smith, and W. Spackman. 1984. The environmental significance of Holocene sediments from the Everglades and saline tidal plain. In: *Environments of South Florida: Present and past II*, edited by P. Gleason, pp. 297–351. Miami Geological Society, Coral Gables, Fla.

Gleason, P. J., and P. Stone. 1994. Age, origin, and landscape evolution of the Everglades peatland. In: *Everglades: The ecosystem and its restoration*, edited by S. M. Davis and J. C. Ogden, pp. 149–197. St. Lucie Press, Delray Beach, Fla.

Gleason, P. J., P. A. Stone, R. Goodrick, G. Guerin, L. Harris, P. A. Stone, R. Goodrick, G. Guerin, and L. Harris. 1975. The significance of paleofloral studies and ecological aspects of floating peat islands to water management in the Everglades conservation areas of the Central and Southern Florida Flood Control District. Unpublished report. South Florida Water Management District, West Palm Beach, Fla. 176 pp. + 4 appendices.

Glenn, J. L. 1982. *My work among the Florida Seminoles*. Edited by H. A. Kersey, Jr. University Presses of Florida, Orlando, Fla. 121 pp.

Goodrick, R. L. 1974. The wet prairies of the northern Everglades. In: *Environments of South Florida: Present and past*, edited by P. Gleason, pp. 47–52. Miami Geological Society, Coral Gables, Fla.

Graham, ? 1907. [Interview with Mr. Graham, February 12, 1907.] Quoted in: Stewart (1907), p. 24.

Graham, W. A. 1951. The Pennsuco sugar experiment. *Tequesta* 11: 27–49.

Gray, W. C. 1903. [Diary entry of Episcopalian Bishop William Crane Gray, Glades Cross Mission, for April 28, 1903.] Quoted in: Kersey (1975), p. 68.

Griffin, J. W., J. T. Milanich, and J. J. Miller. 2002. *Archeology of the Everglades*. Edited by J. Milanich. University Press of Florida, Gainesville, Fla. 432 pp.

Griswold, L. S. 1896. Notes on the geology of southern Florida. Quoted in: Agassiz (1896), pp. 52–59.

Gunderson, L. H. 1989. Historical hydropatterns in wetland communities of Everglades National Park. In: *Freshwater wetlands and wildlife*. CONF-8603101. DOE Symposium Series no. 61, edited by R. R. Sharitz and J. W. Gibbons, pp. 1099–1111. U.S. Dept. of Energy, Office of Scientific and Technical Information, Oak Ridge, Tenn.

———. 1994. Vegetation of the Everglades: Determinants of community composition. In: *Everglades: The ecosystem and its restoration*, edited by S. M. Davis and J. C. Ogden, pp. 323–340. St. Lucie Press, Delray Beach, Fla.

Gunderson, L. H., and J. R. Synder. 1994. Fire patterns in the southern Everglades. In: *Everglades: The ecosystem and its restoration*, edited by S. M. Davis and J. C. Ogden, pp. 291–305. St. Lucie Press, Delray Beach, Fla.

Hagenbuck, W. W., R. Thompson, and D. P. Rodgers. 1974. *A preliminary investigation of the effects of water levels on vegetative communities of Loxahatchee National Wildlife Refuge, Florida*. NTIS PB-231–611. U.S. Department of Interior, Atlanta, Ga. 25 pp.

Hammar, H. E. 1929. The chemical composition of Florida Everglades peat soils, with special reference to their inorganic constituents. *Soil Sci* 28: 1–11.

Hanna, A. J., and K. A. Hanna. 1948. *Lake Okeechobee: Wellspring of the Everglades*. Bobbs-Merrill, Indianapolis, Ind. 380 pp.

Harney, W. S. 1848. Letter from Gen. William S. Harney, United States Army, to Buckingham Smith, Esq.; dated January 23, 1848. Quoted in: U.S. Congress (1911), pp. 56–57.

Harney, W. W. 1884. The drainage of the Everglades. *Harper's New Monthly Magazine* 68: 598–605.

Harper, R. M. 1910a. A preliminary report on the Florida peat deposits. In: *Third Annual Report 1909–1910*, pp. 197–375. Florida Geological Survey, Tallahassee, Fla.

———. 1910b. Tramping and camping on the southeastern rim of the Everglades. *Florida Review*, July, pp. 44–55; August, pp. 147–157.

———. 1925. *Generalized soil map of Florida*. "Florida State Geological Survey, Herman Gunther, State Geologist, in cooperation with Florida Department of Agriculture, Nathan Mayo, Commissioner." 32 × 33 inches. 1:1,000,000. Florida Geological Survey, Tallahassee, Fla.

———. 1927. *Natural resources of southern Florida*. 18th Annual Report. Florida Geological Survey, Tallahassee, Fla. 207 pp.

Harshberger, J. W. 1913. *Phytogeographic map of South Florida*. (hand drawn). 24 × 26 inches. 1:500,000. Wagner Free Institute of Science, Philadelphia.

———. 1914. The vegetation of South Florida, south of 27°30' north, exclusive of the Florida Keys. *Trans Wagner Free Institute of Science* 7: 49–190.

Hartwell, J. H. 1970. *Some aspects of availability of water from the Everglades for Everglades National Park, Florida*. Open-file report 70007. U.S. Geological Survey, Tallahassee, Fla.

Harvey, J. W., R. W. Schaffranek, G. B. Noe, L. G. Larsen, D. J. Nowacki, and B. L. O'Conner. 2009. Hydroecological factors governing surface water flow on a low-gradient floodplain. *Water Resources Research* 45: W03421, doi:10.1029/2008WR007129.

Hassan, A. F. 1915. *[Map of the] State of Florida*. 44 × 60 inches. 1:500,000. U.S. Geological Survey.

Havens, K. E., and D. E. Gawlik. 2005. Lake Okeechobee conceptual ecological model. *Wetlands* 25: 908–925.

Heilprin, A. 1887. *Explorations on the west coast of Florida and in the Okeechobee wilderness*. Paleontological Research Institution, Philadelphia; repr. 1972, Wagner Free Institute of Science, Ithaca, N.Y. 134 pp.

Heisler, L., D. T. Towles, L. A. Brandt, and R. T. Pace. 2002. Tree island vegetation and water management in the central Everglades. In: *Tree islands of the Everglades*, edited by F. H. Sklar and A. van der Valk, pp. 283–309. Kluwer Academic Publishers, Dordrecht, Netherlands.

Henderson, J. R. 1939. *The soils of Florida*. Tech. Bull. #334. University of Florida, Agricultural Experiment Station, Gainesville, Fla. 67 pp.

———. 1942. From the standpoint of research and extension. 1: Problems in soil surveying. *Soil Sci Soc Fl* 4-B: 13–19.

Hendrix, A. N., and W. F. Loftus. 2000. Distribution and relative abundance of the crayfishes *Procambarus alleni* (Faxon) and *P. fallax* (Hagen) in southern Florida. *Wetlands* 20: 194–199.

Hendry, F. A. 1896. [Communication to Board of the Internal Improvement Fund replying to letters of Mr. Disston and J. M. Kreamer regarding closing canals, July 26, 1895.] In: *Minutes of the Trustees, Vol. IV (1904)*, pp. 370–372. Trustees of the Internal Improvement Fund, Tallahassee, Fla.

———. 1897. [Communication to Board of the Internal Improvement Fund regarding closing of canal leading from Hicpochee Lake into the Caloosahatchee river, June 13, 1897.] In: *Minutes of the Trustees, Vol. IV (1904)*, pp. 403–405. Trustees of the Internal Improvement Fund, Tallahassee, Fla.

———. 1907. [Interview with Captain Hendry, Labelle, Florida, March 8, 1907.] Quoted in: Stewart (1907), p. 41.

Henshall, J. A. 1883. [Excerpts from nine articles in *Forest and Stream* describing a voyage from Titusville to Key West.] Quoted in: Reiger (1971), pp. 43–66.

———. 1884. *Camping and cruising in Florida*. Robert Clarke, Cincinnati, Ohio; repr. 1991, Florida Classics Library, Port Salermo, Fla. 168 pp.

Herr, B. 1943. *A report to the Board of Commissioners of Okeechobee Flood Control District on the activities of the District and on Lake Okeechobee, 1929–1943*. Okeechobee Flood Control District, West Palm Beach, Fla. 99 pp.

Hetherington, A. P. 1980. *The river of the long water*. Mickler House Publishers, Chuluota, Fla. 237 pp.

Hills, G. B. 1931. The Everglades of Florida. *Florida Engineering and Construction* 8: 3–13.

Hoffman, W., G. T. Bancroft, and R. J. Sawicki. 1994. Foraging habitat of wading birds in the water conservation areas of the Everglades. In: *Everglades: The ecosystem and its restoration*, edited by S. M. Davis and J. C. Ogden, pp. 585–614. St. Lucie Press, Delray Beach, Fla.

Hofstetter, R. H. 1984. The effect of fire on the pineland and sawgrass communities of southern Florida. In: *Environments of South Florida: Present and past II*, edited by P. Gleason, pp. 465–476. Miami Geological Society, Coral Gables, Fla.

Hopkins, C. F. 1884a. [Letter from C. F. Hopkins, Agent of the Atlantic and Gulf Coast Canal and Okeechobee Land Company, to Board of Trustees of the Internal Improvement Fund, regarding progress of drainage, dated Aug. 1884.] In: *Minutes of the Trustees, Vol. III (1904)*, pp. 324–325. Trustees of the Internal Improvement Fund, Tallahassee, Fla.

———. 1884b. [Letter and report to J. M. Kreamer, Chief Engineer Okeechobee Drainage Co., concerning the *Times-Democrat* expedition of Nov.–Dec. 1883, dated Feb. 4, 1884.] Quoted in: Stewart (1907), pp. 100–109.

Howell, A. H. 1932. *Florida bird life*. Florida Department of Game and Fresh Water Fish in cooperation with Bureau of Biological Survey, U.S. Department of Agriculture. Coward-McCann, Publisher's Agents, New York. 579 pp.

Huser, T. 1989. *Into the fifth decade: The first forty years of the South Florida Water Management District*. Special Publication. South Florida Water Management District, West Palm Beach, Fla. 154 pp.

Ingraham, J. E. 1918. Typescript letter from J. E. Ingraham, Land Department, Model Land Company, to Mr. F. S. Morse, dated St. Augustine, Fla., June 24, 1918. Model Lands Collection, box 18, folder 467, University of Miami.

Interbureau Committee. 1930. *Memorandum regarding agricultural conditions in the Everglades of Florida, and the effect on those conditions of proposed navigation and flood-protection improvements*. House Doc. no. 71-47. 71st Congress, 2nd Session, Washington, D.C.

Ives, J. C. 1856a. *Memoir to accompany a military map of the peninsula of Florida, south of Tampa Bay*. M. B. Wynkoop, Book and Job Printer, New York. 42 pp.

———. 1856b. *Military map of the peninsula of Florida south of Tampa Bay*. Accompanies Ives (1856a). [U.S.] War Department, New York.

Jackson, J. 1847. Handwritten letter from J. Jackson to Col. R. Butler, Surveyor General, State of Florida, dated "Hunting ground near Key biscayne," April 5, 1847. Letters and Reports to Surveyor General, vol. 1, 1825–1847, p. 811. Florida Department of Environmental Protection, Division of State Lands, Bureau of Survey and Mapping, Tallahassee, Fla.

Jansen, J. D., and G. C. Nanson. 2004. Anabranching and maximum flow efficiency in Magela Creek, northern Australia. *Water Resources Res* 40: W04503, doi: 10.1029/2003WR002408.

Jennings, W. S. 1903. Message of Governor W. S. Jennings to the legislature of Florida relative to reclamation of Everglades; dated April 7. Quoted in: U.S. Congress (1911), pp. 84–90.

———. 1907. [Report on a tour of inspection of the drainage canals by Hon. W. S. Jennings, General Counsel. Jacksonville, Fla., Nov. 19, 1907.] In: *Minutes of the Trustees, Vol. VII (1909)*, pp. 122–125. Trustees of the Internal Improvement Fund, Tallahassee, Fla.

Johnson, L. 1947. Stage of Lake Okeechobee, 1915 to 1947. Graph, July 1947; "Showing stage above or below minimum prescribed stage, and average ground elevation near lake. Elevations refer to Lake Okeechobee datum." Everglades Drainage District, West Palm Beach, Fla.

———. 1950. A report on the water resources of Everglades National Park, Florida. Unpublished typescript dated May 22, 1950. Central and Southern Florida Flood Control District, Engineering Department, West Palm Beach, Fla. 16 pp. + 1 table.

———. 1958. A survey of the water resources of Everglades National Park, Florida. Typescript, July, 1958. Lamar Johnson, Consulting Engineer, Lake Worth, Fla. 38(?) pp.

———. 1974. *Beyond the fourth generation*. University Presses of Florida, Gainesville, Fla. 230 pp.

Jones, A. H. 1847. Letter from A. H. Jones, Esq., United States Surveyor, to Buckingham Smith, Esq.; dated November 12, 1847. Quoted in: U.S. Congress (1911), pp. 61–62.

Jones, L. A., R. V. Allison, G. D. Ruehle, J. R. Neller, J. R. Henderson, R. D. Marsden, J. G. Steele, C. K. Davis, B. S. Clayton, J. C. Stephens, et al. 1948a. *[Map of] Florida, Everglades Drainage District, physical land conditions*. Accompanies

Tech. Bull. #442. 18 × 22 inches; 39 sheets. 1:63,360. University of Florida, Agricultural Experiment Station, Gainesville, Fla.

———. 1948b. *[Map of] Florida, Everglades region, districts organized for water control.* Accompanies Tech. Bull. #442. "Compiled by Soil Conservation Service. 1947." 18 × 22 inches. 1:510,000 (approx.). University of Florida, Agricultural Experiment Station, Gainesville, Fla.

———. 1948c. *[Map of] Florida. Everglades region, generalized land conditions.* Accompanies Tech. Bull. #442. "Compiled by Soil Conservation Service. 1947." 18 × 22 inches. 1:510,000 (approx.). University of Florida, Agricultural Experiment Station, Gainesville, Fla.

———. 1948d. *[Map of] Florida, Everglades region, water control.* Accompanies Tech. Bull. #442. "Compiled by Soil Conservation Service. 1947." 18 × 22 inches. 1:510,000 (approx.). University of Florida, Agricultural Experiment Station, Gainesville, Fla.

———. 1948e. *Soils, geology and water control in the Everglades region.* Tech. Bull. #442. University of Florida, Agricultural Experiment Station, Gainesville, Fla. 168 pp.

Jordan, F. 1996. *Spatial ecology of decapods and fishes in a northern Everglades wetland mosaic.* PhD diss., University of Florida, Gainesville, Fla. 155 pp.

Jordan, F., H. L. Jelks, and W. M. Kitchens. 1994. Habitat use by the fishing spider *Dolomedes triton* in a northern Everglades wetland. *Wetlands* 14: 239–242.

———. 1997. Habitat structure and plant community composition in a northern Everglades wetland landscape. *Wetlands* 17: 275–283.

Kelly, H. A. 1931. The Everglades National Park. *J Maryland Acad Sciences* 2: 39–43.

Kersey, H. A. 1975. *Pelts, plumes, and hides: White traders among the Seminole Indians, 1870–1930.* University Presses of Florida, Gainesville, Fla. 158 pp.

King, J. W. 1917a. [Map of] Contemplated location and dimension of the main drainage canals in the Southern Drainage District. Unpublished blueprint to accompany report dated Sept. 17, 1917. 20 × 39 inches. 1:60,000 (approx.). Jaudon Collection, box 16, Historical Museum of Southern Florida, Miami.

———. 1917b. Map showing results of examination of the Tamiami Trail lands in Dade County Florida. Unpublished blueprint to accompany report dated Mar. 23, 1917 (Report item #18). 35 × 35 inches. 1:15,840. Jaudon Collection, box 16, Historical Museum of Southern Florida, Miami.

———. 1917c. Map showing results of exploration and examinations in Lee, Monroe and Dade counties, Florida. Unpublished blueprint to accompany report dated Mar. 23, 1917 (Report item #16). 24 × 76 inches. 1:80,000. Jaudon Collection, box 16, Historical Museum of Southern Florida, Miami.

———. 1917d. Profiles showing soil conditions on Tamiami Trail lands. Unpublished blueprint to accompany report dated Mar. 23, 1917 (Report item #17). 10 × 42 inches. 1:9,700 (Horiz.); 1:48 (Vert.). Jaudon Collection, box 16, Historical Museum of Southern Florida, Miami.

———. 1917e. Report of exploration, examination and reconnaissance of the eastern portion of the property of A. W. Hopkins in Monroe and Dade counties, Florida. Typescript dated Mar. 23, 1917. Jaudon Collection, box 16, Historical Museum of Southern Florida, Miami. 20 pp.

———. 1917f. Report of exploration: Examination and reconnoissance of the lands of the Tamiami Trail in Dade County, Florida. Typescript dated Mar. 23, 1917. Jaudon Collection, box 16, folder 2, Historical Museum of Southern Florida, Miami. 26 pp.

———. 1917g. Report of progress upon preliminary survey for reclaiming lands within the Southern Drainage District. Typescript dated Sept. 17, 1917. Jaudon Collection, box 16, Historical Museum of Southern Florida, Miami. 10 pp.

———. 1918. [Quote from J. W. King from his trip through the Everglades.] Quoted in: Larned (1918c), pp. 148–150, 185–186.

Knetsch, J. 1989. William Cooley explores the Everglades. *Broward Legacy* 12: 40–44.

———. 1992. A surveyor's life: John Jackson in south Florida. *Sunland Tribune: J Tampa Hist Soc* 18: 3–8.

———. 1993. A well-connected man: The career of Marcellus A. Williams. *Broward Legacy* 16: 2–10.

———. 1994. The surveys of George MacKay: A drawer of lines on the map of south Florida. *Florida Surveyor* 2: 14–17.

———. 1995. Filling in the skeleton. *S Fla Hist Mag* 23: 24–29.

———. 1999. Southeast Florida in the Third Seminole War: Roads, scouts and expeditions, part I. *Broward Legacy* 22: 38–45.

———. 2006. *Faces on the frontier: Florida surveyors and developers in the 19th century.* Florida Historical Society Press, Cocoa, Fla. 214 pp.

Kohout, F. A. 1966. Submarine springs: A neglected phenomenon of coastal hydrology. In: *Central Treaty Organization Symp. on Hydrology and Water Resources Development, Feb. 5–12, 1966*, pp. 391–413. U.S. Geological Survey, Washington, D.C.

Kohout, F. A., and J. H. Hartwell. 1967. *Hydrologic effects of Area B flood control plan on urbanization of Dade County, Florida.* Report of Investigations #47. Florida Geological Survey, Tallahassee, Fla. 61 pp.

Kohout, F. A., and M. C. Kolipinski. 1967. Biological zonation related to groundwater discharge along the shore of Biscayne Bay, Miami, Florida. In: *Estuaries,* Publ. no. 83, edited by G. Lauff, pp. 488–499. American Association for the Advancement of Science, Washington, D.C.

Kohout, F. A., M. C. Kolipinski, and A. L. Higer. 1973. Remote sensing of submarine springs: Floridan Plateau and Jamaica, West Indies. In: *Proceedings of the second International Symposium on Groundwater,* Palermo, Italy, April 28–May 2, 1973, pp. 571–578.

Kolipinski, M. C., and A. L. Higer. 1969. *Some aspects of the effects of the quantity and quality of water on biological com-*

munities in Everglades National Park. Open File Report FL-69007. U.S. Geological Survey, Tallahassee, Fla. 96 pp.

Kreamer, J. M. 1890. [Presentation to the Board of the Internal Improvement Fund regarding petition by residents of the Caloosahatchee Valley. May 24, 1890.] In: *Minutes of the Trustees, Vol. IV (1904)*, pp. 94–97. Trustees of the Internal Improvement Fund, Tallahassee, Fla.

———. 1891. [Letter to Dr. H. W. Wiley regarding state of drainage of sawgrass lands, ca. July 1891.] Quoted in: Wiley (1891), p. 166.

———. 1892. *Map of Hic-po-chee [Lake Hicpochee] and Okeechobee sugar lands, Lee and De Soto Co's, Florida, embracing 175,000 acres of land available for sugar cultivation*. 36 × 52 inches. 1:47,500. Atlantic and Gulf Coast Canal and Okeechobee Land Company.

———. 1895. [Letter to Hamilton Disston respecting closing the canals leading from Lake Okeechobee into the upper Caloosahatchee country, dated June 4, 1895.] In: *Minutes of the Trustees, Vol. IV (1904)*, pp. 337–341. Trustees of the Internal Improvement Fund, Tallahassee, Fla.

———. 1897. [Letter to the Board of the Internal Improvement Fund relative to the expediency of placing a dam in the Drainage Canal at Lake Okeechobee or Lake Hicpochee. July 19, 1897.] In: *Minutes of the Trustees, Vol. IV (1904)*, pp. 415–417. Trustees of the Internal Improvement Fund, Tallahassee, Fla.

Krome, W. J. 1902a. Handwritten letter to E. Ben Carter, F.E.C. Railway, dated Camp at Cutler Extension, October 1, 1902. William J. Krome letters collection (Krome 1902a–k; 1903a–m), Historical Association of Southern Florida, Miami, but see also transcriptions by J. Knetsch from letters in the Archives of the Florida East Coast Railway, St. Augustine.

———. 1902b. Handwritten letter to E. Ben Carter, F.E.C. Railway, dated Miami, October 4, 1902.

———. 1902c. Handwritten letter to C. T. McCrimmon, West Palm Beach, dated Cutler, October 20, 1902.

———. 1902d. Handwritten letter to E. Ben Carter, F.E.C. Railway, dated Camp #3, 3 mi. W of Cutler, October 20, 1902.

———. 1902e. Handwritten letter to E. Ben Carter, F.E.C. Railway, dated Miami, October 24, 1902.

———. 1902f. Handwritten letter to E. Ben Carter, F.E.C. Railway, dated Miami, November 2, 1902.

———. 1902g. Handwritten letter to E. Ben Carter, F.E.C. Railway, dated Miami, November 16, 1902. Transcribed by J. Knetsch. Archives of the Florida East Coast Railway, St. Augustine, Fla.

———. 1902h. Handwritten letter to E. Ben Carter, F.E.C. Railway, dated Miami, November 23, 1902.

———. 1902i. Handwritten letter to E. Ben Carter, F.E.C. Railway, dated Miami, December [10?], 1902.

———. 1902j. Handwritten letter to E. Ben Carter, F.E.C. Railway, dated Camp Sherritt, December 14, 1902.

———. 1902k. Handwritten letter to E. Ben Carter, F.E.C. Railway, dated Camp Longview, December 30, 1902.

———. 1903a. Handwritten letter to E. Ben Carter, F.E.C. Railway, dated Camp Seminole, January 12, 1903.

———. 1903b. Handwritten letter to E. Ben Carter, F.E.C. Railway, dated Cape Sable Exploration Field Camp, January 28, 1903. Transcribed by J. Knetsch. Archives of the Florida East Coast Railway, St. Augustine, Fla.

———. 1903c. Handwritten letter to E. Ben Carter, F.E.C. Railway, dated Camp Oak Grove, February 11, 1903.

———. 1903d. Handwritten letter to E. Ben Carter, F.E.C. Railway, dated Field Camp, Cape Sable Exploration, February 20, 1903.

———. 1903e. Handwritten letter to E. Ben Carter, F.E.C. Railway, dated Field Camp, Long Key, March 5, 1903.

———. 1903f. Handwritten letter to E. Ben Carter, F.E.C. Railway, dated Camp Nemo, Long Key, March 10, 1903.

———. 1903g. Handwritten letter to E. Ben Carter, F.E.C. Railway, dated Camp Nemo, Long Key, March 13, 1903.

———. 1903h. Handwritten letter to E. Ben Carter, F.E.C. Railway, dated East Cape Sable, May 18, 1903. Transcribed by J. Knetsch. Archives of the Florida East Coast Railway, St. Augustine, Fla.

———. 1903i. Handwritten letter to E. Ben Carter, F.E.C. Railway, dated Camp Sherritt, September 12, 1903.

———. 1903j. Handwritten letter to E. Ben Carter, F.E.C. Railway, dated Camp Sherritt, September 15, 1903.

———. 1903k. Handwritten letter to E. Ben Carter, F.E.C. Railway, dated Camp Jackson, October 28, 1903.

———. 1903l. Letters from William J. Krome, Asst. Eng. during the Cutler and Cape Sable Railway survey expeditions, Florida East Coast rail, Oct. 1, 1902 to Nov. 20, 1903. Historical Association of Southern Florida, Miami.

———. 1903m. [Map of] Florida East Coast RY [Railway]. From Miami to Cape Sable. Showing location through Dade and Monroe Counties, Fla. June 1903. 28 × 36 inches. 1:120,000. Unpublished. Archives of the Florida East Coast Railway, St. Augustine, Fla.

———. 1904. Railway location in the Florida Everglades. Quoted in: Taylor (1979), pp. 7–16.

Kushlan, J. A. 1976. Environmental stability and fish community diversity. *Ecology* 57: 821–825.

———. 1986. Responses of wading birds to seasonally fluctuating water levels: Strategies and their limits. *Colonial Waterbirds* 9: 155–162.

———. 1987. External threats and internal management: The hydrologic regulation of the Everglades, Florida. *Environ Manag* 11: 109–119.

Kushlan, J. A., and M. S. Kushlan. 1980. Population fluctuations of the prawn, *Palaemonetes paludosus*, in the Everglades. *Am Midl Nat* 103: 401–403.

Kushlan, J. A., J. C. Ogden, and A. L. Higer. 1975. *Relation of water level and fish availability to Wood Stork reproduction in the southern Everglades, Florida*. Open File Report 75-434. U.S. Geological Survey, Tallahassee, Fla. 56 pp.

Lal, A. M. W., R. Van Zee, and M. Belnap. 2005. Case study:

Model to simulate regional flow in South Florida. *J. Hydr. Engrg.* (131: 247–258).

Lamb, J. S. 1975. Typescript letter to Tom Martin, Refuge Manager, Loxahatchee National Wildlife Refuge, regarding the author's observations of the refuge portion of the Everglades in its "natural state" (1925–1950s), 13 pp. . Records of Loxahatchee National Wildlife Refuge.

Larned, W. L. 1917. Lost in the wastes of the Everglades. Part 1: The adventures of an engineering party exploring the unknown water wastes of southernmost Florida—A desperate fight against thirst and starvation. *Forest and Stream*, December, 585–589.

———. 1918a. Lost in the Everglades. Part 1: The water wastes. *Forest and Stream*, January, 22–23, 57.

———. 1918b. Lost in the Everglades. Part 3: In Seminole land. *Forest and Stream*, February, 86–87, 110–113.

———. 1918c. Lost in the Everglades. Part 4: Along Tamiami Trail. *Forest and Stream*, March, 148–150, 185–187.

———. 1918d. Lost in the Everglades. Part 5: The "sloughs" of despond. *Forest and Stream*, April, 210–212, 251, 254.

———. 1918e. Lost in the Everglades. Part 6: The aeroplane search. *Forest and Stream*, May, 276–278, 307–310.

———. 1918f. Lost in the Everglades. Part 7: When the supplies gave out. *Forest and Stream*, June, 340–343, 380.

———. 1918g. Lost in the Everglades. Part 8: Cypress trails. *Forest and Stream*, July, 402–404, 437–438.

———. 1918h. Lost in the Everglades. Part 9: Adventure's last lap. *Forest and Stream*, August, 466–469.

Larsen, L. G., J. W. Harvey, and J. P. Crimaldi. 2007. A delicate balance: Ecohydrological feedbacks governing landscape morphology in a lotic peatland. *Ecol Monogr* 77: 591–614.

———. 2009a. Morphological and transport properties of natural organic floc. *Water Resources Research* 45: W01410, doi:10.1029/2008WR006990.

———. 2009b. Predicting bed shear stress and its role in sediment dynamics and restoration potential of the Everglades and other vegetated flow systems. *Ecol. Eng.* 35: 1773–1785.

Larsen, L. G., J. W. Harvey, G. B. Noe, and J. P. Crimaldi. 2009c. Predicting organic floc transport dynamics in shallow aquatic ecosystems: Insights from the field, the laboratory, and numerical modeling. *Water Resources Research* 45: W01411, doi:10.1029/2008WR007221.

Layne, J. N. 1984. The land mammals of South Florida. In: *Environments of South Florida: Present and past II*, edited by P. Gleason, pp. 269–296. Miami Geological Society, Coral Gables, Fla.

Leach, S. D., H. Klein, and E. R. Hampton. 1972. *Hydrologic effects of water control and management of southeastern Florida.* Report of Investigations #60. U.S. Geological Survey, Tallahassee, Fla. 115 pp.

Leighty, R. G., M. H. Gallatin, J. L. Malcolm, and F. B. Smith. 1954. *[Map of] Soil associations of Dade County, Florida.* Cir. S-77. 30 × 36 inches. No scale. University of Florida, Agricultural Experiment Station, Gainesville, Fla.

Leonard, L., A. Croft, D. Childers, S. Mitchell-Bruker, H. Solo-Gabriele, and M. Ross. 2006. Characteristics of surface-water flows in the ridge and slough landscape of Everglades National Park: Implications for particulate transport. *Hydrobiologia* 569: 5–22.

Light, S. S., and W. J. Dineen. 1994. Water control in the Everglades: A historical perspective. In: *Everglades: The ecosystem and its restoration*, edited by S. M. Davis and J. C. Ogden, pp. 47–84. St. Lucie Press, Delray Beach, Fla.

Lilly, R., J. L. Cutler, D. S. Colby, F. M. Park, and other residents. 1890. [Petition by residents of the Caloosahatchee Valley to the Trustees.] In: *Minutes of the Trustees, Vol. IV (1904)*, pp. 87–89. Trustees of the Internal Improvement Fund, Tallahassee, Fla.

Lloyd, E. R. 1929. Report on the possibilities of agricultural development in the Everglades District of Florida [prepared for Col. Mark Brooke, Division Engineer of the Chief of Engineers; New Orleans, August 28, 1929]. Quoted in: Mayo (1940), pp. 51–79.

Lodge, T. E. 1994. *The Everglades handbook: Understanding the ecosystem.* St. Lucie Press, Delray Beach, Fla. 228 pp.

———. 2004. *The Everglades Handbook: Understanding the ecosystem.* 2nd. ed. CRC Press, Boca Raton, Fla. 286 pp.

Loftus, W. F., J. D. Chapman, and R. Conrow. 1990. Hydroperiod effects on Everglades marsh food webs, with relation to marsh restoration efforts. In: *Proc. 4th Triennial Conf. on Science in National Parks and Equivalent Reserves*, edited by G. Larson and M. Soukup, pp. 1–22. U.S. National Park Service and George Wright Society, Fort Collins, Colo.

Loftus, W. F., and A. Eklund. 1994. Long-term dynamics of an Everglades small-fish assemblage. In: *Everglades: The ecosystem and its restoration*, edited by S. M. Davis and J. C. Ogden, pp. 461–483. St. Lucie Press, Delray Beach, Fla.

Loftus, W. F., and J. A. Kushlan. 1987. Freshwater fishes of southern Florida. *Bull Fla State Mus Biol Sci* 31: 147–344.

Love, R. 1920. *Location map of a portion of Dade Co. Florida.* "Compiled from government and general survey plats and from the records of Dade Co." Gilbert Blueprinting, Miami.

Loveless, C. M. 1956. *Recommended program for Conservation Area 2.* Florida Game and Fresh Water Fish Commission, Tallahassee. [120 pp.]

———. 1959a. A study of the vegetation in the Florida Everglades. *Ecology* 40: 1–9.

———. 1959b. *The Everglades deer herd life history and management.* Tech. Bull. #6. Florida Game and Fresh Water Fish Commission, Tallahassee, Fla. 104 pp.

Loveless, C. M., G. W. Cornwell, R. L. Downing, A. R. Marshall, and J. N. Layne. 1970. *Report of the Special Study Team on the Florida Everglades.* Central and Southern Florida Flood Control District, West Palm Beach, Fla. 42 pp.

Lucas, S. J. E. 1917?a. [Map of the] A. W. Hopkins Property. Monroe and Dade Counties, Florida. Survey, Map and Cruise. Unpublished blueprint. Accompanies Lucas (1917b).

30 × 52 inches. 1:31,680. Jaudon Collection, box 18, Historical Museum of Southern Florida, Miami.

———. 1917?b. Report of a timber survey of the A. W. Hopkins tract, including descriptions of townships T54–55S R31–34E. Unpublished typescript. Jaudon Collection, box 18, Historical Museum of Southern Florida, Miami. 9 pp.

Lupfer, S. L. 1906. The drainage of the Florida Everglades. *Engineering News* 55: 373–376.

MacGonigle, J. N. 1896. The geography of the southern peninsula of the United States. *Nat Geog Mag* 7: 381–394.

MacKay, C. J., and L. J. E. Blake. 1839. *Map of the seat of war in Florida.* Compiled by order of Bvt. Brigr. Genl. Z. Taylor, principally from the surveys and reconnaissance of officers of the U.S. Army. 20 × 41 inches (approx.). 1:710,000 (approx.). Head Quarters, Army of the South, Tampa Bay, Fla.

MacKay, G. 1846. Handwritten letter to Robert Butler, Surveyor General of Florida, dated Lake Monroe, Feb. 15, 1846. Letters and Reports to Surveyor General, vol. 1, 1825–1847, p. 475. Transcribed by J. Knetsch, Feb. 1993, Florida Department of Environmental Protection, Division of State Lands, Bureau of Survey and Mapping, Tallahassee, Fla.

———. 1847. Extracts from letter of George MacKay, Esq., United States Surveyor, to Buckingham Smith, Esq.; dated December 6, 1847. Quoted in: U.S. Congress (1911), pp. 65–66.

Mallory, S. R. 1847. Memorandum from S. R. Mallory, Esq., Collector of Customs of the United States at Key West, to B. Smith, Esq.; dated September 1847. Quoted in: U.S. Congress (1911), pp. 62–63.

Marchman, W. P. 1947. The Ingraham Everglades exploring expedition, 1892. *Tequesta* 7: 3–43.

———. 1949. A dash through the Everglades. *Tequesta* 9: 13–41.

Marshall, A. R. 1980. *A critique of water management in South Florida.* Presented to South Florida Water Management District, Nov. 20–21, 1980. Florida International University, Everglades Digital Library, Miami. 4 pp.

Marston, A., S. H. McCrory, and G. B. Hills. 1927. *Report of Everglades Engineering Board of Review to Board of Commissioners of Everglades Drainage District.* May, 1927. Everglades Drainage District, Tallahassee, Fla. 86 pp.

Matlack, C. 1939. Claude Matlack Photographs. Series 50: Everglades. Historical Museum of Southern Florida, Miami.

Matson, G. C., F. G. Clapp, S. Sanford, and T. W. Vaughn. 1913a. *Geologic and topographic map of Florida.* "Plate 1 of Geology and ground waters of Florida. Water Supply Paper 319." 22 × 31 inches. 1:1,000,000. U.S. Geological Survey, Washington, D.C.

Matson, G. C., and S. Sanford. 1913b. *Geology and ground waters of Florida.* Water Supply Paper #319. U.S. Geological Survey, Washington, D.C. 445 pp.

Mayo, N. 1940. *Possibilities of the Everglades (Revised).* Bulletin #61 new series. Florida Dept. of Agriculture, Tallahassee, Fla. 91 pp.

McCall, G. A. 1868. *Letters from the frontiers.* J. B. Lippincott, Philadelphia; facsimile ed. 1974, University Press of Florida, Gainesville, Fla. 539 pp.

McCollum, S. H., V. W. Carlisle, and B. G. Volk. 1976. Historical and current classification of organic soils in the Florida Everglades. *Soil Crop Sci Soc Fl* 35: 173–177.

McCollum, S. H., O. E. Cruz, L. T. Stem, W. H. Wittstruck, R. D. Ford, and F. C. Watts. 1978. *Soil survey of Palm Beach County area, Florida.* U.S. Department of Agriculture, Soil Conservation Service, 96 pp.

McCormick, P. V., S. Newman, S. L. Miao, D. E. Gawlik, D. Marley, K. R. Reddy, and T. D. Fontaine. 2002. Effects of anthropogenic phosphorus inputs on the Everglades. In: *The Everglades, Florida Bay, and coral reefs of the Florida Keys: An ecosystem sourcebook*, edited by J. W. Porter and K. G. Porter, pp. 83–126. CRC Press, Boca Raton, Fla.

McCutcheon, R. J., J. E. Crane, S. W. Clarke, C. L. Leggett, H. C. Peeples, C. L. Knowles, and Wm. A. Russell. 1907. Report of the Joint Committee of the Florida Legislature for the year 1907, on the drainage of the Everglades. Quoted in: U.S. Congress (1911), pp. 109–113.

McIntosh, W. M. 1893. *Official report of the Board of Trustees of the Internal Improvement Fund to the Legislature of Florida, relative to the drainage operations of the Atlantic and Gulf Coast Canal and Okeechobee Land Company.* May 24, 1893. Trustees of the Internal Improvement Fund, Tallahassee, Fla. 15 pp.

———. 1895. [Letter to Hamilton Disston regarding closing of canal from Lake Okeechobee to the Caloosahatchee River, dated April 20, 1895.] In: *Minutes of the Trustees, Vol. IV (1904)*, pp. 313–314. Trustees of the Internal Improvement Fund, Tallahassee, Fla.

———. 1898. [Filing of petition of H. S. English requesting closure of canal between Lakes Okeechobee and Hicpochee.] In: *Minutes of the Trustees, Vol. IV (1904)*, p. 465. Trustees of the Internal Improvement Fund, Tallahassee, Fla.

McLaughlin, J. T. 1841. *Chart of the S. western extremity of Florida.* June 3, 1841. 10 × 13 inches. No scale.

McPherson, B. F. 1973a. *Vegetation in relation to water depth in Conservation Area 3, Florida.* Open File Report 73-025. U.S. Geological Survey, Tallahassee, Fla. 62 pp.

———. 1973b. *Vegetation map of southern parts of Subareas A and C, Big Cypress Swamp, Florida.* Included with Duever et al. 1996; originally published by the U.S. Geological Survey in 1973 under same title, Hydrologic Investigations Atlas HA-492. 31 × 28 in. 1;120,000 (approx.). National Audubon Society, New York.

McPherson, B. F., and R. Halley. 1996. *The south Florida environment: A region under stress.* Circular 1134. U.S. Geological Survey, Branch of Information Services, Denver, Colo. 61 pp.

McVoy, C. W. 2006. Unpublished synoptic field measurements of slough water depths in Water Conservation Areas 3A and 3B and Everglades National Park, measured in August 2005, December 2005 and March 2006. South Florida Water Management District, West Palm Beach, Fla.

McVoy, C. W. 2007. Unpublished measurements of relative sawgrass ridge and slough elevations (microtopography), two kilometer-long transects in Water Conservation Areas 3A and 3B, measured 2004 to 2007. South Florida Water Management District, West Palm Beach, Fla.

Meigs, J. L. 1879. Examination of Caloosahatchee River, Florida. In: *Annual report of the Chief of Engineers, 1879, Appendix J 17*, pp. 863–870. U.S. Army Corps of Engineers, Mobile, Ala.

Meindl, C. F. 1996. *Environmental perception and the historical geography of the great American wetland: Florida's Everglades, 1895 to 1930*. PhD diss., University of Florida, Gainesville, Fla. 257 pp.

Menge, C. 1947. *Early dredging in the Lake Okeechobee region*. Typescript report, collections of the Clewiston Museum, Clewiston, Fla. 12 pp.

Menge, J. F. 1894a. [Letter from Capt. J. F. Menge to Hamilton Disston, dated Sept. 23, 1894.] In: *Minutes of the Trustees, Vol. IV (1904)*, pp. 333–334. Trustees of the Internal Improvement Fund, Tallahassee, Fla.

———. 1894b. [Letter from Capt. J. F. Menge to Hamilton Disston, dated Oct. 7, 1894.] In: *Minutes of the Trustees, Vol. IV (1904)*, p. 334. Trustees of the Internal Improvement Fund, Tallahassee, Fla.

———. 1894c. [Letter from Capt. J. F. Menge to Hamilton Disston, dated Oct. 22, 1894.] In: *Minutes of the Trustees, Vol. IV (1904)*, p. 334. Trustees of the Internal Improvement Fund, Tallahassee, Fla.

———. 1894d. [Letter from Capt. J. F. Menge to Hamilton Disston, dated Nov. 2, 1894.] In: *Minutes of the Trustees, Vol. IV (1904)*, p. 335. Trustees of the Internal Improvement Fund, Tallahassee, Fla.

———. 1894e. [Letter from Capt. J. F. Menge to Hamilton Disston, dated Nov. 13, 1894.] In: *Minutes of the Trustees, Vol. IV (1904)*, p. 335. Trustees of the Internal Improvement Fund, Tallahassee, Fla.

———. 1907. [Interview with Capt. Fred Menge, Fort Myers, Fla., March 3, 1907.] Quoted in: Stewart (1907), pp. 40–41.

Miami Daily Metropolis. 1921. Walks over proposed Tamiami Trail route from Gulf to ocean: J. F. Jaudon and George Storter made notable jaunt to investigate road-building conditions. Aug. 6.

Miami Herald. 1921. See Tamiami Trail on 70-mile trudge: Jaudon reports highway will be passable by automobile with completion of twelve more miles. Aug. 6.

Miami Metropolis. 1905. Governor's party passed through Miami last night after having come down from Palm Beach in the yacht Linnet. Feb. 17.

———. 1907. A visit to the Everglades drainage operations with Governor Broward and investigating committee. Aug. 30.

Mickler, W. 1859. [Handwritten letter to Col. F. L. Dancy, Surveyor General, State of Florida, dated Lake Arbucel (Arbuckle), March 20, 1859.] Letters to the Surveyor General, vol. 3, 1857–1861, p. 63; transcribed by J. Knetsch, Florida Department of Environmental Protection, Division of State Lands, Bureau of Survey and Mapping, Tallahassee, Fla.

———. 1861. [Handwritten letter to Surveyor General, dated Tallahassee, Jan. 30, 1861.] Letters to the Surveyor General, vol. 3, 1857–1861, p. 136; transcribed by J. Knetsch, Florida Department of Environmental Protection, Division of State Lands, Bureau of Survey and Mapping, Tallahassee, Fla.

———. 1885. [Handwritten letter to Hon. William W. Hicks, U.S. Surveyor General, State of Florida, dated Tallahassee, July 1, 1885.] Drawer: "U.S. Deputy Surveyors: H-N" File: "U.S. Deputy Surveyor: William Mickler"; transcribed by J. Knetsch, Jan. 1993, Florida Department of Environmental Protection, Division of State Lands, Bureau of Survey and Mapping, Tallahassee, Fla.

Miller, C. F. 1918. Inorganic composition of a peat and of the plant from which it was formed. *J Agr Res* 13: 605–609.

Mitchell, A. J., and M. R. Ensign. 1928. *The climate of Florida*. Tech. Bull. #200. University of Florida, Agricultural Experiment Station, Gainesville, Fla. 207 pp.

Mitchell, R. T. 1932. *Map of area adjacent to North New River, South New River, and Dania Canals in Twps 50 & 51 S, Ranges 40, 41 & 42 E*. S.N.R. 2-1-11. 41 × 75 inches. 1:24,000. Everglades Drainage District, Office of Engineer, West Palm Beach, Fla.

Mitchell, R. T., Lee, Page, and W. D. Brockway. 1928. *North New River Canal plan and profile*. 17 Sheets. "EN-7." 24 × 48 inches. Along canal 1 in. = 400 ft.; perpendicular to canal 1 in. = 40 ft. Everglades Drainage District, Office of Chief Assistant Engineer, West Palm Beach, Fla.

Mitsch, W. J., and J. G. Gosselink. 1993. *Wetlands*. Van Nostrand Reinhold Company, New York. 722 pp.

Mock, Roos, and Associates. 1990. *[Map of] Lake Worth Drainage District: Control structures and maintenance elevations*. "P.A. No. 90–180. Sheet 3 of 8." 26 × 36 inches. 1:63,360. Mock, Roos and Associates, West Palm Beach, Fla.

Model Land Company. 1918. Receipt no. 9415, St. Augustine, Fla. June 11, 1918, regarding Contract no. 01561, covering N 1/2 of SE 1/4 of SW 1/4 of SW 1/4 in [Sec.] 3 [Twp] 55 [Rge] 40. Land Department, Model Land Company. Model Lands Collection, box 18, folder 467, University of Miami.

Moller, J. 1998. *Pers. comm., 24 Dec. 1998.*

Molnar, G. 1990. *Successional dynamics of a tropical hardwood hammock on the Miami Rockridge*. Master's thesis, Florida International University, Miami. 198 pp.

Morse, F. S. 1918. Typescript letter to Mr. J. E. Ingraham, President, Model Land Company, dated St. Augustine, Fla., July 8, 1918. Land Department, Model Land Company. Model Lands Collection, box 18, folder 467, University of Miami.

Moses, W. R. 1892. The journal of the Everglades exploring expedition, March 14 to April 16, 1892. Quoted in: Marchman (1947), pp. 4–40.

———. 1907. [Interview with W. R. Moses, Secy of Ingraham Expedition, 1892, West Palm Beach, Fla.] Quoted in: Stewart (1907), p. 43.

Motte, J. R. 1838a. [Journal of Assistant Surgeon Jacob Rhett Motte.] Quoted in: Sunderman (1950), pp. 26–33.

———. 1838b. *Journey into wilderness: An army surgeon's account of life in camp and field during the Creek and Seminole wars*

1836–1838. Edited by J. F. Sunderman. Repr. 1963, University of Florida Press, Gainesville, 326 pp.

Munroe, A. 1890? [Description of Arch Creek.] Quoted in: Parks (1977), p. 54.

Munroe, R. M., and V. Gilpin. 1930. *The commodore's story*. Ives Washburn, New York. 384 pp.

Neller, J. R. 1943. Nature and extent of the surface subsidence of the organic soils of the Everglades. Part II. Natural oxidation and crop requirements as factors in soil losses. *Soil Sci Soc Fl* 5-A: 120–123.

Newell, W. 1937. Annual Report [of the University of Florida Everglades Station]. Quoted in: Mayo (1940), pp. 7–47.

Newman, J. W. 1907. [Interview with J. Newman, drainage engineer for State of Florida.] Quoted in: Stewart (1907), p. 46.

Newman, S., J. B. Grace, and J. W. Koebel. 1996. Effects of nutrients and hydroperiod on *Typha, Cladium*, and *Eleocharis*: Implications for Everglades restoration. *Ecol Appl* 6: 774–783.

Niblack, S. L. 1882. [Communication dated December 18, 1882 regarding report on the boundary of the Okeechobee Drainage District. Lake City, Fla., June 27, 1882.] In: *Minutes of the Trustees, Vol. III (1904)*, pp. 190–195. Trustees of the Internal Improvement Fund, Tallahassee, Fla.

Noe, G. B., J. W. Harvey, and J. E. Saiers. 2007. Characterization of suspended particles in Everglades wetlands. *Limnology and Oceanography* 52: 1166–1178.

Ober, F. A. 1874. Lake Okeechobee. II. *Appletons' Journal* 12: 591–594.

———. 1887. *The Knockabout Club in the Everglades: The adventures of the club in exploring Lake Okeechobee*. Estes and Lauriat, Boston. 213 pp.

Ogden, J. C. 1994. A comparison of wading bird nesting colony dynamics (1931–1946 and 1974–1989) as an indication of ecosystem conditions in the southern Everglades. In: *Everglades: The ecosystem and its restoration*, edited by S. M. Davis and J. C. Ogden, pp. 533–570. St. Lucie Press, Delray Beach, Fla.

Okeechobee Flood Control District of Florida. 1930. *Caloosahatchee River and Lake Okeechobee Drainage Areas Florida. A Report to the US District Engineer, Jacksonville, Florida*. Senate Doc. no. 71–225. 71st Congress, 3rd Session, Washington, D.C. 96 pp.

Olmsted, I. C., and T. V. Armentano. 1997. *Vegetation of Shark Slough, Everglades National Park*. Rep. 97-001. U.S. National Park Service, South Florida Research Center, Homestead, Fla. 41 pp.

Olmsted, I. C., and L. L. Loope. 1984. Plant communities of Everglades National Park. In: *Environments of South Florida: Present and past II*, edited by P. Gleason, pp. 167–184. Miami Geological Society, Coral Gables, Fla.

Olmsted, I. C., L. L. Loope, and R. E. Rintz. 1980. *A survey and baseline analysis of aspects of the vegetation of Taylor Slough, Everglades National Park*. Report T-586. U.S. National Park Service, South Florida Research Center, Homestead, Fla. 71 pp.

Padgett, N. 1998. Conservation Area Number One: Testimony taken August 17, 1998 during hearings for renewal of lease of Water Conservation Area 1 as the Arthur R. Marshall Loxahatchee National Wildlife Refuge. Unpublished typescript. Files of the A. R. Marshall Loxahatchee National Wildlife Refuge, Boynton Beach, Fla. 10 pp.

Parker, G. G. 1974. Hydrology of the pre-drainage system of the Everglades in south Florida. In: *Environments of South Florida: Present and past*, edited by P. Gleason, pp. 18–27. Miami Geological Society, Coral Gables, Fla.

Parker, G. G., G. E. Ferguson, S. K. Love, and Others. 1955. *Water resources of southeastern Florida, with special reference to the geology and ground water of the Miami area*. Water Supply Paper #1255. U.S. Geological Survey, Washington, D.C. 965 pp.

Parker, G. G., and N. D. Hoy. 1943. Further studies of geological relationships affecting soil and water conservation and use in the Everglades. Part I: Additional notes on the geology and ground water of southern Florida. *Soil Sci Soc Fl* 5-A: 33–55.

Parker, Mrs. G. G. 1940. "Muck Fires," nos. A-90, A-91, and A-92 [photographs]. Private collection of Garald G. Parker. Files of C. McVoy, South Florida Water Management District, West Palm Beach, Fla.

Parks, A. M. 1977. *The forgotten frontier*. Banyan Books, Miami. 178 pp.

Peoples, M. D., and E. A. Davis. 1951. Across south central Florida in 1882: The account of the first New Orleans Times-Democrat Exploring Expedition. *Tequesta* 11: 63–92.

Pepper, F. J. 1912. Typescript letter to Mr. F. S. Morse, dated Apr. 27, 1912. Land Department, Model Land Company. Model Lands Collection, box 2, folder 58, University of Miami.

———. 1918. Typescript letter to Mr. J. E. Ingraham, President, Model Land Company, dated June 7, 1918. Land Department, Model Land Company. Model Lands Collection, box 18, folder 467, University of Miami.

Phillips, W. S. 1940. A tropical hammock on the Miami (Florida) limestone. *Ecology* 21: 166–175.

Pierce, C. W. 1962. The cruise of the Bonton. *Tequesta* 22: 3–63.

———. 1970. *Pioneer life in southeast Florida*. Edited by D. Curl. University of Miami Press, Coral Gables, Fla. 264 pp.

Potter, G. W. 1907. [Interview with George W. Potter, Pres. of Dade Co. Bank, West Palm Beach, Fla.] Quoted in: Stewart (1907), p. 44.

Powell, L. M. 1848. Letter from Commander L. M. Powell, United States Navy, to Hon. J. D. Westcott, Jr., United States Senate; dated March 1, 1848. Quoted in: U.S. Congress (1911), p. 60.

Preble, G. H. 1842. The diary of a canoe expedition into the Everglades in 1842. Quoted in: Anonymous (1946), pp. 33–51.

Purcell, A. D. 2001. Plumb lines, politics, and projections: The Florida Everglades and the Wright report controversy. *Fla Hist Q*. 80: 161–197.

Reiger, J. F. 1971. Sailing in South Florida waters in the early 1880s. Part I. *Tequesta* 31: 43–66.

Restrepo, J. I., A. M. Montoya, and J. Obeysekera. 1998. A wetland simulation module for the MODFLOW ground water model. *Ground Water* 36: 764–770.

Richards, R. W. 1903. Reminiscences of the days of Miami. Quoted in: Dieterich (1987), p. 55.

Richardson, J. R., W. L. Bryant, W. M. Kitchens, J. E. Mattson, and K. R. Pope. 1990. An evaluation of refuge habitats and relationships to water quality, quantity, and hydroperiod. Florida Cooperative Fish and Wildlife Research Unit Tech. Report no. 41. 168 pp.

Richmond, S. R. 1907. [Interview with S. R. Richmond, Agent for Florida Land Grant, Miami, Fla., March 16, 1907.] Quoted in: Stewart (1907), p. 45.

Roberts, K. L. 1926. *Florida*. Harper and Brothers Publishers, New York. 324 pp.

Roberts, W. A. 1907. [Interview with W. A. Roberts, timber estimator and land locator, Fort Myers, Fla., March 3, 1907.] Quoted in: Stewart (1907), pp. 39–40.

Robertson, W. B. 1953. *A survey of the effects of fire in Everglades National Park*. U.S. Department of Interior, National Park Service, Everglades National Park, Homestead, Fla. 169 pp.

———. 1954. Everglades fires—past, present and future. *Everglades Nat Hist* 2: 11–16.

———. 1962. Ornithology of the cruise of the Bonton. *Tequesta* 22: 65–77.

Rodgers, C. R. P. 1848. Letter from Lieut. C. R. P. Rodgers, United States Navy, to Hon. James D. Westcott, Jr., United States Senate, dated February 14, 1848. Quoted in: U.S. Congress (1911), pp. 60–61.

Romans, B. 1775. *A concise natural history of east and west Florida*. Printed for the author, New York. Repr. 1961, with an introduction by L. Richardson. Pelican Publishing, New Orleans, La. 316 pp.

Rose, P. W., and P. C. Rosendahl. 1983. Classification of Landsat data for hydrologic application, Everglades National Park. *Photogramm Eng and Remote Sens* 49: 505–511.

Rose, R. E. 1898. [Extract of report to the Florida East Coast Drainage and Sugar Company regarding proposed system of drainage. St. Augustine, Fla., Dec. 16, 1898.] In: *Minutes of the Trustees, Vol. IV (1904)*, pp. 454–457. Trustees of the Internal Improvement Fund, Tallahassee, Fla.

Rosenau, J. C., G. L. Faulkner, J. C. W. Hendry, and R. W. Hull. 1977. *Springs of Florida*. Geology Bulletin #31 (revised). U.S. Geological Survey, Tallahassee, Fla. 461 pp.

Rosendahl, P. C., and P. W. Rose. 1981. Freshwater flow rates and distribution within the Everglades marsh. In: *Proc. Nat. Symp. on Freshwater Inflow to Estuaries*, edited by R. D. Cross and D. L. Williams, pp. 385–401. U.S. Fish and Wildlife Service, Office of Biological Services, Washington, D.C.

Ross, M. S., D. L. Reed, J. P. Sah, P. L. Ruiz, and M. P. Lewin. 2003. Vegetation, environment relationships and water management in Shark Slough, Everglades National Park. *Wetl Ecol Manag* 11: 291–303.

Ross, M. S., P. Ruiz, J. Sah, S. Stoffella, N. Timilsina, and E. Hanan. 2006. *Marl Prairie/Slough Gradients: Patterns and trends in Shark Slough and adjacent marl prairies*. First Annual Report (2005). Department of Environmental Studies, Southeast Environmental Research Center, Florida International University, Miami. 16 pp.

Ross, M. S., J. Sah, P. Ruiz, D. Jones, H. Cooley, R. Travieso, F. Tobias, J. Snyder, and D. Hagyari. 2006. *Effect of hydrologic restoration on the habitat of the Caple Sable Seaside Sparrow*. Annual Report of 2004–2005—Third Annual Report to Everglades National Park. Department of Environtal Studies, Southeast Environmental Research Center, Florida International University, Miami. 46 pp.

Rutchey, K., L. Vilchek, and M. Love. 2005. *Development of a vegetation map for water conservation area 3*. Technical Publication ERA #421. South Florida Water Management District, West Palm Beach, Fla. 25 pp.

Rutchey, K., T. Schall, C. Coronado, M. Nungesser, J. Volin, D. Owen and F. Sklar. 2009. Chapter 6: Ecology of the Everglades Protection Area. Relative Marsh and Tree Island Elevation: Spatial Patterns in Water Conservation Area 3A and 3B. In: *South Florida Environmental Report*, pp. 6-87 to 6-90. South Florida Water Management District, West Palm Beach, Fla.

Sackett, J. W. 1888. *Report of J. W. Sackett, Assistant Engineer, U.S.A. on survey of Caloosahatchee River, Florida*. June 30, 1888. U.S. Army Corps of Engineers, Engineer Office, St. Augustine, Fla. 7 pp.

Said, W., and M. C. Brown. 2010. Natural system regional simulation model v3.3. Implementation, results, and evaluation report, September 2010 (draft). South Florida Water Management District, Hydrologic and Environmental Systems Modeling Dept., West Palm Beach, Fla.

Said, W., M. C. Brown, and T. E. Newton. 2006. Natural system regional simulation model v2.0. Implementation, results, and evaluation report, December 2006 (draft). South Florida Water Management District, Hydrologic and Environmental Systems Modeling Dept., West Palm Beach, Fla. 78 + 27 pp., App. A–G.

Schaffranek, R. W. 2004. *Sheet-flow velocities and factors affecting sheet-flow behavior of importance to restoration of the Florida Everglades*. Fact Sheet 2004-3123. U.S. Geological Survey, Reston, Va. 4 pp.

Schortemeyer, J. L. 1980. *An evaluation of water management practices for optimum wildlife benefits in Conservation Area 3A*. Florida Game and Fresh Water Fish Commission, Fort Lauderdale, Fla. 76 pp.

Schroeder, M. C., H. Klein, and N. D. Hoy. 1958. *Biscayne aquifer of Dade and Broward counties, Florida*. Report of Investigations #17. Florida Geological Survey, Tallahassee, Fla. 56 pp.

Schrontz, C. C. 1942. The principal characteristics of the Kissimmee–Lake Okeechobee Watershed. A: Lake Okeechobee. *Soil Sci Soc Fl* 4-A: 14–23.

Science Coordination Team. 2003. *The role of flow in the Everglades ridge and slough landscape*. Jan. 14, 2003. South Florida Ecosystem Restoration Working Group, Miami. 62 pp.

Scott, R. H. 1939. *Map of the Lake Worth Drainage District. Palm Beach County, Florida*. Ref. no. DM2. 1:31,680. Office of Chief Engineer, Lake Worth Drainage District.

Sellards, E. H. 1912. The soils and other surface residual materials of Florida. In: *Fourth Annual Report, 1910–1911*, pp. 7–86. Florida Geological Survey, Tallahassee, Fla.

Senarath, S., R. Novoa, J. Niedzialek, S. Nair, F. Khatun, J. Morshed, M. Hellinger, and N. Islam. 2008. Glades-LECSA RSM Model (Beta Version) Calibration and Verification. Hydrologic and Environmental Systems Modeling Dept., Project BZ844, South Florida Water Management District, West Palm Beach, Fla. 599 pp.

Senarath, S. U. S., R. J. Novoa, F. Khatun, S. Nair, T. Dessalegne, and V. R. Karri. 2010. Calibration and validation of the Glades–Lower East Coast Service Area hydrologic model. South Florida Water Management District, Hydrologic and Environmental Systems Modeling Dept., West Palm Beach, Fla.

Sewell, J. 1933. *John Sewell's memoirs and history of Miami*. Franklin Press, Miami. 236 pp.

SFWMD, Hydrologic Systems Modeling Division. 1998. *Natural System Model version 4.5 documentation (draft), Randy Van Zee, principal investigator*. February 1998. South Florida Water Management District, Planning Department, West Palm Beach, Fla. 49 pp.

Shaler, N. S. 1890. The topography of Florida. *Bull Mus Comp Zool* 16: 139–158.

Shands, J. F. 1907. [Interview with J. F. Shands, ex-County surveyor and superintendent of Lee County Schools, Fort Myers, Fla., March 2, 1907.] Quoted in: Stewart (1907), pp. 37–39.

Shih, S. F., B. Glaz, and R. E. Barnes. 1997. *Subsidence lines revisited in the Everglades Agricultural Area, 1997*. Tech. Bull. #902. University of Florida, Agricultural Experiment Station, Gainesville, Fla. 38 pp.

Shih, S. F., J. W. Mishoe, J. W. Jones, and D. L. Myhre. 1978. Modeling the subsidence of Everglades organic soil. *Trans Am Soc Agric Eng* 21: 1151–1156.

———. 1979b. Subsidence related to land use in Everglades agricultural area. *Trans Am Soc Agric Eng* 22: 560–563, 568.

Shih, S. F., E. H. Stewart, L. H. Allen, and J. W. Hilliard. 1979a. *An investigation of subsidence lines in 1978*. Belle Glade AREC Research Report EV-1979-1. University of Florida, Agricultural Research and Educational Center, Belle Glade, Fla. 28 pp.

———. 1979c. Variability of depth to bedrock in Everglades organic soil. *Soil Crop Sci Soc Fl* 38: 66–71.

Simmons, G., and L. Ogden. 1998. *Gladesmen: Gator hunters, moonshiners, and skiffers*. University Press of Florida, Gainesville, Fla. 197 pp.

Simpson, C. T. 1920. *In lower Florida wilds*. G. P. Putnam's Sons, New York. 404 pp.

———. 1930. [Letter to H. A. Kelly describing flammability of the peat soil after drainage, dated Nov. 10, 1930.] Quoted in: Kelly (1931), pp. 41–42.

Singletary, I. S. 1907. [Interview with I. S. Singletary, timber estimator and agent of U.S. Dept. of Agric.] Quoted in: Stewart (1907), pp. 4–16.

Sklar, F. H., L. Brandt, D. DeAngelis, C. Fitz, D. Gawlik, S. Krupa, C. Madden, F. Mazzotti, C. McVoy, S. Miao, et al. 2000. Chapter 2: Hydrologic needs—Effects of hydrology on the Everglades. In: *Everglades Consolidated Report*, pp. 2-1 to 2-62. South Florida Water Management District, West Palm Beach, Fla.

Sklar, F. H., M. Cook, E. Call, R. Shuford, M. Kobza, R. Johnson, S. Miao, M. Korvela, C. Coronado, L. Bauman, et al. 2006. Chapter 6: Ecology of the Everglades Protection Area. In: *South Florida Environmental Report*, pp. 6-1 to 6-65. South Florida Water Management District, West Palm Beach, Fla.

Sklar, F. H., M. J. Chimmey, S. Newman, P. V. McCormick, D. Gawlik, S. L. Miao, C. W. McVoy, S. W. Newman, C. Coronado, G. Crozier, M. Korvela, and K. Rutchey. 2005. The ecological-societal underpinnings of Everglades restoration. *Frontiers Ecol Environ* 3: 161–169.

Sklar, F. H., C. Coronado, G. Crozier, M. Darwish, B. Garrett, D. Gawlik, A. Huffman, M. Korvela, J. Leeds, C. J. Madden, et al. 2003. Chapter 6: Ecological effects of hydrology on the Everglades Protection Area. In: *Everglades Consolidated Report*, pp. 6-1 to 6-70. South Florida Water Management District, West Palm Beach, Fla.

Sklar, F. H., C. Coronado-Molina, A. Gras, K. Rutchey, D. Gawlik, and G. Crozier. 2004. Chapter 6: Ecological effects of hydrology. In: *Everglades Consolidated Report*, pp. 6-1 to 6-58. South Florida Water Management District, West Palm Beach, Fla.

Sklar, F. H., and A. van der Valk. 2002. *Tree islands of the Everglades*. Edited by F. H. Sklar and A. van der Valk. Kluwer Academic Publishers, Dordrecht, Netherlands. 541 pp.

Slattery, C. J. R. 1913. *Drainage map, Kissimmee and Caloosahatchee rivers and Lake Okeechobee, Florida*. 20 × 28 inches. 1:530,000 (approx.). Part of House Doc. no. 137, 63rd Congress, 1st Session, Washington, D.C.

Small, J. K. 1914. Exploration in the Everglades and on the Florida Keys. *J New York Bot Gard* 15: 69–79.

———. 1916. Royal Palm Hammock. *J New York Bot Gard* 17: 165–172.

———. 1917. "Water neck-deep in summer," Sm0533 [photograph, slough with hammock islands, southern Everglades, May 1917]. John K. Small Collection, State Archives of Florida,Tallahassee, Fla.

———. 1918. Narrative of a cruise to Lake Okeechobee. *Am Mus J* 18: 684–700.

———. 1921. Photo Sm1528 [photograph, slough with herons, ibis, and other birds, east of Royal Palm Hammock, December 1921]. J. K. Small Collection, State Archives of Florida, Tallahassee, Fla.

———. 1923. Green deserts and dead gardens: A record of exploration in Florida in spring of 1921. *J New York Bot Gard* 24: 193–247.

———. 1929a. The Everglades. *Scientific Monthly* 28: 80–87.

———. 1929b. *From Eden to Sahara: Florida's tragedy*. Science Press Printing, Lancaster, Pa. 123 pp.

———. 1934. An Everglades cypress swamp. *J New York Bot Gard* 34: 261–267.

Smith, B. 1848a. Report of Buckingham Smith, Esq., on his reconnaissance of the Everglades in 1848. Quoted in: U.S. Congress (1911), pp. 46–55.

———. 1848b. *Report of Buckingham Smith, Esq. to the Secretary of the Treasury, June 1, 1848 on Reconnaissance of the Everglades*. Senate Rep. Com. no. 242. Aug. 12, 1848. 30th Congress, 1st Session, Washington, D.C. 141 pp.

Smith, S. R. 1909. *Reclamation of the Florida Everglades*. Privately published, Miami. 32 pp.

Snyder, G. H. 1994. Soils of the EAA. In: *Everglades Agricultural Area (EAA): Water, soil, crop, and environmental management*, edited by A. B. Bottcher and F. T. Izuno, pp. 27–41. University Press of Florida, Gainesville, Fla.

Snyder, G. H., and J. A. Davidson. 1994. Everglades agriculture: Past, present, and future. In: *Everglades: The ecosystem and its restoration*, edited by S. M. Davis and J. C. Ogden, pp. 85–115. St. Lucie Press, Delray Beach, Fla.

South Florida Water Management District. 1991. *Color infrared aerial photography of the South Florida Water Management District*. 1:24,000. Reference no. 40 90-1293. West Palm Beach, Fla.

———. 2005. Documentation of the South Florida Water Management Model (v. 5.5). South Florida Water Management District and the Interagency Modeling Center, West Palm Beach, Fla. November 2005.

———. 2006. *Regional simulation model (RSM) hydrologic simulation engine (HSE) user's manual: RSM version 2.2.9*. South Florida Water Management District, Hydrologic and Environmental Systems Modeling Dept., West Palm Beach, Fla.

Southern Engineering & Construction Co. 1918. *[Profiles] to accompany report of the Plan of Reclamation of Southern Drainage District by the Southern Engineering & Construction Co.* June 1918. 12 × 83 inches. 1:15,840 (horiz.); 1:48 (vert.). Jaudon Collection, box 18, Historical Museum of Southern Florida, Miami.

Spackman, W., A. D. Cohen, P. H. Given, and D. J. Casagrande. 1976. *The comparative study of the Okefenokee Swamp and the Everglades-mangrove swamp-marsh complex of southern Florida: A field guidebook*. Pennsylvania State University, Coal Research Section, University Park, Pa. 403 pp.

Sprague, J. T. 1848. *The origin, progress, and conclusion of the Florida War; to which is appended a record of officers, non-commissioned officers, musicians, and privates of the U.S. Army, Navy, and Marine Corps, who were killed in battle or died of disease. As also the names of officers who were distinguished by brevets, and the names of others recommended. Together with orders for collecting the remains of the dead in Florida, and the ceremony of interment at St. Augustine, East Florida, on the fourteenth day of August, 1842*. D. Appleton and Company, New York. 555 pp.

St. Augustine News. 1841. Notes on the passage across the Everglades [December 4–15, 1840]. Quoted in: Anonymous (1960), pp. 57–65.

Steinberg, B. 1980. Vegetation of the Atlantic Coastal Ridge of Broward County, Florida, based on 1940 imagery. *Fla Sci* 43: 7–12.

Steinman, A. D., K. E. Havens, H. J. Carrick, and R. VanZee. 2002. The past, present, and future hydrology and ecology of Lake Okeechobee and its watersheds. In: *The Everglades, Florida Bay, and coral reefs of the Florida Keys: An ecosystem sourcebook*, edited by J. W. Porter and K. G. Porter, pp. 19–37. CRC Press, Boca Raton, Fla.

Stephens, A. R. 1943. Progress report on a topographic reconnaisance of the Everglades Drainage District. *Soil Sci Soc Fl* 5-A: 24–32.

Stephens, J. C. 1942. The principal characteristics of the Kissimmee–Lake Okeechobee Watershed. B: The Everglades. *Soil Sci Soc Fl* 4-A: 24–28.

———. 1943. Further studies of geological relationships affecting soil and water conservation and use in the Everglades. Part II: The relationship of sub-surface hydrology to water control and land use in the Everglades. *Soil Sci Soc Fl* 5-A: 56–94.

Stephens, J. C., and L. Johnson. 1951a. Subsidence of organic soils in the upper Everglades region of Florida. *Soil Sci Soc Fl* 11: 191–237.

———. 1951b. *Subsidence of peat soils in the Everglades region of Florida*. U.S. Department of Agriculture, Soil Conservation Service. 47 pp.

Stewart, J. T. 1907. *Report on Everglades Drainage Project in Lee and Dade counties, Florida*. U.S. Department of Agriculture, Office of Experimental Stations, Drainage Investigations, Washington, D.C. 109 pp.

Stewart, J. T., and L. Brett. 1907. Statement representing the characteristics of the course of the survey across the Everglades. Quoted in: Wright (1909c), pp. 158–159.

Stone, C. R. 1979. *Forty years in the Everglades*. Atlantic Publishing, Tabor City, N.C. 220 pp.

Stone, P. A., P. J. Gleason, and G. L. Chmura. 2002. Bayhead tree islands on deep peats of the northeastern Everglades. In: *Tree islands of the Everglades*, edited by F. H. Sklar and A. van der Valk, pp. 71–115. Kluwer Academic Publishers, Dordrecht, Netherlands.

Storter, R. 2000. *Crackers in the glade: Life and times in the old Everglades*. University of Georgia Press, Athens, Ga. 127 pp.

Stranahan, F. 1907. [Interview with Frank Stranahan, businessman, banker, and Fort Lauderdale's first postmaster, Fort Lauderdale, Fla.] Quoted in: Stewart (1907), p. 44.

Stuart, V. H. 1891. *Adventures amidst the equatorial forests and rivers of South America; also in the West Indies and the wilds of Florida, to which is added "Jamaica revisited."* John Murray, London. 268 pp.

Sturtevant, W. C. 1953. Chakaika and the "Spanish Indians": Documentary sources compared with Seminole tradition. *Tequesta* 13: 35–73.

———. 1956. A Seminole personal document. *Tequesta* 16: 55–75.

Sunderman, J. F. 1950. Army surgeon reports on lower east coast, 1838. *Tequesta* 10: 25–33.

Tamiami Trail Commissioners. 1928. *History of the Tamiami Trail, and a brief review of the road construction movement in Florida.* Tamiami Trail Commissioners and County Commissioners of Dade County, Florida, Miami. 27 pp.

Taylor, J. C. 1979. Railway location in the Florida Everglades. *Tequesta* 39: 5–16.

Tebeau, C. W. 1968. *Man in the Everglades: 2,000 years of human history in the Everglades National Park.* 2nd., revised ed. University of Miami Press, Coral Gables, Fla. 192 pp.

———. 1973. Past environment from historical sources (typescript). South Florida Environmental Project. U.S. Department of Interior, National Park Service. 58 pp.

Telis, P., and A. Higer. 2006. *The Everglades depth estimation network (EDEN) for support of ecological and biological assessments.* Fact Sheet 2006-3087. U.S. Geological Survey, Jacksonville, Fla. 2 pp.

Thomas, T. M. 1970. *A detailed analysis of climatological and hydrological records of south Florida with reference to man's influence upon ecosystem evolution.* Report to the National Park Service, project period June 1968–June 1970. Rosenstiel School of Marine and Atmospheric Science, University of Miami. 111 pp.

Tomlinson, P. B. 1980. *The biology of trees native to tropical Florida.* Harvard University Printing Office, Boston. 480 pp.

Tooth, S., and G. C. Nanson. 2000. The role of vegetation in the formation of anabranching channels in an ephemeral river, northern plains, arid central Australia. *Hydrological Processes* 14: 3099–3117.

Toth, L. A. 1987. *Effects of hydrologic regimes on lifetime production and nutrient dynamics of sawgrass.* Technical Publication 87-6. South Florida Water Management District, West Palm Beach, Fla. 31 pp.

Trexler, J. C., W. F. Loftus, F. Jordan, J. H. Chick, K. L. Kandl, T. C. McElroy, and O. L. Bass. 2002. Ecological scale and its implications for freshwater fishes of the Florida Everglades. In: *The Everglades, Florida Bay, and coral reefs of the Florida Keys: An ecosystem sourcebook,* edited by J. W. Porter and K. G. Porter, pp. 153–181. CRC Press, Boca Raton, Fla.

Trexler, J. C., W. F. Loftus, and S. Perry. 2005. Disturbance frequency and community structure in a twenty-five-year intervention study. *Oecologia* 145: 140–152.

Tropical Bioindustries. 1990. *Hydroperiod conditions of key environmental indicators of Everglades National Park and adjacent east Everglades area as guide to selection of an optimum water plan for Everglades National Park, Florida.* Final report in fulfillment of Contract #DACW 17-84-C-0031. [Written by Durbin Tabb.] Tropical Bioindustries, Miami. 74 pp.

Trustees of the Internal Improvement Fund. 1881. [Articles of agreement with Hamilton Disston for the reclamation of the overflowed lands in the valley of Lake Okeechobee and Kissimmee river. Feb. 26, 1881.] In: *Minutes of the Trustees, Vol. II (1904),* pp. 463–467. Trustees of the Internal Improvement Fund, Tallahassee, Fla.

———. 1888. [Minutes regarding contract with the Atlantic and Gulf Coast Canal and Okeechobee Land Company.] In: *Minutes of the Trustees, Vol. III (1904),* pp. 500–504. Trustees of the Internal Improvement Fund, Tallahassee, Fla.

———. 1908a. [Instructions to Mr. John Newman concerning dredging of the North New River Canal in the direction of Pine Island. Aug. 12, 1908.] In: *Minutes of the Trustees, Vol. VII (1909),* pp. 280–281. Trustees of the Internal Improvement Fund, Tallahassee, Fla.

———. 1908b. *Official map of canals: Drawn on map of a portion of the State of Florida, showing drainage district.* "Designated in contract with R. J. Bolles and adopted by Trustees of the Internal Improvement Fund, December 23, 1908." 12 × 23 inches. 1:120,000. Jaudon Collection, box 17, Historical Museum of Southern Florida, Miami.

———. 1921. *Map of Lake Okeechobee.* "Soundings refer to lake at elevation 16 ft. Datum = M.L.W. at Punta Rasa, Gulf of Mexico. M.L.W. at Stuart = +1.44 Punta Rasa. Surveyed 1913. Chart revised 1921." 12 × 12 inches. 1:225,280. Trustees of the Internal Improvement Fund, Office of the Chief Drainage Engineer, Tallahassee, Fla.

Turner, A. M., J. C. Trexler, C. F. Jordan, S. J. Slack, P. Geddes, J. H. Chick, and W. F. Loftus. 1999. Targeting ecosystem features for conservation: Standing crops in the Florida Everglades. *Conserv Biol* 14: 898–911.

U.S. Army Corps of Engineers. 1947. *[Map of] Flooded areas—1947.* Drawn on sheet no. 5 of *Central and south Florida comprehensive plan for flood control and other purposes: St. Johns and Kissimmee River valleys and the Everglades, in 21 sheets.* "To accompany report dated Dec. 19, 1947. File No. 3-15-14,599." "FO-SK-23." 15 × 20 inches. 1:980,000. U.S. Army Corps of Engineers, Jacksonville, Fla.

———. 1948. *Comprehensive report on Central and Southern Florida Project for Flood Control and other Purposes.* House Doc. No. 643. 80th Congress, 2nd Session, Washington, D.C. 60 pp.

———. 1954. *Central and Southern Florida Project for Flood Control and other Purposes. Part V: Coastal areas south of St. Lucie Canal, supplement 12, design memorandum, hydrology and hydraulic design, canals in greater Miami area (C-2 through C-9) (Revised), Mar. 23, 1954.* U.S. Army Corps of Engineers, Office of the District Engineer, Jacksonville, Fla. 157 pp.

———. 1957. *Central and Southern Florida Project for Flood Control and Other Purposes. Part I: Agricultural and conservation areas, supplement 25, general design memorandum, plan of regulation for Conservation Area No. 1, Nov. 29, 1957.* U.S. Army Corps of Engineers, Office of the District Engineer, Jacksonville, Fla. 57 pp.

———. 1960a. *[Map of] Central and southern Florida comprehensive plan (below Lake Okeechobee).* "Revised June 1960. File No. 400-25, 255-1.3," "F.C.D. File No. FO-24." 28 × 50 inches. 1:410,000 (approx.). U.S. Army Corps of Engineers, Jacksonville, Fla.

———. 1960b. *Central and Southern Florida Project for Flood Control and other Purposes. Part I: Agricultural and conservation areas, supplement 33, general design memorandum, Conservation Area No. 3, Jun. 22, 1960*. U.S. Army Corps of Engineers, Jacksonville, Fla.

———. 1978. Report on rule curves and key operating criteria. In: *Central and Southern Florida Project for Flood Control and Other Purposes. Master regulation manual*. Vol. 2, Appendix D, pp. 37A–37B. U.S. Army Corps of Engineers, Jacksonville, Fla.

U.S. Bureau of Topographical Engineers. 1846. *[Map of] the State of Florida*. Compiled in the Bureau of Topographical Engineers. 17 × 21 inches. U.S. Bureau of Topographical Engineers. Records of the Office of the Chief of Engineers, Record Group 77, Headquarters Map File, part of map L61. National Archives, Washington, D.C.

———. 1856. *Map of the State of Florida*. "Compiled in the Bureau of Topog'l Eng'rs, from the most recent authorities and prepared by order of the Honorable Jeff. Davis, Sec'y of War, in conformity with a resolution of the Senate of the 11th of Feb'y 1856." 19 × 34 inches. 1:720,000. U.S. Bureau of Topographical Engineers. Records of the Office of the Chief of Engineers, Record Group 77, Headquarters Map File, part of map L84, no. 1. National Archives, Washington, D.C.

U.S. Coast and Geodetic Survey. 1892. *Florida Reefs, from Key Biscayne to Carysfort Reef: Navigation chart 166*. 32 × 40 inches. 1:80,000. U.S. Coast and Geodetic Survey, Rockville, Md.

———. 1928a. *Compilation of aerial photographs. Florida: Cape Sable–Shark River–Oyster Bay*. "Sheet No. 4460. Photographs by U.S. Army Air Corps, Compilation by U.S. Coast and Geodetic Survey. Datum Approx. North American." 28 × 50 inches. 1:20,000. U.S. Coast and Geodetic Survey.

———. 1928b. *Compilation of Aerial Photographs. Florida: Shark River–Tarpon Bay*. "Sheet No. 4459. Photographs by U.S. Army Air Corps, Compilation by U.S. Coast and Geodetic Survey. March 29, 1928." 22 × 35 inches. 1:20,000. U.S. Coast and Geodetic Survey.

———. 1928c. *Compilation of Aerial Photographs. Florida, West Coast: Cape Sable–Shark River–Oyster Bay*. "Sheet No. 4460. Photographs by U.S. Army Air Corps. 11:40 AM March 29, 1928." 28 × 50 inches. 1:20,000. U.S. Coast and Geodetic Survey.

———. 1937a. *Topographic map no. T-5439. Florida: Snake Bight—Fox Lakes*. "This map, without contours, is a reproduction of the original drawing compiled from air photographs taken January 26, 1935, and supplemented by other surveys to April, 1937." 28 × 35 inches. 1:20,000. U.S. Coast and Geodetic Survey.

———. 1937b. *Topographic map no. T-5440. Florida: Madeira Bay–Snake Bight*. "This map, without contours, is a reproduction of the original drawing compiled from air photographs taken January 22 and 26, 1935, and supplemented by other surveys to April 14, 1937." 28 × 28 inches. 1:20,000. U.S. Coast and Geodetic Survey.

———. 1937c. *Topographic map no. T-5441. Florida: Little Madeira Bay*. "This map is a reproduction of the original drawing compiled from Air Photographs taken January 17 and 22, 1935." 26 × 30 inches. 1:20,000. U.S. Coast and Geodetic Survey.

U.S. Congress. 1911. *Everglades of Florida: Acts, reports, and other papers, state and national, relating to the Everglades of the State of Florida and their reclamation*. Senate Doc. No. 89. 62nd Congress, 1st Session, Washington, D.C. 208 pp.

———. 2000. *Comprehensive Everglades Restoration, Public Law No. 106–541, 114 Stat. 2754. Water Resources Development Act: Hearing before the subcommittee on transportation and infrastructure of the Committee on Environment and Public Works, United States Senate, May 23, 2000*. 160th Congress, 2nd Session, Washington, D.C.

U.S. Department of Agriculture. 1938. *Aerial photography index of Dade County, Symbol BUP*. Photographed Dec. 1938 by Aero Service Corp. 11 Sheets. 20 × 24 inches. 1:20,000 (negatives). U.S. Department of Agriculture, Agricultural Adjustment Administration, Washington, D.C.

———. 1940. *Aerial Photography, Everglades Area Florida*. "Photographed 1940 by Aero Service Corp., Philadelphia. Index compiled 6-5-40. Project AIS 20674." 20 × 24 inches (36 index sheets). 1:40,000 (negatives). U.S. Department of Agriculture, Soil Conservation Service, Washington, D.C.

———. 1970. *Detailed soil survey of Palm Beach County, Florida: A preliminary report*. U.S. Department of Agriculture, Soil Conservation Service, Gainesville, Fla. 121 pp.

———. 1980. *Black and white aerial photographs of Broward Co., Florida*. 20 × 24 inches. 1:40,000 (negatives). U.S. Department of Agriculture, Agricultural Stabilization and Conservation Service. Invitation no. ASCS–31–78 SLC, item 4. Flown by Park Aerial Surveys, Inc.

U.S. Department of Interior, Fish and Wildlife Service. 1972. A report on fish and wildlife resources affected by water regulation schedule for Loxahatchee National Wildlife Refuge (Conservation Area No. 1), Florida. Summary letter dated May 30, 1972. U.S. Department of Interior, Fish and Wildlife Service, Atlanta, Ga. 18 pp.

U.S. Geological Survey. 1964. *Aerial phototography, Everglades, Florida [South of Tamiami Trail]*. Uncontrolled mosaic. Approximate scale 1:48,000. 8 sheets, 24 × 30 inches. [Miami, Fla.]

van der Valk, A., and F. H. Sklar. 2002. What we know and should know about tree islands. In: *Tree islands of the Everglades*, edited by F. H. Sklar and A. Van der Valk, pp. 499–522. Kluwer Academic Publishers, Dordrecht, Netherlands.

Vignoles, C. B. 1823a. *Map of Florida*. Accompanies Vignoles (1823b). "Compiled and drawn from various actual surveys and observations." 15 × 17 inches. 1:2,900,000 (approx.). Bliss and White, New York; facsimile ed. 1977, University Presses of Florida, Gainesville.

———. 1823b. *Observations upon the Floridas.* Bliss and White, New York; facsimile ed. 1977, edited by J. Moore, University Presses of Florida, Gainesville. 153 pp.

Voss, G. L. 1970. [Endnotes for *Pioneer Life in Southeast Florida*, provided by Dr. Voss, Chair of Division of Biology, Rosenstiel School of Marine and Atmospheric Science, University of Miami.] Quoted in: Pierce (1970), pp. 256.

Waldin, W. 1912. *Truck farming in the Everglades.* Privately published, Miami. 139 pp.

Walker, R. J. 1847. Extracts from instructions to Buckingham Smith, Esq., by the Secretary of the Treasury; dated June 18, 1847. Quoted in: U.S. Congress (1911), pp. 37–38.

Wallace, H. E. 1955a. *Quarterly progress report for investigations project: Development of conservation areas.* Project No. W-43-D. 1 Oct. 1955. Florida Game and Fresh Water Fish Commission, Vero Beach, Fla.

———. 1955b. *Quarterly progress report for investigations project: Everglades impoundment investigation.* Project No. W-39-R. 1 Apr. 1955. Florida Game and Fresh Water Fish Commission, Vero Beach, Fla.

Wallace, H. E., W. H. Herke, N. F. Schlaack, F. J. Ligas, and C. M. Loveless. 1960. *Recommended program for Conservation Area 3.* Florida Game and Fresh Water Fish Commission, Vero Beach, Fla. 22 pp.

Wanless, H. R., Randall W. Parkinson, and Lenore P. Tedesco. 1994. Sea level control on stability of Everglades wetlands. In: *Everglades: The ecosystem and its restoration*, edited by S. M. Davis and J. C. Ogden, pp. 199–223. St. Lucie Press, Delray Beach, Fla.

Wende, R., and G. C. Nanson. 1998. Anabranching rivers: Ridge-form alluvial channels in tropical northern Australia. *Geomorphology* 22: 205–224.

Werner, H. W. 1975. The biology of the Cape Sable seaside sparrow. Unpublished report prepared for the U.S. Fish and Wildlife Service. Everglades National Park, Homestead, Fla. 215 pp.

West, T., J. R. Miller, J. W. Hatcher, J. H. B. Miller, and A. J. Peaden. 1909. Report of the special joint committee of the Florida Legislature for the year 1909, on the drainage of the Everglades. Quoted in: U.S. Congress (1911), pp. 121–138.

Wetzel, P. R. 2002. Analysis of tree island vegetation communities. In: *Tree islands of the Everglades*, edited by F. H. Sklar and A. van der Valk, pp. 357–389. Kluwer Academic Publishers, Dordrecht, Netherlands.

White, W. A. 1970. *The geomorphology of the Florida peninsula.* Geological Bulletin #51. Florida Department of Natural Resources, Division of Interior Resources, Bureau of Geology, Tallahassee, Fla. 164 pp.

Whitney, A. 1924. Letter to Thomas Will dated 28 September, 1924. Quoted in: Meindl (1996), p. 178.

Whitney, G. C. 1994. *From coastal wilderness to fruited plain: A history of environmental change in temperate North America from 1500 to the present.* Cambridge University Press, Oxford. 451 pp.

Wildeman, M. G. 1893. Aanteekeningen over joodsche geschlachten. *Algemeen Nederlandsch Familieblad, Tijdschrift voor geschiedenis, geschlacht-, Wapen-, Zegelkunde, enz.* X: 21.

Wiley, H. W. 1891. The muck lands of the Florida Peninsula. In: *Annual Report, Secretary of Agriculture*, pp. 163–171. U.S. Department of Agriculture, Bureau of Chemistry, Washington, D.C.

Will, L. E. 1964. *A cracker history of Okeechobee.* Great Outdoors Publishing, St. Petersburg, Fla.; repr. 1977, Glades Historical Society, Belle Glade, Fla. 317 pp.

———. 1965. *Okeechobee boats and skippers.* Great Outdoors Publishing, St. Petersburg, Fla. 165 pp.

Willard, D. A., C. E. Bernhardt, C. W. Holmes, B. Landacre, and M. Marot. 2006. Response of Everglades tree islands to environmental change. *Ecol Monogr* 76: 565–583.

Willard, D. A., and C. W. Holmes. 1997. *Pollen and geochronological data from South Florida: Taylor Creek Site 2.* Open File Report 97-35. U.S. Geological Survey, Reston, Va. 14 pp.

Willard, D. A., C. W. Holmes, M. S. Korvela, D. Mason, J. B. Murray, W. H. Orem, and D. T. Towles. 2002. Paleoecological insights on fixed tree island development in the Florida Everglades. I: Environmental Controls. In: *Tree islands of the Everglades*, edited by F. H. Sklar and A. van der Valk, pp. 117–151. Kluwer Academic Publishers, Dordrecht, Netherlands.

Willard, D. A., C. W. Holmes, and L. M. Weimer. 2001. The Florida Everglades ecosystem: Climatic and anthropogenic impacts over the last two millennia. *Bull Am Paleontol* 361: 41–56.

Willey, D. A. 1911. Reclaiming the Everglades: How the State of Florida is winning millions of acres from the water. *Cassier's Magazine* 39: 418–432.

Williams, A. P. 1882. Across south central Florida in 1882: The account of the first New Orleans Times-Democrat exploring expedition. Quoted in: Peoples and Davis (1951), pp. 63–92.

———. 1884. The Everglades [detailed report of the Times-Democrat expedition, November 4–December 5, 1883]. Quoted in: Wintringham (1964), pp. 65–93.

Williams, J. A., C. T. Culpepper, G. W. Ward, S. A. Robinson, and J. E. Cade. 1911. Report of the joint committee of the Florida Legislature for the year 1911, on the drainage of the Everglades, dated May 31. Quoted in: U.S. Congress (1911), pp. 193–199.

Williams, J. L. 1837. *The territory of Florida: Or sketches of the topography, civil and natural history, of the county, the climate, and the Indian tribes, from the first discovery to the present time.* A. T. Goodrich, New York; facsimile ed. 1962, University of Florida Press, Gainesville. 304 pp.

———. 1847. Extracts from manuscript of John Lee Williams, Esq. Quoted in: U.S. Congress (1911), p. 64.

Willoughby, H. L. 1898a. *Across the Everglades: A canoe journey of exploration*. J. B. Lippincott, Philadelphia; repr. 1992, Florida Classics Library, Port Salerno, Fla. 192 pp.

———. 1898b. *[Map of a] line through the Everglades*. Accompanies Willoughby (1898a). 15 × 10 inches. 1:56,000 (approx.). J. B. Lippincott, Philadelphia.

Wilson, W. 1907. [Interview with William Wilson, Supt. of Travers Orange Grove, Buckingham, Fla., Feb. 27, 1907.] Quoted in: Stewart (1907), pp. 35–36.

Winkler, M. G. 1995. *Progress report for Everglades National Park paleoecology study*. February 1995. Unpublished typescript, Madison, Wis. 25 pp. + appendices.

Wintringham, M. K. 1963. North to south through the Glades in 1883. *Tequesta* 23: 33–59.

———. 1964. North to south through the Everglades in 1883. Part II. *Tequesta* 24: 59–93.

Wood, J. M., and G. W. Tanner. 1990. Gramminoid community composition and structure within four Everglades management areas. *Wetlands* 10: 127–149.

Worth, D. F. 1988. *Environmental response of Water Conservation Area 2A to reduction in regulation schedule and marsh drawdown*. Technical Publication 88-2, DRE-250. South Florida Water Management District, West Palm Beach, Fla. 55 pp.

Wright, H. E., B. A. Coffin, and N. E. Aaseng. 1992. *The peatlands of Minnesota*. University of Minnesota Press, Minneapolis, Minn. 327 pp.

Wright, J. O. 1909a. Extract from a report on the drainage of the Everglades of Florida, by J. O. Wright, supervising drainage engineer, Feb. 25. Quoted in: U.S. Congress (1911), pp. 130–137.

———. 1909b. *Map showing proposed plan for draining the Florida Everglades*. "Prepared from investigations and surveys made in 1907–8. To accompany a report by J. O. Wright, Supervising Drainage Engineer. Drawn by G. F. Pohlers." 24 × 28 inches. 1:570,000. U.S. Department of Agriculture, Office of Experimental Stations, Drainage Investigations, Washington, D.C.

———. 1909c. [Report on the drainage of the Everglades of Florida, by J. O. Wright, supervising drainage engineer, June 25, 1909.] Quoted in: U.S. Congress (1911), pp. 140–180.

———. 1910. [Report of J. O. Wright, Chief Drainage Engineer, on trip of inspection of the drainage work of the Everglades, March 10, 1910.] In: *Minutes of the Trustees, Vol. VIII (1910)*, pp. 332–334. Trustees of the Internal Improvement Fund, Tallahassee, Fla.

———. 1911. [Report of J. O. Wright, Chief Drainage Engineer, relative to water courses on the east side of the Everglades, June 14, 1911.] In: *Minutes of the Trustees, Vol. IX (1913)*, pp. 153–155. Trustees of the Internal Improvement Fund, Tallahassee, Fla.

———. 1912. *The Everglades of Florida: Their adaptability for the growth of sugar cane*. Privately published, Tallahassee, Fla. 83 pp.

Writers Press Association. 1926. *Captain James F. Jaudon: Pioneer in development of Dade County and expert on good roads, Miami, Florida*. Press release J 2276, dated New York, November 2, 1926. Jaudon Collection, Historical Museum of Southern Florida, Miami. 1 p.

Wu, Y., N. Wang, and K. Rutchey. 2006. An analysis of spatial complexity of ridge and slough patterns in the Everglades ecosystem. *Ecological Complexity* 71: 1–10.

Wunderlin, R. P. 1982. *Guide to the vascular plants of central Florida*. University Presses of Florida, Gainesville, Fla. 472 pp.

Yourcenar, M. 1951. *Memoires d'Hadrien*. Librairie Plon, Paris. *Memoirs of Hadrian*. Translated by G. Frick and M. Yourcenar. 1986. Penguin Classics, London. 288 pp.

Zaffke, M. 1983. *Plant communities of Water Conservation Area 3A: Base-line documentation prior to the operation of S-339 and S-340*. Technical Memorandum, DRE-164. South Florida Water Management District, West Palm Beach, Fla. 31 + 4 + 9 + 6 pp.

Township Survey References

The field notes and plat maps produced by land surveyors working either for the U.S. General Land Office or for the State of Florida form an important part of the source material used in this study. Observations made by the surveyors are valuable for three reasons: (1) many surveys were made from the 1840s through the 1870s, prior to any drainage, or in the 1880s and 1910s, shortly after drainage; (2) the observations are well geo-referenced, usually within a few hundred feet (100 m); and (3) the surveyors often noted a number of relevant landscape characteristics, including vegetation, soils, water depths, landscape transitions, and widths of water bodies.

Survey field notes and plat maps are not formally published and can be difficult to obtain. Most U.S. surveys and a portion of the State surveys have recently been made available online at www.labins.org by the State of Florida, but some still remain hard to obtain. Prior to online availability, we made extensive use of the South Florida Water Management District's (SFWMD) file copies of township plat maps and survey notes. These notes rarely indicated the volume numbers of the survey field books from which they had been copied, being filed only by township and range. Original page numbers generally were visible on the copies. We have used www.labins.org to attempt to assign volume numbers but were not always successful. Additionally, scanned copies of survey notes made between approximately 1911 and 1918 during township surveys commissioned by the Trustees of the Internal Improvement Fund (TIIF), under the direction of F. C. Elliot, the Chief Drainage Engineer for the TIIF, were made available to us by Howard Ehmke of the SFWMD. We have referred to the two volumes of these as "FCEv01" and "FCEv02," following Elliot's page numbering where available. Table 1.3 in chapter 1 lists the principal surveys by name and areas surveyed; figure 1.4 shows all surveys consulted in this study.

The importance of these surveys to this study and the large number consulted (over 300) necessitated an unambiguous way to cite each survey individually, and in a way that quickly indicates to the reader both the location and date of the work. Surveys are cited in the text with the author's name, survey year, Township, and Range: MacKay 1845-T55 R40, for example. The list that follows is in the same format, ordered alphabetically by surveyor's name, then within a given surveyor by township, and then by range. When they could be identified, the field book volumes and page numbers have been included. Note that these are handwritten field books, often maintained under challenging conditions; as a result, page numberings were sometimes later revised, duplicate numbering systems were introduced, and/or half page numbers were used ("85.5").

Anonymous. 1909-T51 R41. Supplemental Plat of NW 1/4 of NE 1/4 of Sec. 1 for State Survey.

Apthorp, W. L. 1872-T43 R31. Field notes for U.S. Survey. Vol. 220: 274–279.

Apthorp, W. L. 1872-T43 R32. Field notes for U.S. Survey. Vol. 220: 305–308.

Apthorp, W. L. 1872-T44 R31. Field notes for U.S. Survey. Vol. 220: 267–273.

Apthorp, W. L. 1872-T44 R32. Field notes for U.S. Survey. Vol. 220: 298–304.

Apthorp, W. L. 1872-T44 R33. Field notes for U.S. Survey. Vol. 220: 293–295.

Apthorp, W. L. 1872-T45 R33. Field notes for U.S. Survey. Vol. 220: 285–287, 290–292.

Apthorp, W. L. 1872-T45 R34. Field notes for U.S. Survey. Vol. 220: 284–285.

Apthorp, W. L. 1872-T46 R33. Field notes for U.S. Survey. Vol. 220: 36–38, 282–283.

Apthorp, W. L. 1872-T46 R34. Field notes for U.S. Survey. Vol. 220: 38–39.

Apthorp, W. L. 1872-T47 R33. Field notes for U.S. Survey. Vol. 220: 217–221.

Apthorp, W. L. 1872-T47 R34. Field notes for U.S. Survey.

Broward, N. B. 1918-T47 R38. Plat map for State Survey.

Broward, N. B. 1918-T47 R39. Plat map for State Survey.

Broward, N. B. 1918-T48 R38. Plat map for State Survey.

Broward, N. B. 1918-T48 R39. Plat map for State Survey.
Broward, N. B. 1918-T49 R38. Plat map for State Survey.
Broward, N. B. 1918-T49 R39. Plat map for State Survey.
Broward, N. B. 1917-T49 R40. Plat map for State Survey.
Broward, N. B. 1918-T50 R38. Plat map for State Survey.
Broward, N. B. 1918-T50 R39. Plat map for State Survey.
Broward, N. B. 1918-T50 R40. Plat map for State Survey.
Broward, N. B. 1918-T51 R38. Plat map for State Survey.
Broward, N. B. 1918-T52 R38. Plat map for State Survey.
Broward, N. B. 1918-T52 R39. Plat map for State Survey.
Broward, N. B. 1918-T53 R38. Plat map for State Survey.
Broward, N. B. 1918-T53 R39. Plat map for State Survey.
DuVal, H. S. 1881-T51 R33. Field notes for U.S. Survey. Vol. 242: 147–161.
Elliot, F. C. 1911-T43 R32. Field notes and plat map for State Survey.
Elliot, F. C. 1912-T43 R33. Field notes and plat map for State Survey.
Elliot, F. C. 1912-T43 R34. Field notes for State Survey.
Elliot, F. C. 1912-T44 R33. Field notes for State Survey.
Elliot, F. C. 1911-T44 R34. Field notes for State Survey.
Elliot, F. C. 1911-T45 R33. Field notes for State Survey.
Elliot, F. C. 1912-T45 R34. Field notes for State Survey.
Elliot, F. C. 1911-T46 R33. Field notes for State Survey.
Elliot, F. C. 1911-T46 R34. Field notes for State Survey.
Elliot, F. C. 1911-T47 R34. Field notes for State Survey.
Ensey, R. F. 1911-T44 R41. Field notes for State Survey. Vol. FCEv01: 135–143.
Ensey, R. F. 1911-T44–50 R41. Index (only) to field notes for State Survey. Vol. FCEv01: 85.5.
Ensey, R. F. 1911-T45 R36. Field notes for State Survey. Vol. FCEv01: 173–174.
Ensey, R. F. 1911-T45 R37. Field notes for State Survey. Vol. FCEv01: 172–173.
Ensey, R. F. 1911-T45 R38. Field notes for State Survey. Vol. FCEv01: 171–172.
Ensey, R. F. 1911-T45 R39. Field notes for State Survey. Vol. FCEv01: 166–170.
Ensey, R. F. 1911-T45 R41. Field notes and plat map for State Survey. Vol. FCEv01: 144–154.
Ensey, R. F. 1911-T45–46 R36–40. Field notes for State Survey. Vol. FCEv01: 161.5–176.
Ensey, R. F. 1911-T46 R40. Field notes for State Survey. Vol. FCEv01: 161.5–166.
Ensey, R. F. 1911-T46 R41. Field notes and plat map for State Survey. Vol. FCEv01: 155–161.
Ensey, R. F. 1911-T47 R41. Field notes and plat map for State Survey. Vol. FCEv01: 122–134.
Ensey, R. F. 1911-T48 R41. Field notes and plat map for State Survey. Vol. FCEv01: 118–127.
Ensey, R. F. 1911-T49 R41. Field notes and plat map for State Survey. Vol. FCEv01: 109–117.
Ensey, R. F. 1911-T50 R41. Field notes and plat map for State Survey. Vol. FCEv01: 86–90, 100–108.
Ensey, R. F. 1912-T51 R41. Field notes for State Survey. Vol. FCEv01: 80–85, 91–99.
Ensey, R. F. 1912-T52 R41. Field notes and plat map for State Survey. Vol. FCEv01: 68–79.
Ensey, R. F. 1912-T52 R42. Field notes for State Survey. Vol. FCEv01: 71–72.
Ensey, R. F. 1912-T54 R40. Field notes for State Survey. Vol. FCEv01: 61.5–67.
Franklin, L. D. 1914-T43 R35. Field notes and plat map for State Survey. Vol. FCEv01: 0.5–13.
Franklin, L. D. 1914-T44 R35. Field notes and plat map for State Survey. Vol. FCEv01: 1e–2e.
Franklin, L. D. 1912-T45 R41. Field notes and plat map for State Survey. Vol. FCEv01: 177–192.
Franklin, L. D. 1912-T46 R41. Field notes and plat map for State Survey. Vol. FCEv01: 193–214.
Franklin, L. D. 1912-T47 R41. Field notes and plat map for State Survey. Vol. FCEv01: 214.5–240.
Franklin, L. D. 1912-T48 R41. Field notes and plat map for State Survey. Vol. FCEv01: 240.5–266.
Franklin, L. D. 1912-T49 R41. Field notes and plat map for State Survey. Vol. FCEv01: 267–297.
Franklin, L. D. 1914-T50 R38. Field notes and plat map for State Survey. Vol. FCEv01: 1f–5f.
Franklin, L. D. 1914-T50 R39. Field notes and plat map for State Survey. Vol. FCEv01: 1a–6a.
Franklin, L. D. 1914-T50 R40. Field notes and plat map for State Survey. Vol. FCEv01: 1b–17b.
Franklin, L. D. 1914-T51 R39. Field notes and plat map for State Survey. Vol. FCEv01: 1c–17c.
Frederick, E. S. 1903-T49 R42. Field notes and plat map for U.S. Survey.
Frederick, J. S. 1901-T43 R43. Field notes and plat map for U.S. Survey. Vol. 255: 145–155.
Frederick, J. S. 1901-T44 R43. Field notes and plat map for U.S. Survey. Vol. 255: 156–171.
Frederick, T. E. 1914-T44 R35. Field notes for State Survey. Vol. W: ?.
Frederick, T. E. 1914-T44 R36. Field notes for State Survey. Vol. Y: ?.
Frederick, T. E. 1914-T45 R35. Field notes for State Survey. Vol. X: ?.
Frederick, T. E. 1914-T45 R36. Field notes for State Survey. Vol. V: ?.
Frederick, T. E. 1914-T52 R39. Field notes (partial) for State Survey. Vol. BB: ?.

Frederick, T. E. 1912-T52 R40. Field notes and plat map for State Survey. Vol. AA: ?.

Frederick, T. E. 1912-T53 R40. Field notes and plat map for State Survey. Vol. FCEv01: 14–61.

Fries, J. O. 1898-T50 R40. Field notes and plat map for partial U.S. Survey. Vol. 251: 54a–64a.

Fries, J. O. 1898-T50 R41. Field notes and plat map for partial U.S. Survey. Vol. 251: 16A–21A, 24A, 29A–54A.

Fries, J. O. 1898-T51 R41. Field notes and plat map for partial U.S. Survey. Vol. 251: 22A–23A, 25A–28A.

Fries, J. O. 1898-T51 R42. Field notes for partial U.S. Survey. Vol. 251: 1A–16A.

Hamblen, S. G. 1873-T43 R31. Field notes and plat map for U.S. Survey. Vol. 221: 300–327.

Hamblen, S. G. 1873-T43 R32. Field notes for U.S. Survey. Vol. 221: 453–475.

Hamblen, S. G. 1873-T44 R31. Field notes and plat map for U.S. Survey. Vol. 221: 333–392.

Hamblen, S. G. 1873-T44 R32. Field notes and plat map for U.S. Survey. Vol. 221: 481–535.

Hamblen, S. G. 1873-T44 R33. Field notes and plat map for U.S. Survey. Vol. 221: 638–658.

Hamblen, S. G. 1873-T45 R33. Field notes and plat map for U.S. Survey. Vol. 221: 665–718.

Hamblen, S. G. 1873-T46 R33. Field notes and plat maps for U.S. Survey. Vol. 221: 719–767.

Hancock, J. T. 1907-T38 R35. Field notes for U.S. Survey. Vol. 258: 366–425 (1–60). [Notes describe Lake Okeechobee shoreline meanders from T40 R32 to T38 R35.]

Hancock, J. T. 1907-T40 R32. Field notes and plat map for U.S. Survey. Vol. 258: 390–397 (25–32).

Hancock, J. T. 1907-T40 R33. Field notes and plat map for U.S. Survey. Vol. 258: 398–400 (33–35).

Hancock, J. T. 1907-T41 R32. Field notes for U.S. Survey.

Hardin, O. A. 1916-T42 R36. Field notes and plat map for State Survey. Vol. U: ?.

Hardin, O. A. 1916-T42 R37. Field notes and plat map for State Survey. Vol. D: ?.

Hardin, O. A. 1916-T42 R38. Field notes and plat map for State Survey. Vol. E: 4E–5E, 11E–14E; FCEv02: 421–436.

Hardin, O. A. 1916-T42 R39. Field notes and plat map for State Survey. Vol. R: ?.

Hardin, O. A. 1916-T43 R34. Field notes and plat map for State Survey. Vol. M: ?.

Hardin, O. A. 1916-T43 R36. Field notes and plat map for State Survey. Vol. A: ?.

Hardin, O. A. 1916-T43 R37. Field notes and plat map for State Survey. Vol. B/C/E/F/Z: ?.

Hardin, O. A. 1916-T43 R38. Field notes and plat map for State Survey. Vol. E: ?; FCEv02: 457–473.

Hardin, O. A. 1916-T43 R39. Field notes and plat map for State Survey. Vol. E: 6E, 9E; O: 1–2; FCEv02: 437–456.

Hardin, O. A. 1916-T44 R33. Field notes and plat map for State Survey. Vol. P: ?.

Hardin, O. A. 1916-T44 R34. Field notes for State Survey. Vol. T: ?.

Hardin, O. A. 1915-T44 R35. Field notes and plat map for State Survey. Vol. N: ?.

Hardin, O. A. 1915-T44 R36. Field notes and plat map for State Survey. Vol. N: ?.

Hardin, O. A. 1916-T45 R34. Field notes and plat map for State Survey. Vol. I: ?.

Hardin, O. A. 1915-T45 R35. Field notes for State Survey. Vol. L: ?.

Hardin, O. A. 1915-T45 R36. Field notes for State Survey. Vol. K: ?.

Hardin, O. A. 1916-T46 R34. Field notes and plat map for State Survey. Vol. G: ?.

Hardin, O. A. 1915-T46 R35. Field notes for State Survey. Vol. J: ?.

Hardin, O. A. 1915-T46 R36. Field notes for State Survey. Vol. I: ?.

Hardin, O. A. 1916-T47 R34. Field notes and plat map for State Survey. Vol. Q: ?.

Hardin, O. A. 1915-T47 R35. Field notes for State Survey. Vol. S: ?.

Hardin, O. A. 1918-T53 R37. Field notes and plat map for State Survey. Vol. FCEv02: 274–278.

Hopkins, C. F. 1853-T36 R37. Field notes and plat map for U.S. Survey.

Hopkins, C. F. 1884-T51 R33. Field notes and plat map for U.S. Survey. Vol. 246: 169–249.

Hopkins, C. F. 1884-T51 R34. Field notes and plat map for U.S. Survey. Vol. 246: 254–319.

Hopkins, C. F. 1884-T52 R34. Field notes and plat map for U.S. Survey. Vol. 246: 93–163.

Horne, C. R. 1914-T44 R37. Field notes and plat map for State Survey. Vol. FCEv02: 306–308, 311–334, 392–406.

Horne, C. R. 1914-T44 R38. Field notes and plat map for State Survey. Vol. FCEv02: 303–305, 335–338, 380–391.

Horne, C. R. 1914-T44 R39. Field notes and plat map for State Survey. Vol. FCEv02: 300–302, 339–347, 369–371.

Horne, C. R. 1914-T44 R40. Field notes and plat map for State Survey. Vol. FCEv02: 284–288, 296–299, 342–347.

Horne, C. R. 1915-T45 R37. Field notes and plat map for State Survey. Vol. FCEv02: 358–364, 377–379, 418–420.

Horne, C. R. 1915-T45 R38. Field notes and plat map for State Survey. Vol. FCEv02: 309–330, 358–361, 373–376, 415–417.

Horne, C. R. 1915-T45 R39. Field notes and plat map for State Survey. Vol. FCEv02: 348–357, 369–372, 412–414.

Horne, C. R. 1915-T45 R40. Field notes and plat map for State Survey. Vol. FCEv02: 290–295, 348–353, 365–368, 407–411.

Houston, G. 1845-T40 R40. Field notes for U.S. Survey. Vol. 80: 525–530.

Houston, G. 1845-T40 R41. Field notes and plat map for U.S. Survey. Vol. 80: 583–588; 128: 1–60.

Houston, G. 1845-T40 R42. Field notes for U.S. Survey. Vol. 80: 616–620; 85: 140, 145.

Houston, G. 1845-T40 R43. Field notes for U.S. Survey. Vol. 85: 141–142.

Houston, G. 1845-T41 R43. Field notes for U.S. Survey. Vol. 85: 143.

Jackson, J. 1859-T40 R31. Field notes for U.S. Survey. Vol. 198: 523–540.

Jackson, J. 1860-T40 R31. Field notes for U.S. Survey. Vol. 211: 215–225.

Jackson, J. 1860-T40 R32. Field notes for U.S. Survey. Vol. 211: 213–214.

Jackson, J. 1860-T41 R31. Field notes for U.S. Survey. Vol. 211: 194–205.

Jackson, J. 1860-T41 R32. Field notes for U.S. Survey. Vol. 211: 206–212.

Jackson, J. 1860-T42 R29. Field notes for U.S. Survey. Vol. 211: 254–265.

Jackson, J. 1860-T42 R30. Field notes for U.S. Survey. Vol. 211: 227–232, 248–253.

Jackson, J. 1860-T42 R31. Field notes for U.S. Survey.

Jackson, J. 1860-T42 R32. Field notes for U.S. Survey.

Jackson, J. 1860-T43 R31. Field notes for U.S. Survey. Vol. 211: 233–241.

Jackson, J. 1847-T55 R39. Field notes and plat map for U.S. Survey. Vol. 79: 287–346; 90: 15–26.

Jackson, J. 1847-T55 R40. Field notes and plat map for U.S. Survey. Vol. 79: 493–552.

Jackson, J. 1847-T55 R41. Field notes and plat map for U.S. Survey. Vol. 79: 623–629.

Jackson, J. 1847-T55 R42. Field notes and plat map for U.S. Survey. Vol. 79: ?.

Jackson, J. 1847-T56 R38. Field notes and plat map for U.S. Survey. Vol. 79: 107–165; 90: 33–43.

Jackson, J. 1847-T56 R39. Field notes and plat map for U.S. Survey. Vol. 79: 357–416; 90: 3–14.

Jackson, J. 1847-T56 R40. Field notes and plat map for U.S. Survey. Vol. 79: 563–608; 90: 27–32, 57–59.

Jackson, J. 1847-T57 R38. Field notes and plat map for U.S. Survey. Vol. 79: 171–231; 90: 50–55, 60–65.

Jackson, J. 1847-T57 R39. Field notes and plat map for U.S. Survey. Vol. 79: 421–478; 90: 44–49, 71–74.

Jackson, J. 1847-T58 R38. Field notes and plat map for U.S. Survey. Vol. 79: 237–273; 90: 66–70.

Jones, A. H. 1845-T40 R40. Field notes for U.S. Survey. Vol. 75: 505–510.

Jones, A. H. 1845-T41 R37. Field notes for U.S. Survey. Vol. 75: 532.

Jones, A. H. 1845-T41 R38. Field notes for U.S. Survey. Vol. 75: 529–531.

Jones, A. H. 1845-T41 R39. Field notes for U.S. Survey. Vol. 75: 525–528, 534–536.

Jones, A. H. 1845-T41 R40. Field notes and plat map for U.S. Survey. Vol. 75: 512–514, 538–542; 111: 298–357.

Jones, A. H. 1845-T41 R41. Field notes and plat map for U.S. Survey. Vol. 75: 515–517, 549–552; 111: 378–437.

Jones, A. H. 1845-T41 R42. Field notes and plat map for U.S. Survey. Vol. 75: 518–522; 111: 594–656.

Jones, A. H. 1845-T41 R43. Field notes for U.S. Survey. Vol. 75: 585–588.

Jones, A. H. 1845-T42 R39. Field notes for U.S. Survey. Vol. 75: 556–558.

Jones, A. H. 1845-T42 R40. Field notes for U.S. Survey. Vol. 75: 543–545, 553–555.

Jones, A. H. 1845-T42 R41. Field notes and plat map for U.S. Survey. Vol. 75: 546–562; 111: 443–503.

Jones, A. H. 1845-T42 R42. Field notes for U.S. Survey. Vol. 75: 589–592.

Jones, A. H. 1845-T42 R43. Field notes and plat map for U.S. Survey. Vol. 75: 593–595, 609–612.

Jones, A. H. 1845-T43 R40. Field notes for U.S. Survey. Vol. 75: 365–568.

Jones, A. H. 1845-T43 R41. Field notes and plat map for U.S. Survey. Vol. 75: 563–565, 569–571; 111: 525–584.

Jones, A. H. 1845-T43 R42. Field notes for U.S. Survey. Vol. 75: 596–598.

Jones, A. H. 1845-T43 R43. Field notes for U.S. Survey. Vol. 75: 599–601, 614.

Jones, A. H. 1845-T44 R40. Field notes for U.S. Survey. Vol. 75: 575.

Jones, A. H. 1845-T44 R41. Field notes for U.S. Survey. Vol. 75: 572–574, 576–578.

Jones, A. H. 1845-T44 R42. Field notes for U.S. Survey. Vol. 75: 579–581.

Jones, A. H. 1845-T44 R43. Field notes for U.S. Survey. Vol. 75: 582–584, 602–604.

Kackley, W. J. 1912-T51 R40. Field notes and plat map for State Survey. Vol. BB: ?.

Kackley, W. J. 1912-T51 R41. Field notes and plat map for State Survey. Vol. BB: ?.

King, J. W. 1917-T54 R39. Field notes and plat map for Private Survey.

King, J. W. 1917-T54 R40. Field notes and plat map for Private Survey.

King, J. W. 1917-T54 R41. Field notes and plat map for Private Survey.

King, J. W. 1917-T55 R39. Field notes and plat map for Private Survey.

King, J. W. 1917-T55 R40. Field notes and plat map for Private Survey.

King, J. W. 1917-T55 R41. Field notes and plat map for Private Survey.

Livingston, R. W. and W. E. Heister. 1919-T41 R37. Field notes for U.S. Resurvey of North Boundary of Secs. 1 and 2 and South and East Boundarys of Sec. 33. Vol. 263(?): 001–012. May 1919.

MacKay, G. 1845-T41 R42. Field notes for U.S. Survey. Vol. 84: 220–225.

MacKay, G. 1845-T42 R42. Field notes for U.S. Survey. Vol. 84: 226–231.

MacKay, G. 1845-T43 R32. Field notes for U.S. Survey. Vol. 84: 232–237.

MacKay, G. 1845-T44 R41. Field notes for U.S. Survey. Vol. 84: 244–247.

MacKay, G. 1845-T44 R42. Field notes for U.S. Survey. Vol. 84: 238–243, 248–253.

MacKay, G. 1845-T44 R43. Field notes for U.S. Survey. Vol. 84: 254–258.

MacKay, G. 1845-T45 R42. Field notes for U.S. Survey. Vol. 84: 259–264.

MacKay, G. 1845-T45 R43. Field notes for U.S. Survey. Vol. 84: 344–353.

MacKay, G. 1845-T46 R41. Field notes for U.S. Survey. Vol. 84: 265–266.

MacKay, G. 1845-T46 R42. Field notes and plat map for U.S. Survey. Vol. 70: 3–62; 84: 267–272, 332–343.

MacKay, G. 1845-T46 R43. Field notes for U.S. Survey. Vol. 84: 329–331.

MacKay, G. 1845-T47 R41. Field notes for U.S. Survey. Vol. 84: 273–274.

MacKay, G. 1845-T47 R42. Field notes and plat map for U.S. Survey. Vol. 70: 65–124; 84: 275–280, 317–328.

MacKay, G. 1845-T47 R43. Field notes for U.S. Survey. Vol. 84: 315–316.

MacKay, G. 1845-T48 R41. Field notes for U.S. Survey. Vol. 84: 281–283.

MacKay, G. 1845-T48 R42. Field notes for U.S. Survey. Vol. 84: 284–289, 303–314.

MacKay, G. 1845-T49 R42. Field notes for U.S. Survey. Vol. 84: 290–295.

MacKay, G. 1845-T49 R43. Field notes for U.S. Survey. Vol. 84: 296–302.

MacKay, G. 1845-T50 R42. Field notes for U.S. Survey. Vol. 84: 001–015; 85: 286–289.

MacKay, G. 1845-T51 R42. Field notes for U.S. Survey. Vol. 84: 16–27.

MacKay, G. 1845-T52 R42. Field notes for U.S. Survey. Vol. 84: 28–42.

MacKay, G. 1845-T53 R41. Field notes and plat map for U.S. Survey. Vol. 70: 134–194; 84: 43–54.

MacKay, G. 1845-T53 R42. Field notes for U.S. Survey. Vol. 84: 39–42, 66–71.

MacKay, G. 1846-T54 R40. Field notes and plat map for U.S. Survey. Vol. 70: 238–264; 84: 61–65.

MacKay, G. 1845-T54 R41. Field notes and plat map for U.S. Survey. Vol. 70: 97–235; 84: 55–60, 75–83, 118; 85: 264–265.

MacKay, G. 1845-T55 R40. Field notes for U.S. Survey. Vol. 84: 88–99, 110–114.

MacKay, G. 1845-T55 R41. Field notes for U.S. Survey. Vol. 84: 84–87.

MacKay, G. 1845-T55 R42. Field notes and plat map for U.S. Survey. Vol. 85: ?.

MacKay, G. 1845-T56 R38. Field notes for U.S. Survey. Vol. 84: 106–109.

MacKay, G. 1845-T56 R39. Field notes for U.S. Survey. Vol. 84: 100–105.

Merriam, L. B. 1918-T41 R32. Field notes and plat map for State Survey. Vol. FCEv02: 16–23.

Merriam, L. B. 1918-T42 R32. Field notes and plat map for State Survey. Vol. FCEv02: 24–26.

Merriam, L. B. 1918-T42 R33. Field notes and plat map for State Survey. Vol. FCEv02: 85–91.

Merriam, L. B. 1918-T42 R34. Field notes and plat map for State Survey. Vol. FCEv02: 138–140.

Merriam, L. B. 1918-T43 R34. Field notes and plat map for State Survey. Vol. FCEv02: 141.

Mickler, W. 1885-T52 R35. Field notes for U.S. Survey. Vol. 245: 288–293.

Mickler, W. 1886-T52 R36. Field notes for U.S. Survey. Vol. 245: 367–373, 376–380.

Mickler, W. 1885-T53 R34. Field notes for U.S. Survey. Vol. 245: 252–259.

Mickler, W. 1885-T53 R35. Field notes for U.S. Survey. Vol. 245: 193–202, 267–273, 277–285.

Mickler, W. 1886-T53 R36. Field notes for U.S. Survey. Vol. 245: 347–354, 357–363.

Mickler, W. 1885-T54 R33. Field notes for U.S. Survey. Vol. 245: 17–26, 112–120.

Mickler, W. 1885-T54 R34. Field notes for U.S. Survey. Vol. 245: 229–236, 239–247.

Mickler, W. 1885-T54 R35. Field notes for U.S. Survey. Vol. 245: 187–193, 195–201.

Mickler, W. 1885-T54 R36. Field notes for U.S. Survey. Vol. 245: 327–333, 336–343.

Mickler, W. 1885-T55 R32. Field notes for U.S. Survey. Vol. 245: 80, 135–144, 147–148.

Mickler, W. 1885-T55 R33. Field notes for U.S. Survey. Vol. 245: 28–36, 73–80, 87–95, 99–108.

Mickler, W. 1885-T55 R34. Field notes for U.S. Survey. Vol. 245: 52–57, 207–214, 217–225.

Mickler, W. 1885-T55 R35. Field notes for U.S. Survey. Vol. 245: 59–63, 169–176, 179–185.

Mickler, W. 1885-T55 R36. Field notes for U.S. Survey. Vol. 245: 64–69, 298–304, 307–314.

Mickler, W. 1885-T55 R37. Field notes for U.S. Survey. Vol. 245: 71.

Mickler, W. 1886-T56 R33. Field notes for U.S. Survey. Vol. 245: 37–42, 573–579.

Mickler, W. 1886-T56 R34. Field notes for U.S. Survey. Vol. 245: 415–421.

Mickler, W. 1886-T56 R35. Field notes for U.S. Survey. Vol. 245: 451–457.

Mickler, W. 1886-T56 R36. Field notes for U.S. Survey. Vol. 245: 487–493.

Mickler, W. 1886-T57 R33. Field notes for U.S. Survey. Vol. 245: 44–49, 547–553, 555–557, 559–569.

Mickler, W. 1886-T57 R34. Field notes for U.S. Survey. Vol. 245: 389–404, 407–413.

Mickler, W. 1886-T57 R35. Field notes for U.S. Survey. Vol. 245: 425–438, 441–447.

Mickler, W. 1886-T57 R36. Field notes for U.S. Survey. Vol. 245: 461–467, 469–475, 478–484.

Mickler, W. 1886-T58 R35. Field notes for U.S. Survey. Vol. 245: 497–520.

Mickler, W. 1886-T58 R36. Field notes for U.S. Survey. Vol. 245: 524–543.

Newman, J. W. 1908-T50 R41. Plat map for State Survey.

Newman, J. W. 1908-T51 R41. Field notes and plat map for State Survey. Vol. 259: 463–475.

Newman, J. W. 1908-T54 R40. Field notes and plat map for State Survey. Vol. 259: 477–501.

Pidgeon, C. M. 1913-T43 R35. Field notes and plat map for State Survey. Vol. 261: ?.

Pidgeon, C. M. 1913-T43 R36. Field notes and plat map for State Survey. Vol. 261: ?.

Potter, G. W. 1889-T42 R43. Field notes and plat map for U.S. Survey. Vol. 248: 85–101.

Potter, G. W. 1885-T43 R43. Field notes for U.S. Survey. Vol. 243: 254–263.

Reyes, W. J. 1855-T40 R37. Field notes and plat map for U.S. Survey. Vol. 191: 74–106.

Reyes, W. J. 1855-T40 R38. Field notes and plat map for U.S. Survey. Vol. 191: 105–183.

Reyes, W. J. 1855-T40 R39. Field notes and plat map for U.S. Survey. Vol. 191: 187–255.

Reyes, W. J. 1855-T40 R40. Field notes and plat map for U.S. Survey. Vol. 191: 271–330.

Reyes, W. J. 1855-T40 R42. Field notes and plat map for U.S. Survey. Vol. 191: 336–418.

Reyes, W. J. 1855-T40 R43. Field notes and plat map for U.S. Survey. Vol. 191: 435–447.

Reyes, W. J. 1855-T41 R37. Field notes and plat map for U.S. Survey. Vol. 212: 332–356.

Reyes, W. J. 1855-T41 R38. Field notes and plat map for U.S. Survey. Vol. 191: 459–485.

Reyes, W. J. 1855-T41 R39. Field notes and plat map for U.S. Survey. Vol. 191: 513–583.

Reyes, W. J. 1855-T41 R43. Field notes and plat map for U.S. Survey. Vol. 191: 586–626.

Reyes, W. J. 1855-T42 R39. Field notes and plat maps for U.S. Survey. Vol. 212: 356–414.

Reyes, W. J. 1856-T42 R40. Field notes and plat map for U.S. Survey. Vol. 212: 415–457.

Reyes, W. J. 1858-T42 R40. Field notes and plat map for U.S. Survey. Vol. 208: 131–154.

Reyes, W. J. 1858-T42 R42. Field notes and plat map for U.S. Survey. Vol. 208: 247–322.

Reyes, W. J. 1858-T42 R43. Field notes and plat map for U.S. Survey. Vol. 208: 447–489.

Reyes, W. J. 1858-T43 R40. Field notes and plat map for U.S. Survey. Vol. 208: 157–191.

Reyes, W. J. 1858-T43 R42. Field notes and plat map for U.S. Survey. Vol. 208: 323–383.

Reyes, W. J. 1858-T43 R43. Field notes and plat map for U.S. Survey. Vol. 208: 491–534.

Reyes, W. J. 1858-T44 R41. Field notes and plat map for U.S. Survey. Vol. 208: 196–233.

Reyes, W. J. 1858-T44 R42. Field notes and plat map for U.S. Survey. Vol. 208: 387–445.

Reyes, W. J. 1858-T44 R43. Field notes and plat map for U.S. Survey. Vol. 208: 535–579.

Richards, W. H. 1915-T44 R41. Field notes and plat map for State Survey. Vol. 262: ?.

Richards, W. H. 1915-T44 R42. Field notes and plat map for State Survey. Vol. 262: 446–455.

Richards, W. H. 1915-T44 R43. Field notes and plat map for State Survey. Vol. 262: ?.

Richards, W. H. 1915-T45 R42. Field notes and plat map for State Survey. Vol. 262: 455–464.

Richards, W. H. 1915-T45 R43. Field notes and plat map for State Survey. Vol. 262: 481–490; 263: 530–534.

Shearer, N. 1955-T58 R35. U.S. Survey. 376–379.

Sollie, W. C. 1884-T50 R33. Field notes and plat map for U.S. Survey. Vol. 247: 139–147, 318–395.

Sollie, W. C. 1885-T51 R34. Field notes for U.S. Survey. Vol. 247: 150–181.

Sollie, W. C. 1884-T51 R35. Field notes for U.S. Survey. Vol. 247: 182–205.

Sollie, W. C. 1884-T52 R33. Field notes and plat map for U.S. Survey. Vol. 247: 206–214, 217–225, 401–474.

Sollie, W. C. 1884-T52 R34. Field notes for U.S. Survey. Vol. 247: 227–232, 235–242.

Southern Engineering & Construction Co. 1918-T54 R37. Field notes and plat map for State Survey. Vol. FCEv02: 202–222.

Southern Engineering & Construction Co. 1918-T54 R38. Field notes and plat map for State Survey. Vol. FCEv02: 223–242.

Southern Engineering & Construction Co. 1918-T54 R39. Field notes and plat map for State Survey. Vol. FCEv02: 243–266.

Steams, J. H. 1871-T42 R29. Field notes and plat map for U.S. Survey. Vol. 216: 501–563.

Steams, T. S. 1874-T47 R33. Field notes and plat map for U.S. Survey. Vol. 225: 177–237.

Steams, T. S. 1874-T47 R34. Field notes and plat map for U.S. Survey. Vol. 225: 238–263.

Stearns, J. S. 1874-T47 R34. Plat map and field notes for U.S. Survey. Vol. 225: 238–265.

Tannehill, J. D. 1871-T40 R31. Field notes and plat map for U.S. Survey. Vol. 217: 164–259.

Tannehill, J. D. 1871-T40 R32. Field notes and plat map for U.S. Survey. Vol. 217: 384–468.

Tannehill, J. D. 1871-T40 R33. Field notes and plat map for U.S. Survey. Vol. 217: 412, 619, 620.

Tannehill, J. D. 1871-T41 R31. Field notes and plat map for U.S. Survey. Vol. 217: 265–333.

Tannehill, J. D. 1871-T41 R32. Field notes and plat map for U.S. Survey. Vol. 217: 347–374.

Tannehill, J. D. 1871-T42 R30. Field notes and plat map for U.S. Survey. Vol. 217: 3–73.

Tannehill, J. D. 1871-T42 R31. Field notes and plat map for U.S. Survey. Vol. 217: 88–142.

Tannehill, J. D. 1871-T42 R32. Field notes and plat map for U.S. Survey. Vol. 217: 154–155.

Wallis, W. T. 1925-T57 R40. Field notes and plat map for U.S. Survey.

White, E. L. 1886-T45 R43. Field notes and plat map for U.S. Survey. Vol. 243: 289–300.

Williams, M. A. 1872-T45 R41. Field notes and plat map for U.S. Survey. Vol. 219: 330–340.

Williams, M. A. 1872-T45 R42. Field notes and plat map for U.S. Survey. Vol. 219: 248–296.

Williams, M. A. 1872-T45 R43. Field notes and plat map for U.S. Survey. Vol. 219: 348–384.

Williams, M. A. 1872-T46 R41. Field notes and plat map for U.S. Survey. Vol. 219: 305–324.

Williams, M. A. 1870-T46 R43. Field notes and plat map for U.S. Survey. Vol. 215: 253–273.

Williams, M. A. 1870-T47 R43. Field notes and plat map for U.S. Survey. Vol. 215: 283–302.

Williams, M. A. 1875-T48 R33. Field notes and plat map for U.S. Survey. Vol. 228: 480–524.

Williams, M. A. 1875-T48 R34. Field notes and plat map for U.S. Survey. Vol. 228: 565–600.

Williams, M. A. 1870-T48 R42. Field notes and plat map for U.S. Survey. Vol. 215: 44–65.

Williams, M. A. 1870-T48 R43. Field notes and plat map for U.S. Survey. Vol. 215: 311–324.

Williams, M. A. 1875-T49 R33. Field notes and plat map for U.S. Survey. Vol. 228: 525–564.

Williams, M. A. 1875-T49 R34. Field notes and plat map for U.S. Survey. Vol. 228: 601–639.

Williams, M. A. 1870-T49 R42. Field notes and plat map for U.S. Survey. Vol. 215: 66–68, 79–107.

Williams, M. A. 1870-T49 R43. Field notes and plat map for U.S. Survey. Vol. 215: 69–71.

Williams, M. A. 1870-T50 R42. Field notes and plat map for U.S. Survey. Vol. 215: 115–166.

Williams, M. A. 1870-T50 R43. Field notes for U.S. Survey. Vol. 215: 168.

Williams, M. A. 1870-T51 R42. Field notes and plat map for U.S. Survey. Vol. 215: 178–211.

Williams, M. A. 1870-T52 R41. Field notes and plat map for U.S. Survey. Vol. 215: 1–31.

Williams, M. A. 1870-T52 R42. Field notes and plat map for U.S. Survey. Vol. 215: 208–243.

Williams, M. A. 1874-T53 R42. Field notes and plat map for U.S. Survey. Vol. 226: 1–27.

Williams, M. A. 1874-T57 R39. Field notes and plat map for U.S. Survey. Vol. 226: 34–36.

Williams, M. A. 1847-T57 R40. Field notes for U.S. Survey. Vol. 226: 31–33.

Williams, M. A. 1874-T58 R39. Field notes for U.S. Survey. Vol. 226: 68–109.

Williams, M. A. 1874-T58 R40. Field notes and plat map for U.S. Survey. Vol. 226: 38–68.

Williams, M. A. 1872-T58 R41. Field notes and plat map for U.S. Survey. Vol. 219: 26–79.

Williams, M. A. 1874-T59 R39. Field notes and plat map for U.S. Survey. Vol. 226: 110–154.

Williams, M. A. 1872-T59 R40. Field notes for U.S. Survey. Vol. 219: 109–135.

Williams, M. A. 1874-T59 R40. Field notes and plat map for U.S. Survey. Vol. 226: 116–119.

Index

Page numbers in *italic* refer to illustrations. Page numbers referencing material on the DVD are preceded by a "D."

Christopher W. McVoy, a graduate of Cornell University and the University of Florida, is a soil physicist and wetland ecologist. Winifred Park Said is a botanist, modeler, and environmental planner. Jayantha Obeysekera is a hydrologist and director of the Hydrologic and Environmental Systems Modeling Department at the South Florida Water Management District. Joel A. VanArman is a biologist, retired after thirty-five years of Everglades studies. Thomas W. Dreschel is an environmental scientist specializing in aquatic systems. All authors are current or former scientists at the South Florida Water Management District.